AF249104

Operational Processes in Mechanical Engineering

Operational Processes in Mechanical Engineering

Michael B. Spektor
Professor Emeritus
Oregon Institute of Technology, USA

World Scientific

NEW JERSEY · LONDON · SINGAPORE · BEIJING · SHANGHAI · HONG KONG · TAIPEI · CHENNAI · TOKYO

Published by

World Scientific Publishing Co. Pte. Ltd.

5 Toh Tuck Link, Singapore 596224

USA office: 27 Warren Street, Suite 401-402, Hackensack, NJ 07601

UK office: 57 Shelton Street, Covent Garden, London WC2H 9HE

British Library Cataloguing-in-Publication Data
A catalogue record for this book is available from the British Library.

OPERATIONAL PROCESSES IN MECHANICAL ENGINEERING

ISBN 978-981-12-7772-6 (hardcover)
ISBN 978-981-12-7773-3 (ebook for institutions)
ISBN 978-981-12-7774-0 (ebook for individuals)

For any available supplementary material, please visit
https://www.worldscientific.com/worldscibooks/10.1142/13451#t=suppl

Desk Editor: Rhaimie Wahap

Typeset by Stallion Press
Email: enquiries@stallionpress.com

To my family

PREFACE

The goal of mechanical engineering is to design and implement machines and systems with intentionally defined behaviors and performance. Having taught Mechanical Engineering in universities within the USA, Israel, and former Soviet Union, I am very familiar with the core courses offered in Mechanical Engineering programs throughout the world. These courses focus on calculating the required strength of a single machine element of general application. While the strength of an individual machine element is an extremely important consideration, equally important are the critical constraints around motion that engineers must also accommodate, such as safe ranges of acceleration and deceleration in an aircraft. Despite this, Mechanical Engineering programs offer very limited emphasis on the mathematical analysis of the operational processes of systems.

The motion of a system can be analyzed by composing and solving corresponding differential equation of motion. Calculus courses teach a methodology based on characteristic equations for solving these equations for one-degree-of-freedom systems. However, this approach does not work for two-degree-of-freedom systems, which represent a large portion of the assemblies that engineers need to design.

I conducted a survey of related engineering textbooks and articles published over the last 75 years. While there are many descriptions of two-degree-of-freedom systems, the survey did not reveal even one complete rigorous mathematical solution for a pair of appropriate differential equations of motion of a two-degree-of-freedom system. Furthermore, the differential equations provided in these examples do not account for the fact that the masses in the systems are in permanent relative motion.

This textbook is the first attempt to systemize, analyze, and present the content of a Mechanical Engineering course related to the fundamentals of operational processes of mechanical structures, for both one- and two-degrees-of-freedom systems. I composed and solved the entire spectrum of differential equations for one-degree-of-freedom systems of all possible structural compositions, relevant loading factors, and initial conditions of motion. Additionally, by accounting for the principles of relative motion and using the Laplace Transform methodology, I was able to obtain solutions for two-degree-of-freedom systems for the first time, for the entire spectrum of two-degree-of-freedom systems. Most of these equations, their mathematically rigorous solutions, and the table of Laplace Transform pairs required to obtain the solutions, are presented for the first time in this book. It should be noted that for three-degree-of-freedom systems and higher orders, there is no published methodology for composing and solving their equations of motion. Hence, this textbook only focuses on one- and two-degree of freedom systems, which also happens to represent the overwhelming majority of systems that engineers design.

My hope is that this book will help students gain a solid foundation in mechanical engineering operational processes of one- and two-degree-of-freedom systems. I also hope that this book will inspire the students to explore further applications and research topics in this field.

I would like to express my sincere gratitude and deepest appreciation to Distinguished Professor I. E. Elishakoff for reading the manuscript and sharing his feedback and support.

In addition, I would like to thank my son, Daron Spektor, for his priceless help while I was writing this book, and specifically for his assistance with using Python for numeric analysis.

I also would like to thank my family for their support and encouragement throughout this project.

Michael Spektor, Professor Emeritus
Oregon Institute of Technology, USA
9 June 2023

CONTENTS

INTRODUCTION

Mechanical engineers must design and implement machines with specifically defined behaviors and performance. To do this, they need to understand and consciously control the basic parameters of the machine's motion — the displacement, velocity, and acceleration. This textbook is the first attempt to address the scientific approach to the analytical fundamentals describing the operational processes of mechanical systems. These systems predominately consist of one- and two-degree-of-freedom structures, as described in the two parts of the textbook.

The author's previous book, *Applied Dynamics in Engineering*, showed that the full spectrum of one-degree-of-freedom systems can be described by differential equations corresponding to all possible combinations of resisting and active loading factors and initial conditions of motion. In this book, the full spectrum of structural compositions of differential equations of motion and their solutions are provided for both one- and two-degree-of-freedom systems, allowing this book to serve as a comprehensive guide to the operational processes of the overwhelming majority of systems that engineers work with.

To solve a differential equation for a one-degree-of-freedom system, most calculus courses use a method based on characteristic equations. Unfortunately, this method does not work for two-degree-of-freedom systems because the characteristic equation cannot handle the two unknowns in each of the two differential equations. This textbook demonstrates for the first time how the

Laplace Transform methodology can be used to solve the pairs of simultaneous differential equations of motion for two-degree-of-freedom systems. This textbook offers for the first time a table of the 101 Laplace Transform pairs that are required to solve the linear differential equations of motion for all structural compositions of one- and two-degree-of-freedom systems in mechanical engineering. With the help of this table a differential equation of motion can be converted to an algebraic equation in the Laplace domain, where one of the two unknown variables can be eliminated and the remaining one solved, and then inverted back to the time domain.

Part 1 of this textbook presents the theoretical fundamentals that allow for the comprehensive analysis of one degree-of freedom systems. A one-degree-of-freedom system may be represented by one element or an assembly of several elements, the motion of which is described by one second order differential equation. Part 1 consists of 17 chapters. Chapter 1 discusses the structures of the mathematical terms and of the differential equations of motion, as well as the use of the Laplace Transform methodology to solve them. It should be noted that the same mechanical engineering system can perform different operational processes. This depends on the combinations of loading factors and of the initial conditions of motion applied to these systems. Finally, Chapter 1 addresses how to handle dry friction forces.

Chapters 2–5 deal with the operational processes of one-degree-of-freedom systems whose motion is caused only by the initial conditions, or by all possible combinations of resisting and active forces including the initial conditions of motion. Chapters 6–9 deal with systems that include a flexible link (spring) connecting the mass to a non-movable support. In Chapter 6 it is shown that in the case of free fall it is necessary to use different springs for people of different weights in order to stay within safe limits of deceleration. Further analysis of this case reveals that if the harmonic force applied to the mass of the system possesses the same frequency as the natural frequency of the system, then the system will be subjected to resonance. Chapters 10–13 deal with systems that have a fluid link instead of a spring.

Chapters 14–17 deal with systems that have flexible and fluid links in parallel. Depending on the values of the forces and initial conditions of motion of the mass, this mass could perform under-damped vibration, critically damped motion, or overdamped motion. Part 1 of this book presents analyses of the motion of the systems on horizontal and inclined surfaces, as well as hanging downward and moving under the force of gravity. It also describes numeric solutions for transcendental expressions and presents their graphs using the Python programming language.

Part 2 of this textbook deals with two-degree-of-freedom systems, the motions of the two masses of which are described by an appropriate pair of differential equations by two independent simultaneous differential equations. The two masses are permanently in relative motion, and, therefore, the absolute values of their velocities are not equal. If the velocities of the two masses were equal, then they would be moving as one mass, and hence modeled as a one-degree-of-freedom system. In a two-degree-freedom system, the kinematic link connecting the two masses exerts forces proportional to the difference between the corresponding parameters of motion of these masses. On the other hand, in one-degree-of-freedom systems, the kinematic link exerts a force proportional to the absolute value of the corresponding parameter of motion. The structure of the mathematical terms describing the forces exerted by the kinematic links in one- and two-degree-of-freedom systems are substantially different. Previous publications in the field of vibration present differential equations with a mixture of terms from one- and two-degree-of-freedom systems and no solutions. This textbook for the first time shows how to compose differential equations that comply with the principle of relative motion, as well as how to solve them using the Laplace Transform methodology.

Two-degree-of-freedom systems can be classified as unrestricted or restricted depending on their structural compositions. Unrestricted systems are those where the two masses are only connected to each other, whereas restricted systems are those where one or both masses are also attached to one or two non-movable supports. It should be noted that the overwhelming majority of surveyed publications only covers restricted systems. However, this book covers both types of systems. It should be stressed that at the time of this writing, the entire field related to the motion of two-degree-of-freedom systems had been largely unexplored.

Chapter 18 presented in Part 2 of this textbook demonstrates that the Laplace Transform methodology is also applicable for solving the corresponding differential equations of motion of two-degree-of-freedom systems.

Chapters 19–21 describe the entire spectrum of operational processes of unrestricted two-degree-of-freedom systems. Chapter 19 addresses systems whose masses are connected by a flexible link. Chapter 20 deals with systems that have a fluid link, and Chapter 21 addresses systems with both types of links in parallel. These chapters present solutions to differential equations for various initial conditions and loading factors. They also present the use of Python to obtain numeric solutions and graphs for transcendental equations.

Chapters 22–24 deal with restricted two-degree-of-freedom systems that have two sequences of links, hence only one mass is attached to a non-movable support. Chapter 22 details systems with flexible links, Chapter 23 deals with fluid links, and Chapter 24 details both types of links in parallel. These chapters present numeric solutions and graphs in Python, showing examples of anti-phase vibration, resonance, underdamped, critically damped, and overdamped motion.

Chapters 25–27 describe restricted systems with three sequences of links, and both masses are attached by these links to non-movable supports. Chapter 25 deals with flexible links, Chapter 26 with fluid links, and Chapter 27 with both in parallel. These chapters also present numerical solutions and plots in Python, demonstrating different characteristics of vibrational motion.

This textbook does not address the aspects of motion of systems with three or more degrees of freedom, such as trucks towing two trailers or railroad trains with multiple cars. The author notes that this topic lacks published sources of fundamental investigations.

It is worth noting that most mathematical expressions describing the motion of one and two degrees of freedom systems are transcendental functions of time, the analysis of which in general terms is currently impossible. However, corresponding numeric computer programs allow to analyze the transcendental expressions with high accuracy. Each chapter of this textbook includes at least one example of such a program using the Python programming language. These examples present the numeric solutions and the graphs of the equations of displacement of the masses. These equations, which represent the laws of motion for the systems, are one of the three basic parameters of motion that are functions of time. By adjusting the input data in the Python program, it is possible to analyze the parameters in question and obtain their desired values. In today's rapidly advancing technological landscape, it is crucial to utilize analytical methods in the development of new operational processes to reduce the cost and time associated with testing the prototypes of new systems.

One of the most important features of an undergraduate university course is its completeness. Evaluating the completeness of a course's content can be challenging. Given that mechanical operational processes are characterized by motion, one way to ensure completeness of a course is the ability to analyze all possible structural combinations of systems, all relevant loading factors and initial conditions of motion that determine a system's movement. This text aims to provide a comprehensive methodology to accomplish this for the entire spectrum of one- and two-degree of freedom systems.

PART 1

CHAPTER 1

INTRODUCTION TO THE STUDY OF OPERATIONAL PROCESSES OF ONE-DEGREE-OF-FREEDOM SYSTEMS

Mankind has created a tremendous number of machines, mechanisms, attachments, structures, devices, and other mechanical engineering systems that are extremely beneficial in the everyday lives of humanity. These systems are performing a variety of operational processes, the common characteristics of which represent the parameters of motion. All types of motion are identified and scientifically described in the university course of dynamics. However, the scope of operational processes is not identified, and it does not provide a systemized description of operational processes. An operational process of a mechanical system could be described by one type of motion or by a combination of several types of motion. In addition, it is very important to emphasize that each operational process should comply with certain predetermined requirements. The desired characteristics of the operational process should be defined. It should be clear how to verify and control these characteristics. All these constitute a system of complex problems, the study of which could represent a special university course. This course should address the entire spectrum of the studies of motion of common one- and two-degree-of-freedom systems. In Part 1 of this text, all issues related to the operational processes of one-degree-of-freedom mechanical systems are addressed. A corresponding description of two-degree-of-freedom mechanical systems is provided in Part 2 of the text.

As mentioned above, the operational processes are characterized by different types of motion of related machine elements or their numerous assemblies. It may be stated that each operational process of a mechanical system can be described by a certain second-order differential equation of motion or by a corresponding combination of second-order differential equations of motion. One of the most important objectives of the above-mentioned university course is to deliver to the graduates the necessary knowledge of composing and solving the related second-order differential equations of motion of one- and two-degree-of-freedom mechanical systems, assuming that these systems represent the vast majority of physical structures that are associated with mechanical operational processes.

In the following chapters of this text, detailed descriptions of all possible common differential equations of motion of the above-mentioned mechanical systems are presented. Hence, since the second-order differential equations of motion of mechanical systems comprise the algebraic sums of loading factors, it is necessary to identify all possible justifiable combinations of these factors that could be taken into consideration.

Accounting for the fact that the structures of differential equations describing translational motion are completely analogous to those of the differential equations of rotational motion, we continue the consideration of all aspects related to the differential equations of motion only for the translational motion. Mechanical engineering educational programs are focused on teaching the graduates the necessary skills to create and maintain efficient systems. These objectives are based on the latest achievements in engineering science that are in continuous development.

Mechanical engineering systems are divided into two groups that consist of non-movable and movable structures. This text deals with movable mechanical structures. In this text, we define a mechanical movable system as any moving mass which consists of a single component or a plurality of elements and assemblies that are in a state of motion.

The non-movable and movable mechanical structures should comply with numerous requirements, the most important of which is that the strength of the structure should be sufficient to withstand the relevant loading conditions. Hence, mechanical engineering educational programs contain a sequence of core courses that reflect the contemporary scientific achievements in the computation of the strength of machine elements for general applications. It should be emphasized that the methodologies of engineering calculations related to the strength of machine elements are well established and grounded. Obviously, sufficiently strong machine elements are necessary to execute the operational processes of movable systems, while the educational programs deliver to the graduates the existing knowledge that, with an acceptable degree of accuracy, allows them to create machine elements possessing the required mechanical strength.

As mentioned above, an overwhelming majority of mechanical engineering systems represent one- or two-degree-of-freedom structures. The first part of this book deals with the operational processes of one-degree-of-freedom systems, the motions of which are completely characterized by one second-order differential equation. This suggests that for the analysis of the operational process of a mechanical system, it is necessary to compose an appropriate differential equation of motion, the solution of which, as is well known, yields the law of motion of the system, representing its displacement as a function of running time. The first and second derivatives from this function with respect to time, respectively, describe the velocity and acceleration of the system as functions of time. The displacement, the velocity, and the acceleration represent the basic parameters of motion of the system and allow us to carry out a complete analysis of its operational process.

Therefore, it is justifiable to emphasize some related aspects of a second-order differential equation of motion.

1.1. Second-Order Linear Differential Equation of Motion

The basics of differential equations are presented in the university course of calculus, which is included in mechanical engineering educational programs. The order of the differential equation is determined by the highest order of the derivative in the equation.

The solution of the second-order differential equations of motion represents the only way to obtain the required engineering and analytical representation of the law of motion and other basic characteristics of motion related to the operational processes of one- and two-degree-of-freedom systems. The mathematical expressions describing the structures of second-order differential equations of motion that are applicable to the descriptions of translational and rotational motions are presented in the university course on calculus and dynamics.

As it is known from the course of dynamics, for movable systems in mechanical engineering, the second-order differential equations describe the motion of the system, while each mathematical term of the equation represents a loading factor — a force or a moment. In addition, the displacement of the mass (the system) represents the law of motion of the system as a function of time, while the first and second derivatives of this function with respect to time are, respectively, the velocity and acceleration of the system. It should be reminded that forces cause the translational type of motion, while moments cause the rotational type of motion; therefore, it is needless to mention that a differential equation cannot contain a mix of forces and moments. The structural composition of a differential equation of motion is characterized by the mathematical members or terms that constitute the equation. The definitions of forces and moments and their features are presented and discussed in detail in the courses of physics, statics, and dynamics. It should be mentioned that in this text, it is accepted to consider the forces and moments that resist the motion of mechanical systems as resisting, reactive, or internal forces and moments, while the forces and moments that cause the motion of these systems are considered active, external, or effective forces and moments. In addition, on the left-hand side of the differential equation (before the equal sign), the resisting (reactive, internal) forces and moments have positive values, while on the right-hand side of the same equation, the active (external) forces and moments have positive values.

As is well known, depending on the characteristics of the loading factors, the second-order differential equations of motion could be linear or nonlinear. Unfortunately, currently, there does not exist a straightforward methodology for solving the nonlinear second-order differential equations of motion. However, the solutions of linear second-order differential equations of motion very often possess sufficient practical accuracy and are extremely helpful in understanding and advancing operational processes. Therefore, this course deals with linear differential processes.

It is acceptable to use schematic diagrams that allow us to visualize the relevant characteristics of the mechanical system under consideration. A typical example of a schematic diagram is presented in Figure 1.1.1, which shows a mechanical system moving on a horizontal surface and subjected to all forces that commonly resist or cause motion.

This schematic diagram is like many published schematic diagrams and is used to clarify the structure of the differential equations of motion of mechanical systems. Since the force of inertia is present in all differential equations of motion, it is generally accepted that this force is not required to be shown in schematic diagrams. In addition, it is acceptable to use in schematic diagrams of one-degree-of-freedom mechanical systems the images of kinematic links (springs, dashpots, or a combination of both links in parallel) that attach the masses to non-movable supports. In a one-degree-of-freedom system, the forces exerted by the connecting links toward the mass

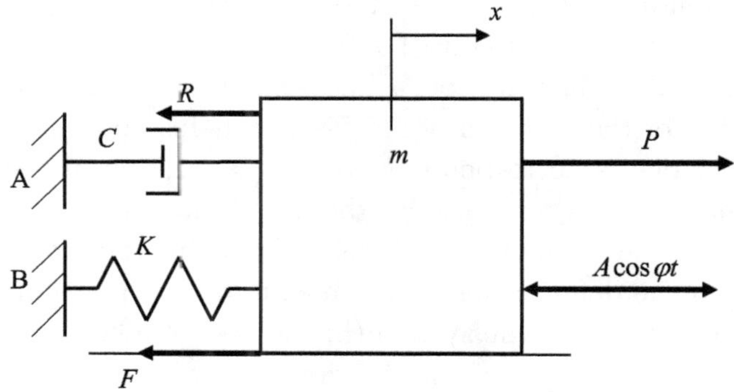

Fig. 1.1.1. Schematic diagram of a mechanical system loaded by all common resisting and active forces.

are proportional to the absolute values of the parameters of motion (displacement and velocity) of the mass. Often, the above-mentioned dashpots and springs symbolically represent viscous or/and elastic media that exert reactive (resisting) forces toward the moving rigid bodies as a reaction to the deformation (distortion) of the media by these masses that represent the mechanical systems. In the case of a viscous medium, the resisting force is proportional to the velocity of the moving body, while in the case of an elastic medium, the resisting force is proportional to the displacement of the moving body. There does also exist a plastic medium; however, the resisting (reactive) force of this medium does not depend on the velocity or displacement of the rigid body and remains constant during its deformation (distortion) by the rigid body. Obviously, the resisting forces of any medium depend on the cross-sectional area and some other characteristics of the rigid body interacting with the medium. Figure 1.1.1 shows a non-movable support, A, attached to a dashpot, while a non-movable support, B, is attached to a spring. The non-movable supports A and B could represent a single support attached to a dashpot or to a spring. The link that is represented by a spring is called a flexible link, while the link represented by a dashpot is called a fluid link. All these are symbolic representations of the features of viscosity and elasticity of media that could be in a state of fluid, gas, or a rigid body. It should be mentioned that in some mechanical systems, real dashpots, springs, or their combinations are used, while the corresponding schematics related to these systems are similar to the schematic shown in Figure 1.1.1

A schematic diagram of a movable system allows for the visualization of the problem of motion and helps accounting for the factors that play a corresponding role in the motion of the system. Among these factors, we consider the interaction between a moving system and a medium. As mentioned above, this interaction is proportional to the deformation of the medium, which is characterized by the displacement of the system (in the case of a flexible link) and proportional to the rate of the deformation, which is characterized by the velocity of the system (in the case of a fluid link).

Equation (1.1.1) shows the mathematical structures describing the resisting forces exerted by a fluid and a flexible link on the moving system.

The following presents the complete version of a second-order differential equation describing the translational motion of a one-degree-of-freedom mechanical system:

$$m\frac{d^2 x}{dt^2} + C\frac{dx}{dt} + Kx + R + F = P + A\cos\varphi t \qquad (1.1.1)$$

where m is the mass of the system, $\frac{d^2 x}{dt^2}$ is the acceleration of the system as a function of time t, and $m\frac{d^2 x}{dt^2}$ is the force of inertia of the mass (system); C is the damping coefficient of the fluid link, dashpot, which symbolically connects the mass to a non-movable support, while $\frac{dx}{dt}$ is the velocity of the mass (system) and is a function of time t; K is the stiffness coefficient of the flexible link, spring, which also symbolically connects the mass to a non-movable support, while x is the displacement of the mass (system) and represents a function of time; R is a constant resisting force that represents a function of time in the form Rt^0; F is a constant dry friction force that represents a function of time in the form Ft^0; P is a constant active force that represents a function of time in the form Pt^0; and $A\cos\varphi t$ is a harmonic force representing a function of time t, where A is the amplitude of the force and φ is the frequency of the harmonic function. It should be noted that in this text, the constant resisting force R could represent a reaction to the deformation of a plastic medium caused by the applied external, active, or effective force, which is denoted by P, or a harmonic force, which is denoted by $A\cos\varphi t$, or caused by a combined action of the forces P and $A\cos\varphi t$. In addition, the constant resisting force R could represent a gravity force in the case of lifting a weight, while the constant active force could represent a gravity force in the case of lowering a weight. The dry friction force F is considered a constant resisting force that is applied to the contacting surfaces of two bodies that perform relative sliding on each other. The damping force $C\frac{dx}{dt}$ represents the product of the multiplication of a constant coefficient C by the velocity of the mass (system) and reflects an interaction between the moving mass and the viscous medium. The resistance that the viscous medium exerts on the moving system in the process of its deformation is proportional to the velocity of the system. The stiffness force Kx equals the product of the multiplication of the constant coefficient K by the displacement of the system x. This force also represents a reaction of the medium to its deformation by the moving system. In this case, as noted above, the resistance that the medium exerts on the moving system is proportional to the displacement of the system.

It should be stressed that the forces involved in equation (1.1.1) represent the entire spectrum of forces that could be applied to a mechanical engineering system. The damping and stiffness forces could be characterized by linear or nonlinear functions of time; however, as mentioned earlier, because of the absence of a universal straightforward methodology for solving nonlinear second-order differential equations, in this text, we consider only linear differential equations of motion. It is allowable to assume that some other forces, as functions of time, could be artificially produced and applied to one-degree-of-freedom systems. In my book *Applied Dynamics in Engineering*, published in 2016 by Industrial Press, an active force was considered a linear function of time; however, to my knowledge, this force has a very limited practical implementation; therefore, it is not considered in this text.

Based on the forces included in equation (1.1.1), 64 primary and an additional 32 secondary linear differential equations of motion of one-degree-of-freedom systems are composed and solved, which represent all possible combinations of forces that could be utilized in corresponding differential equations describing the motions of common mechanical systems. Therefore, it becomes clear that the total number of common mechanical operational processes that could possibly exist consists of $64 + 32 = 96$ processes. Columns A–D of Table 1.6.1 indicate in order the numbers of sections in Part 1 of this book where the solutions and appropriate analyses of the solutions of the 64 primary differential equations are presented, while Table 1.6.2 presents in order the numbers of those sections where the 32 secondary differential equations are given. Hence, the 64 primary and

32 secondary differential equations of motion describe the 96 possible operational processes that represent the entire spectrum of the processes associated with translational motion. Therefore, the scope of the operational processes is determined, and the rigorous solutions of the 92 differential equations describing the possible operational processes are obtained using the Laplace transform methodology, some related details of which are presented in the following. It should be emphasized that the purposeful improvement and development of mechanical systems is based on the mathematical analysis of their operational processes, and it is justifiable to implement the study of operational processes as a core course in engineering educational programs.

As mentioned earlier, since the differential equations of motion and their solutions for translational and rotational motion are completely similar, there is no need to present both types of equations. To compare these two versions, the version of the differential equation for rotational motion is presented as follows:

$$I\frac{d^2\theta}{dt^2} + \sigma\frac{d\theta}{dt} + \kappa\theta + \rho + \mu = \psi + \Delta\cos\Omega t \qquad (1.1.2)$$

where I is the mass moment of inertia; $\frac{d^2\theta}{dt^2}$ is the angular acceleration of the system as a function of time t; σ is the rotational damping coefficient of the fluid link that connects the system to a non-movable support; $\frac{d\theta}{dt}$ is the angular velocity of the system, which is a function of time t; κ is the rotational stiffness coefficient of the flexible link (spiral spring coil) that connects the system to a non-movable support; θ is the angular displacement of the system and is a function of time; ρ is a constant resisting moment that represents a function of time in the form σt^0; μ is a constant dry friction moment that represents a function of time in the form μt^0; ψ is a constant active moment that represents a function of time in the form ψt^0; and $\Delta\cos\Omega t$ is a harmonic moment representing a function of time t, where Δ is the amplitude of the moment and Ω is the frequency of the harmonic function.

1.2. Linearity of Differential Equations of Motion

As is well known, a second-order differential equation of motion consists of mathematical terms representing forces or moments as functions of time, and obviously, the discussion is continued by addressing only the forces.

If even one mathematical term of a differential equation of motion is not linear, the differential equation is not linear as well, and its exact solution is unknown. There are some handbooks and catalogs that contain approximate solutions of certain nonlinear differential equations; however, their applicability to mechanical engineering problems is very limited. The analysis of equation (1.1.1) with regard to the linearity of its particular terms shows that the following three terms could be linear or nonlinear: (1) force of inertia: $m\frac{d^2x}{dt^2}$, (2) damping force: $C\frac{dx}{dt}$, and (3) stiffness force: Kx. The nonlinearity of the term $m\frac{d^2x}{dt^2}$ is usually associated with a decrease or increase in mass during the motion of the system. As an example of the decrease or increase in mass, the mass of a transportation means could be considered, where a gradual decrease in mass due to fuel consumption or a gradual increase due to rain or snow is observed. In more practical cases, the mentioned change in mass is insignificant and ignorable. The damping and stiffness forces according to the second and third terms of equation (1.1.1) play the decisive role in establishing the linearity or nonlinearity of the differential equation. The second term deals with the dependence of the resisting damping force on the velocity of the system, while the third term is associated with the dependence of the resisting stiffness force on the displacement of the system. The study of rheology fundamentally addresses all aspects of the interaction of a rigid body with the medium. The rigid body represents the moving system (mass), and the medium represents the environment in which the body is moving, while the interaction between the system and the medium represents the distortion (deformation and other physical changes) of the medium because of the motion of the system. The distortion is accompanied by the medium exerting resisting forces that are applied to the moving system. The analytical expressions of the nonlinear resisting forces in most cases are not known. Therefore, it is acceptable to plot graphs of the corresponding nonlinear resisting forces (damping or stiffness) and apply to them the appropriate techniques to estimate the level of their linearity. In many cases, the methodology of piece-wise linear approximation allows us to obtain a reasonably accurate solution of a nonlinear differential equation of motion that contains only one nonlinear term — the damping or stiffness resisting force. The appropriate examples can be found in my book *Solving Engineering Problems in Dynamics.* However, it should be stressed that the analysis of the operational processes of mechanical systems as well as most scientific investigations in mechanical engineering could be performed only with the appropriate linear second-order differential equations of motion.

It is necessary to indicate that the forces applied to the system determine the characteristics of its motion, while for each combination of applied forces, a single differential equation of motion could be composed.

1.3. Solving Second-Order Linear Differential Equations of Motion of One-Degree-of-Freedom Systems

As mentioned above, each operational process of a mechanical system could be analytically described by the three basic parameters of motion of the system's mass. As we know, these basic parameters of motion represent analytical expressions describing the displacement, velocity, and acceleration of the system's mass as functions of time. As is well known, the velocity and acceleration, respectively, represent the first and second derivatives of the displacement. The law of motion of the system or the equation describing its displacement can be determined by solving the corresponding second-order differential equation of motion of the system. For each operational process, the corresponding second-order differential equation of motion should be composed and solved. The university courses on calculus in engineering programs offer a methodology for solving these equations by using the so-called characteristic equation. This methodology is not universal and is often associated with certain mathematical difficulties.

However, since this textbook deals not just with one-degree-of-freedom systems but also with two-degree-of-freedom systems, it is necessary to have the methodologies for solving the pairs of simultaneous second-order differential equations of motion. It is obvious that composing and solving simultaneous second-order differential equations of motion of two-degree-of-freedom systems is associated with two unknowns in each differential equation of motion, while it is impossible to apply the methodology with the characteristic equation to a differential equation with two unknowns. However, it becomes clear from the second part of this textbook that the Laplace transform methodology works with the two-degree-of-freedom systems. This justifies the discussion on the applicability of the Laplace transform methodology to one- as well as two-degree-of-freedom systems. Since the delivery of the mathematical fundamentals of this methodology in the framework of higher education requires a relatively long learning cycle, most undergraduate programs avoid offering this methodology. However, this textbook shows that with the availability of the corresponding Laplace transform tables of conversion and inversion pairs of related analytical expressions, it becomes possible without any difficulties and in a short time to solve the differential equations of motion of one- and two-degree-of-freedom systems, while the graduates do not need to take any additional courses in calculus.

The idea of the Laplace transform methodology consists in the conversion of the given differential equation from the time domain into a corresponding algebraic equation in the Laplace domain by using the appropriate table of conversion–inversion pairs. During this step, we apply to the obtained algebraic equation the conventional procedures to solve this equation for the displacement of the system in the Laplace domain (using the Laplace variable). In the next step, we invert the obtained equation expressing the displacement in the Laplace domain expression into the time domain expression by using the appropriate table of conversion–inversion pairs of Laplace transforms. As a result, we obtain the solution of the given differential of motion. In the following, several examples demonstrating in detail the use of the Laplace transform methodology are presented for solving second-order differential equations of motion of one-degree-of-freedom systems. It should be mentioned that in the case of two-degree-of-freedom systems, we apply to the obtained algebraic equation in Laplace domain the method of substitutions and eliminate one unknown from the mentioned algebraic equation. This algebraic procedure is demonstrated in the second part of the textbook.

In Appendix A-1 of this text, a table of 101 Laplace Transform pairs is presented, which allows us to solve all the second-order linear differential equations of motion of all possible common mechanical systems and their operational processes. Many of these pairs were developed by me.

It is important to emphasize that all the solutions of the corresponding differential equations are obtained in general terms.

1.3.1. *Example 1*

The simplest differential equation of motion reads

$$m\frac{d^2x}{dt^2} = 0 \tag{1.3.1}$$

Dividing equation (1.3.1) by m, we have

$$\frac{d^2x}{dt^2} = 0 \tag{1.3.2}$$

The initial conditions of motion are

$$\text{for } t = 0 \quad x = 0; \quad \frac{dx}{dt} = V \tag{1.3.3}$$

Applying Laplace Transform pair 5 from Appendix A-1 to differential equation (1.3.2) with its initial conditions of motion according to expression (1.3.3), we see that term $\frac{d^2x}{dt^2}$ corresponds to the left-hand side of pair 5, $\frac{d^2u}{dt^2}$. Therefore, we must adjust the notations in pair 5 by changing u to x. After that, the right-hand side of pair 5 will have the following form:

$$s^2 x(s) - sV = 0 \tag{1.3.4}$$

where s is the Laplace variable and $x(s)$ is the displacement in the Laplace domain. Substituting into equation (1.3.4) the corresponding values of the initial velocity $\frac{dx}{dt} = V$ and initial displacement $x = 0$ according to expression (1.3.3), we obtain

$$sx(s) - V = 0 \tag{1.3.5}$$

Hence, expression (1.3.5) represents the algebraic equation from which we determine the displacement of the mass in the Laplace domain:

$$x(s) = \frac{1}{s} V \tag{1.3.6}$$

Now, we need to invert equation (1.3.6) from the Laplace domain into the time domain. Applying to equation 1.3.6) Laplace Transform pair 3 from the table in Appendix A-1, we see that the Laplace domain function $\frac{1}{s}$ corresponds to the time t in the time domain. Therefore, the solution of differential equation (1.3.2) with the initial conditions of motion according to expression (1.3.3) reads

$$x = Vt \tag{1.3.7}$$

Taking the first derivative from equation (1.3.7) with respect to t (time), we determine the velocity of the system:

$$\frac{dx}{dt} = V \tag{1.3.8}$$

Assuming in equations (1.3.7) and (1.3.8) that $t = 0$, we obtain $x = 0$ and $\frac{dx}{dt} = V$, as expected according to initial conditions of motion (1.3.3).

1.3.2. *Example 2*

This example demonstrates the derivation of a well-known formula for calculating displacement, which is introduced in high school as part of a course on physics.

The differential equation has the following form:

$$m\frac{d^2x}{dt^2} = P \tag{1.3.9}$$

where P is the constant active force. The initial conditions of motion are

$$\text{for } t = 0 \quad x = 0; \quad \frac{dx}{dt} = 0 \tag{1.3.10}$$

Dividing equation (1.3.9) by m, we write

$$\frac{d^2x}{dt^2} = p \tag{1.3.11}$$

where

$$p = \frac{P}{m} \tag{1.3.12}$$

Applying to equation (1.3.12) Laplace Transform pair 5, we convert this equation with its initial conditions of motion according to expression (1.3.3) from the time domain into an algebraic equation in the Laplace domain:

$$s^2 x(s) = p \tag{1.3.13}$$

Solving equation (1.3.13) for the displacement in the Laplace domain, we obtain

$$x(s) = \frac{1}{s^2}p \tag{1.3.14}$$

Applying to equation (1.3.14) Laplace Transform pair 6, we invert this equation from the Laplace domain into the time domain and obtain the solution of differential equation (1.3.9) with initial conditions of motion (1.3.10) in the time domain:

$$x = \frac{1}{2}pt^2 \tag{1.3.15}$$

Taking the first derivative from equation (1.3.15), we determine the velocity of the system:

$$\frac{dx}{dt} = pt \tag{1.3.16}$$

Assuming in equations (1.3.15) and (1.3.16) that $t = 0$, we obtain $x = 0$ and $\frac{dx}{dt} = 0$, as it should be according to initial conditions of motion (1.3.10).

1.3.3. *Example 3*

The following differential equation describes the free vibratory motion of a system:

$$m\frac{d^2x}{dt^2} + Kx = 0 \tag{1.3.17}$$

Dividing equation (1.3.17) by m, we have

$$\frac{d^2x}{dt^2} + \omega_0^2 x = 0 \tag{1.3.18}$$

where

$$\omega_0^2 = \frac{K}{m} \tag{1.3.19}$$

while ω_0 is the natural frequency of the system. The initial conditions of motion are as follows:

$$\text{for } t = 0 \quad x = S; \quad \frac{dx}{dt} = 0 \tag{1.3.20}$$

Applying to equation (1.3.18) Laplace Transform pairs 5, 2, and 1, we convert this equation with its initial conditions of motion (1.3.20) from the time domain into an algebraic equation in the Laplace domain:

$$s^2 x(s) - s^2 S + \omega_0^2 x(s) = 0 \tag{1.3.21}$$

Solving equation (1.3.21) for the displacement in the Laplace domain, we obtain

$$x(s) = \frac{s^2 S}{s^2 + \omega_0^2} \tag{1.3.22}$$

Applying to equation (1.3.22) the Laplace Transform pair 29, we invert equation (1.3.22) with its initial conditions of motion (1.3.20) from the Laplace domain into the time domain and obtain the solution of differential equation (1.3.18) with its initial conditions of motion (1.3.20):

$$x = S \cos \omega_0 t \tag{1.3.23}$$

Taking the first derivative from equation (1.3.23), we determine the velocity of the system:

$$\frac{dx}{dt} = -S\omega_0 \sin \omega_0 t \tag{1.3.24}$$

Assuming in equations (1.3.23) and (1.3.24) that $t = 0$, we obtain $x = S$ and $\frac{dx}{dt} = 0$, as expected according to initial conditions of motion (1.3.20).

1.3.4. *Example 4*

The following differential equation describes a forced vibrational motion:

$$m\frac{d^2 x}{dt^2} + C\frac{dx}{dt} = A \cos \varphi t \tag{1.3.25}$$

where C is the damping coefficient, A is the amplitude of the harmonic force, and φ is the frequency of the harmonic function.

The initial conditions of motion are taken according to expression (1.3.10).

Dividing equation (1.3.25) by m, we have

$$\frac{d^2 x}{dt^2} + 2n\frac{dx}{dt} = a \cos \varphi t \tag{1.3.26}$$

where n is the damping factor, while

$$2n = \frac{C}{m}; \quad a = \frac{A}{m} \tag{1.3.27}$$

Applying to the equation (1.3.26) Laplace Transform pairs 5, 2, 4, 2, and 29, we convert this equation with initial conditions of motion (1.3.10) from the time domain into an algebraic equation in the Laplace domain:

$$s^2 x(s) + 2nsx(s) = a\frac{s^2}{s^2 + \varphi^2} \tag{1.3.28}$$

Solving equation (1.3.28) for the displacement in the Laplace domain, we obtain

$$x(s) = \frac{sa}{(s + 2n)(s^2 + \varphi^2)} \tag{1.3.29}$$

Applying to equation (1.3.29) Laplace Transform pair 44, we obtain the solution of differential equation (1.3.25) with its initial conditions of motion (1.3.20):

$$x = \frac{a}{\varphi^2 + 4n^2}\left(e^{-2nt} + \frac{2n}{\varphi}\sin\varphi t - \cos\varphi t\right) \tag{1.3.30}$$

Taking the first derivative from equation (1.3.30), we determine the velocity of the system:

$$\frac{dx}{dt} = \frac{a}{\varphi^2 + 4n^2}(-2ne^{-2nt} + 2n\cos\varphi t + \varphi\sin\varphi t) \tag{1.3.31}$$

Supposing that in equations (1.3.30) and (1.3.31), $t = 0$, we obtain that $x = 0$ and $\frac{dx}{dt} = 0$, as expected according to initial conditions of motion (1.3.20).

1.3.5. *Example 5*

The following differential equation describes different kinds of motion depending on the values of the loading factors:

$$m\frac{d^2x}{dt^2} + C\frac{dx}{dt} + Kx + R = P + A\cos\varphi t \tag{1.3.32}$$

This example shows that, depending on the values of parameters C and K, the system could be in a state of underdamped vibration, critical damping, or overdamped motion.

Dividing equation (1.3.32) by m, we may write

$$\frac{d^2x}{dt^2} + 2n\frac{dx}{dt} + \omega_0^2 x + r = p + a\cos\varphi t \qquad (1.3.33)$$

where

$$r = \frac{R}{m} \qquad (1.3.34)$$

The initial conditions of motion are taken according to expression (1.3.10).

Applying Laplace Transform pairs 5, 2, 4, 2, 1, 2, 2, and 29, we convert equation (1.3.33) with the initial conditions of motion according to expression (1.3.10) from the time domain into the Laplace domain and obtain the corresponding algebraic equation in the Laplace domain:

$$s^2 x(s) + 2nsx(s) + \omega_0^2 x(s) + r = p + \frac{as^2}{s^2 + \varphi^2} \qquad (1.3.35)$$

Rearranging equation (1.3.35), we have

$$x(s)(s^2 + 2ns + \omega_0^2) = p - r + \frac{as^2}{s^2 + \varphi^2} \qquad (1.3.36)$$

Solving equation (1.3.36) for the displacement $x(s)$ in the Laplace domain, we obtain

$$x(s) = \frac{p - r}{s^2 + 2ns + \omega_0^2} + \frac{as^2}{(s^2 + 2ns + \omega_0^2)(s^2 + \varphi^2)} \qquad (1.3.37)$$

Rearranging the expression $(s^2 + 2ns + \omega_0^2)$ of the denominators in equation (1.3.37), we may write

$$s^2 + 2ns + \omega_0^2 + n^2 - n^2 = (s + n)^2 + \omega^2 \qquad (1.3.38)$$

where

$$\omega^2 = \omega_0^2 - n^2 \qquad (1.3.39)$$

while ω^2 could be positive, equal to zero, or negative. Therefore, if $\omega^2 > 0$, then the system performs underdamped vibratory motion; in the case where $\omega^2 = 0$, we have critical damping, which is a non-vibratory motion; and in the case where $\omega^2 < 0$, the system performs overdamped non-vibratory motion. Rearranging equation (1.3.37) by combining it with equations (1.3.38) and (1.3.39), we obtain

$$x(s) = \frac{p - r}{(s + n)^2 + \omega^2} + \frac{as^2}{[(s + n)^2 + \omega^2](s^2 + \varphi^2)} \qquad (1.3.40)$$

In the following, all three cases are described.

1.3.5.1. *Underdamped vibration, $\omega^2 > 0$*

Applying to equation (1.3.40) Laplace Transform pairs 1, 21, and 68, we invert this equation from the Laplace domain into the time domain and obtain the solution of differential equation of motion(1.3.32) with the initial conditions of motion according to expression (1.3.10):

$$
\begin{aligned}
x = {} & \frac{p - r}{\omega^2 + n^2}\left[1 - e^{-nt}\left(\cos\omega t + \frac{n}{\omega}\sin\omega t\right)\right] \\
& + \frac{a}{4n^2\varphi^2 + (\omega^2 + n^2 - \varphi^2)^2}[(\omega^2 + n^2 - \varphi^2)(\cos\varphi t - e^{-nt}\cos\omega t) \\
& + 2n\varphi\sin\varphi t - \frac{n}{\omega}(\omega^2 + n^2 + \varphi^2)e^{-nt}\sin\omega t] \qquad (1.3.41)
\end{aligned}
$$

Taking the first derivative from equation (1.3.41), we obtain the equation describing the velocity of the system:

$$
\begin{aligned}
\frac{dx}{dt} = {} & \frac{p - r}{\omega}e^{-nt}\sin\omega t + \frac{a}{4n^2\varphi^2 + (\omega^2 + n^2 - \varphi^2)^2} \\
& \times \left\{ 2n\varphi^2(\cos\varphi t - e^{-nt}\cos\omega t) - \varphi(\omega^2 + n^2 - \varphi^2)\sin\varphi t \right. \\
& \left. + \frac{1}{\omega}e^{-nt}[(\omega^2 + n^2)^2 - \varphi^2(\omega^2 + n^2)]\sin\omega t \right\} \qquad (1.3.42)
\end{aligned}
$$

Assuming that in equations (1.3.41) and (1.3.42), $t = 0$, we get that $x = 0$ and $\frac{dx}{dt} = 0$. as expected according to initial conditions of motion (1.3.10).

The second derivative from equation (1.3.41) yields the acceleration of the system:

$$\frac{d^2 x}{dt^2} = (p - r)e^{-nt}\left(\cos\omega t - \frac{n}{\omega}\sin\omega t\right) + \frac{a}{4n^2\varphi^2 + (\omega^2 + n^2 - \varphi^2)^2}$$

$$\times \left\{[(\omega^2 + n^2)^2 + \varphi^2(3n^2 - \omega^2)]e^{-nt}\cos\omega t - \varphi^2(\omega^2 + n^2 - \varphi^2)\right.$$

$$\times \cos\varphi t - 2\varphi^3 n\sin\varphi t - \frac{n}{\omega}[(\omega^2 + n^2)^2 + \varphi^2(n^2 - 3\omega^2)]e^{-nt}$$

$$\times \left. \sin\omega t - 2\varphi^3 n\sin\varphi t\right\} - \varphi(\omega^2 + n^2 - \varphi^2)\sin\varphi t$$

$$+ \frac{1}{\omega}e^{-nt}[(\omega^2 + n^2)^2 - \varphi^2(\omega^2 + n^2)]\sin\omega t\bigg\} \tag{1.3.43}$$

Equations (1.3.41)–(1.3.43) represent the basic parameters of motion of the system.

1.3.5.2. *Critical damping, $\omega^2 = 0$*

In equation (1.3.40), equating the parameter ω^2 to zero, we determine the displacement $x(s)$ of the system in the Laplace domain:

$$x(s) = \frac{p - r}{(s + n)^2} + \frac{s^2 a}{(s + n)^2(s^2 + \varphi^2)} \tag{1.3.44}$$

By applying Laplace Transform pairs 1, 19, and 71 to equation (1.3.44), we invert this equation from the Laplace domain into the time domain and obtain the solution of differential equation (1.3.34) with its initial conditions of motion according to expression (1.3.10) for the case of critical damping:

$$x = \frac{p - r}{n^2}[1 - e^{-nt}(1 + nt)] + \frac{a}{(\varphi^2 + n^2)^2}\{2n\varphi\sin\varphi t$$

$$+ e^{-nt}[\varphi^2 - n^2 - nt(\varphi^2 + n^2) - (\varphi^2 - n^2)\cos\varphi t\}$$

$$\tag{1.3.45}$$

Taking the first derivative from equation (1.3.45), we determine the velocity of the system:

$$\frac{dx}{dt} = (p - r)te^{-nt} + \frac{a}{(\varphi^2 + n^2)^2}[(\varphi^2 - n^2)(\varphi\sin\varphi t + n^2 te^{-nt})$$
$$- 2\varphi^2 n(\cos\varphi t - e^{-nt})] \tag{1.3.46}$$

Supposing that in equations (1.3.45) and (1.3.46), $t = 0$, we get that $x = 0$ and $\frac{dx}{dt} = 0$, as it should be according to initial conditions of motion (1.3.10).

1.3.5.3. *Overdamped motion, $\omega^2 < 0$*

For this case, we change in equations (1.3.41) the sign of the parameter ω^2 from positive to negative, and we obtain

$$x(s) = \frac{p - r}{(s + n)^2 - \omega^2} + \frac{s^2 a}{[(s + n)^2 - \omega^2](s^2 + \varphi^2)} \tag{1.3.47}$$

Applying Laplace Transform pairs 1, 28, and 69, we invert equation (1.3.47) from the Laplace domain into the time domain and obtain for this case the solution of differential equation of motion (1.3.32) with the initial conditions of motion according to expression (1.3.10):

$$x = \frac{p - r}{n^2 - \omega^2}\left[1 - e^{-nt}\left(\cosh\omega t - \frac{n}{\omega}\sinh\omega t\right)\right]$$
$$+ \frac{a}{4n^2\varphi^2 + (n^2 - \omega^2 - \varphi^2)^2}\left[(n^2 - \omega^2 - \varphi^2)(\cos\varphi t\right.$$
$$\left. - e^{-nt}\cosh\omega t) + 2n\varphi\sin\varphi t - \frac{n}{\omega}(n^2 - \omega^2 + \varphi^2)e^{-nt}\sinh\omega t\right] \tag{1.3.48}$$

The first derivative from equation (1.3.48) yields the velocity of the system:

$$\frac{dx}{dt} = \frac{p-r}{\omega}e^{-nt}\sinh\omega t + \frac{a}{4n^2\varphi^2 + (n^2 - \omega^2 - \varphi^2)^2}\left\{2n\varphi^2(\cos\varphi t\right.$$

$$-\,e^{-nt}\cos\omega t) - \varphi(n^2 - \omega^2 - \varphi^2)\sin\varphi t + \frac{1}{\omega}e^{-nt}[(n^2 - \omega^2)^2$$

$$\left.+\;\varphi^2(\omega^2 + n^2)]\sinh\omega t\right\} \tag{1.3.49}$$

If in equations (1.3.48) and (1.3.49), $t = 0$, then we obtain that $x = 0$ and $\frac{dx}{dt} = 0$, as expected according to initial conditions of motion (1.3.10).

1.4. Understanding the Aspects of the Analysis of the Operational Process of a Movable Mechanical System

1.4.1. *Basic parameters of motion and their analyses*

The operational process of a movable mechanical system represents the motion of the system. Motion can be completely defined by three basic parameters. A basic parameter of motion represents a single defining feature of motion that cannot be decomposed into interrelated components. The basic parameters of motion represent displacement, velocity, and acceleration/deceleration. All parameters of motion represent functions of time. The mathematical composition of the basic parameters of motion should be determined for each operational process. The solution of the corresponding second-order differential equation of motion represents the law of motion of the system. For each operational process, the corresponding second-order differential equation of motion should be composed

As mentioned earlier, there are 64 primary and 32 secondary second-order differential equations of motion that characterize the entire spectrum of one-degree-of-freedom mechanical systems. Hence, the number of these differential equations of motion coincides with the number of possible operational processes associated with the one-degree-of-freedom systems. All these equations are composed and solved. Therefore, we have laws of motion related to all common operational processes. Obviously, we have all necessary derivatives; therefore, we possess all related analytical expressions of the basic parameters of motion of common mechanical systems. However, most of these expressions contain transcendental trigonometric and exponential functions, whose analysis in general terms is, in most cases, impossible. As is known from courses on calculus, a transcendental trigonometric or exponential function could be approximated by certain series of increasing orders of powers of the argument. University graduates and engineering practitioners might be familiar with this mathematical concept, which is associated with Taylor's and Maclaurin's series. However, in the vast majority of practical cases, the achievable accuracy of engineering calculations associated with this concept is limited to the cubic powers of the corresponding parameters, which may be not sufficient for certain calculations related to operational processes. For this reason, in this textbook, the methodology based on numeric solutions is used with the help of the computer programming language Python, which allows us to obtain the necessary accuracy of the analyses of the related parameters of operational processes. In this book for all Python solutions the time is presented in s and the displacement in m. Therefore, in each chapter of the book, at least one example of using Python programs for the analysis of the corresponding operational processes is presented.

1.5. Analysis of the Motion of a System Subjected to a Dry Friction Force in the Case of Vibration or Other Reciprocating Motions

Consider a system moving on a horizontal frictional surface, while the air resistance to the motion is negligible. For this case, we ignore the difference between the static and dynamic friction coefficients. The system is subjected to dry friction and harmonic forces, while the initial velocity of the system equals zero. It is obvious that the motion of the system is possible only in the case where the amplitude of the harmonic force exceeds the dry friction force. However, the amplitude of the harmonic force periodically decreases to zero and becomes smaller than the dry friction force. This leads to periodical stops of the system that are accompanied by changes in the direction of the velocity of the system. A change in the direction of the velocity causes a change in the direction of the dry friction force that is always directed opposite to the direction of the velocity. However, the differential equation of motion of the system as well as the analytical expressions describing the solution of this equation cannot change the direction of the dry friction force by themselves. In the following, we demonstrate a simplified example of how to determine the displacement of a mass before the first stop of the system occurs. To describe the continuation of the motion after the first stop, we have to use the same differential equation of motion that was used before the first stop while applying new initial conditions of motion that indicate that the motion starts from the point of the first stop and the initial velocity equals zero. Let us consider this example that is based on the following differential equation of motion:

$$m\frac{d^2x}{dt^2} + F = A\cos\varphi t \tag{1.5.1}$$

The initial conditions of motion are taken according to expression (1.3.10). Dividing equation (1.5.1) by m, we have

$$\frac{d^2x}{dt^2} + f = a\cos\varphi t \tag{1.5.2}$$

Applying to equation (1.5.2) Laplace Transform pairs 5, 2, 2, and 29, we convert differential equation (1.5.2) with the initial conditions of motion (1.3.10) from the time domain into the Laplace domain and obtain an algebraic equation in the Laplace domain:

$$s^2x(s) + f = \frac{as^2}{s^2 + \varphi^2} \tag{1.5.3}$$

where

$$f = \frac{F}{m} \tag{1.5.4}$$

Solving equation (1.5.3) for the displacement $x(s)$ in the Laplace domain, we write

$$x(s) = \frac{a}{s^2 + \varphi^2} - \frac{f}{s^2} \tag{1.5.5}$$

Applying Laplace Transform pairs 1, 17, 2, 2, and 6 to equation (1.5.5), we obtain the solution of differential equation of motion (1.5.1) with the initial conditions of motion (1.3.10):

$$x = \frac{a}{\varphi^2}(1 - \cos \varphi t) - \frac{1}{2}ft^2 \qquad (1.5.6)$$

The first derivative from equation (1.5.6) yields the velocity of the system:

$$\frac{dx}{dt} = \frac{a}{\varphi} \sin \varphi t - ft \qquad (1.5.7)$$

Taking that in equation (1.5.6) and (1.5.7), $t = 0$, we obtain

$$x = 0; \quad \frac{dx}{dt} = 0 \qquad (1.5.8)$$

as expected according to initial conditions of motion (1.3.10).

Taking the first derivative from equation (1.5.7), we determine the acceleration of the system:

$$\frac{d^2x}{dt^2} = a \cos \varphi t - f \qquad (1.5.9)$$

Combining equations (1.5.2) and (1.5.8), we have

$$a \cos \varphi t - f + f - a \cos \varphi t = 0 \qquad (1.5.10)$$

Hence, expressions (1.5.8) and (1.5.10) confirm that the solution of differential equation (1.5.1) at initial conditions of motion (1.3.10) is correct, and we may continue to analyze the parameters of motion.

The first stop of the system will occur when

$$\frac{dx}{dt} = 0 \qquad (1.5.11)$$

Combining equations (1.5.7) and (1.5.11), we determine the interval of time passed before the first stop:

$$T_1 = \frac{a}{f\varphi} \sin \varphi T_1 \qquad (1.5.12)$$

Combining equations (1.5.6) and (1.5.12), we determine the displacement S_1 of the mass before the first stop:

$$S_1 = \frac{a}{\varphi^2}(1 - \cos \varphi T_1) - \frac{1}{2} f T_1^2. \tag{1.5.13}$$

The expression (1.5.13) allows us to determine the new initial displacement, while the new initial velocity, as was indicated above, equals zero. Hence, we considered the first iteration of the investigational process related to this case. The next iterations are similar. The number of iterations depends on the value of the total displacement that the system is required to perform. It should be clear that the minimum displacement equals S_1, while the maximum displacement equals

$$S_{\max} = S_1 + n_i S_i$$

where n_i is the number of iterations after the first stop and S_i is the value of the displacement during one iteration.

1.6. Comments to Tables 1.6.1 and 1.6.2

Tables 1.6.1 and 1.6.2 are very helpful in achieving the completeness of the analyses of operational processes.

The tables are composed in accordance with the corresponding parameters that characterize the operational processes. These parameters consist of the characteristics of the combinations of resisting and active forces involved in the operational process. The "star" symbol (*) indicates a resisting force or the forces involved in the operational process. In the following, Tables 1.6.1 and 1.6.2 related to Part 1 of this text are presented.

Column A indicates that no active forces are applied to the system. This means that the motion of the system occurs due to its initial conditions of motion. Columns B, C, and D indicate the force or the combination of active forces that are applied to the system. In Table 1.6.1, the numbers in columns A–D indicate the orderly numbers of the differential equations of motion

Table 1.6.1. Primary differential equations of motion.

#	Resisting (Reactive, Internal) Forces				No Force (A)	Active (External) Forces (B, C, D)		
					Number of the Differential Equation of Motion			
1	2	3	4	5	A	B	C	D
1	$C\frac{dx}{dt}$	Kx	F	R	0	P	$A\cos\varphi\,t$	$P+A\cos\varphi\,t$
2					2.1.1	2.2.1	2.3.3	2.4.1
3				*	3.1.1	3.2.1	3.3.1	3.4.1
4			*		4.1.1	4.2.1	4.3.1	4.4.1
5			*	*	5.1.1	5.2.1	5.3.1	5.4.1
6		*			6.1.1	6.2.1	6.3.1	6.4.1
7		*		*	7.1.1	7.2.1	7.3.1	7.4.1
8		*	*		8.1.1	8.2.1	8.3.1	8.4.1
9		*	*	*	9.1.1	9.2.1	9.3.1	9.4.1
10	*				10.1.1	10.2.1	10.3.1	10.4.1
11	*			*	11.1.1	11.2.1	11.3.1	11.4.1
12	*		*		12.1.1	12.2.1	12.3.1	12.4.1
13	*		*	*	13.1.1	13.2.1	13.3.1	13.4.1
14	*	*			14.1.1	14.2.1	14.3.1	14.4.1
15	*	*		*	15.1.1	15.2.1	15.3.1	15.4.1
16	*	*	*		16.1.1	16.2.1	16.3.1	16.4.1
17	*	*	*	*	17.1.1	17.2.1	17.3.1	17.4.1

Table 1.6.2. Secondary differential equations of motion.

#	Resisting (Reactive, Internal) Forces				No Force (A)	Active (External) Forces (B, C, D)			Values of the Square of Compound Frequencies
					Number of Sections Describing Secondary Differential Equations of Motion and Their Solutions				
1	2	3	4	5	A	B	C	D	
1	$C\frac{dx}{dt}$	Kx	R	F	0	P	$A\cos\varphi t$	$P+A\cos\varphi t$	ω^2
2	*	*			14.1.2	14.2.2	14.3.2	14.4.2	$\omega^2=0$
3	*	*			14.1.3	14.2.3	14.3.3	14.4.3	$\omega^2<0$
4	*	*	*		15.1.2	15.2.2	15.3.2	15.4.2	$\omega^2=0$
5	*	*	*		15.1.3	15.2.3	15.3.3	15.4.3	$\omega^2<0$
6	*	*		*	16.1.2	16.2.2	16.3.2	16.4.2	$\omega^2=0$
7	*	*		*	16.1.3	16.2.3	16.3.3	16.4.3	$\omega^2<0$
8	*	*	*	*	17.1.2	17.2.2	17.3.2	17.4.2	$\omega^2=0$
9	*	*	*	*	17.1.3	17.2.3	17.3.3	17.4.3	$\omega^2<0$

related to a particular operational process present in the book. For instance, the number 11.2.1 indicates the number of the differential equation of motion related to the operational process in which the mass is subjected to a resisting force exerted by a dashpot, a constant resisting force R, and a constant active force P.

In Table 1.6.2, the numbers in columns A–D indicate the numbers of sections in which the solutions of the differential equations describing the motion of the masses in certain operational processes are presented. For example, the number 16.3.3 indicates that a mass is subjected to resisting forces exerted by a dashpot and a spring, a dry friction force, and also a harmonic force.

The meanings of the rest of the symbols in Tables 1.6.1 and 1.6.2 are self-explanatory.

MOTION OF A SYSTEM ON A HORIZONTAL FRICTIONLESS SURFACE IN THE ABSENCE OF RESISTING AND ACTIVE FORCES

Investigation of the process of motion of a mechanical system is based on the analysis of the basic parameters of motion, namely displacement, velocity, and acceleration, as functions of time. These functions could be determined by solving the corresponding second-order differential equations of motion. As discussed in Chapter 1, the solution of a second-order differential equation of motion represents an expression that analytically describes the law of motion of a certain mechanical system or its mass as a function of time. The analytical expression of the law of motion characterizes the displacement of the system (or its mass) as a function of running time. Therefore, the main objectives of the analytical investigations of the operational processes of mechanical systems consist of composing and solving the corresponding differential equations of motion of mechanical systems (or masses).

2.1. Motion of a System Due to Its Initial Velocity

We start with composing and solving the simplest linear second-order differential equation of motion of a one-degree-of-freedom mechanical system. This equation describes the operational process of a system moving on a horizontal frictionless surface due to its initial velocity.

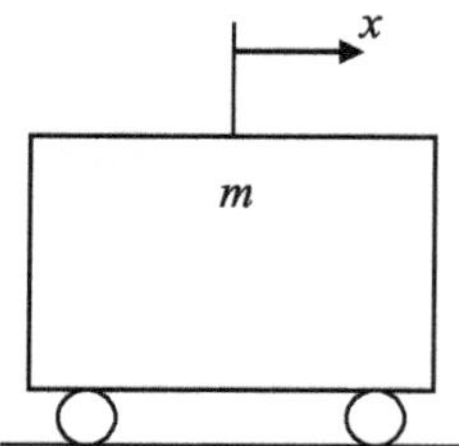

Fig. 2.1.1. Schematic diagram of a system moving on a horizontal frictionless surface.

The system, possessing a mass, m, is moving on a wheeled platform that symbolizes a frictionless interaction between the mass and the surface. The air resistance to the motion of the system is negligible.

Figure 2.1.1 shows a schematic diagram of a mechanical system performing translational motion in the horizontal direction indicated by the abscissa x.

Based on the considerations mentioned above and the schematic diagram shown in Figure 2.1.1, we compose the following differential equation of motion of the mass m of the system:

$$m\frac{d^2x}{dt^2} = 0 \tag{2.1.1}$$

where t is the running time, which is the argument. Hence, the general initial conditions of motion are

$$\text{for } t = 0 \quad x = S; \quad \frac{dx}{dt} = V \tag{2.1.2}$$

In fact, this case has an important practical application. Consider a situation in which a rigid body of a known mass m is located on a horizontal frictionless surface and subjected to an impulse loading directed parallel to this surface. It is known that it is impossible to determine the characteristics of the force applied to the mass during the impulse. This prevents the opportunity of composing the second-order differential equation of motion resulting from the force developed during the impulse. However, since

$$P_i t_i = mV \tag{2.1.3}$$

where P_i is the force exerted during the impulse toward the mass m and t_i is the time duration of the impulse, while V is the instantaneous velocity of the mass m after the impulse.

Contemporary measurement techniques allow measuring the instantaneous velocity of the mass after the impulse. Therefore, by assigning to the mass the initial velocity V according to expression (2.1.2), the value of which was determined by measuring the instantaneous velocity of the mass immediately after the impulse, it becomes possible to compose the appropriate differential equation of motion that is expressed in equation (2.1.1). Dividing equation (2.1.1) by m, we may write

$$\frac{d^2 x}{dt^2} = 0 \tag{2.1.4}$$

As noted above, in this book, the solutions of the linear second-order differential equations of motion are based on the mathematically rigorous, straightforward, and universal Laplace transform methodology, the first step of which is converting the differential equation from the time domain into an algebraic equation in the Laplace domain. The conversion can be performed by applying to equation (2.1.4), with the initial conditions of motion according to expression (2.1.2), Laplace Transform pair 5 from the corresponding table of Laplace Transform pairs presented in Appendix A-1. The following expression represents the above-mentioned conversion, which is an algebraic equation with one unknown, $x(s)$, playing the role of the Laplace domain displacement of the mass, as a function of the Laplace variable s.

$$s^2 x(s) - sV - s^2 S = 0 \tag{2.1.5}$$

Solving the algebraic equation (2.1.5) for the displacement in the Laplace domain $x(s)$, we obtain

$$x(s) = \frac{V}{s} + S \tag{2.1.6}$$

Applying to the algebraic equation (2.1.6) Laplace Transform pairs 1, 3, and 2 listed in Appendix A-1, we invert this equation from the Laplace domain into the time domain and obtain the law of motion that describes the displacement of a mass as a function of time and represents the solution of the differential equation (2.1.1) with the initial conditions of motion according to expression (2.1.2):

$$x = S + Vt \qquad (2.1.7)$$

Equation (2.1.7) indicates that the system performs a uniform translational motion caused by the initial velocity of the mass.

The first derivative from equation (2.1.7) describes the velocity of the system:

$$\frac{dx}{dt} = V \qquad (2.1.8)$$

Assuming for the equations (2.1.7) and (2.1.8) that $t = 0$, we obtain that $x = S$ and $\frac{dx}{dt} = V$, as expected from the initial conditions of motion according to expression (2.1.2).

Hence, the values of the basic parameters of motion of the system can be determined using the expressions (2.1.7), (2.1.8), and (2.1.4). According to equation (2.1.7), the motion will continue indefinitely with the constant velocity V.

2.2. Motion of a System Due to Its Initial Velocity and a Constant Active Force

We consider the operational process of a movable system, the motion of which is predetermined by the initial velocity of the system and by a constant active force applied to it. The system is moving on a horizontal frictionless surface, while the air resistance is negligible. Figure 2.2.1 shows the schematic diagram of the system described

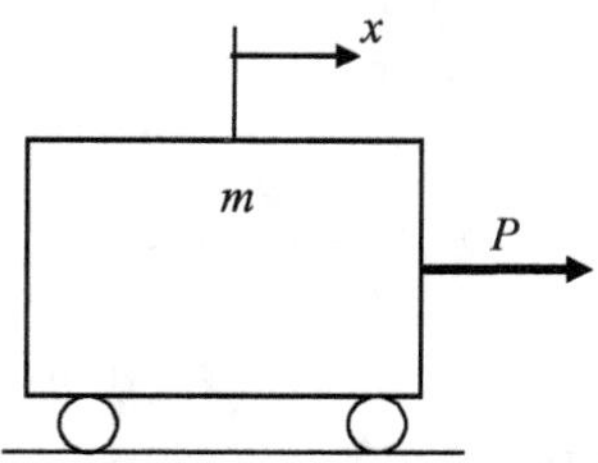

Fig. 2.2.1. Schematic diagram of a system moving on a horizontal frictionless surface, while being subjected to a constant active force.

above. The notations in this figure are self-explanatory.

$$m\frac{d^2x}{dt^2} = P \tag{2.2.1}$$

The general initial conditions of motion for the current system are characterized by expression (2.1.2).

Dividing differential equation (2.2.1) by m, we may write

$$\frac{d^2x}{dt^2} = p \tag{2.2.2}$$

where

$$p = \frac{P}{m} \tag{2.2.3}$$

Applying Laplace Transform pairs 5 and 2, we convert differential equation (2.2.1) with the initial conditions of motion (2.1.2) from the time domain into an algebraic equation in the Laplace domain:

$$s^2x(s) - sV - s^2S = p \tag{2.2.4}$$

Solving equation (2.2.4) for the displacement in the Laplace domain $x(s)$, we obtain

$$x(s) = S + \frac{V}{s} + \frac{p}{s^2} \tag{2.2.5}$$

Applying to equation (2.2.5) Laplace Transform pairs 1, 2, 3, and 6 from Appendix A-1, we invert this equation from the Laplace domain into the time domain and obtain the solution of differential equation (2.2.1) with initial conditions of motion (2.1.2):

$$x = S + Vt + \frac{1}{2}pt^2 \tag{2.2.6}$$

According to this equation, the system is in accelerated translational motion.

Taking the first derivative from equation (2.2.6), we determine the velocity of the system:

$$\frac{dx}{dt} = V + pt \tag{2.2.7}$$

Assuming that in equations (2.2.6) and (2.2.7), $t = 0$, we obtain that $x = S$ and $\frac{dx}{dt} = V$, as it should be according to the initial conditions of motion presented in expression (2.1.2). The first derivative from equation (2.2.7) yields the acceleration of the system:

$$\frac{d^2 x}{dt^2} = p \tag{2.2.8}$$

Hence, equations (2.2.6)–(2.2.8) characterize the three basic parameters of motion of the system.

2.2.1. *Interrelation between velocity, time, and power*

Let us determine the parameters of motion that allow us to obtain a certain velocity in the required interval of time. In this case, we consider the initial displacement and initial velocity of the system to be equal to zero, namely

$$\text{for} \quad t = 0 \quad x = 0; \quad \frac{dx}{dt} = 0 \tag{2.2.9}$$

Hence, based on equation (2.2.6), we obtain

$$x = \frac{1}{2}pt^2 \tag{2.2.10}$$

The first derivative from equation (2.2.10) yields the velocity of the system:

$$\frac{dx}{dt} = pt \tag{2.2.11}$$

Assuming that the desired velocity is V_T and the required time is T, we may write

$$V_T = pT \tag{2.2.12}$$

Based on equation (2.2.12) and recalling notation (2.2.3), we determine the value of force P_T that should be applied to the system:

$$P_T = m\frac{V_T}{T} \qquad (2.2.13)$$

Taking the first derivative from equation (2.2.11), we determine the acceleration of the system:

$$\frac{d^2 x}{dt^2} = p \qquad (2.2.14)$$

Hence, equations (2.2.10), (2.2.11), and (2.2.14) represent the three basic parameters of motion of the system, while the power N is the product of multiplying the force by the velocity, and it reads

$$N = P_T V_T.$$

2.2.2. *Downward motion of a system on an inclined frictionless surface*

Consider the operational process associated with the downward motion of a system having a weight W. The air resistance to the motion is negligible. Figure 2.2.2 shows the schematic diagram of the system, the notations of which are self-explanatory.

Based on the schematic diagram shown in Figure 2.2.2 and the considerations mentioned above, we compose the differential equation

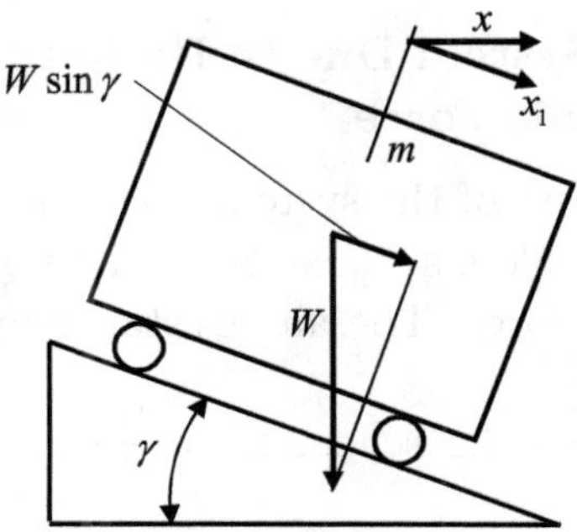

Fig. 2.2.2. Schematic diagram of a system moving downward on an inclined frictionless surface.

of motion of the system along the axis x_1:

$$m\frac{dx_1^2}{dt^2} = W \sin \gamma \tag{2.2.15}$$

Dividing equation (2.2.15) by m, we write

$$\frac{dx_1^2}{dt^2} = g \sin \gamma \tag{2.2.16}$$

where

$$g = \frac{W}{m} \tag{2.2.17}$$

is the acceleration of gravity.

The initial conditions of motion are taken according to expression (2.2.9). The solution of differential equation (2.2.16) with the initial conditions of motion (2.2.9) is similar to the solution of equation (2.2.6) and has the following form:

$$x_1 = \frac{1}{2}gt^2 \sin \gamma \tag{2.2.18}$$

According to Figure 2.2.2, the displacement of the system in the horizontal direction can be determined from the following expression:

$$x = \frac{1}{4}gt^2 \sin 2\gamma \tag{2.2.19}$$

2.3. Motion of a System Due to Its Initial Velocity and a Harmonic Force

The operational process of the system is characterized by its motion on a horizontal frictionless surface due to its initial velocity and the action of a harmonic force. The air resistance applied to the system is negligible.

Figure 2.3.1 shows the schematic diagram of the system mentioned above.

Based on the considerations mentioned above and the schematic diagram shown in Figure 2.3.1, we compose the following differential

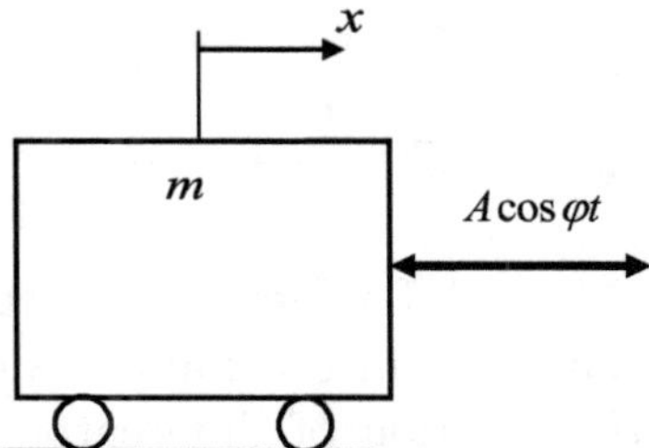

Fig. 2.3.1. Schematic diagram of a system moving on a horizontal frictionless surface while being subjected to a harmonic force.

equation of motion of the system:

$$m\frac{d^2x}{dt^2} = A\cos\varphi t \qquad (2.3.1)$$

where A is the amplitude of the harmonic force and φ is the frequency of the harmonic function. The initial conditions of motion are taken according to expression (2.1.2).

Dividing equation (2.3.1) by m, we may write

$$\frac{d^2x}{dt^2} = a\cos\varphi t \qquad (2.3.2)$$

where

$$a = \frac{A}{m} \qquad (2.3.3)$$

Taking for differential equation (2.3.2) the initial conditions of motion (2.1.2) and applying Laplace Transform pairs 5, 2, and 29 from Appendix A-1, we convert this equation from the time domain into an algebraic equation in the Laplace domain:

$$s^2 x(s) - sV - s^2 S = \frac{as^2}{s^2 + \varphi^2} \qquad (2.3.4)$$

Solving equation (2.3.4) for the displacement $x(s)$, we obtain an algebraic equation describing the displacement as a function of the

Laplace variable:

$$x(s) = S + \frac{V}{s} + \frac{a}{s^2 + \varphi^2} \qquad (2.3.5)$$

Applying to equation (2.3.5) Laplace Transform pairs 1, 2, 3, and 17, we invert this equation from the Laplace domain into the time domain and obtain the solution of differential equation of motion (2.3.1) with the initial conditions of motion according to expression (2.1.2):

$$x = S + Vt + \frac{a}{\varphi^2}(1 - \cos \varphi t) \qquad (2.3.6)$$

Equation (2.3.6) shows that the motion of the system represents a superposition of a uniform motion caused by the initial conditions of motion and a vibratory motion due to the harmonic force.

Taking the first and second derivatives from equation (2.3.6), we obtain the corresponding expressions describing the velocity and acceleration of the system:

$$\frac{dx}{dt} = V + \frac{a}{\varphi} \sin \varphi t \qquad (2.3.7)$$

$$\frac{d^2 x}{dt^2} = a \cos \varphi t \qquad (2.3.8)$$

Assuming for equations (2.3.6) and (2.3.7), $t = 0$, we determine that $x = S$ and $\frac{dx}{dt} = V$, as expected from the initial conditions of motion according to expression (2.1.2).

Hence, equations (2.3.6)–(2.3.8) describe the basic parameters of motion of the system.

Based on equations (2.3.7) and (2.3.8), we determine, respectively, the extreme values of the velocity and acceleration of the system:

$$\frac{dx}{dt} = V \pm \frac{a}{\varphi} \qquad (2.3.9)$$

$$\frac{d^2 x}{dt^2} = \pm a \qquad (2.3.10)$$

2.4. Motion of a System Due to Its Initial Velocity, a Constant Active Force, and a Harmonic Force

The analysis of the operational process of the system moving on a horizontal frictionless surface is presented in the following.

The motion of the system is caused by its initial velocity, a constant active force, and a harmonic force. The influence of air resistance on the motion of the system is negligible. Figure 2.4.1 shows a schematic diagram of a system moving on a frictionless horizontal surface while being subjected to the actions of a constant active force and a harmonic force. The notations in this figure are self-explanatory.

Based on the above-mentioned considerations and the schematic diagram shown in Figure 2.4.1, we compose the differential equation of motion of the system:

$$m\frac{d^2x}{dt^2} = P + A\cos\varphi t \qquad (2.4.1)$$

The initial conditions of motion are taken according to expression (2.1.2).

Dividing equation (2.4.1) by m, we may write

$$\frac{d^2x}{dt^2} = p + a\cos\varphi t \qquad (2.4.2)$$

Applying Laplace Transform pairs 5, 2, 2, and 29, we convert differential equation (2.4.2) with the initial conditions of motion (2.1.2) from the time domain into an algebraic equation in the

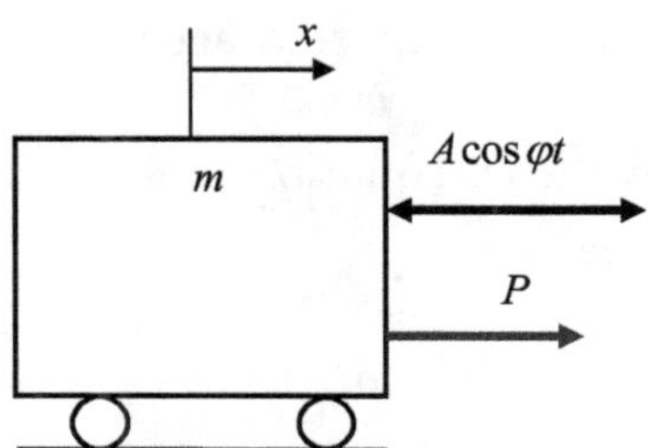

Fig. 2.4.1. Schematic diagram of a system moving on a horizontal frictionless surface while being subjected to a constant active force and a harmonic force.

Laplace domain:

$$s^2 x(s) - sV - s^2 S = p + \frac{as^2}{s^2 + \varphi^2} \qquad (2.4.3)$$

Solving equation (2.4.3) for the displacement $x(s)$ as a function of the Laplace variable s, we obtain an algebraic equation of the displacement in the Laplace domain:

$$x(s) = S + \frac{V}{s} + \frac{p}{s^2} + \frac{a}{s^2 + \varphi^2} \qquad (2.4.4)$$

Inverting this equation using Laplace Transform pairs 1, 2, 3, 6, and 17 from the Laplace domain into the time domain, we obtain the solution of the differential equation of motion (2.4.1) with the initial conditions of motion according to expression (2.1.2):

$$x = S + Vt + \frac{1}{2}pt^2 + \frac{a}{\varphi^2}(1 - \cos \varphi t) \qquad (2.4.5)$$

Equation (2.4.5) indicates that the motion of the system represents a superposition of a uniform motion caused by the initial velocity, an accelerated translational motion caused by the constant active force, and a vibratory motion resulting from the harmonic force.

Taking the first derivative from equation (2.4.6), we determine the velocity of the system:

$$\frac{dx}{dt} = V + pt + \frac{a}{\varphi} \sin \varphi t \qquad (2.4.6)$$

Assuming for equations (2.4.5) and (2.4.6) that $t = 0$, we obtain that $x = S$ and $\frac{dx}{dt} = V$, which is expected according to initial conditions of motion (2.1.2).

The first derivative from equation (2.4.6) yields the acceleration of the system:

$$\frac{d^2 x}{dt^2} = p + a \cos \varphi t \qquad (2.4.7)$$

Hence, equations (2.4.5)–(2.4.7) characterize the basic parameters of motion.

2.4.1. *Motion on a downward inclined frictionless surface*

Consider the motion of a system on a downward inclined frictionless surface. The system is subjected to the action of a component of its weight, $W \sin \gamma$, and a harmonic force, while the air resistance to the motion is negligible. Figure 2.4.2 shows a schematic diagram of the system, the notations in which are self-explanatory.

This schematic diagram shows that the system is moving on a frictionless surface that is inclined at an angle γ, while $W \sin \gamma$ combined with P is the constant active force and $A \cos \varphi t$ is the harmonic force that are applied to the system.

Actually, the motion of the system occurs in the plane xz, and consequently, the description of the motion in this plane should be considered relative to the coordinate axes x and z. This requires composing and solving the two corresponding differential equations of motion. However, from Figure 2.4.2, it can be seen that $x = x_1 \cos \gamma$, and it is more convenient to analyze the motion of the system along the axis x_1 using just one equation.

Hence, based on the schematic diagram shown in Figure 2.4.2 and the considerations mentioned above, we compose the following

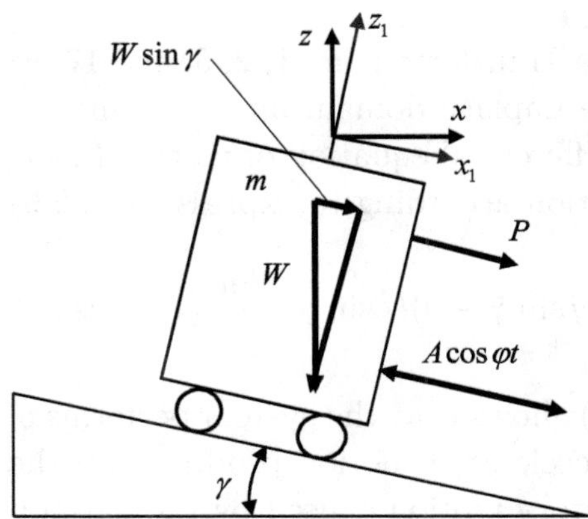

Fig. 2.4.2. Schematic diagram of a system moving downward on an inclined frictionless surface while being subjected to constant active forces and a harmonic force.

differential equation of motion of the system:

$$m\frac{d^2x_1}{dt^2} = W\sin\gamma + P + A\cos\varphi t \tag{2.4.8}$$

The initial conditions of motion are taken according to expression (2.2.9). Dividing equation (2.4.8) by m, we may write

$$\frac{d^2x_1}{dt^2} = g\sin\gamma + p + a\cos\varphi t \tag{2.4.9}$$

Using Laplace Transform pairs 5, 2, 2, 2, and 29, we convert equation (2.4.9) with the initial conditions of motion according to expression (2.2.9) from the time domain into the Laplace domain and obtain an algebraic equation in the Laplace domain:

$$s^2x_1(s) = g\sin\gamma + p + \frac{as^2}{s^2 + \varphi^2} \tag{2.4.10}$$

Solving equation (2.4.10) for $x_1(s)$, we obtain the algebraic equation describing the displacement of the system in the Laplace domain:

$$x_1(s) = \frac{g\sin\gamma + p}{s^2} + \frac{a}{s^2 + \varphi^2} \tag{2.4.11}$$

Applying Laplace Transform pairs 1, 2, 6, and 17, we invert equation (2.4.11) from the Laplace domain into the time domain and obtain the solution of differential equation of motion (2.4.8) with the initial conditions of motion according to expression (2.2.9):

$$x_1 = \frac{1}{2}(g\sin\gamma + p)t^2\sin\gamma + \frac{a}{\varphi^2}(1 - \cos\varphi t) \tag{2.4.12}$$

Equation (2.4.12) shows that the system performs a superposition of a translational accelerated motion produced by the constant active force and a vibratory motion caused by the harmonic force.

Taking the first derivative from equation (2.4.12), we determine the velocity of the system:

$$\frac{dx_1}{dt} = (g\sin\gamma + p)t + \frac{a}{\varphi}\sin\varphi t \qquad (2.4.13)$$

The first derivative from equation (2.4.13) yields the acceleration of the system:

$$\frac{d^2 x_1}{dt^2} = g\sin\gamma + p + a\cos\varphi t \qquad (2.4.14)$$

Hence, equations (2.4.12)–(2.4.14) describe the basic parameters of motion of the system. However, the transcendency of these equations does not allow their analysis in general terms.

2.4.2. *Numerical solution*

Consider case described in Section 2.4.

$$x = S + Vt + \frac{1}{2}pt^2 + \frac{a}{\varphi^2}(1 - \cos\varphi t) \qquad (2.4.5)$$

With the help of a computer program written in Python, we can obtain the numeric solution and the corresponding graph of equation (2.4.5).

```python
from matplotlib. pyplot import plot, show
from numpy import linspace, cos

S = 0.1
V = 0.1
p = 5
a = 200
phi = 30

t = Lin space(0, 2, 1000)
x = S + V*t + (1/2)*p*t**2 + a/(phi**2)*(1 - cos(phi*t))

plot (t, x)
show ()
```

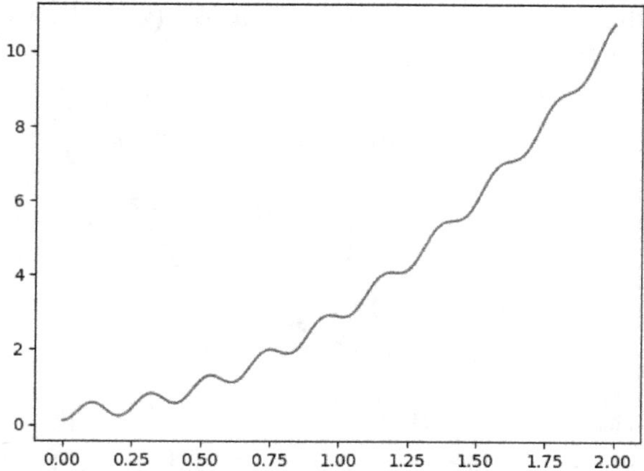

Fig. 2.4.3. Graph of equation (2.4.5).

This equation shows that the displacement continuously increases with time, while the amplitude of vibration is alternating between $\frac{a}{\varphi^2}$ and $\frac{2a}{\varphi^2}$. The graph shown in Figure 2.4.3 reflects the characteristics of equation (2.4.5).

MOTION OF A SYSTEM ON A FRICTIONLESS SURFACE WHILE BEING SUBJECTED TO A CONSTANT RESISTING FORCE

The operational processes analyzed in this chapter deal with constant resisting forces.

These forces could be exerted during deformation of a medium possessing features of plasticity and could be related to the gravitational forces of systems.

3.1. Motion of a System Due to Its Initial Velocity While Being Subjected to a Constant Resisting Force

This section presents the analysis of an operational process of a system moving on a horizontal frictionless surface due to an initial velocity while being subjected to a constant resisting force R. The air resistance to the motion of the system is negligible.

Figure 3.1.1 shows a schematic diagram of the system described above. The notations in this figure are self-explanatory.

According to the considerations mentioned above and the schematic diagram shown in Figure 3.1.1, we compose the differential equation describing the motion of the system:

$$m\frac{d^2x}{dt^2} + R = 0 \qquad (3.1.1)$$

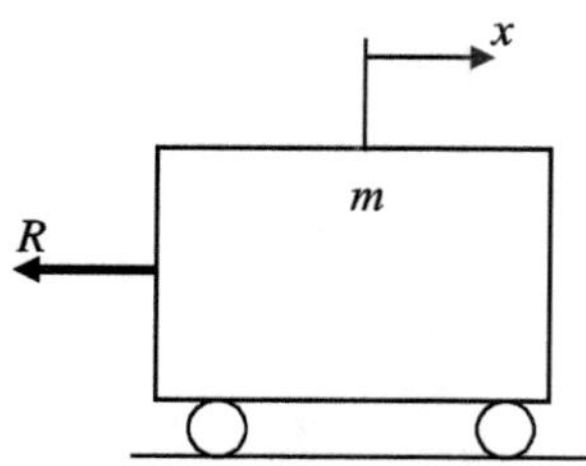

Fig. 3.1.1. Schematic diagram of a system moving on a horizontal frictionless surface while being subjected to a constant resisting force.

The initial conditions of motion are assigned according to expression (2.1.2). Dividing equation (3.1.1) by m, we may write

$$\frac{d^2x}{dt^2} + r = 0 \tag{3.1.2}$$

where

$$r = \frac{R}{m} \tag{3.1.3}$$

Applying Laplace Transform pairs 5 and 2 to equation (3.1.2) with the initial condition of motion according to expression (2.1.2), we convert this equation from the time domain into an algebraic equation in the Laplace domain, where $x(s)$ is the displacement of the system as a function of the Laplace variable s:

$$x(s) - sV - s^2S + r = 0 \tag{3.1.4}$$

Solving equation (3.1.4) for the displacement $x(s)$ in the Laplace domain, we may write

$$x(s) = sV + s^2S - r \tag{3.1.5}$$

Using Laplace Transform pairs 1, 3, 6, 2, and 2, we invert equation (3.1.5) from the Laplace domain into the time domain and obtain the solution of differential equation (3.1.1) with initial conditions of motion (2.1.2):

$$x = S + Vt - \frac{1}{2}rt^2 \tag{3.1.6}$$

Taking the first derivative from equation (3.1.6), we determine the velocity of the system:

$$\frac{dx}{dt} = V - rt \qquad (3.1.7)$$

Supposing for equations (3.1.6) and (3.1.7) that $t = 0$, we have $x = S$ and $\frac{dx}{dt} = V$, which is expected according to initial conditions of motion (2.1.2).

Taking the first derivative from equation (3.1.6), we obtain the equation describing the deceleration of the system:

$$\frac{d^2 x}{dt^2} = -r \qquad (3.1.8)$$

Hence, equations (3.1.6)–(3.1.8) represent the three basic parameters of motion of the system.

Equating equation (3.1.7) to zero, we determine the time T that the motion could last before the system comes to a stop:

$$0 = V - rT \qquad (3.1.9)$$

Solving equation (3.1.9) for T, we obtain

$$T = \frac{V}{r} \qquad (3.1.10)$$

Equation (3.1.8) shows that during the considered braking process, the deceleration represents a constant value that, in certain cases, should not exceed the corresponding allowable value.

3.2. Motion of a System on a Horizontal Frictionless Surface While Being Subjected to a Constant Resisting Force and a Constant Active Force

The description of an operational process of a movable system possessing an initial velocity while being subjected to the actions of a constant resisting force and a constant active force is presented in this section. The motion of the system occurs on a horizontal frictionless surface, while the air resistance is considered negligible. In this case, it is assumed that the active force exceeds the resisting force; therefore, the motion is accelerated.

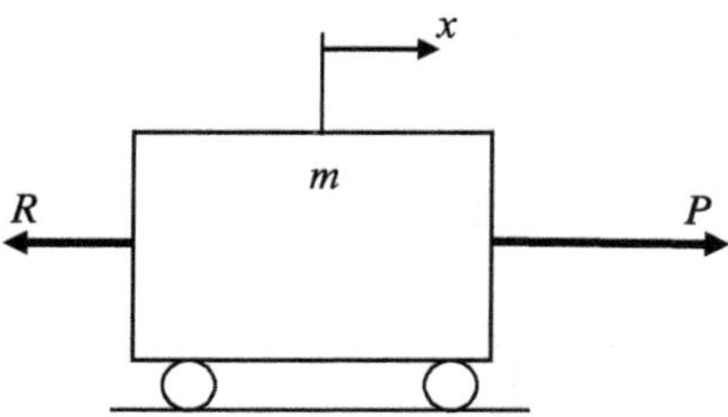

Fig. 3.2.1. Schematic diagram of a system moving on a horizontal frictionless surface while being subjected to constant resisting and active forces.

Figure 3.2.1 shows a schematic diagram that represents a system moving on a horizontal frictionless surface and subjected to a constant resisting force and a constant active force.

Based on the above-mentioned considerations and the schematic diagram shown in Figure 3.2.1, we compose the differential equation of motion of the system:

$$m\frac{d^2x}{dt^2} + R = P \tag{3.2.1}$$

The general initial conditions of motion of the system are taken according to expression (2.1.2).

Dividing equation (3.2.1) by m, we may write

$$\frac{d^2x}{dt^2} + r = p \tag{3.2.2}$$

Applying to equation (3.2.2) Laplace Transform pairs 5, 2, and 2, we convert this equation with initial conditions of motion (2.1.2) from the time domain into an algebraic equation in the Laplace domain:

$$s^2x(s) - sV - s^2S + r = p \tag{3.2.3}$$

Solving the algebraic equation (3.2.3) for the displacement $x(s)$ in the Laplace domain, we may write

$$x(s) = S + \frac{V}{s} + \frac{p - r}{s^2} \tag{3.2.4}$$

With the help of Laplace Transform pairs 1, 2, 3, 2, and 6, we invert equation (3.2.4) from the Laplace domain into the time domain and obtain the solution of differential equation (3.2.1) with the initial conditions of motion according to expression (2.1.2):

$$x = S + Vt + \frac{1}{2}(p - r)t^2 \tag{3.2.5}$$

Taking the first derivative from equation (3.2.5), we determine the velocity of the system:

$$\frac{dx}{dt} = V + (p - r)t \tag{3.2.6}$$

Supposing for equations (3.2.5) and (3.2.6) that $t = 0$, we determine that $x = S$ and $\frac{dx}{dt} = V$, as it is expected according to initial conditions of motion (2.1.2).

The first derivative from equation (3.2.6) allows for obtaining the equation describing the acceleration/deceleration of the system:

$$\frac{d^2 x}{dt^2} = p - r \tag{3.2.7}$$

Analyzing equation (3.2.7), it becomes clear that if $p > r$, then the motion of the system is accelerated.

Hence, equations (3.2.5)–(3.2.7) represent the basic parameters of motion.

3.2.1. *Motion of a system on an inclined frictionless surface*

Consider a case where a system is moving on an upward inclined frictionless surface and is subjected to the action of a constant resisting force and a constant active force, while the air resistance is negligible. We assume that the active force exceeds the resisting force. Figure 3.2.2 shows a schematic diagram of a system moving upward on an inclined frictionless surface and subjected to constant resisting and active forces. Figure 3.2.2 also shows that the coordinate system $x_1 z_1$ is rotated by an angle γ relative to the coordinate system xz, and W is the weight of the system. The rest of the notations are self-explanatory.

The motion of the system occurs in the plane xz; however, it is more convenient to consider the motion of the system along the axis

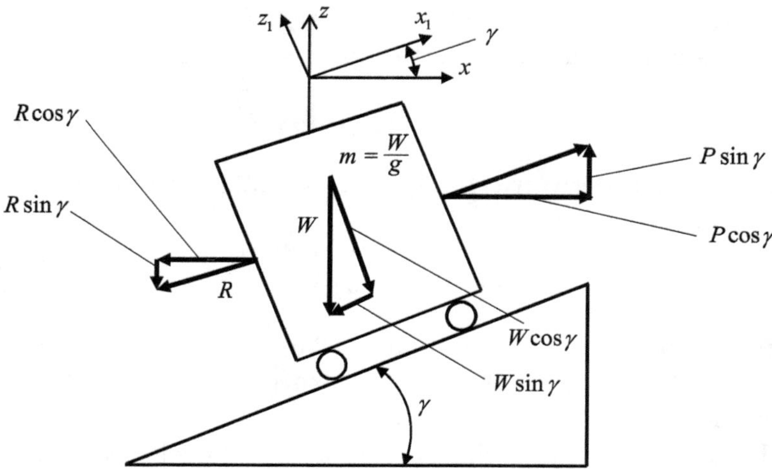

Fig. 3.2.2. Schematic diagram of a system moving on an upward inclined frictionless surface while being subjected to constant resisting and active forces.

x_1 while keeping in mind that

$$x = x_1 \cos\gamma \tag{3.2.8}$$

Hence, based on all these considerations and the schematic diagram in Figure 3.2.2, we compose the differential equation of motion:

$$m\frac{d^2 x_1}{dt^2} + R + W\sin\gamma = P \tag{3.2.9}$$

Dividing equation (3.2.9) by m, we may write

$$\frac{d^2 x_1}{dt^2} + r + g\sin\gamma = p \tag{3.2.10}$$

Using Laplace Transform pairs 5, 2, 2, and 2, we convert differential equation (3.2.10) with its initial conditions of motion from the time domain into the algebraic equation in the Laplace domain:

$$s^2 x_1(s) - sV - s^2 S + r + g\sin\gamma = p \tag{3.2.11}$$

Solving equation (3.2.11) for the displacement in the Laplace domain $x_1(s)$, we have

$$x_1(s) = S + \frac{V}{s} + \frac{1}{s^2}(p - r - g\sin\gamma) \tag{3.2.12}$$

Applying Laplace Transform pairs 1, 2, 3, 6, and 2, we invert equation (3.2.12) from the Laplace domain into the time domain and obtain the solution of differential equation (3.2.9) with its initial conditions of motion:

$$x_1 = S + Vt + \frac{1}{2}t^2(p - r - g\sin\gamma) \qquad (3.2.13)$$

Multiplying equation (3.2.13) by $\cos\gamma$, we may write

$$x = \left[S + Vt + \frac{1}{2}t^2(p - r - g\sin\gamma) \right]\cos\gamma \qquad (3.2.14)$$

Taking the first and second derivatives from equation (3.2.14), we determine, respectively, the velocity and acceleration of the system:

$$\frac{dx_1}{dt} = V + t(p - r - g\sin\gamma) \qquad (3.2.15)$$

$$\frac{d^2x_1}{dt^2} = p - r - g\sin\gamma \qquad (3.2.16)$$

Equations (3.2.14)–(3.2.16) represent the basic parameters of motion of the system.

3.3. Motion of a System on a Horizontal Frictionless Surface While Being Subjected to a Constant Resisting Force and a Harmonic Force

Consider the operational process of a system moving on a horizontal frictionless surface while it is subjected to a constant resisting force and a harmonic force.

The motion of the system is caused by its initial velocity as well as by a harmonic force. The influence of air resistance on the motion of the system is negligible. Figure 3.3.1 shows a schematic diagram of a system moving on a frictionless horizontal surface while being subjected to the actions of a constant resisting force and a harmonic force. The notations in this figure are self-explanatory.

Based on the above-mentioned considerations and the schematic diagram shown in Figure 3.3.1, we compose the differential equation

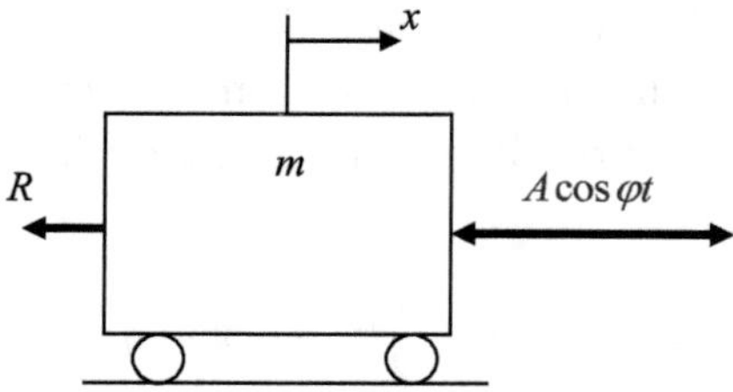

Fig. 3.3.1. Schematic diagram of a system moving on a horizontal frictionless surface, while it is being subjected to a constant resisting force and a harmonic force.

of motion of the system:

$$m\frac{d^2x}{dt^2} + R = A\cos\varphi t \qquad (3.3.1)$$

The initial conditions of motion of the system are characterized by expression (2.1.2).

Dividing equation (3.3.1) by m, we may write

$$\frac{d^2x}{dt^2} + r = a\cos\varphi t \qquad (3.3.2)$$

Applying Laplace Transform pairs 5, 2, 2, and 29, we convert differential equation (3.3.2) with initial conditions of motion (2.1.2) from the time domain into an algebraic equation in the Laplace domain:

$$s^2x(s) - sV - s^2S + r = \frac{as^2}{s^2 + \varphi^2} \qquad (3.3.3)$$

Solving equation (3.3.3) for the displacement $x(s)$ as a function of the Laplace variable s, we obtain an algebraic equation of the displacement in the Laplace domain:

$$x(s) = S + \frac{V}{s} - \frac{r}{s^2} + \frac{a}{s^2 + \varphi^2} \qquad (3.3.4)$$

Inverting this equation using Laplace Transform pairs 1, 2, 3, 6, 2, 2, and 17 from the Laplace domain into the time domain, we obtain the solution of differential equation of motion (3.3.1) with the initial conditions according to expression (2.1.2):

$$x = S + Vt - \frac{1}{2}rt^2 + \frac{a}{\varphi^2}(1 - \cos \varphi t) \qquad (3.3.5)$$

Taking the first derivative from equation (3.3.5), we determine the velocity of the system:

$$\frac{dx}{dt} = V - rt + \frac{a}{\varphi}\sin \varphi t \qquad (3.3.6)$$

Assuming for equations (3.3.5) and (3.3.6) that $t = 0$, we obtain that $x = S$ and $\frac{dx}{dt} = V$, which is expected according to initial conditions of motion (2.1.2).

The first derivative from equation (3.3.6) yields the acceleration of the system:

$$\frac{d^2x}{dt^2} = a \cos \varphi t - r \qquad (3.3.7)$$

Hence, equations (3.3.5)–(3.3.7) characterize the basic parameters of motion.

3.4. Motion of a System on a Horizontal Frictionless Surface Due to a Constant Active Force and a Harmonic Force

The operational process analyzed in the following is associated with the motion of a system on a horizontal frictionless surface while it is subjected to the actions of a constant resisting force, a constant active force, and a harmonic force. The air resistance to the motion of the system is negligible. Figure 3.4.1 shows the schematic diagram of the system described above. The notations in the figure are self-explanatory.

Based on the schematic diagram shown in Figure 3.4.1 and the considerations mentioned above, we compose the following

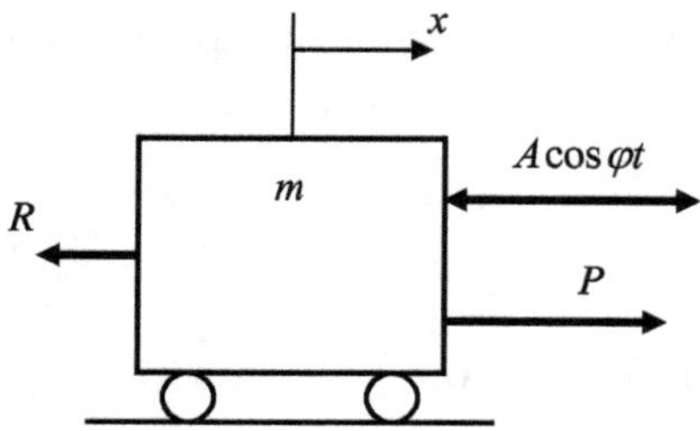

Fig. 3.4.1. Schematic diagram of a system moving on a horizontal frictionless surface while being subjected to a constant resisting force, a constant active force, and a harmonic force.

differential equation of motion:

$$m\frac{d^2x}{dt^2} + R = P + A\cos\varphi t \tag{3.4.1}$$

The initial conditions of motion are taken according to expression (2.1.2). Dividing equation (3.4.1) by m, we have

$$\frac{d^2x}{dt^2} + r = p + a\cos\varphi t \tag{3.4.2}$$

Applying Laplace Transform pairs 5, 2, 2, and 29, we convert equation (3.4.2) with the initial conditions of motion according to expression (2.1.2) from the time domain into the Laplace domain and obtain an algebraic equation in the Laplace domain:

$$s^2x(s) - sV - s^2S + r = p + \frac{as^2}{s^2 + \varphi^2} \tag{3.4.3}$$

Solving equation (3.4.3) for the displacement $x(s)$ as a function of the Laplace variable s, we may write

$$x(s) = S + \frac{V}{s} + \frac{p - r}{s^2} + \frac{a}{s^2 + \varphi^2} \tag{3.4.4}$$

Applying Laplace Transform pairs 1, 2, 3, 6, and 17, we invert equation (3.4.4) from the Laplace domain into the time domain and obtain the solution of differential equation of motion (3.4.1) with the initial conditions of motion according to expression (2.1.2):

$$x = S + Vt + \frac{1}{2}t^2(p - r) + \frac{a}{\varphi^2}(1 - \cos\varphi t) \qquad (3.4.5)$$

Taking the first derivative from equation (3.4.5), we determine the velocity of the system:

$$\frac{dx}{dt} = V + t(p - r) + \frac{a}{\varphi}\sin\varphi t \qquad (3.4.6)$$

Assuming in equations (3.4.5) and (3.4.6) that $t = 0$, we obtain that $x(0) = S$ and $\frac{dx(0)}{dt} = V$, as it should be according to expression (2.1.2).

Taking the first derivative from equation (3.4.6) allows us to determine the acceleration of the system. $\qquad (3.4.7)$

Hence, equations (3.4.5)–(3.4.7) describe the basic parameters of motion of the system.

3.4.1. *Motion of a system on a downward inclined frictionless surface due to its initial velocity, a component of the system's weight, and a harmonic force*

Consider the motion of a system on a downward inclined frictionless surface. The system is subjected to the action of a component of its weight, $W\sin\gamma$, and a harmonic force, while the air resistance to the motion is negligible. Figure 3.4.2 shows a schematic diagram of the system, the notations in which are self-explanatory.

This schematic diagram shows that the system is moving on a frictionless surface that is inclined at an angle γ, while $W\sin\gamma$ is the constant active force and $A\cos\varphi t$ is the harmonic force that are applied to the system.

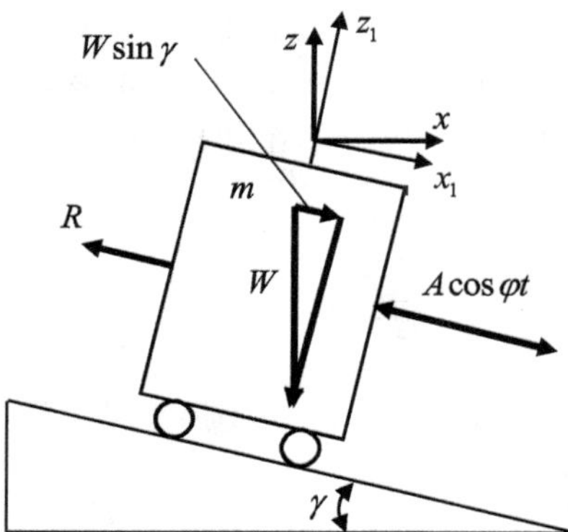

Fig. 3.4.2. Schematic diagram of a system moving downward on an inclined frictionless surface while being subjected to a constant resisting force, a constant component of the system's weight, and a harmonic force.

Actually, the motion of the system occurs in the plane xz; however, from Figure 3.4.2, it can be seen that $x = x_1 \cos\gamma$, and it is convenient to analyze the motion of the system along the axis x_1.

Hence, based on the schematic diagram shown in Figure 3.4.2 and the considerations mentioned above, we compose the following differential equation of motion of the system:

$$m\frac{d^2 x_1}{dt^2} + R = W\sin\gamma + A\cos\varphi t \tag{3.4.8}$$

Dividing equation (3.4.8) by m, we may write

$$\frac{d^2 x_1}{dt^2} + r = g\sin\gamma + a\cos\varphi t \tag{3.4.9}$$

Applying to equation (3.4.9) Laplace Transform pairs 5, 2, 2, 2, and 29, we convert equation (3.4.9) with the initial conditions of motion according to expression (2.1.2) from the time domain into the Laplace domain and obtain an algebraic equation describing the motion of the system in the Laplace domain:

$$s^2(x_1) - sV - s^2 S + r = g\sin\gamma + \frac{as^2}{s^2 + \varphi^2} \tag{3.4.10}$$

Solving equation (3.4.10) for $x_1(s)$, we obtain the algebraic equation describing the displacement of the system in the Laplace domain as a function of the Laplace variable s:

$$x_1(s) = S + \frac{V}{s} + \frac{g\sin\gamma - r}{s^2} + \frac{a}{s^2 + \varphi^2} \qquad (3.4.11)$$

Applying Laplace Transform pairs 1, 2, 3, 6, and 17, we invert equation (3.4.11) from the Laplace domain into the time domain and obtain the solution of differential equation of motion (3.4.8) with the initial conditions of motion according to expression (2.1.2):

$$x_1 = S + Vt + \frac{1}{2}t^2(g\sin\gamma - r) + \frac{a}{\varphi^2}(1 - \cos\varphi t) \qquad (3.4.12)$$

Taking the first derivative from equation (3.4.12), we determine the velocity of the system:

$$\frac{dx_1}{dt} = V + t(g\sin\gamma - r) + \frac{a}{\varphi}\sin\varphi t \qquad (3.4.13)$$

The first derivative from equation (3.4.13) yields the acceleration of the system:

$$\frac{d^2x_1}{dt^2} = g\sin\gamma + a\cos\varphi t \qquad (3.4.14)$$

Hence, equations (3.4.12)–(3.4.14) describe the basic parameters of motion of the system. However, these equations are transcendental, which prevents their analysis in general terms.

3.4.2. *Numerical solution*

Following is a Python program to plot equation (3.4.12).

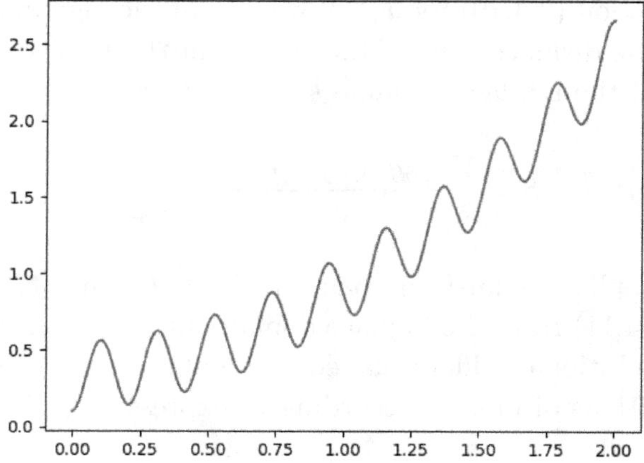

Fig. 3.4.3. Graph of equation (3.4.12).

```python
from matplotlib.pyplot import plot, show
from numpy import linspace, sin, cos

S = 0.1
V = 0.1
g = 9.8
r = 1
a = 200
phi = 30
singamma = 0.2

t = linspace(0, 2, 1000)
x1 = S + V*t + (1/2)*t**2*(g*singamma - r) + (a/phi**2)*(1 - cos(phi*t))

plot(t, x1)
show()
```

Figure 3.4.3 presents the graph of the equation, which is plotted using the above Python program.

CHAPTER 4

MOTION OF A SYSTEM BEING SUBJECTED TO A DRY FRICTION FORCE

4.1. Motion of a System on a Horizontal Surface Due to Its Initial Velocity

We consider the analysis of the operational process of a system moving on a horizontal surface due to its initial velocity. The schematic diagram of the system is shown in Figure 4.1.1, the notations in which are self-explanatory.

Based on the considerations mentioned above and the schematic diagram shown in Figure 4.1.1, we compose the differential equation describing the motion of the system:

$$m\frac{d^2x}{dt^2} + F = 0 \tag{4.1.1}$$

We accept the initial conditions of motion presented in the expression (2.1.2). Dividing equation (4.1.1) by m, we may write

$$\frac{d^2x}{dt^2} + f = 0 \tag{4.1.2}$$

where

$$f = \frac{F}{m} \tag{4.1.3}$$

Applying Laplace Transform pairs 5 and 2 to equation (4.1.2) with the initial condition of motion according to expression (2.1.2), we

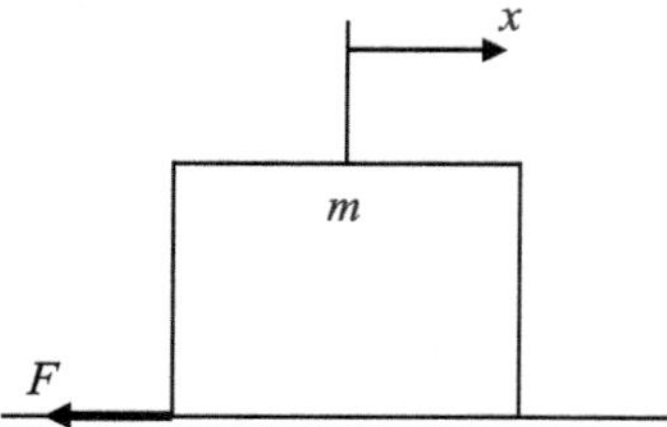

Fig. 4.1.1. Schematic diagram of a system moving on a horizontal surface while being subjected to a dry friction force.

convert this equation from the time domain into the algebraic equation in the Laplace domain:

$$x(s) - sV - s^2 S + f = 0 \qquad (4.1.4)$$

The solution of equation (4.1.4) for the displacement $x(s)$ in the Laplace domain represents the following expression:

$$x(s) = sV + s^2 S - f \qquad (4.1.5)$$

With the help of Laplace Transform pairs 1, 3, 6, and 2, we invert equation (4.1.5) from the Laplace domain into the time domain and obtain the solution of differential equation (4.1.1) with initial conditions of motion (2.1.2):

$$x = S + Vt - \frac{1}{2}ft^2 \qquad (4.1.6)$$

Equation (4.1.6) describes a decelerated translational motion.

The first derivative from equation (4.1.6) yields the velocity of the system:

$$\frac{dx}{dt} = V - ft \qquad (4.1.7)$$

Taking for equations (4.1.6) and (4.1.7) that $t = 0$, we obtain $x = S$ and $\frac{dx}{dt} = V$, as it is expected according to initial conditions of motion (2.1.2).

The first derivative from equation (4.1.7) represents the deceleration of the system:

$$\frac{d^2 x}{dt^2} = -f \tag{4.1.8}$$

It may be noted that, in some cases, the absolute value of the deceleration should not exceed the allowable value.

Equations (4.1.6)–(4.1.8) represent the three basic parameters of motion of the system.

Equating equation (4.1.7) to zero, we determine the time T that the motion could last before the system comes to a stop:

$$0 = V - fT \tag{4.1.9}$$

From equation (4.1.9), we obtain

$$T = \frac{V}{f} \tag{4.1.10}$$

4.2. Motion of a System on a Horizontal Surface While It Is Being Subjected to a Dry Friction Force and a Constant Active Force

In this case, it is assumed that the active force exceeds the dry friction force, while the air resistance to the motion of the system is negligible.

Figure 4.2.1 shows a schematic diagram of the above-mentioned system.

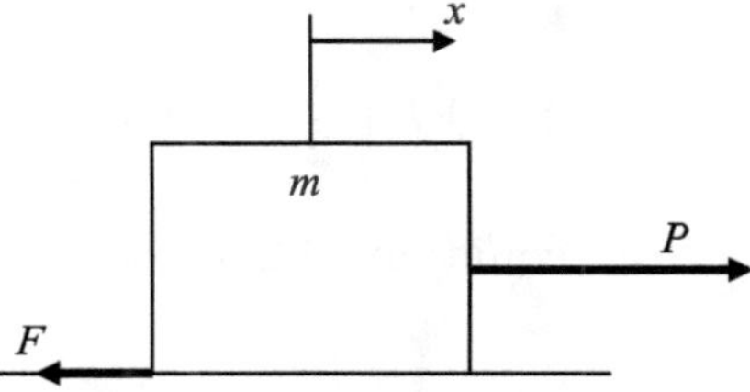

Fig. 4.2.1. Schematic diagram of a system moving on a horizontal surface while being subjected to a dry friction force and to a constant active force.

Based on the above-mentioned considerations and the schematic diagram shown in Figure 4.2.1, we compose the differential equation of motion of the system:

$$m\frac{d^2x}{dt^2} + F = P \qquad (4.2.1)$$

The generalinitial conditions of motion of the system are taken according to expression (2.1.2).

Dividing equation (4.2.1) by m, we may write

$$\frac{d^2x}{dt^2} + f = p \qquad (4.2.2)$$

Applying to equation (4.2.2) Laplace Transform pairs 5, 2, and 2, we convert this equation with initial conditions of motion (2.1.2) from the time domain into an algebraic equation in the Laplace domain:

$$s^2x(s) - sV - s^2S + f = p \qquad (4.2.3)$$

The solution of the algebraic equation (4.2.3) for the displacement $x(s)$ in the Laplace domain is:

$$x(s) = S + \frac{V}{s} + \frac{p-f}{s^2} \qquad (4.2.4)$$

Using Laplace Transform pairs 1, 2, 3, 2, and 6, we invert equation (4.2.4) from the Laplace domain into the time domain and obtain the solution of differential equation (4.2.1) with the initial conditions of motion according to expression (2.1.2):

$$x = S + Vt + \frac{1}{2}(p-f)t^2 \qquad (4.2.5)$$

The first derivative from equation (4.2.5) yields the velocity of the system:

$$\frac{dx}{dt} = V + (p-f)t \qquad (4.2.6)$$

Assuming for equations (4.2.5) and (4.2.6) that $t = 0$, we determine that $x = S$ and $\frac{dx}{dt} = V$, as expected according to initial conditions of motion (2.1.2).

The first derivative from equation (4.2.6) represents the acceleration of the system:

$$\frac{d^2x}{dt^2} = p - f \tag{4.2.7}$$

Since in equation (4.2.7), $p > f$, the motion of the system is accelerated.

Hence, equations (4.2.5)–(4.2.7) represent the basic parameters of motion.

4.2.1. *Motion of a system on a downward inclined frictional surface*

The system is moving due to its weight on a downward inclined frictional surface and is subjected to the action of a dry friction force. The air resistance to the motion of the system is negligible. Figure 4.2.2 shows a schematic diagram of the system moving downward on an inclined frictional surface and subjected to a dry friction force. It is seen from Figure 4.2.2 that the abscissa x_1 is

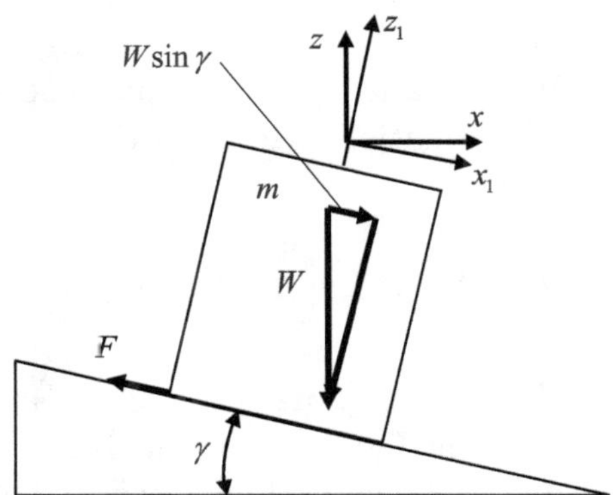

Fig. 4.2.2. Schematic diagram of a system moving downward on an inclined surface and subjected to a dry friction force and a component of the system's weight.

rotated by an angle γ relative to the abscissa x, while the weight of the system is W. The rest of the notations are self-explanatory.

It is reasonable to consider the motion of the system along the axis x_1 while keeping in mind that

$$x = x_1 \cos \gamma \qquad (4.2.8)$$

Accounting for all these considerations and the schematic diagram shown in Figure 4.2.2, we compose the differential equation of motion:

$$m\frac{d^2 x_1}{dt^2} + F = W \sin \gamma \qquad (4.2.9)$$

Obviously, the motion occurs in the case if $F < W \sin \gamma$.

Dividing equation (4.2.9) by m, we may write

$$\frac{d^2 x}{dt^2} + f = g \sin \gamma \qquad (4.2.10)$$

Applying Laplace Transform pairs 5, 2, and 2, we convert differential equation (4.2.10) with its initial conditions of motion from the time domain into an algebraic equation in the Laplace domain:

$$s^2 x_1(s) - sV - s^2 S + f = g \sin \gamma \qquad (4.2.11)$$

The solution of equation (4.2.11) for the displacement in the Laplace domain $x_1(s)$ yields the equation describing this displacement of the mass:

$$x_1(s) = S + \frac{V}{s} + \frac{1}{s^2}(g \sin \gamma - f) \qquad (4.2.12)$$

Based on Laplace Transform pairs 1, 2, 3, 6, and 2, we invert equation (4.2.12) from the Laplace domain into the time domain and obtain the solution of differential equation (4.2.9) with its initial conditions of motion (2.1.2):

$$x_1 = S + Vt + \frac{1}{2}t^2 (g \sin \gamma - f) \tag{4.2.13}$$

Multiplying equation (4.2.13) by $\cos \gamma$, we have

$$x = \left[S + Vt + \frac{1}{2}t^2 (g \sin \gamma - f) \right] \cos \gamma \tag{4.2.14}$$

The first and second derivatives from equation (4.2.13), respectively, yield the velocity and acceleration of the system:

$$\frac{dx_1}{dt} = V + t(g \sin \gamma - f) \tag{4.2.15}$$

$$\frac{d^2 x_1}{dt^2} = g \sin \gamma - f \tag{4.2.16}$$

Equations (4.2.14)–(4.2.16) represent the basic parameters of motion of the system.

4.2.2. Braking of a system subjected to a dry friction force and to a constant active force

This situation may occur when the system is moving on a downward inclined surface.

Equating equation (4.2.14) to zero, we determine the time T that the motion could last until the system stops:

$$T = \frac{V}{f - g \sin \gamma} \tag{4.2.17}$$

Solving together equations (4.2.13) and (4.2.17), we determine the breaking distance $S_{1\mathrm{br}}$:

$$S_{1\mathrm{br}} = S + \frac{V^2}{2(f - g \sin \gamma)} \tag{4.2.18}$$

It is seen from equation (4.2.16) that the deceleration D of the system equals

$$D = -f + g \sin \gamma \tag{4.2.19}$$

The absolute value of the deceleration, according to equation (4.2.19), in certain cases should not exceed the corresponding allowable value.

4.3. Motion of a System on a Horizontal Surface While It Is Being Subjected to a Dry Friction Force and to a Harmonic Force

A system moves due to its initial velocity as well as a harmonic force. The air resistance to the motion of the system is negligible. Figure 4.3.1 shows a schematic diagram of a system moving on a horizontal surface while being subjected to the actions of a dry friction force and a harmonic force. The notations in this figure are self-explanatory.

Based on the above-mentioned considerations and the schematic diagram shown in Figure 4.3.1, we compose the differential equation of motion of the system:

$$m\frac{d^2x}{dt^2} + F = A\cos\varphi t \tag{4.3.1}$$

The initial conditions of motion of the system are characterized by expression (2.1.2).

Dividing equation (4.3.1) by m, we may write

$$\frac{d^2x}{dt^2} + f = a\cos\varphi t \tag{4.3.2}$$

Applying Laplace Transform pairs 5, 2, 2, and 29, we convert differential equation (4.3.2) with initial conditions of motion (2.1.2)

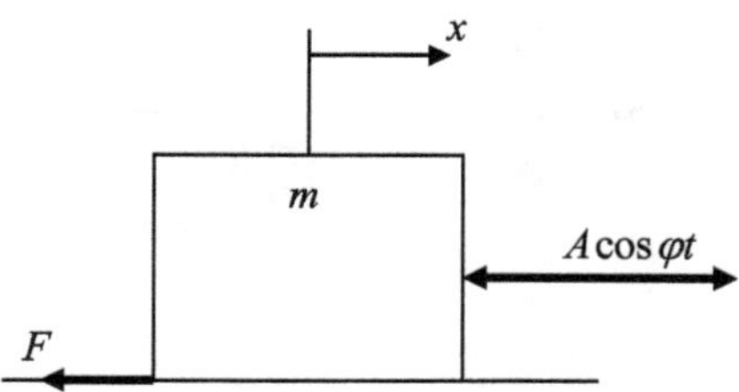

Fig. 4.3.1. Schematic diagram of a system moving on a horizontal surface while being subjected to dry friction and a harmonic force.

from the time domain into an algebraic equation in the Laplace domain:

$$s^2 x(s) - sV - s^2 S + f = \frac{as^2}{s^2 + \varphi^2} \qquad (4.3.3)$$

The solution of equation (4.3.3) for the displacement $x(s)$ as a function of the Laplace variable s represents an algebraic equation of the displacement of the mass in the Laplace domain:

$$x(s) = S + \frac{V}{s} - \frac{f}{s^2} + \frac{a}{s^2 + \varphi^2} \qquad (4.3.4)$$

Applying to equation (4.3.4) the Laplace Transform pairs 1, 2, 3, 6, 2, 2, and 17, we invert this equation from the Laplace domain into the time domain and obtain the solution of differential equation of motion (4.3.1) with the initial conditions according to expression (2.1.2):

$$x = S + Vt - \frac{1}{2} ft^2 + \frac{a}{\varphi^2}(1 - \cos \varphi t) \qquad (4.3.5)$$

The first derivative from equation (4.3.5) yields the velocity of the system:

$$\frac{dx}{dt} = V - ft + \frac{a}{\varphi} \sin \varphi t \qquad (4.3.6)$$

Supposing that in equations (4.3.5) and (4.3.6), $t = 0$, we obtain that $x = S$ and $\frac{dx}{dt} = V$, which is expected according to initial conditions of motion (2.1.2).

It should be kept in mind that, as was addressed earlier, the dry friction force instantaneously changes its direction with the change in the direction of the velocity; however, the equations describing the motion cannot perform this change by themselves. Therefore, these equations remain valid only for the instance when the change in the direction of the velocity occurs. For it to continue to describe the motion of the system after the change in the direction of the velocity, it is necessary to revise the differential equation of motion and assign new initial conditions of motion. Often, it is required to get related numeric data to determine the values of certain parameters of motion.

Taking the first derivative from equation (4.3.6), we determine the acceleration of the system:

$$\frac{d^2x}{dt^2} = a\cos\varphi t - f \tag{4.3.7}$$

Hence, equations (4.3.5)–(4.3.7) characterize the basic parameters of motion.

4.4. Motion of a System on a Horizontal Surface Due to a Constant Active Force and a Harmonic Force

We consider the motion of a system on a horizontal surface, while the system is subjected to the actions of dry friction, a constant active force, and a harmonic force. The air resistance to the motion of the system is negligible. Figure 4.4.1 shows a schematic diagram of the system described above. The notations in the figure are self-explanatory.

Based on the schematic diagram shown in Figure 4.4.1 and the considerations mentioned above, we compose the following differential equation of motion:

$$m\frac{d^2x}{dt^2} + F = P + A\cos\varphi t \tag{4.4.1}$$

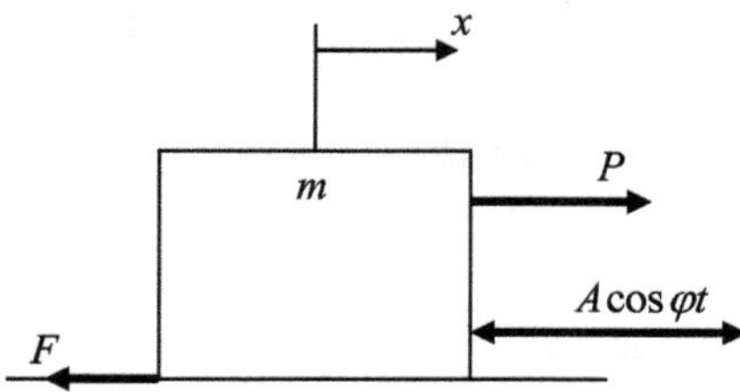

Fig. 4.4.1. Schematic diagram of a system moving on a horizontal surface while being subjected to dry friction, a constant active force, and a harmonic force.

The initial conditions of motion are assigned according to expression (2.1.2). Dividing equation (4.4.1) by m, we have

$$\frac{d^2x}{dt^2} + f = p + a\cos\varphi t \tag{4.4.2}$$

Applying Laplace Transform pairs 5, 2, 2, and 29, we convert equation (4.4.2) with the initial conditions of motion according to expression (2.1.2) from the time domain into the Laplace domain and obtain an algebraic equation in the Laplace domain:

$$s^2 x(s) - sV - s^2 S + f = p + \frac{as^2}{s^2 + \varphi^2} \tag{4.4.3}$$

Solving equation (4.4.3) for the displacement $x(s)$ as a function of Laplace variable s, we obtain

$$x(s) = S + \frac{V}{s} + \frac{p - f}{s^2} + \frac{a}{s^2 + \varphi^2} \tag{4.4.4}$$

Applying Laplace Transform pairs 1, 2, 3, 6, and 17, we invert equation (4.4.4) from the Laplace domain into the time domain and obtain the solution of differential equation of motion (4.4.1) with the initial conditions of motion according to expression (2.1.2):

$$x = S + Vt + \frac{1}{2}t^2(p - f) + \frac{a}{\varphi^2}(1 - \cos\varphi t) \tag{4.4.5}$$

The first derivative from equation (4.4.5) yields the velocity of the system:

$$\frac{dx}{dt} = V + t(p - f) + \frac{a}{\varphi}\sin\varphi t \tag{4.4.6}$$

Assuming that in equations (4.4.5) and (4.4.6), $t = 0$, we obtain that $x = S$ and $\frac{dx}{dt} = V$, as it should be according to expression (2.1.2).

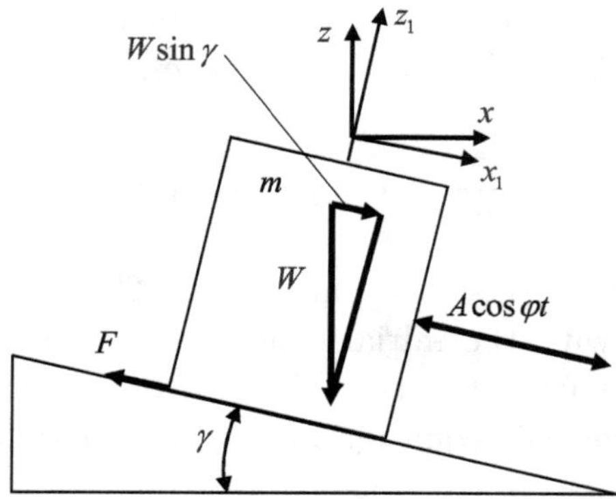

Fig. 4.4.2. Schematic diagram of a system moving downward on an inclined surface while being subjected to a dry friction force, a component of the system's weight, and a harmonic force.

Taking the first derivative from equation (4.4.6) allows us to determine the acceleration of the system:

$$\frac{d^2x}{dt^2} = p - f + a\cos\varphi t \qquad (4.4.7)$$

Hence, equations (4.4.5)–(4.4.7) describe the basic parameters of motion of the system.

4.4.1. *Motion of a system on a downward inclined surface due to its initial velocity, a component of the system's weight, and a harmonic force*

Consider the motion of a system on a downward inclined surface. The system is subjected to the actions of a dry friction force, a component of its weight, $W\sin\gamma$, and a harmonic force, while the air resistance to the motion is negligible. Figure 4.4.2 shows a schematic diagram of the system, the notations in which are self-explanatory.

Actually, the motion of the system occurs in the plane xz; however, from Figure 4.4.2, it can be seen that $x = x_1\cos\gamma$, and it is convenient to analyze the motion of the system along the axis x_1.

Hence, based on the schematic diagram shown in Figure 4.4.2 and the considerations mentioned above, we compose the following differential equation of motion of the system:

$$m\frac{d^2x_1}{dt^2} + F = W\sin\gamma + A\cos\varphi t \qquad (4.4.8)$$

Dividing equation (4.4.8) by m, we have

$$\frac{d^2x_1}{dt^2} + f = g\sin\gamma + a\cos\varphi t \qquad (4.4.9)$$

Applying to equation (4.4.9) Laplace Transform pairs 5, 2, 2, 2, and 29, we convert equation (4.4.9) with the initial conditions of motion according to expression (2.1.2) from the time domain into the Laplace domain and obtain an algebraic equation describing the motion of the system in the Laplace domain:

$$s^2(x_1) - sV - s^2S + f = g\sin\gamma + \frac{as^2}{s^2 + \varphi^2} \qquad (4.4.10)$$

The solution of equation (4.4.10) for $x_1(s)$ leads to the algebraic equation describing the displacement of the system in the Laplace domain as a function of the Laplace variable s:

$$x_1(s) = S + \frac{V}{s} + \frac{g\sin\gamma - f}{s^2} + \frac{a}{s^2 + \varphi^2} \qquad (4.4.11)$$

Using Laplace Transform pairs 1, 2, 3, 6, and 17, we invert equation (4.4.11) from the Laplace domain into the time domain and obtain the solution of differential equation of motion (4.4.8) with the initial conditions of motion according to expression (2.1.2):

$$x_1 = S + Vt + \frac{1}{2}t^2(g\sin\gamma - f) + \frac{a}{\varphi^2}(1 - \cos\varphi t) \qquad (4.4.12)$$

The first derivative from equation (4.4.12) yields the velocity of the system:

$$\frac{dx_1}{dt} = V + t(g\sin\gamma - f) + \frac{a}{\varphi}\sin\varphi t \qquad (4.4.13)$$

Taking the first derivative from equation (4.4.13), we determine the acceleration of the system:

$$\frac{d^2 x_1}{dt^2} = g \sin \gamma - f + a \cos \varphi t \qquad (4.4.14)$$

Hence, equations (4.4.12)–(4.4.14) describe the basic parameters of motion of the system.

4.4.2. *Numerical solution*

Following is a Python program to plot the graph of equation (4.4.13), which allows us to determine when the velocity of the mass will reach zero for the first time.

```python
from matplotlib.pyplot import plot, show, text
from numpy import linspace, sin
from scipy.optimize import fsolve

S = 0.1
V = 0.1
g = 9.8
f = 0.8
a = 20
singamma = 0.2
phi = 30

t = linspace(0, 0.5, 1000)
V1 = lambda t: V + t*(g*singamma - f) + (a/phi)*sin(phi*t)

[root] = fsolve(V1, 0.1)
root = round(root, 3)

plot(t, V1(t))
plot(root, 0, 'bo')
text(0.045, -0.05, f't={root}')
show()
```

The graph of equation (4.4.13) shown in Figure 4.4.3 indicates that the first stop of the mass occurs after 0.117 s from the start.

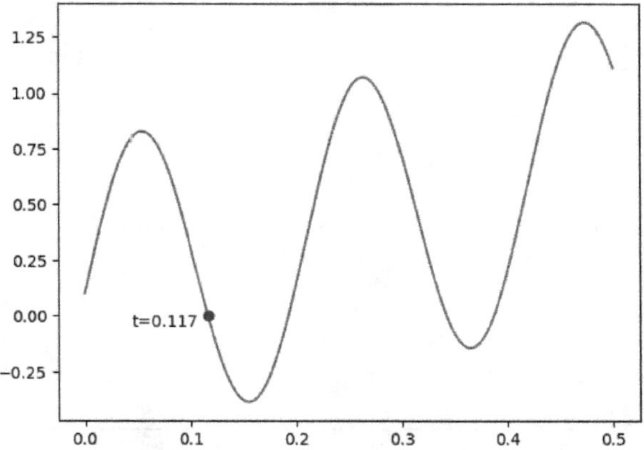

Fig. 4.4.3. Graph of equation (4.4.13).

The following shows a Python program to plot the displacement using equation (4.4.12) from $t = 0$ s to $t = 0.117$ s.

```python
from matplotlib.pyplot import plot, show, text
from numpy import linspace, cos

S = 0.1
V = 0.3
g = 9.8
f = 0.8
a = 20
singamma = 0.2
phi = 30

x1 = lambda t: \
    S + V*t + (1/2)*t**2*(g*singamma - f) + (a/phi**2)*(1 - cos(phi*t))

stopTime = .117
t = linspace(0, stopTime, 1000)
plot(t, x1(t))

stopPosition = round(x1(stopTime), 3);
plot(stopTime, stopPosition, 'bo')
text(0.104, 0.177, f'x1={stopPosition}')

show()
```

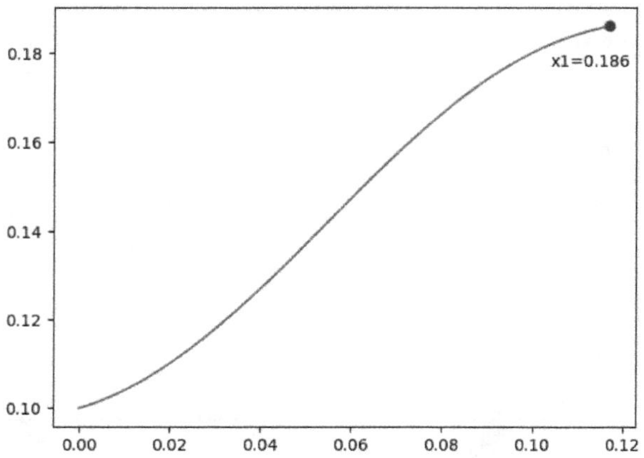

Fig. 4.4.4. Graph of equation (4.4.12).

As shown in Figure 4.4.4, the value of $x1$ is 0.186 m at time $t = 0.117$ s, and the next iteration should be analyzed by applying differential equation (4.4.8) with the following initial conditions:

$$\text{for} \quad t = 0 \quad x_1 = 0.186; \quad \frac{dx_1}{dt} = 0$$

For more details, see Section 1.5.

MOTION OF A SYSTEM ON A HORIZONTAL SURFACE BEING SUBJECTED TO A DRY FRICTION FORCE AND A CONSTANT RESISTING FORCE

5.1. Motion of a System on a Horizontal Surface Due to Its Initial Velocity While the System Is Subjected to Dry Friction and Constant Resisting Forces

We consider the analysis of the operational process of a system moving on a horizontal surface due to the initial velocity of the system that is subjected to the actions of a dry friction force F and a constant resisting force R in the absence of active forces. The air resistance to the motion of the system is ignorable.

Figure 5.1.1 shows a schematic diagram of the system mentioned above. The notations in this figure are self-explanatory. The air resistance to the motion of the system is negligible.

Accounting for the considerations mentioned above and based on the schematic diagram shown in Figure 5.1.1, we compose the differential equation describing the motion of the system:

$$m\frac{d^2x}{dt^2} + F + R = 0 \tag{5.1.1}$$

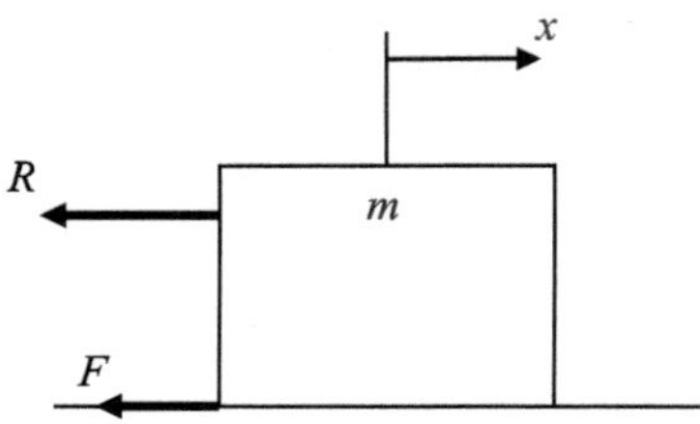

Fig. 5.1.1. Schematic diagram of a system moving on a horizontal surface while being subjected to a dry friction force and a constant resisting force.

We accept the initial conditions of motion presented in expression (2.1.2). Dividing equation (5.1.1) by m, we may write

$$\frac{d^2x}{dt^2} + f + r = 0 \tag{5.1.2}$$

Applying Laplace Transform pairs 5, 2, and 2 to equation (5.1.2) and accounting for the initial condition of motion according to expression (2.1.2), we convert this equation from the time domain into an algebraic equation in the Laplace domain:

$$s^2(x)s - sV - s^2S + f + r = 0 \tag{5.1.3}$$

Solving equation (5.1.3) for the displacement $x(s)$ in the Laplace domain, we obtain the equation that describes the displacement as a function of the Laplace variable s:

$$s^2(x)s = sV + s^2S - f - r \tag{5.1.4}$$

With the help of Laplace Transform pairs 1, 3, 6, 2, and 2, we invert equation (5.1.4) from the Laplace domain into the time domain and obtain the solution of differential equation (5.1.1) with initial conditions of motion (2.1.2):

$$x = S + Vt - \frac{1}{2}t^2(f + r) \tag{5.1.5}$$

Taking the first derivative from equation (5.1.5), we determine the velocity of the system:

$$\frac{dx}{dt} = V - t(f + r) \tag{5.1.6}$$

Taking for equations (5.1.5) and (5.1.6) that $t = 0$, we obtain $x = S$ and $\frac{dx}{dt} = V$, as expected to be according to initial conditions of motion (2.1.2).

The first derivative from equation (5.1.6) represents the deceleration of the system:

$$\frac{d^2x}{dt^2} = -f - r \tag{5.1.7}$$

It may be noted that, in some cases, the absolute value of the deceleration should not exceed the corresponding allowable value.

Equations (5.1.5)–(5.1.7) represent the three basic parameters of motion of the system.

Equating equation (5.1.6) to zero, we determine the time T that the motion could last before the system comes to a stop:

$$0 = V - T(f + r) \tag{5.1.8}$$

From equation (5.1.8), we obtain

$$T = \frac{V}{f + r} \tag{5.1.9}$$

5.2. Motion of a System on a Horizontal Surface, While the System Is Being Subjected to Dry Friction, Constant Resisting, and Constant Active Forces

In this case, it is assumed that the active force exceeds the sum of the dry friction and constant resisting forces, while the air resistance to the motion of the system is negligible.

Figure 5.2.1 shows a schematic diagram of the system mentioned above.

Based on the above-mentioned considerations and the schematic diagram shown in Figure 5.2.1, we compose the differential equation of motion of the system:

$$m\frac{d^2x}{dt^2} + F + R = P \tag{5.2.1}$$

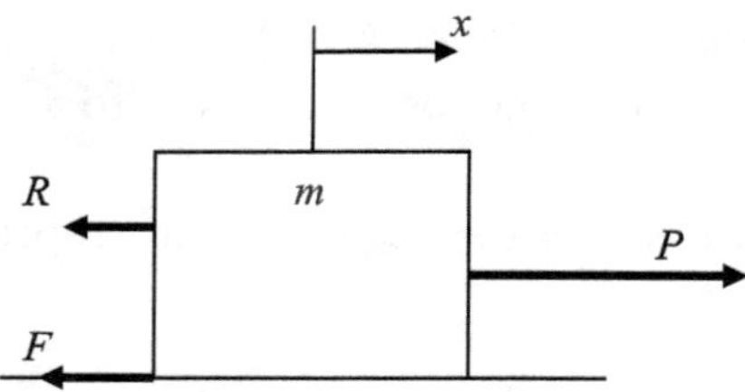

Fig. 5.2.1. Schematic diagram of a system moving on a horizontal surface while being subjected to dry friction, constant resisting, and constant active forces.

The general initial conditions of motion of the system are taken according to expression (2.1.2).

Dividing equation (5.2.1) by m, we have

$$\frac{d^2x}{dt^2} + f + r = p \qquad (5.2.2)$$

Applying to equation (5.2.2) Laplace Transform pairs 5, 2, 2, and 2, we convert this equation with initial conditions of motion (2.1.2) from the time domain into an algebraic equation in the Laplace domain:

$$s^2 x(s) - sV - s^2 S + f + r = p \qquad (5.2.3)$$

Solving equation (5.2.3) for the displacement $x(s)$ in the Laplace domain, we may write

$$x(s) = S + \frac{V}{s} + \frac{p - f - r}{s^2} \qquad (5.2.4)$$

Applying Laplace Transform pairs 1, 2, 3, 2, and 6, we invert equation (5.2.4) from the Laplace domain into the time domain and obtain the solution of differential equation (5.2.1) with the initial conditions of motion according to expression (2.1.2):

$$x = S + Vt + \frac{1}{2}t^2(p - f - r) \qquad (5.2.5)$$

The first derivative from equation (5.2.5) represents the velocity of the system:

$$\frac{dx}{dt} = V + t(p - f - r) \qquad (5.2.6)$$

Assuming for equations (5.2.5) and (5.2.6) that $t = 0$, we determine that $x = S$ and $\frac{dx}{dt} = V$, as it is expected according to initial conditions of motion (2.1.2).

The first derivative from equation (5.2.6) yields the acceleration of the system:

$$\frac{d^2x}{dt^2} = p - f - r \tag{5.2.7}$$

Since it is taken that in equation (5.2.7), $p > f + r$, the motion of the system is accelerated.

Hence, equations (5.2.5)–(5.2.7) represent the basic parameters of motion.

5.2.1. *Motion of a system on a downward inclined frictional surface*

The system is moving due to its weight on a downward inclined surface while being subjected to the actions of dry friction and constant resisting forces. The air resistance to the motion of the system is negligible. Figure 5.2.2 shows the schematic diagram of the system mentioned above. It is seen from Figure 5.2.2 that the abscissa x_1 is rotated by an angle γ relative to the abscissa x, while W is the weight of the system. The rest of the notations in this figure are self-explanatory.

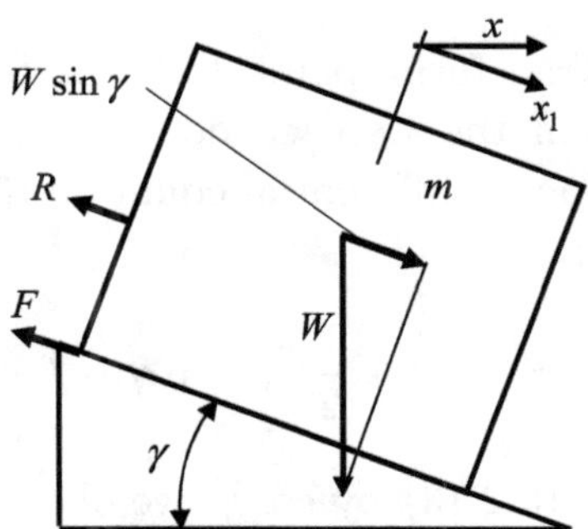

Fig. 5.2.2. Schematic diagram of a system moving downward on an inclined surface while being subjected to dry friction and constant resisting forces and a component of the system's weight.

Considering the motion of the system along the axis x_1, we compose the following differential equation:

$$m\frac{d^2 x_1}{dt^2} + F + R = W \sin\gamma \qquad (5.2.8)$$

Obviously, the motion occurs in the case if

$$F + R < W \sin\gamma \qquad (5.2.9)$$

Dividing equation (5.2.8) by m, we may write

$$\frac{d^2 x}{dt^2} + f + r = g \sin\gamma \qquad (5.2.10)$$

Applying Laplace Transform pairs 5, 2, 2, and 2, we convert differential equation (5.2.10) with its initial conditions of motion from the time domain into an algebraic equation in the Laplace domain:

$$s^2 x_1(s) - sV - s^2 S + f + r = g \sin\gamma \qquad (5.2.11)$$

Solving equation (5.2.11) for the displacement in the Laplace domain $x_1(s)$, we have

$$x_1(s) = S + \frac{V}{s} + \frac{1}{s^2}(g \sin\gamma - f - r) \qquad (5.2.12)$$

Applying Laplace Transform pairs 1, 2, 3, 6, and 2, we invert equation (5.2.12) from the Laplace domain into the time domain and obtain the solution of differential equation (5.2.8) with its initial conditions of motion:

$$x_1 = S + Vt + \frac{1}{2}t^2(g \sin\gamma - f - r) \qquad (5.2.13)$$

Multiplying equation (5.2.13) by $\cos\gamma$, we obtain

$$x = \left[S + Vt + \frac{1}{2}t^2(g \sin\gamma - f - r)\right]\cos\gamma \qquad (5.2.14)$$

Taking the first and second derivatives from equation (5.2.13), we determine the velocity and acceleration of the system, respectively:

$$\frac{dx_1}{dt} = V + t(g\sin\gamma - f - r) \tag{5.2.15}$$

$$\frac{d^2x_1}{dt^2} = g\sin\gamma - f - r \tag{5.2.16}$$

Equations (5.2.13), (5.2.15), and (5.2.16) represent the basic parameters of motion of the system.

5.2.2. *Braking of a system subjected to a dry friction force, a constant resisting force, and a component of the system's weight*

This situation may occur when the system is moving downhill, where the constant active force represents a component of the system's weight that cannot be controlled by the operator.

Equating equation (5.2.15) to zero, we determine the time T that the motion could last until the system stops:

$$T = \frac{V}{f + r - g\sin\gamma} \tag{5.2.17}$$

Solving together equations (5.2.13) and (5.2.17), we determine the breaking distance $S_{1\text{br}}$:

$$S_{1\text{br}} = S + \frac{V^2}{2(f + r - g\sin\gamma)} \tag{5.2.18}$$

It is seen from equation (5.2.16) that the deceleration D of the system equals

$$D = -f - r + g\sin\gamma \tag{5.2.19}$$

The absolute value of the deceleration according to equation (5.2.19) in certain cases should not exceed the corresponding allowable value.

5.3. Motion of a System on a Horizontal Surface While the System Is Being Subjected to Dry Friction, Constant Resisting, and Harmonic Forces

The system moves due to its initial velocity as well as due to a harmonic force. The air resistance to the motion of the system is negligible. Figure 5.3.1 shows a schematic diagram of a system moving on a horizontal surface while being subjected to the actions of a dry friction force, a constant resisting force, and a harmonic force. The notations in this figure are self-explanatory.

Based on the above- mentioned considerations and the schematic diagram shown in Figure 5.3.1, we compose the differential equation of motion of the system:

$$m\frac{d^2x}{dt^2} + F + R = A\cos\varphi t \qquad (5.3.1)$$

The initial conditions of motion of the system are characterized by expression (2.1.2).

Dividing equation (5.3.1) by m, we may write

$$\frac{d^2x}{dt^2} + f + r = a\cos\varphi t \qquad (5.3.2)$$

Applying Laplace Transform pairs 5, 2, 2, 2, and 29, we convert differential equation (5.3.2) with initial conditions of motion (2.1.2) from the time domain into an algebraic equation in the Laplace

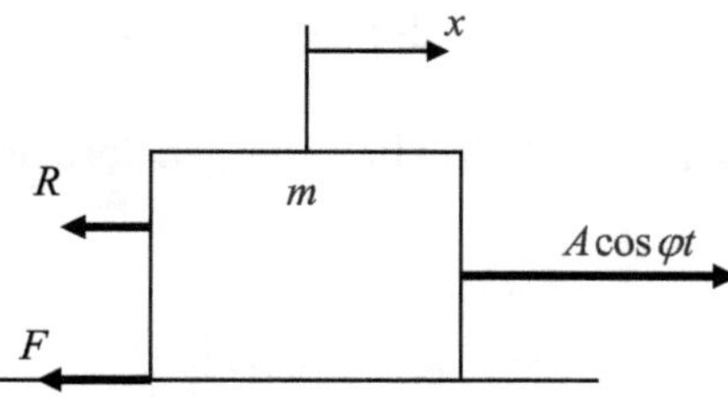

Fig. 5.3.1. Schematic diagram of a system moving on a horizontal surface while being subjected to a dry friction force, a constant resisting force, and to a harmonic force.

domain:

$$s^2 x(s) - sV - s^2 S + f + r = \frac{as^2}{s^2 + \varphi^2} \qquad (5.3.3)$$

Solving equation (5.3.3) for the displacement $x(s)$ as a function of the Laplace variable s, we obtain an algebraic equation of the displacement of the system in the Laplace domain:

$$x(s) = S + \frac{V}{s} - \frac{f+r}{s^2} + \frac{a}{s^2 + \varphi^2} \qquad (5.3.4)$$

Applying to equation (5.3.4) the Laplace Transform pairs 1, 2, 3, 6, 2, 2, and 17, we invert this equation from the Laplace domain into the time domain and obtain the solution of the differential equation of motion (5.3.1) with the initial conditions according to expression (2.1.2):

$$x = S + Vt - \frac{1}{2}t^2(f + r) + \frac{a}{\varphi^2}(1 - \cos \varphi t) \qquad (5.3.5)$$

Taking the first derivative from equation (5.3.5), we determine the velocity of the system:

$$\frac{dx}{dt} = V - t(f + r) + \frac{a}{\varphi} \sin \varphi t \qquad (5.3.6)$$

Taking in equations (5.3.5) and (5.3.6) that $t = 0$, we have $x = S$ and $\frac{dx}{dt} = V$, which is expected according to the initial conditions of motion given in expression (2.1.2).

The first derivative from equation (5.3.6) yields the acceleration of the system:

$$\frac{d^2 x}{dt^2} = a \cos \varphi t - f - r \qquad (5.3.7)$$

Hence, equations (5.3.5)–(5.3.7) characterize the basic parameters of motion of the system.

5.4. Motion of a System on a Horizontal Surface Being Subjected to Dry Friction, Constant Resisting, Constant Active, and Harmonic Forces

We consider the motion of a system on a horizontal surface, while the system is subjected to the actions of dry friction, constant resisting, and harmonic forces. The air resistance to the motion of the system is negligible. Figure 5.4.1 shows a schematic diagram of the system described above. The notations in the figure are self-explanatory.

Based on the schematic diagram shown in Figure 5.4.1 and the considerations mentioned above, we compose the following differential equation of motion:

$$m\frac{d^2x}{dt^2} + F + R = P + A\cos\varphi t \tag{5.4.1}$$

The initial conditions of motion are taken according to expression (2.1.2). Dividing equation (5.4.1) by m, we have

$$\frac{d^2x}{dt^2} + f + r = p + a\cos\varphi t \tag{5.4.2}$$

Applying Laplace Transform pairs 5, 2, 2, 2, and 29, we convert equation (5.4.2) with the initial conditions of motion according to expression (2.1.2) from the time domain into the Laplace domain and obtain an algebraic equation in the Laplace domain:

$$s^2x(s) - sV - s^2S + f + r = p + \frac{as^2}{s^2 + \varphi^2} \tag{5.4.3}$$

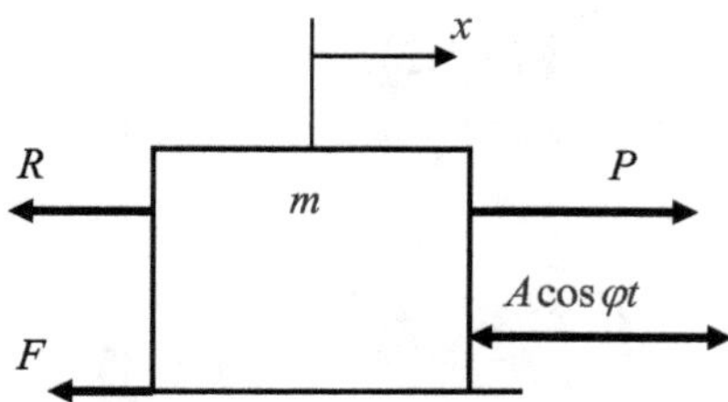

Fig. 5.4.1. Schematic diagram of a system moving on a horizontal surface while being subjected to dry friction, constant resisting, constant active, and harmonic forces.

Solving equation (5.4.3) for the displacement $x(s)$ as a function of the Laplace variable s, we obtain

$$x(s) = S + \frac{V}{s} + \frac{p - f - r}{s^2} + \frac{a}{s^2 + \varphi^2} \qquad (5.4.4)$$

Applying Laplace Transform pairs 1, 2, 3, 6, and 17, we invert equation (5.4.4) from the Laplace domain into the time domain and obtain the solution of differential equation of motion (5.4.1) with the initial conditions of motion according to expression (2.1.2):

$$x = S + Vt + \frac{1}{2}t^2(p - f - r) + \frac{a}{\varphi^2}(1 - \cos \varphi t) \qquad (5.4.5)$$

The first derivative from equation (5.4.5) yields the velocity of the system:

$$\frac{dx}{dt} = V + t(p - f - r) + \frac{a}{\varphi}\sin \varphi t \qquad (5.4.6)$$

Assuming in equations (5.4.5) and (5.4.6) that $t = 0$, we obtain that $x = S$ and $\frac{dx}{dt} = V$, as it should be according to expression (2.1.2).

The first derivative from equation (5.4.6) yields the acceleration of the system:

$$\frac{d^2x}{dt^2} = p - f - r + a \cos \varphi t \qquad (5.4.7)$$

Hence, equations (5.4.5)–(5.4.7) describe the basic parameters of motion of the system.

5.4.1. *Motion of a system on a downward inclined surface due to its initial velocity, a component of the system's weight, and a harmonic force*

Consider the motion of a system on a downward inclined surface while the system is subjected to the actions of a dry friction force, a constant resisting force, a component of the system's weight, $W \sin \gamma$, and a harmonic force, while the air resistance to the motion is negligible. Figure 5.4.2 shows a schematic diagram of the system, the notations in which are self-explanatory.

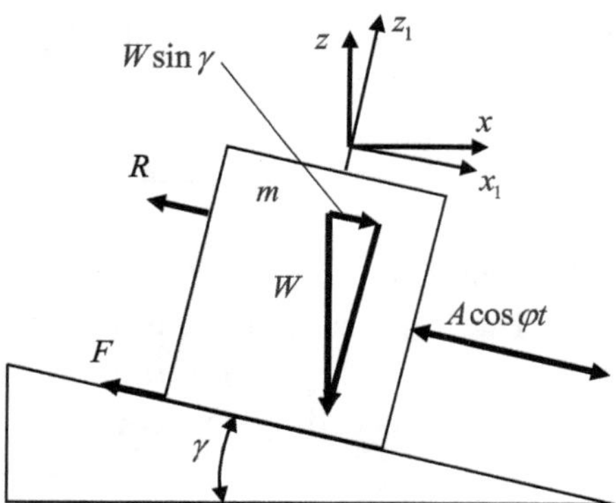

Fig. 5.4.2. Schematic diagram of a system moving downward on an inclined surface while being subjected to a dry friction force, a constant resisting force, a component of the system's weight, and a harmonic force.

Figure 5.4.2 shows that $x = x_1 \cos \gamma$, which makes it reasonable to analyze the motion of the system along the axis x_1.

Hence, based on the schematic diagram shown in Figure 5.4.2 and the considerations mentioned above, we compose the following differential equation of motion of the system:

$$m\frac{d^2 x_1}{dt^2} + F + R = W \sin \gamma + A \cos \varphi t \tag{5.4.8}$$

Dividing equation (5.4.8) by m, we have

$$\frac{d^2 x_1}{dt^2} + f + r = g \sin \gamma + a \cos \varphi t \tag{5.4.9}$$

Applying to equation (5.4.9) Laplace Transform pairs 5, 2, 2, 2, 2, and 29, we convert this equation with the initial conditions of motion according to expression (2.1.2) from the time domain into the Laplace domain and obtain an algebraic equation describing the motion of the system in the Laplace domain:

$$s^2(x_1) - sV - s^2 S + f + r = g \sin \gamma + \frac{a s^2}{s^2 + \varphi^2} \tag{5.4.10}$$

Solving equation (5.4.10) for $x_1(s)$ allows us to obtain an expression describing the displacement of the system in the Laplace domain as a function of the Laplace variable s:

$$x_1(s) = S + \frac{V}{s} + \frac{g\sin\gamma - f - r}{s^2} + \frac{a}{s^2 + \varphi^2} \tag{5.4.11}$$

Using Laplace Transform pairs 1, 2, 3, 6, and 17, we invert equation (5.4.11) from the Laplace domain into the time domain and obtain the solution of differential equation of motion (5.4.8) with the initial conditions of motion according to expression (2.1.2):

$$x_1 = S + Vt + \frac{1}{2}t^2(g\sin\gamma - f - r) + \frac{a}{\varphi^2}(1 - \cos\varphi t) \tag{5.4.12}$$

The first derivative from equation (5.4.12) allows us to determine the velocity of the system:

$$\frac{dx_1}{dt} = V + t(g\sin\gamma - f - r) + \frac{a}{\varphi}\sin\varphi t \tag{5.4.13}$$

The first derivative from equation (5.4.13) yields the acceleration of the system:

$$\frac{d^2x_1}{dt^2} = g\sin\gamma - f - r + a\cos\varphi t \tag{5.4.14}$$

Hence, equations (5.4.12)–(5.4.14) describe the basic parameters of motion of the system.

5.4.2. *Numerical solution*

Following is a Python program to plot the velocity represented by equation (5.4.13), which is solved for when the velocity first reaches zero.

```python
from matplotlib.pyplot import plot, show, text
from numpy import linspace, sin
from scipy.optimize import fsolve

S = 0.1
V = 0.3
g = 9.8
f = 0.8
r = 2
a = 20
singamma = 0.2
phi = 30

t = linspace(0, 0.5, 1000)
V1 = lambda t: V + t*(g*singamma - f - r) + (a/phi)*sin(phi*t)

[root] = fsolve(V1, 0.1)
root = round(root, 3)

plot(t, V1(t))
plot(root, 0, 'bo')
text(0.04, -0.055, f't={root}')
show()
```

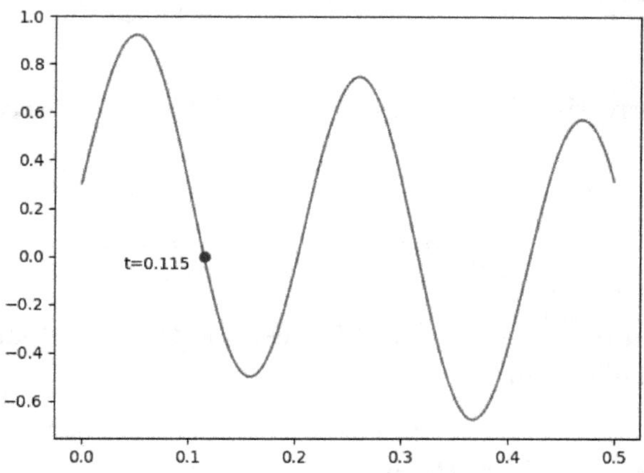

Fig. 5.4.3. Graph of equation (5.4.13).

The graph of equation (5.4.13) shown in Figure 5.4.3 indicates that the first stop of the mass occurs 0.115 s after the start. Following is a Python program to plot the displacement using equation (5.4.12) from $t = 0$ s to $t = 0.115$ s.

```python
from matplotlib.pyplot import plot, show, text
from numpy import linspace, cos

S = 0.1
V = 0.3
g = 9.8
f = 0.5
r = 2
a = 20
phi = 30
singamma = 0.2

x1 = lambda t: \
    S + V*t + (1/2)*t**2*(g*singamma - f - r) + (a/phi**2)*(1 - cos(phi*t))

stopTime = .115
t = linspace(0, stopTime, 1000)
plot(t, x1(t))

stopPosition = round(x1(stopTime), 3);
plot(stopTime, stopPosition, 'bo')
text(0.101, 0.167, f'x1={stopPosition}')

show()
```

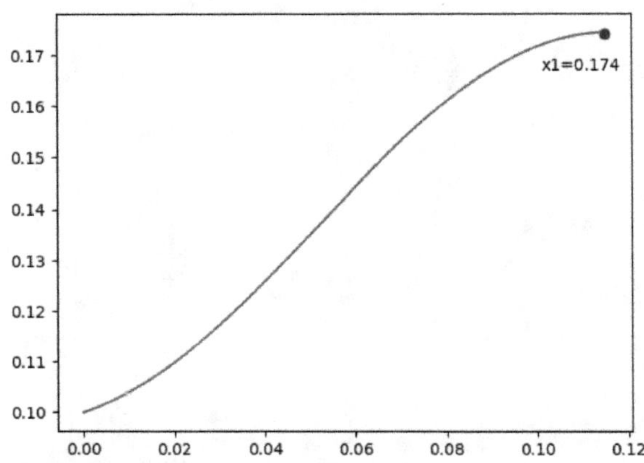

Fig. 5.4.4. Graph of equation (5.4.12).

Hence, the next iteration should be analyzed using differential equation (5.4.8) with the following initial conditions of motion:

$$\text{for} \quad t = 0 \quad x_1 = 0.174; \quad \frac{dx_1}{dt} = 0.$$

For more details, see Section 1.5.

CHAPTER 6

MOTION OF A SYSTEM RESTRICTED BY A FLEXIBLE LINK

A flexible link exerts a force that is proportional to the displacement of a system. In this text, we consider just forces of linear elasticity. In the physical sense, the flexible link may represent a spring, an elastic medium, a magnetic interaction, etc. A spring or an elastic medium absorbs the kinetic energy of the mechanical system and converts it into potential energy through deformation of the spring or medium. This potential energy could be returned to the system in the form of kinetic energy.

It is acceptable to schematically present the flexible link as a spring.

6.1. Motion of a System Restricted by a Flexile Link and Possessing an Initial Displacement and Initial Velocity

This section is focused on the analysis of an operational process associated with the free vibratory motion of a system on a frictionless horizontal surface. The system possesses an initial displacement and an initial velocity that cause an elastic deformation of the link, which, due to the initial displacement, accumulates a certain amount of potential energy. The air resistance in the motion of the system is negligible. Figure 6.1.1 shows a schematic diagram of this system, where K is the stiffness coefficient of the spring connecting the

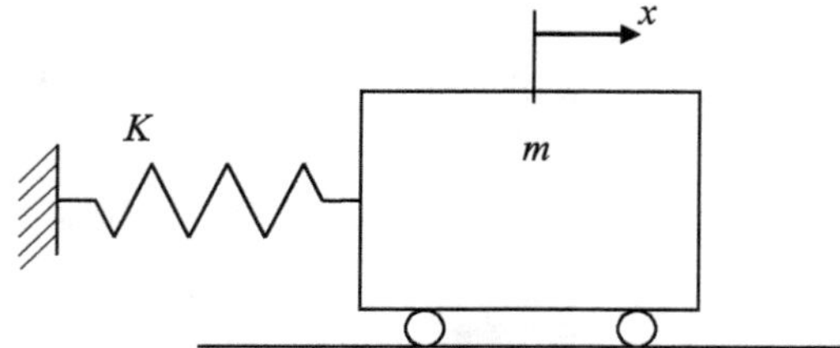

Fig. 6.1.1. Schematic diagram of a system moving on a horizontal frictionless surface while being restricted by a flexible link.

system to a non-movable support. The notations in this figure are self-explanatory.

Based on considerations mentioned above and the schematic diagram shown in Figure 6.1.1, we compose the differential equation of motion of the system:

$$m\frac{d^2x}{dt^2} + Kx = 0 \qquad (6.1.1)$$

The initial conditions of motion are presented in expression (2.1.2). Dividing equation (6.1.1) by m, we may write

$$\frac{d^2x}{dt^2} + \omega^2 x = 0 \qquad (6.1.2)$$

where

$$\omega^2 = \frac{K}{m} \qquad (6.1.3)$$

while ω is the natural frequency of the system.

Applying to differential equation (6.1.2) Laplace Transform pairs 1, 5, 2, and 1, we convert this equation with the initial conditions of motion according to expression (2.1.2) from the time domain into an algebraic equation in the Laplace domain:

$$s^2x(s) - sV - s^2S + \omega^2 x(s) = 0 \qquad (6.1.4)$$

Solving equation (6.1.4) for the displacement $x(s)$ in the Laplace domain, we may write

$$x(s) = \frac{sV}{s^2 + \omega^2} + \frac{s^2S}{s^2 + \omega^2} \qquad (6.1.5)$$

Using Laplace Transform pairs 1, 23, 2, 29, and 2, we invert equation
(6.1.5) from the Laplace domain into the time domain and obtain the
solution of differential equation (6.1.1) with the initial conditions of
motion according to expression (2.1.2):

$$x = \frac{V}{\omega}\sin\omega t + S\cos\omega t \tag{6.1.6}$$

In order to simplify the analysis of the operational process, we apply
to equation (6.1.6) some conventional procedures:

$$x = \frac{V}{\omega}\sin\omega t + S\cos\omega t = \frac{\sqrt{V^2 + S^2\omega^2}}{\omega}$$
$$\times \left(\frac{V\sin\omega t}{\sqrt{V^2 + S^2\omega^2}} + \frac{S\omega\cos\omega t}{\sqrt{V^2 + S^2\omega^2}} \right) \tag{6.1.7}$$

Denoting

$$\cos\delta_0 = \frac{V}{\sqrt{V^2 + S^2\omega^2}}; \quad \sin\delta_0 = \frac{S\omega}{\sqrt{V^2 + S^2\omega^2}} \tag{6.1.8}$$

we may write

$$x = \frac{\sqrt{V^2 + S^2\omega^2}}{\omega}\sin(\omega t + \delta_0) \tag{6.1.9}$$

Taking the first derivative from equation (6.1.9), we determine the
velocity of the system:

$$\frac{dx}{dt} = \sqrt{V^2 + S^2\omega^2}\cos(\omega t + \delta_0) \tag{6.1.10}$$

Supposing that in equations (6.1.9) and (6.1.10), $t = 0$, we determine
that $x = S$ and $\frac{dx}{dt} = V$, as it should be according to the initial conditions of motion presented in expression (2.1.2). The first derivative
from equation (6.1.10) yields the acceleration of the system:

$$\frac{d^2x}{dt^2} = -\omega\sqrt{V^2 + S^2\omega^2}\sin(\omega t + \delta_0) \tag{6.1.11}$$

Hence, equations (6.1.9)–(6.1.11) represent the basic parameters of motion of the system.

6.2. Motion of a System on a Horizontal Frictionless Surface and Being Restricted by a Flexible Link and Subjected to a Constant Active Force

The operational process under consideration is associated with the motion of a system moving on a horizontal frictionless surface due to the initial displacement and velocity as well as a constant active force. The system is restricted by a flexible link. The air resistance to the motion of the system is negligible. Figure 6.2.1 shows a schematic diagram of the system. The notations in this figure are self-explanatory.

Accounting for the considerations mentioned above and the schematic diagram shown in Figure 6.2.1, we compose the differential equation of motion of the system:

$$m\frac{d^2x}{dt^2} + Kx = P \tag{6.2.1}$$

The initial conditions of motion are taken according to expression (2.1.2). Dividing equation (6.2.1) by m, we have

$$\frac{d^2x}{dt^2} + \omega^2 x = p \tag{6.2.2}$$

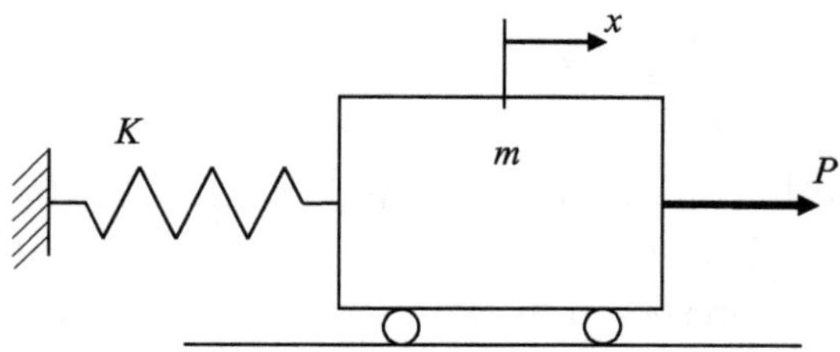

Fig. 6.2.1. Schematic diagram of a system moving on a horizontal frictionless surface while being restricted by a flexible link and subjected to a constant active force.

Using Laplace Transform pairs 5, 2, 1, and 2, we convert differential equation (6.2.2) with the initial conditions of motion according to expression (2.1.2) from the time domain into an algebraic equation in the Laplace domain:

$$s^2 x(s) - sV - s^2 S + \omega^2 x(s) = p \tag{6.2.3}$$

Solving equation (6.2.3) for the displacement $x(s)$ in the Laplace domain, we have

$$x(s) = \frac{p}{s^2 + \omega^2} + \frac{sV}{s^2 + \omega^2} + \frac{s^2 S}{s^2 + \omega^2} \tag{6.2.4}$$

Using Laplace Transform pairs 1, 17, 23, and 29, we invert equation (6.2.4) from the Laplace domain into the time domain and obtain the solution of differential equation (6.2.1) with the initial conditions of motion according to expression (2.1.2):

$$x = \frac{p}{\omega^2}(1 - \cos \omega t) + \frac{V}{\omega} \sin \omega t + S \cos \omega t \tag{6.2.5}$$

Applying to equation (6.2.5) the conventional procedures, we may write

$$x = \frac{p}{\omega^2} - \frac{p}{\omega^2} \cos \omega t + \frac{V}{\omega} \sin \omega t + S \cos \omega t$$

$$= \frac{p}{\omega^2} + \frac{1}{\omega^2}[(S\omega^2 - p) \cos \omega t + V\omega \sin \omega t]$$

$$= \frac{p}{\omega^2} + \frac{\sqrt{(S\omega^2 - p)^2 + V^2 \omega^2}}{\omega^2} \left[\frac{V\omega}{\sqrt{(S\omega^2 - p^2)^2 + V^2 \omega^2}} \sin \omega t \right.$$

$$\left. + \frac{S\omega^2 - p}{\sqrt{(p - S\omega^2)^2 + V^2 \omega^2}} \cos \omega t \right]$$

Based on the presented expressions above, we obtain

$$x = \frac{1}{\omega^2}[p + \sqrt{V^2 \omega^2 + (S\omega^2 - p)^2} \sin(\omega t + \delta)] \tag{6.2.6}$$

where

$$\cos\delta = \frac{p - S\omega^2}{\sqrt{V^2\omega^2 + (S\omega^2 - p)^2}}; \quad \sin\delta = \frac{V\omega}{\sqrt{V^2\omega^2 + (S\omega^2 - p)^2}}$$

$$(6.2.7)$$

The first derivative from equation (6.2.6) allows us to get the expression for computing the velocity of the system:

$$\frac{dx}{dt} = \frac{1}{\omega}\sqrt{V^2\omega^2 + (S\omega^2 - p)^2}\cos(\omega t + \delta)] \qquad (6.2.8)$$

Assuming that in equations (6.2.6) and (6.2.8), $t = 0$, we determine that $x = S$ and $\frac{dx}{dt} = V$, as expected according to the initial conditions of motion presented in expression (2.1.2). Taking the first derivative from equation (6.2.8), we obtain the equation for the acceleration of the system:

$$\frac{d^2x}{dt^2} = -\sqrt{V^2\omega^2 + (S\omega^2 - p)^2}\sin(\omega t + \delta)] \qquad (6.2.9)$$

It should be emphasized that the deformation of the spring or the elastic medium is equivalent to the displacement of the system. Therefore, the force exerted by the link toward the system is proportional to this displacement and is causing the system to stop. Hence, equating to zero the velocity according to equation (6.2.8), we obtain

$$\cos(\omega t + \delta) = 0 \qquad (6.2.10)$$

Therefore, taking $\sin(\omega t + \delta) = 1$, we determine from equations (6.2.6) and (6.2.9) the maximum values of the displacement $S_{\max}$ and acceleration $a_{\max}$, respectively, at the moment when the system stops:

$$S_{\max} = \frac{1}{\omega^2}\left[p + \sqrt{V^2\omega^2 + (S\omega^2 - p^2)^2}\right] \qquad (6.2.11)$$

$$a_{\max} = \left|-\sqrt{V^2\omega^2 + (S\omega^2 - p)^2}\right| \qquad (6.2.12)$$

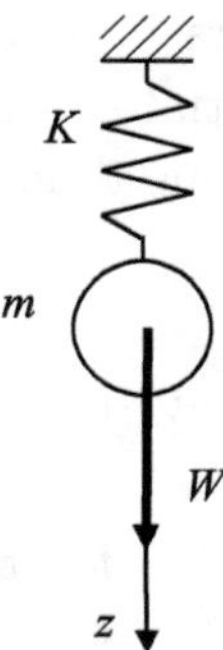

Fig. 6.2.2. Schematic diagram of a system restricted in the vertical direction by a flexible link and subjected to the action of a force of gravity.

Hence, equations (6.2.6), (6.2.8), and (6.2.9) are the basic parameters of motion of the system.

6.2.1. *Downward motion of a system restricted by a flexible link*

Figure 6.2.2 shows a schematic diagram of a system suspended from a spring with a stiffness coefficient K. The displacement of the system occurs along the vertical coordinate z. The air resistance to the motion of the mass is negligible.

The objective of this investigation is to determine the peculiarities of the deceleration and braking distance of the system.

Based on this schematic diagram and related considerations, we compose the differential equation of motion of the mass:

$$m\frac{d^2z}{dt^2} + Kz = W \qquad (6.2.13)$$

The initial conditions of motion at $t = 0$ are

$$z = 0; \quad \frac{dz}{dt} = V \qquad (6.2.14)$$

Dividing equation (6.2.13) by m, we have

$$\frac{d^2z}{dt^2} + \omega^2 z = g \qquad (6.2.15)$$

Using Laplace Transform pairs 5, 1, and 2, we convert differential equation (6.2.15) with the initial conditions of motion according to expression (6.2.14) from the time domain into an algebraic equation in the Laplace domain:

$$s^2 z(s) - sV + \omega^2 z(s) = g \tag{6.2.16}$$

The solution of equation (6.2.16) for the displacement $z(s)$ has the following expression:

$$z(s) = \frac{sV}{s^2 + \omega^2} + \frac{g}{s^2 + \omega^2} \tag{6.2.17}$$

Applying to equation (6.2.17) Laplace Transform pairs 1, 23, and 17, we invert this equation from the Laplace domain into the time domain and obtain the solution of the differential equation of motion (6.2.13) with the initial conditions of motion according to expression (6.2.14):

$$z = \frac{g}{\omega^2}(1 - \cos \omega t) + \frac{V}{\omega} \sin \omega t \tag{6.2.18}$$

Rearranging equation (6.2.18), we may write

$$z = \frac{1}{\omega^2}(g + V\omega \sin \omega t - g \cos \omega t) = \frac{1}{\omega^2}\left\{ g + \left[\sqrt{V^2\omega^2 + g^2} \right.\right.$$

$$\times \left. \left. \left(\sin \omega t \frac{V\omega}{\sqrt{V^2\omega^2 + g^2}} - \cos \omega t \frac{g}{\sqrt{V^2\omega^2 + g^2}} \right) \right] \right\} \tag{6.2.19}$$

Denoting

$$\cos \delta_1 = \frac{g}{\sqrt{V^2\omega^2 + g^2}} \quad \text{and} \quad \sin \delta_1 = \frac{V\omega}{\sqrt{V^2\omega^2 + g^2}} \tag{6.2.20}$$

we write

$$z = \frac{1}{\omega^2} \left\{ g + \left[\sqrt{V^2\omega^2 + g^2}(\sin \omega t \cos \delta_1 - \cos \omega t \sin \delta_1) \right] \right\} \quad (6.2.21)$$

Applying to equation (6.2.20) the sine rule of the sum of angles, we obtain

$$z = \frac{1}{\omega^2} \left[g + \sqrt{V^2\omega^2 + g^2} \sin(\omega t - \delta_1) \right] \quad (6.2.22)$$

The first and second derivatives from equation (6.2.22), respectively, represent the velocity and acceleration/deceleration of the mass:

$$\frac{dz}{dt} = \frac{1}{\omega}\sqrt{V^2\omega^2 + g^2} \cos(\omega t - \delta_1) \quad (6.2.23)$$

$$\frac{d^2 z}{dt^2} = -\sqrt{V^2\omega^2 + g^2} \sin(\omega t - \delta_1) \quad (6.2.24)$$

As the velocity becomes equal to zero, we obtain the maximum absolute value of the acceleration/deceleration of the mass. Therefore, by equating expression (6.2.23) to zero, we determine the time T when the mass stops, and we have

$$\cos(\omega T - \delta_1) = 0 \quad (6.2.25)$$

and therefore,

$$\sin(\omega T - \delta_1) = 1 \quad (6.2.26)$$

Substituting the value of expression (6.2.26) into equation (6.2.24) and recalling that $g = \frac{W}{m}$, we determine the deceleration D of the mass:

$$D = -\sqrt{V^2\frac{Kg}{W} + g^2} \quad (6.2.27)$$

It should be mentioned that, according to the available corresponding information related to health and safety requirements, the deceleration of a human body during a free fall should not exceed $3g$. Applying this analysis to a human body, it is justifiable to assume that the allowable level of deceleration is $3g$, and according to equation (6.2.27), we may write

$$9g^2 = V^2 \frac{Kg}{W} + g^2 \tag{6.2.28}$$

From equation (6.2.28), we determine

$$V^2 \frac{K}{W} = 8g \tag{6.2.29}$$

Considering a case where the weight of the human body is $0.5W$ and combining equations (6.2.29) and (6.2.27), we have $D = -\sqrt{V^2 \frac{Kg}{0.5W} + g^2} = -\sqrt{\frac{8g^2}{0.5} + g^2} = -4.2g$, which exceeds the allowable deceleration of $|3g|$. This leads to the conclusion that it is impossible to comply with the health requirements for different weights of human bodies using the same spring.

6.3. Motion of a System Restricted by a Flexible Link and Subjected to a Harmonic Force

The operational process is related to the system that is moving on a horizontal frictionless surface and is restricted by a flexible link while being subjected to a harmonic force. The air resistance to the motion of the system is negligible. Figure 6.3.1 shows a schematic diagram that represents the system described above. The notations in this figure are self-explanatory.

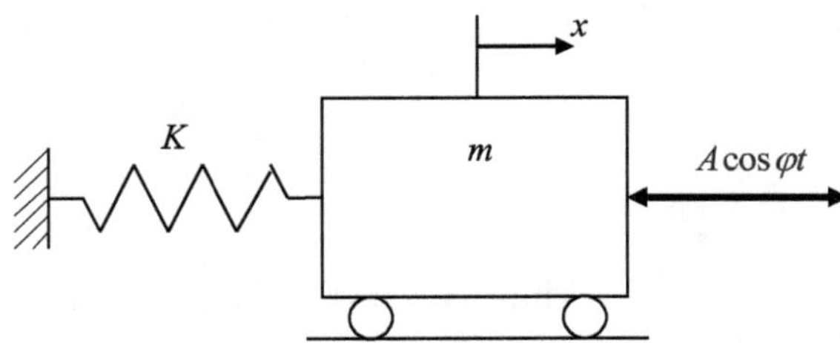

Fig. 6.3.1. Schematic diagram of a system restricted by a flexible link while moving on a frictionless surface and subjected to a harmonic force.

Based on this schematic diagram and the related considerations mentioned above, we compose the differential equation of motion of the system:

$$m\frac{d^2x}{dt^2} + Kx = A\cos\varphi t \tag{6.3.1}$$

The initial conditions of motion are described in expression (2.1.2). Dividing equation (6.3.1) by m, we may write

$$\frac{d^2x}{dt^2} + \omega^2 x = a\cos\varphi t \tag{6.3.2}$$

Applying Laplace Transform pairs 5, 2, 1, 2, and 29 to equation (6.3.2) with the initial conditions of motion according to expression (2.1.2), we convert this equation from the time domain into an algebraic equation in the Laplace domain:

$$s^2x(s) - Vs - s^2S + \omega^2x(s) = \frac{as^2}{s^2 + \varphi^2} \tag{6.3.3}$$

Applying to equation (6.3.3) the conventional algebraic procedures, we obtain

$$x(s)(s^2 + \omega^2) = s^2S + sV + \frac{as^2}{s^2 + \varphi^2} \tag{6.3.4}$$

The solution of equation (6.3.4) for the displacement $x(s)$ in the Laplace domain allows us to obtain an algebraic equation describing the displacement $x(s)$ of the system as a function of the Laplace variable s:

$$x(s) = \frac{s^2S}{s^2 + \omega^2} + \frac{sV}{s^2 + \omega^2} + \frac{as^2}{(s^2 + \varphi^2)(s^2 + \omega^2)} \tag{6.3.5}$$

Based on Laplace Transform pairs 1, 29, 23, and 72, we invert equation (6.3.5) from the Laplace domain into the time domain and obtain the solution of differential equation (6.3.1) with the initial

conditions of motion according to expression (2.1.2):

$$x = S\cos\omega t + \frac{V}{\omega}\sin\omega t + \frac{a(\cos\omega t - \cos\varphi t)}{\varphi^2 - \omega^2} \tag{6.3.6}$$

Taking the first derivative from equation (6.3.6), we determine the velocity of the system:

$$\frac{dx}{dt} = -S\omega\sin\omega t + V\cos\omega t + \frac{a(\varphi\sin\varphi t - \omega\sin\omega t)}{\varphi^2 - \omega^2} \tag{6.3.7}$$

Assuming in equations (6.3.6) and (6.3.7) that $t = 0$, we obtain $x = S$ and $\frac{dx}{dt} = V$, which is expected according to the initial conditions of motion given by expression (2.1.2). The first derivative from equation (6.3.7) yields the acceleration of the system:

$$\frac{d^2x}{dt^2} = -S\omega^2\cos\omega t - V\omega\sin\omega t + \frac{a(\varphi^2\cos\varphi t - \omega^2\cos\omega t)}{\varphi^2 - \omega^2} \tag{6.3.8}$$

Hence, equations (6.3.6)–(6.3.8) describe the basic parameters of motion of the system.

6.3.1. *Resonance*

Analyzing equation (6.3.6), it can be seen that in the case when $\varphi^2 = \omega^2$, we obtain an expression with an indeterminate of form $\frac{0}{0}$, which could cause resonance:

$$x = S\cos\omega t + \frac{V}{\omega}\sin\omega t + \frac{0}{0} \tag{6.3.9}$$

Obviously, the first two terms of equations (6.3.6) and (6.3.9) represent harmonic functions that cannot have any influence on resonance. For this reason, these two terms could be excluded from the verification of equation (6.3.6) on the possibility of being subjected to resonance. Furthermore, in the case of resonance, these two terms are negligible and, obviously, may not be included in the equation that is associated with resonance. Therefore, it is necessary to investigate just the term that represents indeterminacy, which is the last term of equation (6.3.6). Equating this term to X, we may write

$$X = \frac{a(\cos \omega t - \cos \varphi t)}{\varphi^2 - \omega^2} \qquad (6.3.10)$$

Applying to equation (6.3.10) L'Hopital's rule, we evaluate the indeterminate term:

$$X = \lim_{\varphi \to \omega} \frac{a(\cos \omega t - \cos \varphi t)}{\varphi^2 - \omega^2} \qquad (6.3.11)$$

Differentiating the numerator and denominator of the fraction in equation (6.3.11) with respect to ω, we may write

$$X = \lim_{\varphi \to \omega} \frac{a(-t \sin \omega t)}{-2\omega} \qquad (6.3.12)$$

Completing the conventional procedures related to the limit, we evaluate the above-mentioned indeterminate term:

$$X = \frac{at}{2\omega} \sin \omega t \qquad (6.3.13)$$

Equation (6.3.13) contains in the numerator the multiplier t that constantly increases the amplitude of the harmonic function, causing the amplitude to rise to infinity, bringing the system to a state of resonance.

The numeric solution of equation (6.3.13) obtained with the help of a computer program in Python is presented as follows:

```python
from matplotlib.pyplot import plot, show
from numpy import linspace, sin

a = 20
omega = 60

t = linspace(0, 0.5, 1000)
x = (a*t)/(2*omega)*sin(omega*t)

plot(t, x)
show()
```

Figure 6.3.2 represents the graph of equation (6.3.13), which shows a tendency of the amplitude of vibrations of the system to increase toward infinity.

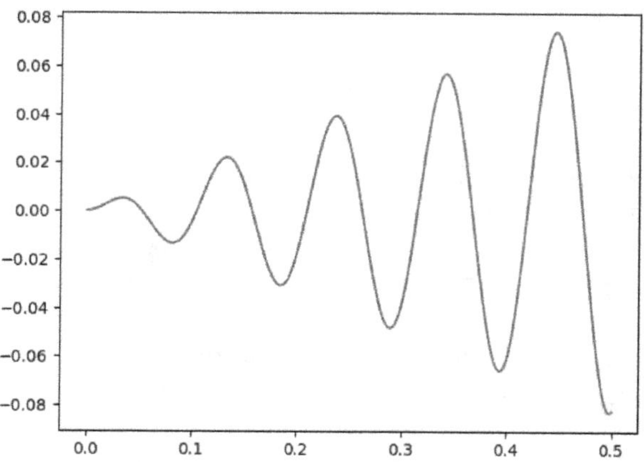

Fig. 6.3.2. Graph demonstrating the tendency of vibrational motion in resonance.

6.4. Motion of a System Restricted by a Flexible Link and Subjected to a Constant Active Force and a Harmonic Force

We consider the motion of a system on a horizontal frictionless surface, while the system is restricted by a flexible link and subjected to the actions of a constant active force and a harmonic force. The air resistance to the motion of the system is negligible. Figure 6.4.1 shows a schematic diagram that represents the system described above. The notations in the figure are self-explanatory.

Based on the above considerations and the schematic diagram shown in Figure 6.4.1, we compose the differential equation of motion of the system:

$$m\frac{d^2x}{dt^2} + Kx = P + A\cos\varphi t \qquad (6.4.1)$$

The initial conditions of motion are taken according to expression (2.1.2). Dividing equation (6.4.1) by m, we have

$$\frac{d^2x}{dt^2} + \omega^2 x = p + a\cos\varphi t \qquad (6.4.2)$$

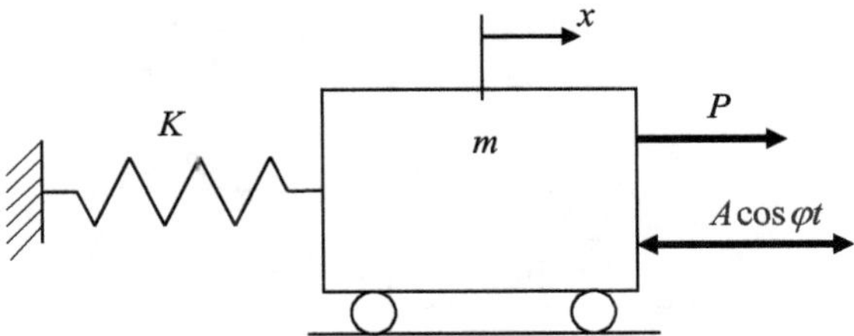

Fig. 6.4.1. Schematic diagram of a system restricted by a flexible link and subjected to a constant active force and a harmonic force.

Applying Laplace Transform pairs 5, 2, 1, 2, 2, and 29 to equation (6.4.2) with the initial conditions of motion according to expression (2.1.2), we convert this equation from the time domain into an algebraic equation in the Laplace domain:

$$s^2 x(s) - Vs - s^2 S + \omega^2 x(s) = p + \frac{as^2}{s^2 + \varphi^2} \tag{6.4.3}$$

Applying to equation (6.4.3) some conventional algebraic procedures, we may write

$$x(s)(s^2 + \omega^2) = s^2 S + sV + p + \frac{as^2}{s^2 + \varphi^2} \tag{6.4.4}$$

The solution of equation (6.4.4) for the displacement $x(s)$ in the Laplace domain yields an algebraic equation describing the displacement of the system as a function of Laplace variable s:

$$x(s) = \frac{s^2 S}{s^2 + \omega^2} + \frac{sV}{s^2 + \omega^2} + \frac{p}{s^2 + \omega^2} + \frac{as^2}{(s^2 + \varphi^2)(s^2 + \omega^2)} \tag{6.4.5}$$

Applying Laplace Transform pairs 1, 29, 23, 17, and 72, we invert equation (6.4.5) from the Laplace domain into the time domain and obtain the solution of differential equation (6.4.1) with the initial conditions of motion according to expression (2.1.2):

$$x = S \cos \omega t + \frac{V}{\omega} \sin \omega t + \frac{p}{\omega^2}(1 - \cos \omega t)$$

$$+ \frac{a(\cos \omega t - \cos \varphi t)}{\varphi^2 - \omega^2} \tag{6.4.6}$$

The first derivative from equation (6.4.6) yields the velocity of the system:

$$\frac{dx}{dt} = -S\omega \sin \omega t + V \cos \omega t + \frac{p}{\omega} \sin \omega t$$

$$+ \frac{a(\varphi \sin \varphi t - \omega \sin \omega t)}{\varphi^2 - \omega^2} \tag{6.4.7}$$

Assuming that in equations (6.4.6) and (6.4.7) $t = 0$, we determine that $x = S$ and $\frac{dx}{dt} = V$, as it should be according to the initial conditions of motion given by expression (2.1.2). The first derivative from equation (6.4.7) represents the acceleration of the system:

$$\frac{d^2x}{dt^2} = -S\omega^2 \cos \omega t - V\omega \sin \omega t + p \cos \omega t$$

$$+ \frac{a(\varphi^2 \cos \varphi t - \omega^2 \cos \omega t)}{\varphi^2 - \omega^2} \tag{6.4.8}$$

Equations (6.4.6)–(6.4.8) represent the basic parameters of motion of the system. In addition, it should be mentioned that the fraction on the right-hand side of equation (6.4.6), as was shown earlier for equation (6.3.9), leads to the resonance of the system in the case when $\varphi = \omega$.

6.4.1. *Numerical solution*

Following is a Python program to plot the graph of equation (6.4.6). The graph is plotted in Figure 6.4.2.

```python
from matplotlib.pyplot import plot, show
from numpy import linspace, sin, cos

S = 0.1
p = 25
V = 0.1
a = 10
phi = 30
omega = 15

t = linspace(0, 1, 1000)
x = S*cos(omega*t) + (V/omega)*sin(omega*t) + (p/omega**2)*(1 -
cos(omega*t)) \
  + a*(cos(omega*t) - cos(phi*t))/(phi**2 - omega**2)

plot(t, x)
show()
```

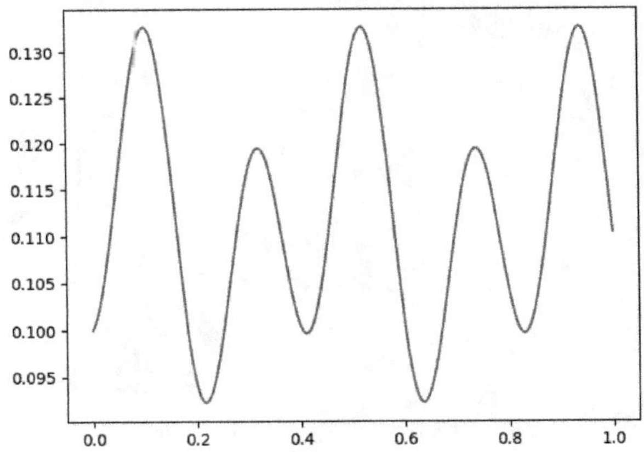

Fig. 6.4.2. Graph of equation.

CHAPTER 7

MOTION OF A SYSTEM RESTRICTED BY A FLEXIBLE LINK AND SUBJECTED TO A CONSTANT RESISTING FORCE

7.1. Motion of a System Restricted by a Flexible Link and Subjected to a Constant Resisting Force while Moving Due to the Initial Displacement and Velocity

The current operational process is related to a mechanical system restricted by a flexible link and moving on a horizontal frictionless surface due to initial displacement and initial velocity, while the system is subjected to a resisting force R. Due to the initial displacement, the flexible link becomes deformed, resulting in storing a corresponding amount of potential energy, which is proportional to the displacement of the system. The potential energy will be returned by the link to the system in the form of kinetic energy. The air resistance to the motion of the system is negligible.

Figure 7.1.1 shows the schematic diagram of the system that is described above. The notations in the figure are self-explanatory.

Based on the schematic diagram shown in Figure 7.1.1 and the considerations mentioned above, we compose the differential equation of motion that reads

$$m\frac{d^2x}{dt^2} + Kx + R = 0 \tag{7.1.1}$$

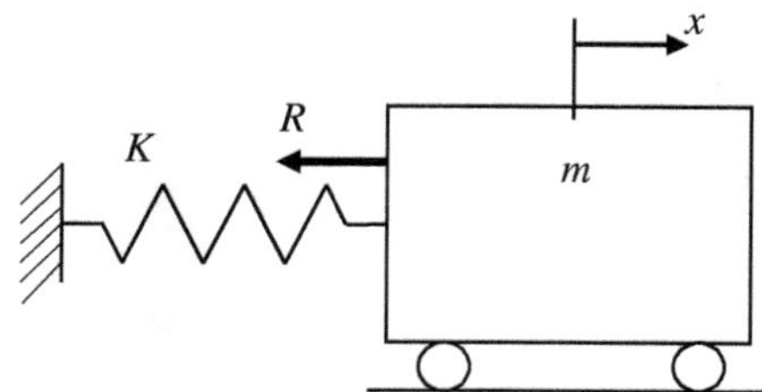

Fig. 7.1.1. Schematic diagram of a system restricted by a flexible link and subjected to a constant restricting force.

The initial conditions of motion are taken according to expression (2.1.2).

Dividing equation (7.1.1) by m, we may write

$$\frac{d^2x}{dt^2} + \omega^2 x + r = 0 \tag{7.1.2}$$

Applying Laplace Transform pairs 5, 2, 1, and 2, we convert equation (7.1.2) with initial conditions of motion according to expression (2.1.2) from the time domain into the Laplace domain and obtain a corresponding algebraic equation in the Laplace domain:

$$s^2 x(s) - sV - s^2 S + \omega^2 x(s) + r = 0 \tag{7.1.3}$$

Rearranging equation (7.1.3), we have

$$x(s)(s^2 + \omega^2) = sV + s^2 S - r \tag{7.1.4}$$

The solution of equation (7.1.4) for the displacement $x(s)$ in the Laplace domain reads

$$x(s) = \frac{sV}{s^2 + \omega^2} + \frac{s^2 S}{s^2 + \omega^2} - \frac{r}{s^2 + \omega^2} \tag{7.1.5}$$

With the help of Laplace Transform pairs 1, 23, 29, and 17, we invert equation (7.1.5) from the Laplace domain into the time domain and obtain the solution of differential equation (7.1.1) with the initial conditions of motion according to expression (2.1.2):

$$x = \frac{V}{\omega} \sin \omega t + S \cos \omega t - \frac{r}{\omega^2}(1 - \cos \omega t) \tag{7.1.6}$$

Taking the first derivative from equation (7.1.6), we determine the velocity of the system:

$$\frac{dx}{dt} = V \cos \omega t - S\omega \sin \omega t - \frac{r}{\omega} \sin \omega t \qquad (7.1.7)$$

Supposing that in equations (7.1.6) and (7.1.7), $t = 0$, we determine that $x = S$ and $\frac{dx}{dt} = V$, as it should be according to the initial conditions of motion according to expression (2.1.2).

The first derivative from equation (7.1.7) yields the acceleration of the system:

$$\frac{d^2 x}{dt^2} = -V\omega \sin \omega t - S\omega^2 \cos \omega t - r \cos \omega t \qquad (7.1.8)$$

Hence, equations (7.1.6)–(7.1.8) represent the basic parameters of motion of the system.

7.2. Motion of a System Restricted by a Flexible Link and Subjected to Constant Resisting and to a Constant Active Force

The operational process of the system represents its motion on a horizontal frictionless surface due to the initial displacement and the initial velocity and a constant active force. The system is restricted by a flexible link and is subjected to the action of a constant resisting force. The air resistance to the motion of the system is negligible. Figure 7.2.1 shows the schematic diagram of the system described above. The notations in this figure are self-explanatory.

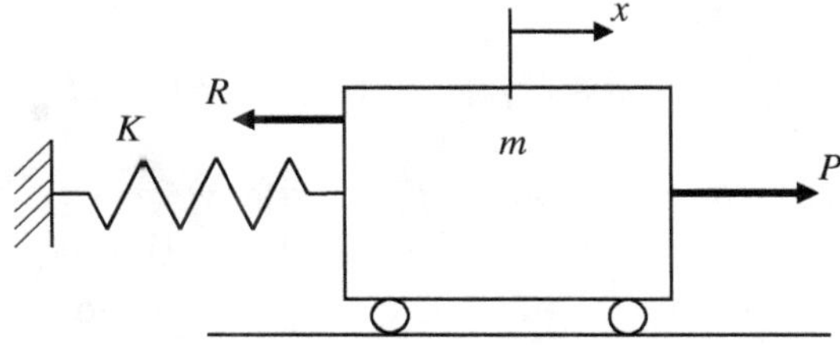

Fig. 7.2.1. Schematic diagram of a system moving on a horizontal frictionless surface while being restricted by a flexible link and subjected to a constant resisting force and a constant active force.

Accounting for the considerations mentioned above and accounting for the schematic diagram shown in Figure 7.2.1, we compose the differential equation of motion of the system:

$$m\frac{d^2x}{dt^2} + Kx + R = P \qquad (7.2.1)$$

The initial conditions of motion are taken according to expression (2.1.2).

Dividing equation (7.2.1) by m, we may write

$$\frac{d^2x}{dt^2} + \omega^2 x + r = p \qquad (7.2.2)$$

Applying Laplace Transform pairs 5, 2, 1, 2, and 2, we convert differential equation (7.2.2) with the initial conditions of motion according to expression (2.1.2) from the time domain into the algebraic equation in Laplace domain:

$$s^2 x(s) - sV - s^2 S + \omega^2 x(s) + r = p \qquad (7.2.3)$$

Solving equation (7.2.3) for the displacement $x(s)$ in the Laplace domain, we have

$$x(s) = \frac{p-r}{s^2+\omega^2} + \frac{sV}{s^2+\omega^2} + \frac{s^2 S}{s^2+\omega^2} \qquad (7.2.4)$$

Applying Laplace Transform pairs 1, 17, 23, and 29, we invert equation (7.2.4) from the Laplace domain into the time domain and obtain the solution of differential equation (7.2.1) with the initial conditions of motion according to expression (2.1.2):

$$x = \frac{p-r}{\omega^2}(1 - \cos\omega t) + \frac{V}{\omega}\sin\omega t + S\cos\omega t \qquad (7.2.5)$$

It is assumed that in equation (7.2.5), we have $p - r > 0$.

It may be noted that in cases where the equation of the displacement represents a sum of sine and cosine functions, it is possible with the help of conventional procedures to represent this equation as a function of a sine or cosine of two angles. As shown in Chapter 6, this allows us to simplify the analysis of the equation.

Hence, applying to equation (7.2.5) the conventional procedures, we may write

$$x = \frac{p-r}{\omega^2} + \frac{1}{\omega^2}[(r + S\omega^2 - p)\cos\omega t + V\omega\sin\omega t]$$

$$= \frac{p-r}{\omega^2} + \frac{\sqrt{(r + S\omega^2 - p)^2 + V^2\omega^2}}{\omega^2}$$

$$\times \left[\frac{r + S\omega^2 - p}{\sqrt{(r + S\omega^2 - p)^2 + V^2\omega^2}}\cos\omega t \right.$$

$$\left. + \frac{V\omega}{\sqrt{(r + S\omega^2 - p)^2 + V^2\omega^2}}\sin\omega t \right]$$

Based on the presented above expressions, we obtain

$$x = \frac{1}{\omega^2}[p - r + \sqrt{V^2\omega^2 + (r + S\omega^2 - p)^2}\sin(\omega t + \lambda_0)] \qquad (7.2.6)$$

where

$$\cos\lambda_0 = \frac{r + S\omega^2 - p}{\sqrt{V^2\omega^2 + (r + S\omega^2 - p)^2}};$$

$$\sin\lambda_0 = \frac{V\omega}{\sqrt{V^2\omega^2 + (r + S\omega^2 - p)^2}} \qquad (7.2.7)$$

The first derivative from equation (7.2.6) yields the velocity of the system:

$$\frac{dx}{dt} = \frac{1}{\omega}\sqrt{V^2\omega^2 + (r + S\omega^2 - p)^2}\cos(\omega t + \lambda_0)] \qquad (7.2.8)$$

Assuming that in equations (7.2.6) and (7.2.8) $t = 0$, we determine that $x = S$ and $\frac{dx}{dt} = V$, as expected according to the initial conditions of motion presented in expression (2.1.2).

Taking the first derivative from equation (7.2.8), we obtain the acceleration of the system:

$$\frac{d^2x}{dt^2} = -\sqrt{V^2\omega^2 + (r + S\omega^2 - p)^2}\sin(\omega t + \lambda_0)] \qquad (7.2.9)$$

Hence, equations (7.2.6), (7.2.8), and (7.2.9) are the basic parameters of motion of the system.

7.3. Motion of a System Restricted by a Flexible Link and Subjected to a Resisting and a Harmonic Force

Consider the operational process of a system restricted by a flexible link, while moving on a horizontal frictionless surface. The system is subjected to a constant resistance and to a harmonic force. The air resistance to the motion of the system is negligible.

Figure 7.3.1 shows the schematic diagram of the system described above. The notations in this figure are self-explanatory.

Based on the schematic diagram presented in Figure 7.3.1 and on the above-mentioned considerations, we compose the differential equation of motion of the system:

$$m\frac{d^2x}{dt^2} + Kx + R = A\cos\varphi t \tag{7.3.1}$$

The initial conditions of motion are taken according to expression (2.1.2).

Dividing equation (7.3.1) by m, we have

$$\frac{d^2x}{dt^2} + \omega^2 x + r = a\cos\varphi t \tag{7.3.2}$$

Using the equation (7.3.2) with the initial conditions of motion according to expression (2.1.2), the Laplace Transform pairs 5, 2, 1, 2, 2, and 29, we convert this equation from the time domain into

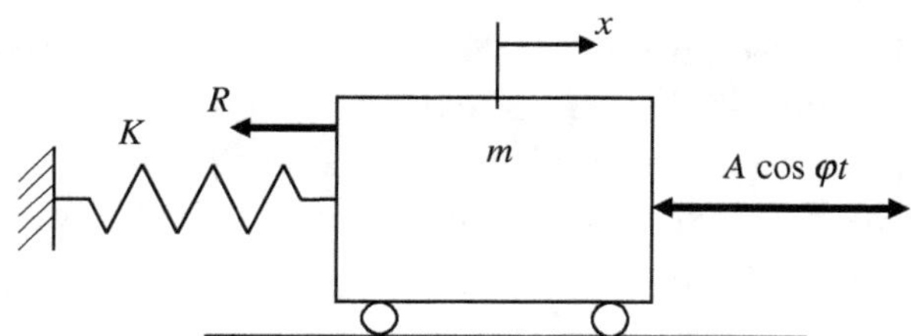

Fig. 7.3.1. Schematic diagram of a system restricted by a flexible link and moving on a frictionless surface while being subjected to a constant resistance and a harmonic force.

an algebraic equation in the Laplace domain:

$$s^2 x(s) - Vs - s^2 S + \omega^2 x(s) + r = \frac{as^2}{s^2 + \varphi^2} \qquad (7.3.3)$$

Rearranging equation (7.3.3), we may write

$$x(s)(s^2 + \omega^2) = s^2 S + sV - r + \frac{as^2}{s^2 + \varphi^2} \qquad (7.3.4)$$

The solution of equation (7.3.4) for the displacement $x(s)$ in the Laplace domain represents an algebraic equation describing a function of Laplace variable s:

$$x(s) = \frac{s^2 S}{s^2 + \omega^2} + \frac{sV}{s^2 + \omega^2} - \frac{r}{s^2 + \omega^2} + \frac{as^2}{(s^2 + \varphi^2)(s^2 + \omega^2)}$$

$$(7.3.5)$$

Applying Laplace Transform pairs 1, 29, 23, 17, and 72, we invert equation (7.3.5) from the Laplace domain into the time domain and obtain the solution of differential equation (7.3.1) with the initial conditions of motion according to expression (2.1.2):

$$x = S \cos \omega t + \frac{V}{\omega} \sin \omega t - \frac{r}{\omega^2}(1 - \cos \omega t) + \frac{a(\cos \omega t - \cos \varphi t)}{\varphi^2 - \omega^2}$$

$$(7.3.6)$$

It may be noted that, as shown earlier, in the case when $\varphi = \omega$, this equation leads to resonance.

Taking the first derivative from equation (7.3.6), we determine the velocity of the system:

$$\frac{dx}{dt} = -S\omega \sin \omega t + V \cos \omega t - \frac{r}{\omega} \sin \omega t + \frac{a(\varphi \sin \varphi t - \omega \sin \omega t)}{\varphi^2 - \omega^2}$$

$$(7.3.7)$$

Supposing that in equations (7.3.6) and (7.3.7) we have $t = 0$, then we obtain that $x = S$ and $\frac{dx}{dt} = V$, which is expected according to initial conditions of motion according to expression (2.1.2).

The first derivative from equation (7.3.7) yields the acceleration of the system:

$$\frac{d^2x}{dt^2} = -S\omega^2 \cos\omega t - V\omega \sin\omega t - r\cos\omega t$$

$$+ \frac{a(\varphi^2 \cos\varphi t - \omega^2 \cos\omega t)}{\varphi^2 - \omega^2} \tag{7.3.8}$$

Hence, equations (7.3.6)–(7.3.8) describe the basic parameters of motion of the system.

7.4. Motion of a System Restricted by a Flexible Link and Subjected to a Constant Resistance, a Constant Active and a Harmonic Force

The operational process is related to the motion of the system on a horizontal frictionless surface while being restricted by a flexible link. The system is subjected to the action of a constant resistance, a constant active and a harmonic force. The air resistance causes negligible action on the motion of the system. Figure 7.4.1 shows the schematic diagram of the system described above. The notations in the figure are self-explanatory.

Based on the schematic diagram shown in Figure 7.4.1 and on the above-mentioned considerations, we compose the differential equation of motion of the system:

$$m\frac{d^2x}{dt^2} + Kx + R = P + A\cos\varphi t \tag{7.4.1}$$

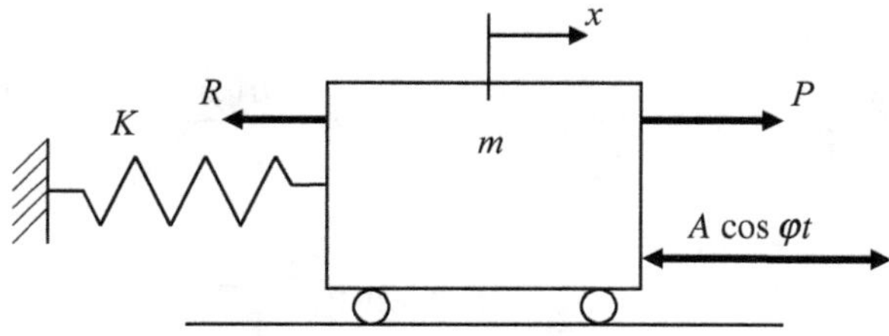

Fig. 7.4.1. Schematic diagram of a system restricted by a flexible link and moving on a frictionless surface while being subjected to a constant resistance, a constant active, and a harmonic force.

The initial conditions of motion are taken according to expression (2.1.2).

Dividing equation (7.4.1) by m, we may write

$$\frac{d^2x}{dt^2} + \omega^2 x + r = p + a\cos\varphi t \tag{7.4.2}$$

With the help of Laplace Transform pairs 5, 2, 1, 2, 2, and 29, we convert equation (7.4.2) with the initial conditions of motion according to expression (2.1.2) from the time domain into the Laplace domain and obtain an algebraic equation in the Laplace domain:

$$s^2 x(s) - Vs - s^2 S + \omega^2 x(s) + r = p + \frac{as^2}{s^2 + \varphi^2} \tag{7.4.3}$$

Applying to equation (7.4.3) the conventional algebraic procedures, we may write

$$x(s)(s^2 + \omega^2) = s^2 S + sV + p - r + \frac{as^2}{s^2 + \varphi^2} \tag{7.4.4}$$

Solving equation (7.4.4) for the displacement $x(s)$ in the Laplace domain, we obtain an algebraic equation describing the displacement $x(s)$ of the system as a function of Laplace variable s:

$$x(s) = \frac{s^2 S}{s^2 + \omega^2} + \frac{sV}{s^2 + \omega^2} + \frac{p - r}{s^2 + \omega^2} + \frac{as^2}{(s^2 + \varphi^2)(s^2 + \omega^2)} \tag{7.4.5}$$

Using Laplace Transform pairs 1, 29, 23, 17, and 72, we invert equation (7.4.5) from the Laplace domain into the time domain and obtain the solution of differential equation (7.4.1) with the initial conditions of motion according to expression (2.1.2):

$$x = S\cos\omega t + \frac{V}{\omega}\sin\omega t + \frac{p - r}{\omega^2}(1 - \cos\omega t) + \frac{a(\cos\omega t - \cos\varphi t)}{\varphi^2 - \omega^2} \tag{7.4.6}$$

The first derivative from equation (7.4.6) yields the velocity of the system:

$$\frac{dx}{dt} = -S\omega\sin\omega t + V\cos\omega t + \frac{p - r}{\omega}\sin\omega t + \frac{a(\varphi\sin\varphi t - \omega\sin\omega t)}{\varphi^2 - \omega^2} \tag{7.4.7}$$

Assuming that in equations (7.4.6) and (7.4.7) $t = 0$, we determine that $x = S$ and $\frac{dx}{dt} = V$ as it should be according to initial conditions of motion presented in expression (2.1.2).

Taking the first derivative from equation (7.4.7), we determine the acceleration of the system

$$\frac{d^2x}{dt^2} = -S\omega^2 \cos \omega t - V\omega \sin \omega t + (p - r) \cos \omega t$$

$$+ \frac{a(\varphi^2 \cos \varphi t - \omega^2 \cos \omega t)}{\varphi^2 - \omega^2} \tag{7.4.8}$$

Equations (7.4.6)–(7.4.8) represent the basic parameters of motion of the system.

In addition, it should be mentioned that equation (7.4.6) applies in the case when $\varphi = \omega$ that leads to the resonance phenomenon.

7.4.1. *Numerical solution*

Following is a Python program to plot the graph of equation (7.4.6) (Figure 7.4.2):

```python
from matplotlib.pyplot import plot, show
from numpy import linspace, sin, cos

S = 0.1
p = 25
V = 0.1
a = 10
r = 3
phi = 30
omega = 15

t = linspace(0, 1, 1000)
x = S * cos(omega*t) + (V/omega)*sin(omega*t) \
    + (p-r)/(omega**2)*(1 - cos(omega*t)) \
    + (a*(cos(omega*t) - cos(phi*t)))/(phi**2 - omega**2)

plot(t, x)
show()
```

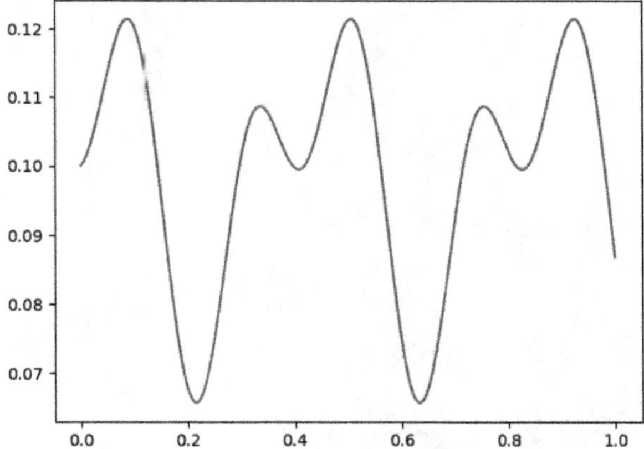

Fig. 7.4.2. Graph of equation (7.4.6).

MOTION OF A SYSTEM ATTACHED BY A FLEXIBLE LINK TO A NON-MOVABLE SUPPORT AND SUBJECTED TO A DRY FRICTION FORCE

The operational process considered in this chapter is related to a system connected by a flexible link to a non-movable support while being subjected to a dry friction force and moving due to initial conditions of motion.

8.1. A System Restricted by a Flexible Link and Subjected to a Dry Friction Force, Possessing an Initial Displacement and Velocity While Moving on a Horizontal Surface

It should be stressed that the flexible link accumulates a corresponding amount of potential energy due to its deformation, and this energy will be returned to the system in the form of kinetic energy. The air resistance to the motion of the system is negligible.

Figure 8.1.1 show schematic diagram of the system described above. The notations in the figure are self-explanatory.

Based on the schematic diagram shown in Figure 8.1.1 and the considerations mentioned above, we compose the differential equation

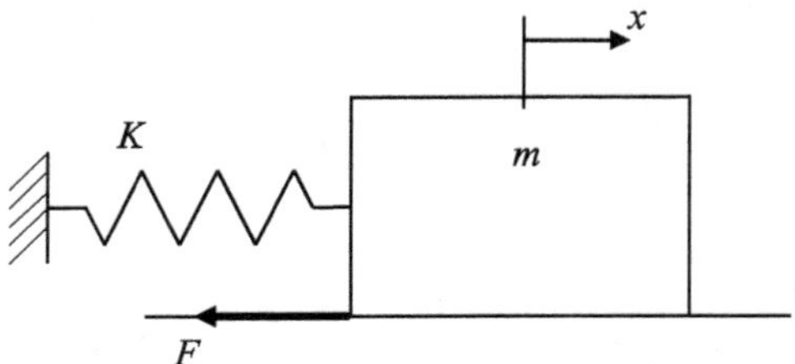

Fig. 8.1.1. Schematic diagram of a system moving on a horizontal surface while being restricted by a flexible link and subjected to a dry friction force.

of motion that reads:

$$m\frac{d^2x}{dt^2} + Kx + F = 0 \tag{8.1.1}$$

The initial conditions of motion are taken according to expression (2.1.2).

Dividing equation (8.1.1) by m, we may write

$$\frac{d^2x}{dt^2} + \omega^2 x + f = 0 \tag{8.1.2}$$

With the help of Laplace Transform pairs 5, 2, 1, and 2, we convert equation (8.1.2) with initial conditions of motion according to expression (2.1.2) from the time domain into the Laplace domain and obtain a corresponding algebraic equation in the Laplace domain:

$$s^2 x(s) - sV - s^2 S + \omega^2 x(s) + f = 0 \tag{8.1.3}$$

Rearranging equation (8.1.3), we have

$$x(s)(s^2 + \omega^2) = sV + s^2 S - f \tag{8.1.4}$$

Solving equation (8.1.4) for the displacement $x(s)$ in the Laplace domain, we have

$$x(s) = \frac{sV}{s^2 + \omega^2} + \frac{s^2 S}{s^2 + \omega^2} - \frac{f}{s^2 + \omega^2} \tag{8.1.5}$$

Applying Laplace Transform pairs 1, 23, 29, and 17, we invert equation (8.1.5) from the Laplace domain into the time domain and obtain the solution of differential equation (8.1.1) with the initial conditions of motion according to expression (2.1.2):

$$x = \frac{V}{\omega} \sin \omega t + S \cos \omega t - \frac{f}{\omega^2}(1 - \cos \omega t) \qquad (8.1.6)$$

The first derivative from equation (8.1.6) yields the velocity of the system:

$$\frac{dx}{dt} = V \cos \omega t - S\omega \sin \omega t - \frac{f}{\omega} \sin \omega t \qquad (8.1.7)$$

Assuming that in equations (8.1.6) and (8.1.7), $t = 0$, we determine that $x = S$ and $\frac{dx}{dt} = V$, as it should be according to the initial conditions of motion according to expression (2.1.2).

It should be stressed that the velocity given by equation (8.1.7) is decreasing as a function of time; sooner or later, it will become equal to zero, and this will result in an instantaneous change in the direction of the velocity. Unfortunately, as mentioned earlier, this change in the direction of velocity cannot occur in equation (8.1.7) by itself; therefore, differential equation (8.1.1) will not describe the motion of the system anymore. This means that, to continue describing the motion of the system, it is necessary to compose and solve a new differential equation with the initial conditions of motion that existed at the instance of the change in the direction of the velocity. More related details are presented in the following.

The first derivative from equation (8.1.7) allows us to determine the acceleration of the system:

$$\frac{d^2 x}{dt^2} = -V\omega \sin \omega t - S\omega^2 \cos \omega t - f \cos \omega t \qquad (8.1.8)$$

Hence, equations (8.1.6)–(8.1.8) represent the basic parameters of motion of the system.

8.1.1. *The phenomenon of dry friction force in vibratory motion*

Consider the analysis of an operational process associated with the presence of a dry friction force in a system performing vibratory motion. The system is moving on a horizontal surface and is attached by a flexible link to a non-movable support. The system is subjected to a dry friction force that is always directed oppositely to the velocity of the system. The air resistance to the motion of the system is negligible. The notations in this figure are self-explanatory. In order to simplify the current analysis of the vibratory process, we equate the initial velocity of the system to zero. We are keeping in mind that differential equation (8.1.1) is applicable to this example. The initial conditions of motion for the second half of the period of vibratory motion are

$$\text{for } t = 0 \; x = S; \quad \frac{dx}{dt} = 0 \tag{8.1.9}$$

Therefore, by applying Laplace Transform pairs 5, 2, 1, and 2 to differential equation (8.1.2) with the initial conditions of motion according to expression (8.1.9), we convert this differential equation with the initial conditions of motion from the time domain into an algebraic equation in the Laplace domain:

$$s^2 x(s) - s^2 S + \omega^2 x(s) + f = 0 \tag{8.1.10}$$

Rearranging equation (8.1.10), we may write

$$x(s)(s^2 + \omega^2) = s^2 S - f \tag{8.1.11}$$

Solving equation (8.1.11) for the displacement $x(s)$ in the Laplace domain, we have

$$x(s) = \frac{s^2 S}{s^2 + \omega^2} - \frac{f}{s^2 + \omega^2} \tag{8.1.12}$$

Using Laplace Transform pairs 1, 29, and 17, we invert equation (8.1.12) from the Laplace domain into the time domain and obtain the solution of differential equation (8.1.1) with the initial conditions of motion according to expression (8.1.9):

$$x = S \cos \omega t - \frac{f}{\omega^2}(1 - \cos \omega t) \tag{8.1.13}$$

The first derivative from equation (8.1.13) yields the velocity of the system:

$$\frac{dx}{dt} = -S\omega \sin \omega t - \frac{f}{\omega} \sin \omega t \qquad (8.1.14)$$

Assuming that in equations (8.1.13) and (8.1.14), $t = 0$, we determine that $x = S$ and $\frac{dx}{dt} = 0$, as it should be according to the initial conditions of motion given by expression (8.1.9).

Taking the first derivative from equation (8.1.14), we determine the deceleration of the system:

$$\frac{d^2 x}{dt^2} = -(S\omega^2 + f) \cos \omega t \qquad (8.1.15)$$

Hence, equations (8.1.13)–(8.1.15) represent the basic parameters of motion of the system.

Continuing with the analysis related to dry friction and vibration, we may emphasize that when the system stops, it immediately changes the direction of its velocity to the opposite value, while equation of velocity (8.1.14) cannot by itself reflect this change. Therefore, for the second half of the period of the vibrational motion, the original differential equation should be used with new values of the initial conditions of motion. More details are presented in the following.

Equation (8.1.14) shows that the first stop occurs when $\sin \omega t = 0$, which could happen at the end of the first half of the period of vibration, while

$$\omega t = \pi \qquad (8.1.16)$$

and consequently,

$$\cos \omega t = -1 \qquad (8.1.17)$$

Combining equations (8.1.13) and (8.1.17), we determine the displacement of the system at its first stop, or, in other words, at the end of the first half of the period of vibrations of the system:

$$S_1 = -S - \frac{2f}{\omega^2} \qquad (8.1.18)$$

Hence, based on the analysis carried out, we compose the differential equation of motion for the second half of period of vibration, where the function is denoted by x_1:

$$m\frac{d^2 x_1}{dt^2} + Kx_1 + F = 0 \qquad (8.1.19)$$

and we take the initial displacement according to equation (8.1.18), while the initial velocity equals zero. Therefore, the initial conditions of motion for the second half of the period of motion of the system are

$$x_1 = S_1; \frac{dx_1}{dt} = 0 \qquad (8.1.20)$$

For more details, see Section 1.5.

8.2. Motion of a System Restricted by a Flexible Link and Subjected to a Dry Friction Force and a Constant Active Force

The analysis of the operational process of the system is associated with the study related to the methodology of clarifying the basic parameters of motion, the control of which allows us to achieve the required performance of the system. Consider the motion of a system restricted by a flexible link while moving on a horizontal surface and subjected to a dry friction force and a constant active force. The air resistance to the motion of the system is negligible. Figure 8.2.1 shows a schematic diagram of the system mentioned above. The notations in this figure are self-explanatory.

Accounting for the considerations mentioned above and the schematic diagram shown in Figure 8.2.1, we compose the differential

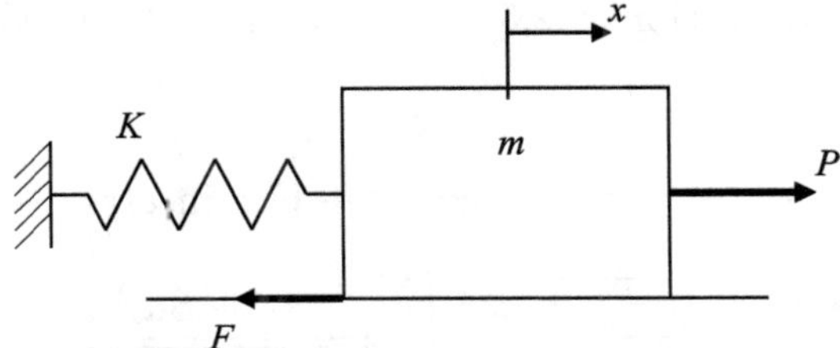

Fig. 8.2.1. Schematic diagram of a system moving on a horizontal surface while being restricted by a flexible link and subjected to a dry friction force and to a constant active force.

equation of motion of the system:

$$m\frac{d^2x}{dt^2} + Kx + F = P \tag{8.2.1}$$

The initial conditions of motion are taken according to expression (2.1.2). Dividing equation (8.2.1) by m, we may write

$$\frac{d^2x}{dt^2} + \omega^2 x + f = p \tag{8.2.2}$$

With the help of Laplace Transform pairs 5, 2, 1, 2, and 2, we convert differential equation (8.2.2) with the initial conditions of motion according to expression (2.1.2) from the time domain into an algebraic equation in the Laplace domain:

$$s^2 x(s) - sV - s^2 S + \omega^2 x(s) + f = p \tag{8.2.3}$$

Solving equation (8.2.3) for the displacement $x(s)$ in the Laplace domain, we have

$$x(s) = \frac{p - f}{s^2 + \omega^2} + \frac{sV}{s^2 + \omega^2} + \frac{s^2 S}{s^2 + \omega^2} \tag{8.2.4}$$

Applying Laplace Transform pairs 1, 17, 23, and 29, we invert equation (8.2.4) from the Laplace domain into the time domain and obtain the solution of differential equation (8.2.1) with the initial conditions of motion according to expression (2.1.2):

$$x = \frac{p - f}{\omega^2}(1 - \cos \omega t) + \frac{V}{\omega}\sin \omega t + S \cos \omega t \tag{8.2.5}$$

Obviously, it is assumed that in equation (8.2.5), $p - f > 0$.

Applying to equation (8.2.5) the conventional procedures, we may write

$$x = \frac{p-f}{\omega^2} + \frac{1}{\omega^2}[(f + S\omega^2 - p)\cos\omega t + V\omega\sin\omega t] = \frac{p-f}{\omega^2}$$

$$+ \frac{\sqrt{(f + S\omega^2 - p)^2 + V^2\omega^2}}{\omega^2}\left[\frac{f + S\omega^2 - p}{\sqrt{(f + S\omega^2 - p)^2 + V^2\omega^2}}\cos\omega t\right.$$

$$\left.+ \frac{V\omega}{\sqrt{(f + S\omega^2 - p)^2 + V^2\omega^2}}\sin\omega t\right]$$

Based on the expressions presented above, we obtain

$$x = \frac{1}{\omega^2}[p - f + \sqrt{V^2\omega^2 + (f + S\omega^2 - p)^2}\sin(\omega t + \lambda_1)] \qquad (8.2.6)$$

where

$$\cos\lambda_1 = \frac{f + S\omega^2 - p}{\sqrt{V^2\omega2 + (f + S\omega^2 - p)^2}};$$

$$\sin\lambda_1 = \frac{V\omega}{\sqrt{V^2\omega^2 + (f + S\omega^2 - p)^2}} \qquad (8.2.7)$$

The first derivative from equation (8.2.6) yields the velocity of the system:

$$\frac{dx}{dt} = \frac{1}{\omega}\sqrt{V^2\omega^2 + (f + S\omega^2 - p)^2}\cos(\omega t + \lambda_1)] \qquad (8.2.8)$$

Assuming that in equations (8.2.6) and (8.2.8), $t = 0$, we determine that $x = S$ and $\frac{dx}{dt} = V$, as expected according to the initial conditions of motion presented in expression (2.1.2). Taking the first derivative from equation (8.2.8), we determine the acceleration of the system:

$$\frac{d^2x}{dt^2} = -\sqrt{V^2\omega^2 + (f + S\omega^2 - p)^2}\sin(\omega t + \lambda_1)] \qquad (8.2.9)$$

Hence, equations (8.2.6),(8.2.8), and (8.2.9) are the basic parameters of motion of the system.

8.3. Motion of a System Restricted by a Flexible Link and Subjected to a Dry Friction Force and a Harmonic Force

Consider the operational process of a system restricted by a flexible link while moving on a horizontal surface. The system is subjected to a dry friction force and a harmonic force. The air resistance to the motion of the system is negligible. Figure 8.3.1 shows a schematic diagram of the system described above. The notations in this figure are self-explanatory.

Based on the schematic diagram presented in Figure 8.3.1 and the above-mentioned considerations, we compose the differential equation of motion of the system:

$$m\frac{d^2x}{dt^2} + Kx + F = A\cos\varphi t \qquad (8.3.1)$$

The initial conditions of motion are taken according to expression (2.1.2). Dividing equation (8.3.1) by m, we may write

$$\frac{d^2x}{dt^2} + \omega^2 x + f = a\cos\varphi t \qquad (8.3.2)$$

By applying Laplace Transform pairs 5, 2, 1, 2, and 29 to equation (8.3.2) with the initial conditions of motion according to expression (2.1.2), we convert this equation from the time domain

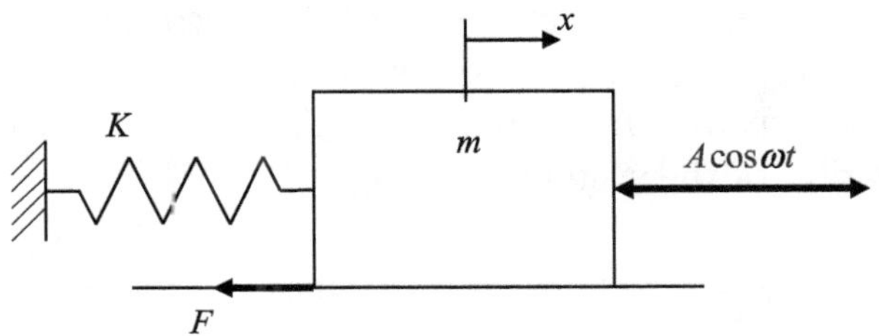

Fig. 8.3.1. Schematic diagram of a system moving on a horizontal surface while being subjected to a dry friction force and a harmonic force.

into an algebraic equation in the Laplace domain:

$$s^2 x(s) - Vs - s^2 S + \omega^2 x(s) + f = \frac{as^2}{s^2 + \varphi^2} \tag{8.3.3}$$

Rearranging equation (8.3.3), we may write

$$x(s)(s^2 + \omega^2) = s^2 S + sV - f + \frac{as^2}{s^2 + \varphi^2} \tag{8.3.4}$$

Solving equation (8.3.4) for the displacement $x(s)$ in the Laplace domain, we obtain an algebraic equation describing the displacement as a function of Laplace variable s:

$$x(s) = \frac{s^2 S}{s^2 + \omega^2} + \frac{sV}{s^2 + \omega^2} - \frac{f}{s^2 + \omega^2} + \frac{as^2}{(s^2 + \varphi^2)(s^2 + \omega^2)} \tag{8.3.5}$$

Applying Laplace Transform pairs 1, 29, 23, 17, and 72, we invert equation (8.3.5) from the Laplace domain into the time domain and obtain the solution of differential equation (8.3.1) with the initial conditions of motion according to expression (2.1.2):

$$x = S\cos\omega t + \frac{V}{\omega}\sin\omega t - \frac{f}{\omega^2}(1 - \cos\omega t) + \frac{a(\cos\omega t - \cos\varphi t)}{\varphi^2 - \omega^2} \tag{8.3.6}$$

The first derivative from equation (8.3.6) yields the velocity of the system:

$$\frac{dx}{dt} = -S\omega\sin\omega t + V\cos\omega t - \frac{f}{\omega}\sin\omega t + \frac{a(\varphi\sin\varphi t - \omega\sin\omega t)}{\varphi^2 - \omega^2} \tag{8.3.7}$$

Assuming that in equations (8.3.6) and (8.3.7), $t = 0$, we obtain that $x = S$ and $\frac{dx}{dt} = V$, which is expected according to the initial conditions of motion given by expression (2.1.2).

The first derivative from equation (8.3.7) allows us to determine the acceleration of the system:

$$\frac{d^2 x}{dt^2} = -S\omega^2\cos\omega t - V\omega\sin\omega t - f\cos\omega t$$

$$+ \frac{a(\varphi^2\cos\varphi t - \omega^2\cos\omega t)}{\varphi^2 - \omega^2} \tag{8.3.8}$$

Hence, equations (8.3.6)–(8.3.8) describe the basic parameters of motion of the system.

8.4. A System Moving on a Horizontal Surface Being Restricted by a Flexible Link and Subjected to a Dry Friction Force, a Constant Active Force, and a Harmonic Force

The operational process is related to the motion of a system on a horizontal surface, while the system is restricted by a flexible link. The system is subjected to the action of a dry friction force, a constant active force, and a harmonic force. The air resistance causes negligible action on the motion of the system. Figure 8.4.1 shows the schematic diagram that represents the system described above. The notations in the figure are self-explanatory.

Based on the schematic diagram shown in Figure 8.4.1 and the above-mentioned considerations, we compose the differential equation of motion of the system:

$$m\frac{d^2x}{dt^2} + Kx + F = P + A\cos\varphi t \tag{8.4.1}$$

The initial conditions of motion are taken according to expression (2.1.2).

Dividing equation (8.4.1) by m, we may write

$$\frac{d^2x}{dt^2} + \omega^2 x + f = p + a\cos\varphi t \tag{8.4.2}$$

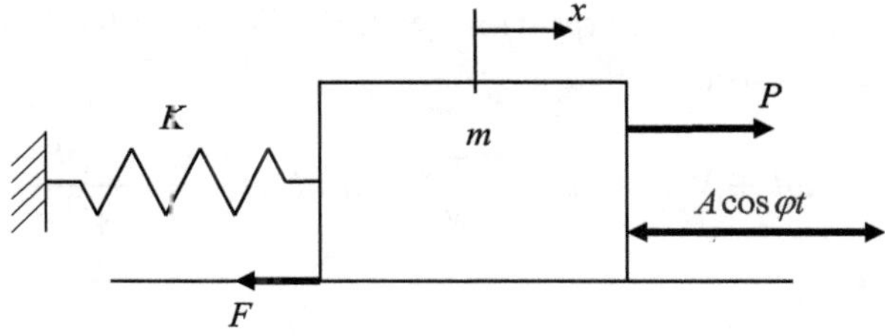

Fig. 8.4.1. Schematic diagram of a system moving on a horizontal surface while being subjected to a dry friction force, a constant active force, and a harmonic force.

Applying Laplace Transform pairs 5, 2, 1, 2, 2, and 29, we convert equation (8.4.2) with the initial conditions of motion according to expression (2.1.2) from the time domain into the Laplace domain and obtain an algebraic equation in the Laplace domain:

$$s^2 x(s) - Vs - s^2 S + \omega^2 x(s) + f = p + \frac{as^2}{s^2 + \varphi^2} \tag{8.4.3}$$

Applying to equation (8.4.3) the conventional algebraic procedures, we may write

$$x(s)(s^2 + \omega^2) = s^2 S + sV + p - f + \frac{as^2}{s^2 + \varphi^2} \tag{8.4.4}$$

The solution of equation (8.4.4) for the displacement $x(s)$ in the Laplace domain allows us to obtain an algebraic equation describing the displacement $x(s)$ of the system as a function of Laplace variable s:

$$x(s) = \frac{s^2 S}{s^2 + \omega^2} + \frac{sV}{s^2 + \omega^2} + \frac{p - f}{s^2 + \omega^2} + \frac{as^2}{(s^2 + \varphi^2)(s^2 + \omega^2)} \tag{8.4.5}$$

Applying Laplace Transform pairs 29, 23, 17, and 72, we invert equation (8.4.5) from the Laplace domain into the time domain and obtain the solution of differential equation (8.4.1) with the initial conditions of motion according to expression (2.1.2):

$$x = S \cos \omega t + \frac{V}{\omega} \sin \omega t + \frac{p - f}{\omega^2}(1 - \cos \omega t) + \frac{a(\cos \omega t - \cos \varphi t)}{\varphi^2 - \omega^2} \tag{8.4.6}$$

The first derivative from equation (8.4.6) allows us to determine the velocity of the system:

$$\frac{dx}{dt} = -S\omega \sin \omega t + V \cos \omega t + \frac{p - f}{\omega} \sin \omega t + \frac{a(\varphi \sin \varphi t - \omega \sin \omega t)}{\varphi^2 - \omega^2} \tag{8.4.7}$$

Assuming that in equations (8.4.6) and (8.4.7), $t = 0$, we determine that $x = S$ and $\frac{dx}{dt} = V$, as it should be according to the initial conditions of motion presented in expression (2.1.2). The first derivative from equation(8.4.7) yields the acceleration of the system:

$$\frac{d^2x}{dt^2} = -S\omega^2 \cos\omega t - V\omega \sin\omega t + (p - f)\cos\omega t$$

$$+ \frac{a(\varphi^2 \cos\varphi t - \omega^2 \cos\omega t)}{\varphi^2 - \omega^2} \tag{8.4.8}$$

Equations (8.4.6)–(8.4.8) represent the basic parameters of motion of the system.

8.4.1. *Numerical solution*

Following is a Python program to plot equation (8.4.7) and solve it for when velocity first reaches zero.

```python
from matplotlib.pyplot import plot, show, text
from numpy import linspace, sin, cos
from scipy.optimize import fsolve

S = 0.1
p = 2
V = 8
a = 20
f = 6
phi = 30
omega = 15

t = linspace(0, 0.5, 1000)
V1 = lambda t: \
    - S*omega*sin(omega*t) + V*cos(omega*t) \
    + ((p - f)/omega)*sin(omega*t) \
    + (a*(phi*sin(phi*t) - omega*sin(omega*t)))/(phi**2 - omega**2)

[root] = fsolve(V1, 0.1)
root = round(root, 3)

plot(t, V1(t))
plot(root, 0, 'bo')
text(0.02, -0.5, f't={root}')
show()
```

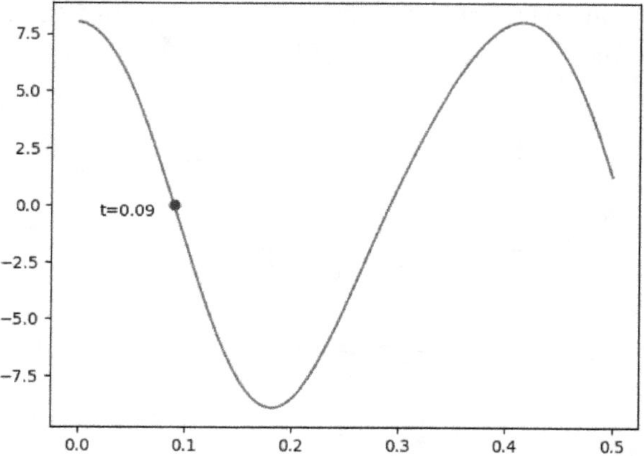

Fig. 8.4.2. Graph of equation (8.4.7).

Following is a Python program to plot the displacement represented by equation (8.4.6) from $t = 0$ to when the system first changes direction at $t = 0.09$ (Figure 8.4.2).

```python
from matplotlib.pyplot import plot, show, text
from numpy import linspace, sin, cos

S = 0.1
p = 2
V = 8
a = 20
f = 6
phi = 30
omega = 15

x = lambda t: \
  S*cos(omega*t) + (V/omega)*sin(omega*t) \
  + (p - f)/(omega**2)*(1 - cos(omega*t)) \
  + a*(cos(omega*t) - cos(phi*t))/(phi**2 - omega**2)

stopTime = 0.09
t = linspace(0, stopTime, 1000)
plot(t, x(t))

stopPosition = round(x(stopTime), 3)
plot(stopTime, stopPosition, 'bo')
text(0.08, 0.52, f'x={stopPosition}')

show()
```

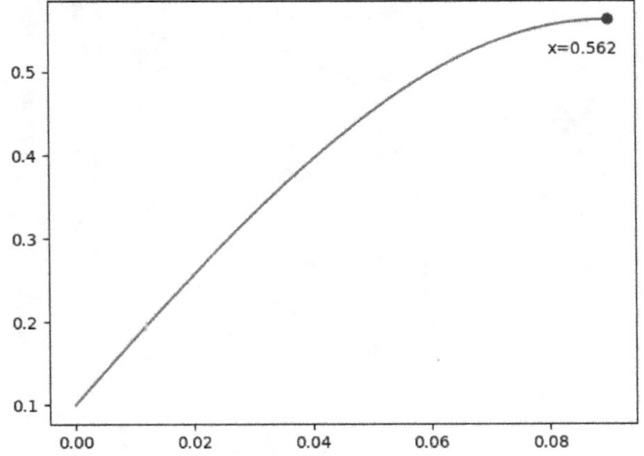

Fig. 8.4.3. Graph of equation 8.4.6)

For the next iteration, apply equation (8.4.1) with the following initial conditions of motion (Figure 8.4.3):

$$\text{for} \quad t = 0 \quad x = 0.562; \quad \frac{dx}{dt} = 0.$$

For more details, see Section 1.5.

CHAPTER 9

MOTION OF A SYSTEM RESTRICTED BY A FLEXIBLE LINK AND SUBJECTED TO A CONSTANT RESISTING FORCE AND A DRY FRICTION FORCE

9.1. Motion of a System Restricted by a Flexible Link While Subjected to a Dry Friction Force and a Constant Resisting Force and Moving Due to Its Initial Displacement and Velocity

The current operational process is related to a movable system restricted by a flexible link while moving on a horizontal surface under the actions of an initial displacement and initial velocity. The system is subjected to a constant resisting force and a dry friction force. Due to the initial conditions of motion, the flexible link becomes deformed, and this results in the storage of some potential energy that is proportional to the displacement of the system. The potential energy will be returned by the link to the system in the form of kinetic energy. The air resistance to the motion of the system is negligible.

Figure 9.1.1 shows a schematic diagram of the system described above. The notations in the figure are self-explanatory.

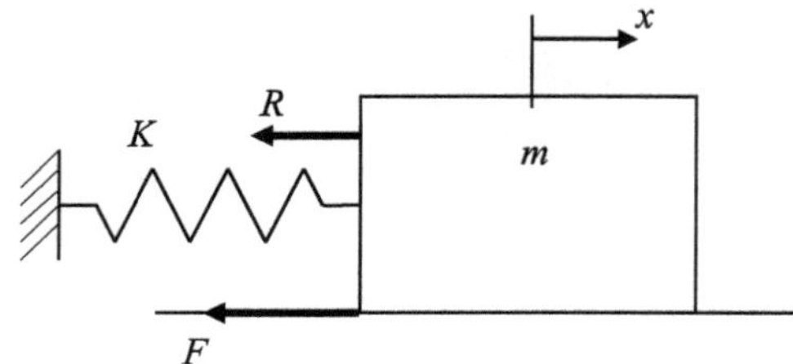

Fig. 9.1.1. Schematic diagram of a system moving on a horizontal surface while being restricted by a flexible link and subjected to a dry friction force and a constant resisting force.

Based on the schematic diagram shown in Figure 9.1.1 and the considerations mentioned above, we compose the differential equation of motion that reads

$$m\frac{d^2x}{dt^2} + Kx + F + R = 0 \tag{9.1.1}$$

The initial conditions of motion are taken according to expression (2.1.2).

Dividing equation (9.1.1) by m, we may write

$$\frac{d^2x}{dt^2} + \omega^2 x + f + r = 0 \tag{9.1.2}$$

Applying Laplace Transform pairs 5, 2, 1, 2, and 2, we convert equation (9.1.2) with the initial conditions of motion according to expression (2.1.2) from the time domain into the Laplace domain and obtain a corresponding algebraic equation in the Laplace domain:

$$s^2x(s) - sV - s^2S + \omega^2 x(s) + f + r = 0 \tag{9.1.3}$$

Rearranging equation (9.1.3), we have

$$x(s)(s^2 + \omega^2) = sV + s^2S - f - r \tag{9.1.4}$$

The solution of equation (9.1.4) for the displacement $x(s)$ in the Laplace domain reads

$$x(s) = \frac{sV}{s^2 + \omega^2} + \frac{s^2S}{s^2 + \omega^2} - \frac{f + r}{s^2 + \omega^2} \tag{9.1.5}$$

With the help of Laplace Transform pairs 1, 23, 29, and 17, we invert equation (9.1.1) from the Laplace domain into the time domain and obtain the solution of differential equation (9.1.1) with the initial conditions of motion according to expression (2.1.2):

$$x = \frac{V}{\omega} \sin \omega t + S \cos \omega t - \frac{f+r}{\omega^2}(1 - \cos \omega t) \tag{9.1.6}$$

Taking the first derivative from equation (9.1.6), we determine the velocity of the system:

$$\frac{dx}{dt} = V \cos \omega t - S\omega \sin \omega t - \frac{f+r}{\omega} \sin \omega t \tag{9.1.7}$$

Supposing that in equations (9.1.6) and (9.1.7), $t = 0$, we determine that $x = S$ and $\frac{dx}{dt} = V$, as it should be according to the initial conditions of motion according to expression (2.1.2).

The first derivative from equation (9.1.7) yields the acceleration of the system:

$$\frac{d^2 x}{dt^2} = -V\omega \sin \omega t - S\omega^2 \cos \omega t - (f+r) \cos \omega t \tag{9.1.8}$$

Hence, equations (9.1.6)–(9.1.8) represent the basic parameters of motion of the system.

9.2. Motion of a System Restricted by a Flexible Link and Subjected to a Constant Resisting Force, a Dry Friction Force, and a Constant Active Force

The system is moving on a horizontal surface due to an initial displacement and initial velocity and to a constant active force. The system is restricted by a flexible link and subjected to the actions of a constant resisting force and a dry friction force. The air resistance to the motion of the system is negligible. Figure 9.2.1 shows a schematic diagram that describes the system. The notations in this figure are self-explanatory.

Accounting for the considerations mentioned above and the schematic diagram shown in Figure 9.2.1, we compose the differential

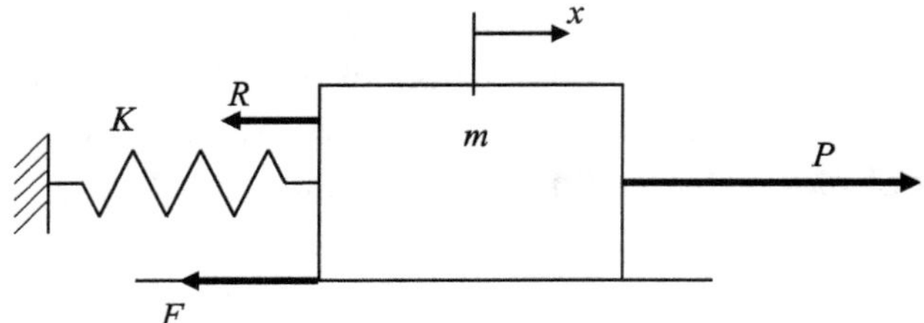

Fig. 9.2.1. Schematic diagram of a system moving on a horizontal surface while being restricted by a flexible link and subjected to a dry friction force, a constant resisting force, and a constant active force.

equation of motion of the system:

$$m\frac{d^2x}{dt^2} + Kx + F + R = P \tag{9.2.1}$$

The initial conditions of motion are taken according to expression (2.1.2).

Dividing equation (9.2.1) by m, we may write

$$\frac{d^2x}{dt^2} + \omega^2 x + f + r = p \tag{9.2.2}$$

Applying Laplace Transform pairs 5, 2, 1, 2, 2, and 2, we convert differential equation (9.2.2) with the initial conditions of motion according to expression (2.1.2) from the time domain into an algebraic equation in the Laplace domain:

$$s^2 x(s) - sV - s^2 S + \omega^2 x(s) + f + r = p \tag{9.2.3}$$

Solving equation (9.2.3) for the displacement $x(s)$ in the Laplace domain, we have

$$x(s) = \frac{p - f - r}{s^2 + \omega^2} + \frac{sV}{s^2 + \omega^2} + \frac{s^2 S}{s^2 + \omega^2} \tag{9.2.4}$$

Applying Laplace Transform pairs 1, 17, 23, and 29, we invert equation (9.2.4) from the Laplace domain into the time domain and obtain the solution of differential equation (9.2.1) with the initial

conditions of motion according to expression (2.1.2):

$$x = \frac{p-f-r}{\omega^2}(1 - \cos\omega t) + \frac{V}{\omega}\sin\omega t + S\cos\omega t \qquad (9.2.5)$$

Obviously, it is assumed that in equation (9.2.5), $p - f - r > 0$.

Applying to equation (9.2.5) the conventional procedures, we may write

$$x = \frac{p-f-r}{\omega^2} + \frac{1}{\omega^2}[(f + r + S\omega^2 - p)\cos\omega t + V\omega\sin\omega t]$$

$$= \frac{p-f-r}{\omega^2} + \frac{\sqrt{(f + r + S\omega^2 - p)^2 + V^2\omega^2}}{\omega^2}$$

$$\times \left[\frac{f + r + S\omega^2 - p}{\sqrt{(f + r + S\omega^2 - p)^2 + V^2\omega^2}}\cos\omega t \right.$$

$$\left. + \frac{V\omega}{\sqrt{(f + r + S\omega^2 - p)^2 + V^2\omega^2}}\sin\omega t \right]$$

Based on the expressions presented above, we obtain

$$x = \frac{1}{\omega^2}\left[p - f - r + \sqrt{V^2\omega^2 + (f + r + S\omega^2 - p)^2}\,\sin(\omega t + \lambda_2) \right]$$
$$(9.2.6)$$

where

$$\cos\lambda_2 = \frac{f + r + S\omega^2 - p}{\sqrt{V^2\omega^2 + (f + r + S\omega^2 - p)^2}};$$

$$\sin\lambda_2 = \frac{V\omega}{\sqrt{V^2\omega^2 + (f + r + S\omega^2 - p)^2}} \qquad (9.2.7)$$

The first derivative from equation (9.2.6) yields the velocity of the system:

$$\frac{dx}{dt} = \frac{1}{\omega}\sqrt{V^2\omega^2 + (f + r + S\omega^2 - p)^2}\,\cos(\omega t + \lambda_2)] \qquad (9.2.8)$$

Assuming that in equations (9.2.6) and (9.2.8), $t = 0$, we determine that $x = S$ and $\frac{dx}{dt} = V$, as expected according to the initial conditions of motion presented in expression (2.1.2).

Taking the first derivative from equation (9.2.8), we obtain the equation for the acceleration of the system:

$$\frac{d^2x}{dt^2} = -\sqrt{V^2\omega^2 + (f + r + S\omega^2 - p)^2}\sin(\omega t + \lambda_2)] \qquad (9.2.9)$$

Hence, equations (9.2.6), (9.2.8), and (9.2.9) are the basic parameters of motion of the system.

9.3. Motion of a System Restricted by a Flexible Link and Subjected to a Constant Resisting Force, a Dry Friction Force, and a Harmonic Force

Consider the operational process of a system restricted by a flexible link while moving on a horizontal surface. The system is subjected to a constant resisting force, a dry friction force, and a harmonic force. The air resistance to the motion of the system is negligible.

Figure 9.3.1 shows a schematic diagram that represents the system described above. The notations in this figure are self-explanatory.

Based on the schematic diagram presented in Figure 9.3.1 and the above-mentioned considerations, we compose the differential

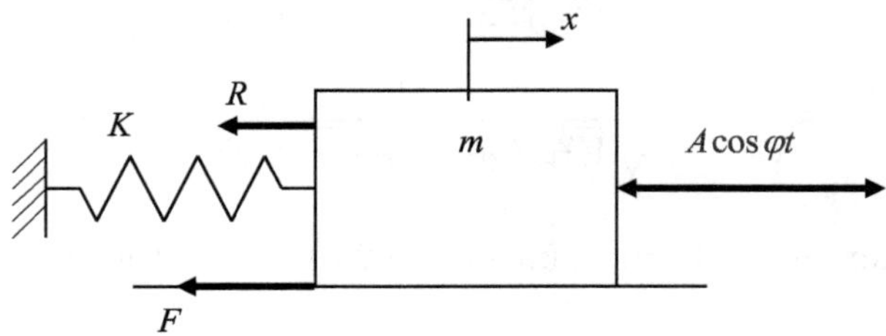

Fig. 9.3.1. Schematic diagram of a system moving on a horizontal surface while being restricted by a flexible link and subjected to a dry friction force, a constant resisting force, and a harmonic force.

equation of motion of the system:

$$m\frac{d^2x}{dt^2} + Kx + F + R = A\cos\varphi t \qquad (9.3.1)$$

The initial conditions of motion are taken according to expression (2.1.2).

Dividing equation (9.3.1) by m, we have

$$\frac{d^2x}{dt^2} + \omega^2 x + f + r = a\cos\varphi t \qquad (9.3.2)$$

Applying Laplace Transform pairs 5, 2, 1, 2, 2, and 29 to equation (9.3.2) with the initial conditions of motion according to expression (2.1.2), we convert this equation from the time domain into an algebraic equation in the Laplace domain:

$$s^2 x(s) - Vs - s^2 S + \omega^2 x(s) + f + r = \frac{as^2}{s^2 + \varphi^2} \qquad (9.3.3)$$

Rearranging equation (9.3.3), we may write

$$x(s)(s^2 + \omega^2) = s^2 S + sV - f - r + \frac{as^2}{s^2 + \varphi^2} \qquad (9.3.4)$$

The solution of equation (9.3.4) for the displacement $x(s)$ in the Laplace domain represents an algebraic equation describing a function of the Laplace variable s:

$$x(s) = \frac{s^2 S}{s^2 + \omega^2} + \frac{sV}{s^2 + \omega^2} - \frac{f + r}{s^2 + \omega^2} + \frac{as^2}{(s^2 + \varphi^2)(s^2 + \omega^2)} \qquad (9.3.5)$$

Applying Laplace Transform pairs 1, 29, 23, 17, and 72, we invert equation (9.3.5) from the Laplace domain into the time domain and obtain the solution of differential equation (9.3.1) with the initial conditions of motion according to expression (2.1.2):

$$x = S\cos\omega t + \frac{V}{\omega}\sin\omega t - \frac{f + r}{\omega^2}(1 - \cos\omega t) + \frac{a(\cos\omega t - \cos\varphi t)}{\varphi^2 - \omega^2}$$

$$(9.3.6)$$

Taking the first derivative from equation (9.3.6), we determine the velocity of the system:

$$\frac{dx}{dt} = -S\omega \sin \omega t + V \cos \omega t - \frac{f+r}{\omega} \sin \omega t + \frac{a(\varphi \sin \varphi t - \omega \sin \omega t)}{\varphi^2 - \omega^2}$$

$$(9.3.7)$$

Supposing that in equations (9.3.6) and (9.3.7), we have $t = 0$, then we obtain that $x = S$ and $\frac{dx}{dt} = V$, which is expected according to the initial conditions of motion given by expression (2.1.2).

The first derivative from equation (9.3.7) yields the acceleration of the system:

$$\frac{d^2 x}{dt^2} = -S\omega^2 \cos \omega t - V\omega \sin \omega t - (f+r) \cos \omega t$$

$$+ \frac{a(\varphi^2 \cos \varphi t - \omega^2 \cos \omega t)}{\varphi^2 - \omega^2} \qquad (9.3.8)$$

Hence, equations (9.3.6)–(9.3.8) describe the basic parameters of motion of the system.

9.4. A System Moving on a Horizontal Surface is Restricted by a Flexible Link and Subjected to Constant Resisting, Dry Friction, Constant Active, and Harmonic Forces

The operational process is related to the motion of a system on a horizontal surface, while the system is restricted by a flexible link. The system is subjected to the actions of a constant resisting force, a dry friction force, a constant active force, and a harmonic force. The air resistance causes negligible action on the motion of the system. Figure 9.4.1 shows a schematic diagram of the system described above. The notations in the figure are self-explanatory.

Based on the schematic diagram shown in Figure 9.4.1 and the above-mentioned considerations, we compose the differential equation of motion of the system:

$$m\frac{d^2 x}{dt^2} + Kx + F + R = P + A \cos \varphi t \qquad (9.4.1)$$

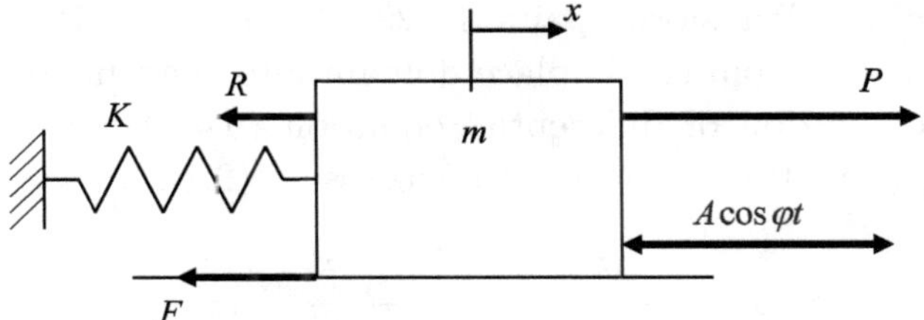

Fig. 9.4.1. Schematic diagram of a system moving on a horizontal surface while being restricted by a flexible link and subjected to a dry friction force, a constant resisting force, a constant active force, and a harmonic force.

The initial conditions of motion are taken according to expression (2.1.2).

Dividing equation (9.4.2) by m, we may write

$$\frac{d^2x}{dt^2} + \omega^2 x + f + r = p + a\cos\varphi t \qquad (9.4.2)$$

With the help of Laplace Transform pairs 5, 2, 1, 2, 2, 2, and 29, we convert equation (9.4.2) with the initial conditions of motion according to expression (2.1.2) from the time domain into the Laplace domain and obtain an algebraic equation in the Laplace domain:

$$s^2 x(s) - Vs - s^2 S + \omega^2 x(s) + f + r = p + \frac{as^2}{s^2 + \varphi^2} \qquad (9.4.3)$$

Applying to equation (9.4.3) the conventional algebraic procedures, we may write

$$x(s)(s^2 + \omega^2) = s^2 S + sV + p - f - r + \frac{as^2}{s^2 + \varphi^2} \qquad (9.4.4)$$

Solving equation (9.4.4) for the displacement $x(s)$ in the Laplace domain, we obtain an algebraic equation describing the displacement $x(s)$ of the system as a function of Laplace variable s:

$$x(s) = \frac{s^2 S}{s^2 + \omega^2} + \frac{sV}{s^2 + \omega^2} + \frac{p - f - r}{s^2 + \omega^2} + \frac{as^2}{(s^2 + \varphi^2)(s^2 + \omega^2)} \qquad (9.4.5)$$

Using Laplace Transform pairs 1, 29, 23, 17, and 72, we invert equation (9.4.5) from the Laplace domain into the time domain and obtain the solution of differential equation (9.4.1) with the initial conditions of motion according to expression (2.1.2):

$$x = S\cos\omega t + \frac{V}{\omega}\sin\omega t + \frac{p - f - r}{\omega^2}(1 - \cos\omega t)$$

$$+ \frac{a(\cos\omega t - \cos\varphi t)}{\varphi^2 - \omega^2} \tag{9.4.6}$$

The first derivative from equation (9.4.6) yields the velocity of the system:

$$\frac{dx}{dt} = -S\omega\sin\omega t + V\cos\omega t + \frac{p - f - r}{\omega}\sin\omega t$$

$$+ \frac{a(\varphi\sin\varphi t - \omega\sin\omega t)}{\varphi^2 - \omega^2} \tag{9.4.7}$$

Assuming that in equations (9.4.6) and (9.4.7), $t = 0$, we determine that $x = S$ and $\frac{dx}{dt} = V$, as it should be according to the initial conditions of motion presented in expression (2.1.2).

Taking the first derivative from equation (9.4.7), we determine the acceleration of the system:

$$\frac{d^2x}{dt^2} = -S\omega^2\cos\omega t - V\omega\sin\omega t + (p - f - r)\cos\omega t$$

$$+ \frac{a(\varphi^2\cos\varphi t - \omega^2\cos\omega t)}{\varphi^2 - \omega^2} \tag{9.4.8}$$

Equations (9.4.6)–(9.4.8) represent the basic parameters of motion of the system.

9.4.1. *Numerical solution*

The following Python program plots equation (9.4.7) and solves it for the time when velocity first reaches zero.

```python
from matplotlib.pyplot import plot, show, text
from numpy import linspace, sin, cos
from scipy.optimize import fsolve


S = 0.1
V = 8
p = 2
r = 1
f = 6
a = 20
phi = 30
omega = 10


t = linspace(0, 1, 1000)

V1 = lambda t: \
    - S*omega*sin(omega*t) + V*cos(omega*t) \
    + ((p - f - r)/omega) * sin(omega*t) \
    + (a*(phi*sin(phi*t) - omega*sin(omega*t)))/(phi**2 - omega**2)

[root] = fsolve(V1, 0.1)
root = round(root, 3)

plot(t, V1(t))
plot(root, 0, 'bo')
text(0.16, 0, f't={root}')
show()
```

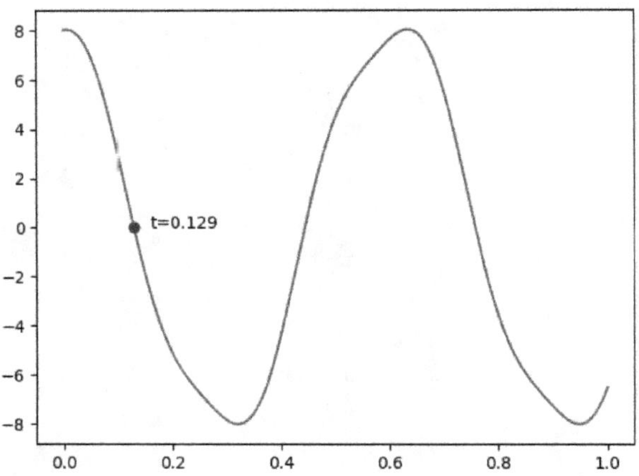

Fig. 9.4.2. Graph of equation (9.4.7).

As shown in Figure 9.4.2, the velocity first reaches zero at $t = 0.129$ s. The following is a Python program to plot equation (9.4.6) from $t = 0$ s to $t = 0.129$ s.

```python
from matplotlib.pyplot import plot, show, text
from numpy import linspace, sin, cos

S = 0.1
V = 8
p = 2
r = 1
f = 6
a = 20
phi = 30
omega = 10

x = lambda t: \
  S*cos(omega*t) + (V/omega)*sin(omega*t) \
  + ((p - f - r)/(omega**2))*(1 - cos(omega*t)) \
  + a*(cos(omega*t) - cos(phi*t))/(phi**2 - omega**2)

stopTime = 0.129
t = linspace(0, stopTime, 1000)
plot(t, x(t))

stopPosition = round(x(stopTime), 3)
plot(stopTime, stopPosition, 'bo')
text(0.115, 0.73, f'x={stopPosition}')

show()
```

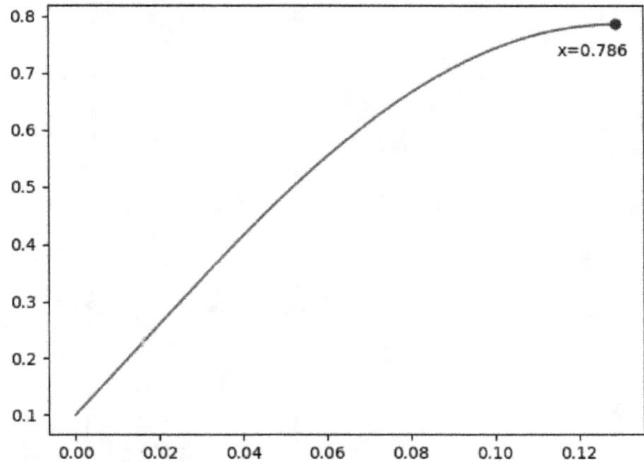

Fig. 9.4.3. Graph of equation (9.4.6).

For the next iteration, apply differential equation of motion (9.4.1) with the following initial conditions of motion:

$$\text{for} \quad t = 0 \quad x = 0.786; \quad \frac{dx}{dt} = 0.$$

For more details, see Section 1.5.

CHAPTER 10

MOTION OF A SYSTEM RESTRICTED BY A FLUID LINK

As mentioned earlier, a fluid link represents a dashpot or a fluid medium that exerts a resisting force toward the system as a result of interactions between the system and the link. The magnitude of this resisting force is proportional to the velocity of the system. It is acceptable to represent the fluid link on related schematic diagrams as a dashpot.

10.1. A System Connected to a Fluid Link Is Moving on a Horizontal Frictionless Surface Due to Its Initial Velocity

Figure 10.1.1 shows a schematic diagram of a system moving on a frictionless horizontal surface while being restricted by a fluid link with the damping coefficient C. The rest of the notations in this figure are self-explanatory.

Based on the schematic diagram shown in Figure 10.1.1 and the considerations mentioned above, we compose the differential equation of motion of the system:

$$m\frac{d^2x}{dt^2} + C\frac{dx}{dt} = 0 \qquad (10.1.1)$$

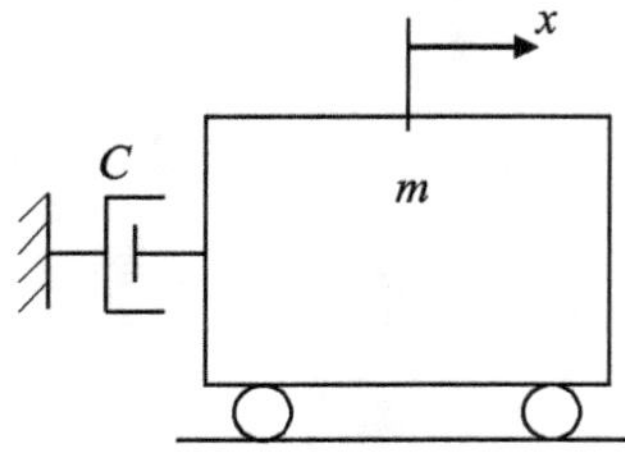

Fig. 10.1.1. Schematic diagram of a system moving on a horizontal frictionless surface while being restricted by a fluid link.

The initial conditions of motion are presented in expression (2.1.2).

Dividing equation (10.1.1) by m, we may write

$$\frac{d^2x}{dt^2} + 2n\frac{dx}{dt} = 0 \tag{10.1.2}$$

where

$$2n = \frac{C}{m} \tag{10.1.3}$$

while n is the damping factor of the system.

The initial conditions of motion are taken according to expression (2.1.2).

Using Laplace Transform pairs 5, 2, and 4, we convert differential equation (10.1.2) with the initial conditions of motion according to expression (2.1.2) from the time domain into an algebraic equation in the Laplace domain:

$$s^2x(s) - sV - s^2S + 2nsx(s) - 2nsS = 0 \tag{10.1.4}$$

The solution of equation (10.1.4) for the displacement $x(s)$ in the Laplace domain reads:

$$x(s) = \frac{V + 2nS}{s + 2n} + \frac{sS}{s + 2n} \tag{10.1.5}$$

Appling Laplace Transform pairs 1, 8, and 14, we invert equation (10.1.5) from the Laplace domain into the time domain and obtain the solution of differential equation (10.1.1) with the initial

conditions of motion according to expression (2.1.2):

$$x = \frac{V + 2nS}{2n}(1 - e^{-2nt}) + Se^{-2nt} \tag{10.1.6}$$

Simplifying equation (10.1.6), we obtain

$$x = S + \frac{V}{2n}(1 - e^{-2nt}) \tag{10.1.7}$$

The first derivative from equation (10.1.6) yields the velocity of the system:

$$\frac{dx}{dt} = Ve^{-2nt} \tag{10.1.8}$$

Assuming that in equations (10.1.7) and (10.1.8), $t = 0$, we determine that $x = S$ and $\frac{dx}{dt} = V$, as it should be according to the initial conditions of motion given by expression (2.1.2).

Taking the first derivative from equation (10.1.8), we determine the acceleration of the system:

$$\frac{d^2x}{dt^2} = -2nVe^{-2nt} \tag{10.1.9}$$

Hence, equations (10.1.7)–(10.1.9) represent the basic parameters of motion of the system.

10.2. Motion of a System Restricted by a Fluid Link and Subjected to a Constant Active Force

The operational process is related to the motion of a system on a horizontal frictionless surface while being restricted by a fluid link and subjected to a constant active force. Figure 10.2.1 shows a schematic diagram of the system. The notations in this figure are self-explanatory.

Based on the considerations mentioned above and the schematic diagram shown in Figure 10.2.1, we compose the differential equation of motion of the system:

$$m\frac{d^2x}{dt^2} + C\frac{dx}{dt} = P \tag{10.2.1}$$

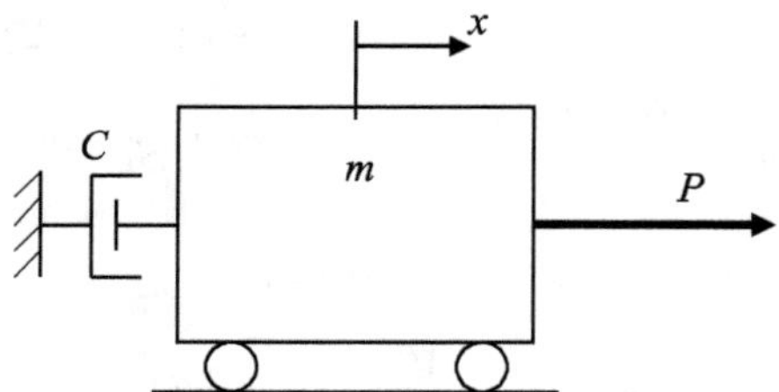

Fig. 10.2.1. Schematic diagram of a system moving on a horizontal frictionless surface while being restricted by a fluid link and subjected to a constant active force.

The initial conditions of motion are taken according to expression (2.1.2). Dividing equation (10.1.1) by m, we have

$$\frac{d^2x}{dt^2} + 2n\frac{dx}{dt} = p \tag{10.2.2}$$

Applying Laplace Transform pairs 5, 2, 4, and 2, we convert differential equation (10.2.2) with the initial conditions of motion according to expression (2.1.2) from the time domain into the algebraic equation in the Laplace domain:

$$s^2x(s) - sV - s^2S + 2nsx(s) - 2nsS = p \tag{10.2.3}$$

The solution of equation (10.2.3) for the displacement $x(s)$ in the Laplace domain reads:

$$x(s) = \frac{V + 2nS}{s + 2n} + \frac{sS}{s + 2n} + \frac{p}{s(s + 2n)} \tag{10.2.4}$$

With the help of Laplace Transform pairs 1, 8, 14, and 16, we invert equation (10.2.4) from the Laplace domain into the time domain and obtain the solution of differential equation (10.2.1) with the initial conditions of motion according to expression (2.1.2):

$$x = \frac{V + 2nS}{2n}(1 - e^{-2nt}) + Se^{-2nt} + \frac{p}{2n}\left[t + \frac{1}{2n}(e^{-2nt} - 1)\right] \tag{10.2.5}$$

Applying to equation (10.2.5) the conventional procedures, we obtain

$$x = S + \frac{V}{2n}(1 - e^{-2nt}) + \frac{p}{2n}\left[t + \frac{1}{2n}(e^{-2nt} - 1)\right] \tag{10.2.6}$$

The first derivative from equation (10.2.6) yields the velocity of the system:

$$\frac{dx}{dt} = V e^{-2nt} + \frac{p}{2n}(1 - e^{-2nt}) \qquad (10.2.7)$$

Supposing that in equations (10.2.6) and (10.2.7), $t = 0$, we determine that $x = S$ and $\frac{dx}{dt} = V$, as expected according to the initial conditions of motion presented in expression (2.1.2).

Taking the first derivative from equation (10.2.7), we determine the acceleration of the system:

$$\frac{d^2 x}{dt^2} = e^{-2nt}(p - 2nV) \qquad (10.2.8)$$

Hence, equations 10.2.6)–(10.2.8) describe the basic parameters of motion of the system.

10.2.1. *Landing with a parachute*

Assuming that the air resistance is a linear function of velocity, we can clarify some important natural phenomena of the operational process of landing with a parachute. Usually, before opening the parachute, the parachutist performs a free fall and gains a certain initial velocity.

However, for this case, we suppose that the gained initial velocity is negligible, and we accept that the initial displacement as well as the initial velocity equal zero.

Figure 10.2.2 shows a schematic diagram related to landing with a parachute. The notations in this figure are self-explanatory.

Based on the schematic diagram shown in Figure 10.2.2 and the considerations mentioned above, we compose the differential equation of motion of the body:

$$m\frac{d^2 z}{dt^2} + C\frac{dz}{dt} = W \qquad (10.2.9)$$

Dividing equation (10.2.9) by m, we have

$$\frac{d^2 z}{dt^2} + 2n\frac{dz}{dt} = g \qquad (10.2.10)$$

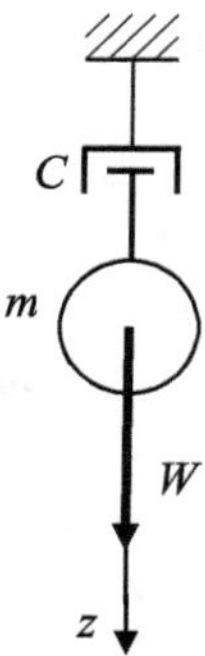

Fig. 10.2.2. Schematic diagram of a system restricted in the vertical direction by a fluid link while being subjected to the force of gravity.

With the help of Laplace Transform pairs 5, 2, 4, and 2, we convert differential equation (10.2.10) with the above-mentioned initial conditions of motion from the time domain into an algebraic equation in the Laplace domain:

$$s^2 z(s) + 2ns z(s) = g \tag{10.2.11}$$

Solving equation (10.2.11) for the displacement $x(s)$ in the Laplace domain, we may write

$$z(s) = \frac{g}{s(s + 2n)} \tag{10.2.12}$$

Applying Laplace Transform pairs 1, 16, and 2, we invert equation (10.2.12) from the Laplace domain into the time domain and obtain the solution of differential equation (10.2.9):

$$z = \frac{g}{2n} \left[t + \frac{1}{2n}(e^{-2nt} - 1) \right] \tag{10.2.13}$$

Taking the first derivative from equation (10.2.13), we determine the velocity of the body:

$$\frac{dz}{dt} = \frac{g}{2n}(1 - e^{-2nt}) \tag{10.2.14}$$

Assuming that in equations (10.2.13) and (10.2.14), $t = 0$, we determine that $z = 0$ and $\frac{dz}{dt} = 0$, as expected according to the above-mentioned initial conditions of motion.

The first derivative from equation (10.2.14) yields the acceleration of the body:

$$\frac{d^2 z}{dt^2} = g e^{-2nt} \tag{10.2.15}$$

In this case, at the moment of time when the acceleration becomes equal to zero, the velocity reaches its maximum value, and the body continues its motion at a constant velocity, performing a uniform motion. This motion will continue regardless of the remaining height to landing. Hence, by equating equation (10.2.15) to zero, we obtain

$$e^{-2nt} \to 0 \tag{10.2.16}$$

Combining equations (10.2.14) and (10.2.16), we conclude that the maximum value of the velocity that a body can reach by landing with a parachute equals

$$V_{\max} = \frac{g}{2n} \tag{10.2.17}$$

It should be emphasized that according to equation (10.2.16), we obtain that the time T tends to infinity:

$$T \to \infty \tag{10.2.18}$$

However, in reality, the actual velocity could be close to the maximum value, but it cannot be reached.

Hence, with the appropriate selection of the damping factor, the landing with a parachute is considered safe regardless of the height.

10.2.2. *Maximum velocity of a ship*

Here, we determine the value of the maximum velocity that could be achieved by a ship possessing a certain source of energy that is enabled to develop a constant active force P. The motion resistance forces of a ship represent the sum of air and water resistance forces that are proportional to the velocity of the ship. The combined damping coefficient of the ship C equals the sum of the related damping coefficients of air and water. The schematic diagram of the system is presented in Figure 10.2.1. Accepting that the initial displacement and velocity of the ship are equal to zero, we determine the law of motion of the ship with the help of equation (10.2.6):

$$x = \frac{p}{2n}\left[t + \frac{1}{2n}(e^{-2nt} - 1)\right] \qquad (10.2.19)$$

The first and second derivatives from equation (10.2.19) yield the velocity and the acceleration of the system, respectively:

$$\frac{dx}{dt} = \frac{p}{2n}(1 - e^{-2nt}) \qquad (10.2.20)$$

$$\frac{d^2x}{dt^2} = pe^{-2nt} \qquad (10.2.21)$$

Equation (10.2.20) shows that the maximum value of the velocity of the ship tends to

$$\left(\frac{dx}{dt}\right)_{\max} \to \frac{p}{2n} \qquad (10.2.22)$$

According to equation (10.2.21), the maximum acceleration occurs at the beginning of the process when $t = 0$; therefore,

$$\left(\frac{d^2x}{dt^2}\right)_{\max} = p \qquad (10.2.23)$$

10.3. Motion of a System Restricted by a Fluid Link and Subjected to a Harmonic Force

This operational process is associated with a system that is moving on a horizontal frictionless surface and is restricted by a fluid link while being subjected to a harmonic force.

Figure 10.3.1 shows a schematic diagram of the system described above. The notations in this figure are self-explanatory.

Based on the considerations mentioned above and the schematic diagram shown in Figure 10.3.1, we compose the differential equation of motion of the system:

$$m\frac{d^2x}{dt^2} + C\frac{dx}{dt} = A\cos\varphi t \qquad (10.3.1)$$

The initial conditions of motion are taken according to expression (2.1.2).

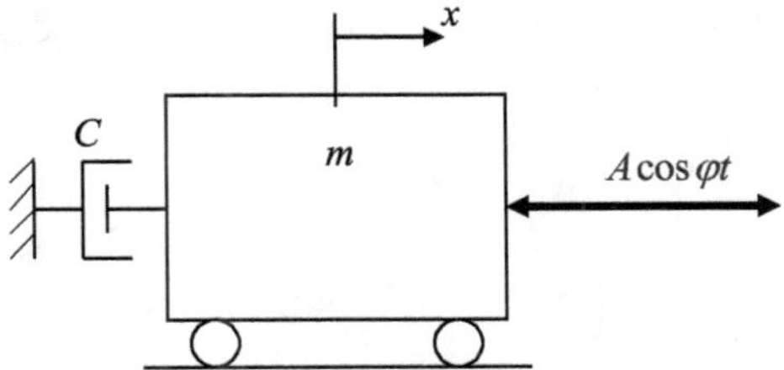

Fig. 10.3.1. Schematic diagram of a system moving on a horizontal frictionless surface while being restricted by a fluid link and subjected to a harmonic force.

Dividing equation (10.3.1) by m, we may write

$$\frac{d^2x}{dt^2} + 2n\frac{dx}{dt} = a\cos\varphi t \tag{10.3.2}$$

Applying Laplace Transform pairs 5, 2, 4, and 29 to equation (10.3.2) with the initial conditions of motion according to expression (2.1.2), we convert this equation from the time domain into an algebraic equation in the Laplace domain:

$$s^2x(s) - Vs - s^2S + 2nsx(s) - 2nsS = \frac{as^2}{s^2 + \varphi^2} \tag{10.3.3}$$

Using for equation (10.3.3) certain conventional algebraic procedures, we have

$$x(s)s(s + 2n) = s^2S + sV + \frac{as^2}{s^2 + \varphi^2} + 2nsS \tag{10.3.4}$$

The solution of equation (10.3.4) for the displacement $x(s)$ in the Laplace domain has the following expression:

$$x(s) = \frac{sS}{s + 2n} + \frac{V + 2nS}{s + 2n} + \frac{as}{(s + 2n)(s^2 + \varphi^2)} \tag{10.3.5}$$

With the help of Laplace Transform pairs 1, 14, 8, and 44, we invert equation (10.3.5) from the Laplace domain into the time domain and obtain the solution of differential equation (10.3.1) with the initial

conditions of motion according to expression (2.1.2):

$$x = Se^{-2nt} + \frac{V + 2nS}{2n}(1 - e^{-2nt})$$

$$+ \frac{a}{\varphi^2 + 4n^2}\left(e^{-2nt} + \frac{2n}{\varphi}\sin\varphi t - \cos\varphi t\right) \qquad (10.3.6)$$

Rearranging equation (10.3.6), we may write

$$x = S + \frac{V}{2n} + e^{-2nt}\left(\frac{a}{\varphi^2 + 4n^2} - \frac{V}{2n}\right)$$

$$+ \frac{a}{\varphi^2 + 4n^2}\left(\frac{2n}{\varphi}\sin\varphi t - \cos\varphi t\right) \qquad (10.3.7)$$

Further simplification of equation (10.3.7) yields the following expression:

$$x = S + \frac{V}{2n}(1 - e^{-2nt}) + \frac{a}{\varphi^2 + 4n^2}e^{-2nt} - \frac{a\cos(\varphi t + \lambda)}{\varphi\sqrt{\varphi^2 + 4n^2}}$$

$$(10.3.8)$$

where

$$\cos\lambda = \frac{\varphi}{\sqrt{\varphi^2 + 4n^2}}; \quad \sin\lambda = \frac{2n}{\sqrt{\varphi^2 + 4n^2}} \qquad (10.3.9)$$

Taking the first derivative from equation (10.3.8), we determine the velocity of the system:

$$\frac{dx}{dt} = Ve^{-2nt} - \frac{2nae^{-2nt}}{\varphi^2 + 4n^2} + \frac{a\sin(\varphi t + \lambda)}{\sqrt{\varphi^2 + 4n^2}} \qquad (10.3.10)$$

Supposing that in equations (10.3.8) and (10.3.10), we have $t = 0$, we obtain $x = S$ and $\frac{dx}{dt} = V$, which is expected according to the initial conditions presented in expression (2.1.2).

The first derivative from equation (10.3.10) yields the acceleration of the system:

$$\frac{d^2x}{dt^2} = -2nVe^{-2nt} + \frac{4n^2ae^{-2nt}}{\varphi^2 + 4n^2} + \frac{a\varphi\cos(\varphi t + \lambda)}{\sqrt{\varphi^2 + 4n^2}} \qquad (10.3.11)$$

Hence, equations (10.3.8), (10.3.10), and (10.3.11) describe the basic parameters of motion of the system.

10.4. Motion of a System Restricted by a Fluid Link and Subjected to a Constant Active Force and a Harmonic Force

We consider an operational process that is characterized by the motion of a system on a horizontal frictionless surface, while the system is restricted by a fluid link. The system is subjected to the actions of a constant active force and a harmonic force. Figure 10.4.1 shows the schematic diagram of the system described above. The notations in the figure are self-explanatory.

Based on the schematic diagram shown in Figure 10.4.1 and the above-mentioned considerations, we compose the differential equation of motion of the system:

$$m\frac{d^2x}{dt^2} + C\frac{dx}{dt} = P + A\cos\varphi t \qquad (10.4.1)$$

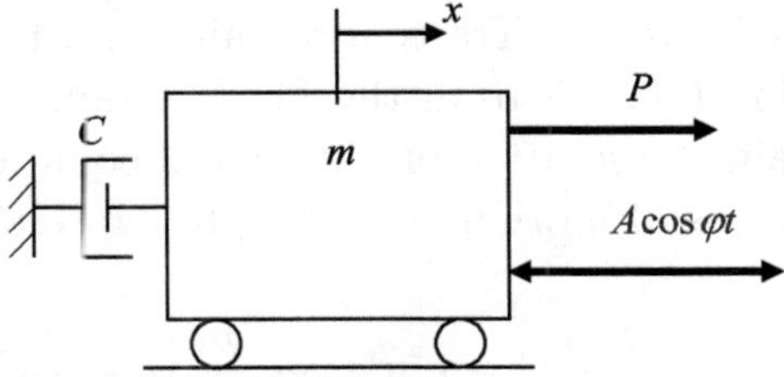

Fig. 10.4.1. Schematic diagram of a system restricted by a fluid link and moving on a horizontal frictionless surface while being subjected to a constant active force and a harmonic force.

The initial conditions of motion are taken according to expression (2.1.2):

Dividing equation (10.4.1) by m, we have

$$\frac{d^2x}{dt^2} + 2n\frac{dx}{dt} = p + a\cos\varphi t \qquad (10.4.2)$$

With the help of Laplace Transform pairs 5, 2, 4, 2, and 29, we convert equation (10.4.2) with the initial conditions of motion according to expression (2.1.2) from the time domain into an algebraic equation in the Laplace domain:

$$s^2x(s) - Vs - s^2S + 2nsx(s) - 2nsS = p + \frac{as^2}{s^2 + \varphi^2} \qquad (10.4.3)$$

Applying to equation (10.4.3) the conventional algebraic procedures, we may write

$$x(s)s(s + 2n) = s^2S + sV + 2nsS + p + \frac{as^2}{s^2 + \varphi^2} \qquad (10.4.4)$$

Solving equation (10.4.4) for the displacement $x(s)$ in the Laplace domain, we obtain

$$x(s) = \frac{sS}{s + 2n} + \frac{V + 2nS}{s + 2n} + \frac{p}{s(s + 2n)} + \frac{as}{(s + 2n)(s^2 + \varphi^2)} \qquad (10.4.5)$$

With the help of Laplace Transform pairs 1, 14, 8, 16, and 44, we invert equation (10.4.5) from the Laplace domain into the time domain and obtain the solution of differential equation (10.4.1) with the initial conditions of motion according to expression (2.1.2):

$$x = Se^{-2nt} + \frac{V + 2nS}{2n}(1 - e^{-2nt}) + \frac{p}{2n}\left[t + \frac{1}{2n}(e^{-2nt} - 1)\right]$$

$$+ \frac{a}{\varphi^2 + 4n^2}\left(e^{-2nt} + \frac{2n}{\varphi}\sin\varphi t - \cos\varphi t\right) \qquad (10.4.6)$$

Rearranging equation (10.4.6), we may write

$$x = S + \frac{V}{2n}(1 - e^{-2nt}) + \frac{pt}{2n} + \frac{p}{4n^2}e^{-2nt} - \frac{p}{4n^2}$$

$$+ \frac{ae^{-2nt}}{\varphi^2 + 4n^2} + \frac{a}{\varphi^2 + 4n^2}\left(\frac{2n}{\varphi}\sin\varphi t - \cos\varphi t\right) \quad (10.4.7)$$

Simplifying equation (10.4.7), we have

$$x = S + \frac{V}{2n} - \frac{p}{4n^2} + \frac{pt}{2n} - e^{-2nt}\left(\frac{V}{2n} - \frac{p}{4n^2} - \frac{a}{\varphi^2 + 4n^2}\right)$$

$$+ \frac{a}{\varphi^2 + 4n^2}\left(\frac{2n}{\varphi}\sin\varphi t - \cos\varphi t\right) \quad (10.4.8)$$

The first derivative from equation (10.4.8) yields the velocity of the system:

$$\frac{dx}{dt} = Ve^{-2nt} + \frac{p}{2n}(1 - e^{-2nt}) - \frac{2an}{\varphi^2 + 4n^2}(e^{-2nt} - \cos\varphi t)$$

$$+ \frac{a\varphi}{\varphi^2 + 4n^2}\sin\varphi t \quad (10.4.9)$$

Supposing that in equations (10.4.8) and (10.4.9), $t = 0$, we obtain $x = S$ and $\frac{dx}{dt} = V$, which is expected according to the initial conditions of motion given by expression (2.1.2).

Taking the first derivative from equation (10.4.9), we determine the acceleration of the system:

$$\frac{d^2x}{dt^2} = (p - 2nV)e^{-2nt} + \frac{a}{\varphi^2 + 4n^2}$$

$$\times (\varphi^2 \cos\varphi t + 4n^2 e^{-2nt} - 2n\varphi\sin\varphi t) \quad (10.4.10)$$

Hence, equations (10.4.8)–(10.4.10) describe the basic parameters of motion of the system.

10.4.1. *Numerical solution*

The following is a Python program to plot the graph of equation (10.4.8). The graph is plotted in Figure 10.4.2.

```python
from matplotlib.pyplot import plot, show
from numpy import linspace, sin, cos, exp


S = 0.1
V = 0.1
n = 50
p = 5
a = 50
phi = 30

t = linspace(0, 2, 1000)
x = S + V/(2*n) - p/(4*n**2) + (p*t)/(2*n) \
    - exp(-2*n*t)*(V/(2*n) - p/(4*n**2) - a/(phi**2 + 4*n**2)) \
    + (a/(phi**2 + 4*n**2))*(((2*n)/phi)*sin(phi*t) - cos(phi*t))

plot(t, x)
show()
```

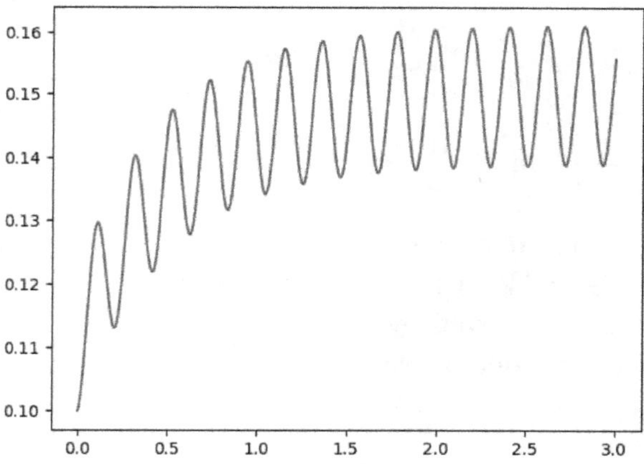

Fig. 10.4.2. Graph of equation (10.4.8).

MOTION OF A SYSTEM RESTRICTED BY A FLUID LINK AND SUBJECTED TO A CONSTANT RESISTING FORCE

11.1. Motion of a System on a Horizontal Frictionless Surface Being Restricted by a Fluid Link and Subjected to a Constant Resisting Force

Consider an operational process associated with a system restricted by a fluid link and moving on a horizontal frictionless surface under the actions of an initial displacement and velocity, while the system is subjected to a constant resisting force.

Figure 11.1.1 shows a schematic diagram of the system described above. The notations in the figure are self-explanatory.

Based on the considerations mentioned above and the schematic diagram shown in Figure 11.1.1, we compose the following differential equation of motion:

$$m\frac{d^2x}{dt^2} + C\frac{dx}{dt} + R = 0 \tag{11.1.1}$$

The initial conditions of motion are taken according to expression (2.1.2). Dividing equation (11.1.1) by m, we have

$$\frac{d^2x}{dt^2} + 2n\frac{dx}{dt} + r = 0 \tag{11.1.2}$$

With the help of Laplace Transform pairs 5, 2, 4, and 2, we convert differential equation (11.1.2) with initial conditions of motion

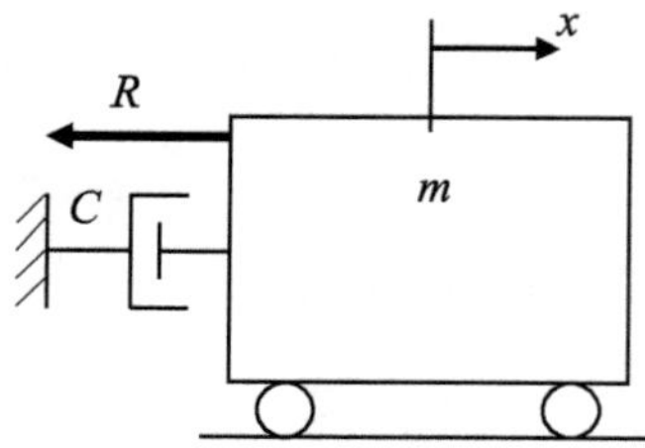

Fig. 11.1.1. Schematic diagram of a system moving on a horizontal frictionless surface while being restricted by a fluid link and subjected to a constant resisting force.

according to expression (2.1.2) from the time domain into the Laplace domain and obtain a corresponding algebraic equation in the Laplace domain:

$$s^2 x(s) - sV - s^2 S + 2nsx(s) - 2nsS + r = 0 \qquad (11.1.3)$$

Rearranging equation (11.1.3), we have

$$x(s)s(s + 2n) = sV + s^2 S + 2nsS - r \qquad (11.1.4)$$

Solving equation (11.1.4) for the displacement $x(s)$ in the Laplace domain, we may write

$$x(s) = \frac{V}{s + 2n} + \frac{sS}{s + 2n} + \frac{2nS}{s + 2n} - \frac{r}{s(s + 2n)} \qquad (11.1.5)$$

Using Laplace Transform pairs 1, 8, 14, 8, and 16, we invert equation (11.1.5) from the Laplace domain into the time domain and obtain the solution of differential equation (11.1.1) with the initial conditions of motion according to expression (2.1.2):

$$x = \frac{V}{2n}(1 - e^{-2nt}) + Se^{-2nt} + S(1 - e^{-2nt})$$
$$- \frac{r}{2n}\left[t + \frac{1}{2n}(e^{-2nt} - 1)\right] \qquad (11.1.6)$$

Rearranging equation (11.1.6), we obtain

$$x = S + \frac{V}{2n} + \frac{r}{4n^2} - \frac{r}{2n}t - e^{-2nt}\left(\frac{V}{2n} + \frac{r}{4n^2}\right) \qquad (11.1.7)$$

Taking the first derivative from equation (11.1.7), we determine the velocity of the system:

$$\frac{dx}{dt} = 2ne^{-2nt}\left(\frac{V}{2n} + \frac{r}{4n^2}\right) - \frac{r}{2n} \tag{11.1.8}$$

Supposing that in equations (11.1.7) and (11.1.8), $t = 0$, we determine that $x = S$ and $\frac{dx}{dt} = V$, as it should be according to the initial conditions of motion according to expression (2.1.2). The first derivative from equation (11.1.8) yields the acceleration of the system:

$$\frac{d^2 x}{dt^2} = -4n^2 e^{-2nt}\left(\frac{V}{2n} + \frac{r}{4n^2}\right) \tag{11.1.9}$$

Hence, equations (11.1.7)–(11.1.9) represent the basic parameters of motion of the system.

11.2. Motion of a System Restricted by a Fluid Link and Subjected to a Constant Resisting Force and a Constant Active Force

Consider the motion of a system restricted by a fluid link, moving on a horizontal frictionless surface, and being subjected to constant resisting and constant active forces. Figure 11.2.1 shows a schematic diagram of the system. The notations in this figure are self-explanatory.

Based on the considerations mentioned above and the schematic diagram shown in Figure 11.2.1, we compose the differential equation

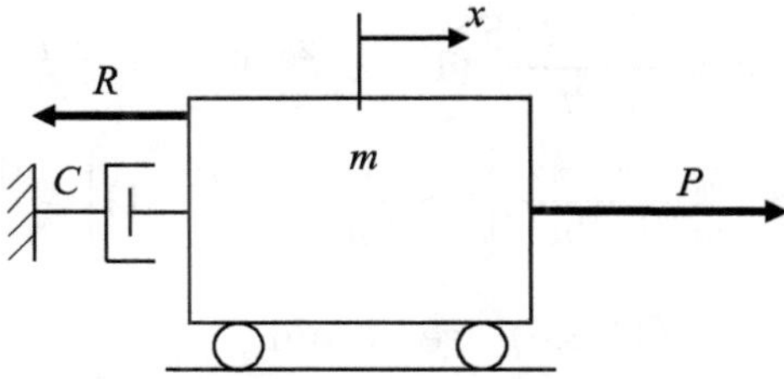

Fig. 11.2.1. Schematic diagram of a system moving on a horizontal frictionless surface, while being restricted by a fluid link and subjected to constant resisting and constant active forces.

of motion of the system:

$$m\frac{d^2x}{dt^2} + C\frac{dx}{dt} + R = P \tag{11.2.1}$$

The initial conditions of motion are taken according to expression (2.1.2).

Dividing equation (11.2.1) by m, we have

$$\frac{d^2x}{dt^2} + 2n\frac{dx}{dt} + r = p \tag{11.2.2}$$

Applying Laplace Transform pairs 5, 2, 4, 2, and 2, we convert differential equation (11.2.2) with the initial conditions of motion according to expression (2.1.2) from the time domain into the algebraic equation in the Laplace domain:

$$s^2x(s) - sV - s^2S + 2nsx(s) - 2nsS + r = p \tag{11.2.3}$$

The solution of equation (11.2.3) for the displacement $x(s)$ in the Laplace domain reads:

$$x(s) = \frac{V + 2nS}{s + 2n} + \frac{sS}{s + 2n} + \frac{p - r}{s(s + 2n)} \tag{11.2.4}$$

Using Laplace Transform pairs 1, 8, 14, and 16, we invert equation (11.2.4) from the Laplace domain into the time domain and obtain the solution of differential equation (11.2.1) with the initial conditions of motion according to expression (2.1.2):

$$x = \frac{V + 2nS}{2n}(1 - e^{-2nt}) + Se^{-2nt}$$

$$+ \frac{p - r}{2n}\left[t + \frac{1}{2n}(e^{-2nt} - 1)\right] \tag{11.2.5}$$

Simplifying equation (11.2.5), we obtain

$$x = S + \frac{V}{2n}(1 - e^{-2nt}) + \frac{p - r}{2n}\left[t + \frac{1}{2n}(e^{-2nt} - 1)\right] \tag{11.2.6}$$

The first derivative from equation (11.2.6) yields the velocity of the system:

$$\frac{dx}{dt} = Ve^{-2nt} + \frac{p-r}{2n}(1 - e^{-2nt}) \qquad (11.2.7)$$

Assuming that in equations (11.2.6) and (11.2.7), $t = 0$, we determine that $x = S$ and $\frac{dx}{dt} = V$, as expected according to the initial conditions of motion presented in expression (2.1.2).

Taking the first derivative from equation (11.2.7), we determine the acceleration of the system:

$$\frac{d^2 x}{dt^2} = e^{-2nt}(p - r - 2nV) \qquad (11.2.8)$$

Hence, equations (11.2.6)–(11.2.8) are the basic parameters of motion of the system.

11.3. Motion of a System Restricted by a Fluid Link and Subjected to a Constant Resisting Force and a Harmonic Force

We consider the operational process of a system that is moving on a horizontal frictionless surface. The system is restricted by a fluid link, and it is subjected to a constant resisting force and a harmonic force.

Figure 11.3.1 shows a schematic diagram that characterizes the system described above. The notations in this figure are self-explanatory.

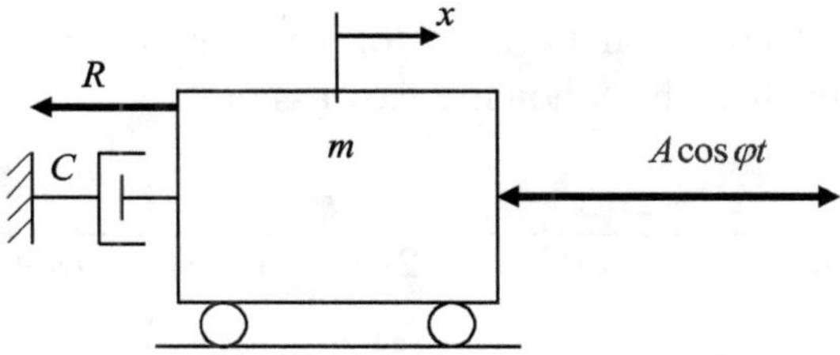

Fig. 11.3.1. Schematic diagram of a system moving on a horizontal frictionless surface, while being restricted by a fluid link and subjected to a constant resisting force and a harmonic force.

Based on the schematic diagram shown in Figure 11.3.1 and on the considerations mentioned above, we compose the differential equation of motion of the system:

$$m\frac{d^2x}{dt^2} + C\frac{dx}{dt} + R = A\cos\varphi t \qquad (11.3.1)$$

The initial conditions of motion are taken according to expression (2.1.2).

Dividing equation (11.3.1) by m, we may write

$$\frac{d^2x}{dt^2} + 2n\frac{dx}{dt} + r = a\cos\varphi t \qquad (11.3.2)$$

Applying Laplace Transform pairs 5, 2, 4, 2, and 29 to equation (11.3.2) with the initial conditions of motion according to expression (2.1.2), we convert this equation from the time domain into an algebraic equation in the Laplace domain:

$$s^2x(s) - Vs - s^2S + 2nsx(s) - 2nsS + r = \frac{as^2}{s^2 + \varphi^2} \qquad (11.3.3)$$

Applying to equation (11.3.3) the conventional algebraic procedures, we may write

$$x(s)s(s + 2n) = s^2S + sV + \frac{as^2}{s^2 + \varphi^2} + 2nsS - r \qquad (11.3.4)$$

The solution of equation (11.3.4) for the displacement $x(s)$ in the Laplace domain has the following expression:

$$x(s) = \frac{sS}{s + 2n} + \frac{V + 2nS}{s + 2n} + \frac{as}{(s + 2n)(s^2 + \varphi^2)} - \frac{r}{s(s + 2n)} \qquad (11.3.5)$$

Applying Laplace Transform pairs 1, 14, 8, 44, and 16, we invert equation (11.3.5) from the Laplace domain into the time domain and obtain the solution of differential equation (11.3.1) with the initial

conditions of motion according to expression (2.1.2):

$$x = Se^{-2nt} + \frac{V + 2nS}{2n}(1 - e^{-2nt}) + \frac{a}{\varphi^2 + 4n^2}$$

$$\times \left(e^{-2nt} + \frac{2n}{\varphi} \sin \varphi t - \cos \varphi t \right) \tag{11.3.6}$$

$$- \frac{r}{2n}\left[t + \frac{1}{2n}(e^{-2nt} - 1) \right]$$

Rearranging equation (11.3.6), we may write

$$x = S + \frac{V}{2n} + e^{-2nt}\left(\frac{a}{\varphi^2 + 4n^2} - \frac{V}{2n} \right)$$

$$+ \frac{a}{\varphi^2 + 4n^2}\left(\frac{2n}{\varphi} \sin \varphi t - \cos \varphi t \right) \tag{11.3.7}$$

$$- \frac{r}{2n}\left[t + \frac{1}{2n}(e^{-2nt} - 1) \right]$$

Combining equation (11.3.7) with expression (10.3.9), we have

$$x = S + \frac{V}{2n}(1 - e^{-2nt}) + \frac{a}{\varphi^2 + 4n^2}e^{-2nt} - \frac{a\cos(\varphi t + \lambda)}{\varphi\sqrt{\varphi^2 + 4n^2}}$$

$$- \frac{r}{2n}\left[t + \frac{1}{2n}(e^{-2nt} - 1) \right] \tag{11.3.8}$$

Taking the first derivative from equation (11.3.8), we determine the velocity of the system:

$$\frac{dx}{dt} = Ve^{-2nt} - \frac{2nae^{-2nt}}{\varphi^2 + 4n^2} + \frac{a\sin(\varphi t + \lambda)}{\sqrt{\varphi^2 + 4n^2}} - \frac{r}{2n}(1 - e^{-2nt}) \tag{11.3.9}$$

Supposing that in equations (11.3.8) and (11.3.9), we have $t = 0$, we obtain $x = S$ and $\frac{dx}{dt} = V$, as is expected according to the initial conditions of motion presented in expression (2.1.2).

The first derivative from equation (11.3.9) yields the acceleration of the system:

$$\frac{d^2x}{dt^2} = -2nVe^{-2nt} + \frac{4n^2ae^{-2nt}}{\varphi^2 + 4n^2} + \frac{a\varphi\cos(\varphi t + \lambda)}{\sqrt{\varphi^2 + 4n^2}} - re^{-2nt} \quad (11.3.10)$$

Hence, equations (11.3.8)–(11.3.10) describe the basic parameters of motion of the system.

11.4. Motion of a System Restricted by a Fluid Link and Subjected to Constant Resisting, Constant Active, and Harmonic Forces

This operational process is characterized by the motion of a system on a horizontal frictionless surface. The system is restricted by a fluid link and is subjected to the action of constant resisting, constant active, and harmonic forces. Figure 11.4.1 shows a schematic diagram of the system described above. The notations in the figure are self-explanatory.

Based on the schematic diagram shown in Figure 11.4.1 and on the above-mentioned considerations, we compose the differential equation of motion of the system:

$$m\frac{d^2x}{dt^2} + C\frac{dx}{dt} + R = P + A\cos\varphi t \quad (11.4.1)$$

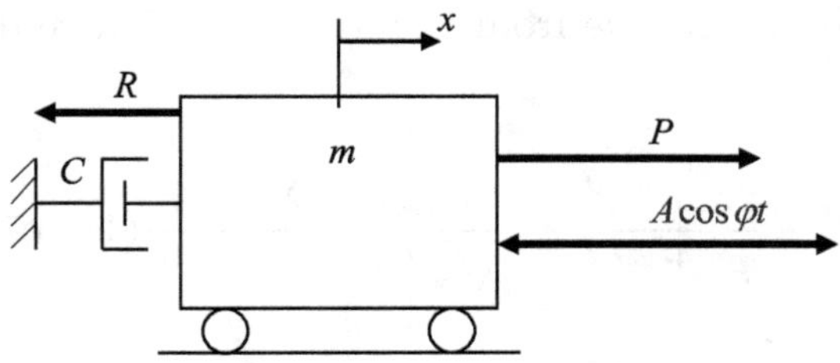

Fig. 11.4.1. Schematic diagram of a system moving on a horizontal frictionless surface, while being restricted by a fluid link and subjected to constant resisting, constant active, and harmonic forces.

The initial conditions of motion are taken according to expression (2.1.2).

Dividing equation (11.4.1) by m, we have

$$\frac{d^2x}{dt^2} + 2n\frac{dx}{dt} + r = p + a\cos\varphi t \qquad (11.4.2)$$

Applying Laplace Transform pairs 5, 2, 4, 2, 2, and 29, we convert equation (11.4.2) with the initial conditions of motion according to expression (2.1.2) from the time domain into an algebraic equation in the Laplace domain:

$$s^2x(s) - Vs - s^2S + 2nsx(s) - 2nsS + r = p + \frac{as^2}{s^2 + \varphi^2} \qquad (11.4.3)$$

Applying to equation (11.4.3) the conventional algebraic procedures, we have

$$x(s)s(s + 2n) = s^2S + sV + 2nsS + p - r + \frac{as^2}{s^2 + \varphi^2} \qquad (11.4.4)$$

Solving equation (11.4.4) for the displacement $x(s)$ in the Laplace domain, we obtain

$$x(s) = \frac{sS}{s + 2n} + \frac{V + 2nS}{s + 2n} + \frac{p - r}{s(s + 2n)} + \frac{as}{(s + 2n)(s^2 + \varphi^2)} \qquad (11.4.5)$$

Applying Laplace Transform pairs 1, 14, 8, 16, and 44, we invert equation (11.4.5) from the Laplace domain into the time domain and obtain the solution of differential equation (11.4.1) with the initial conditions of motion according to expression (2.1.2):

$$x = Se^{-2nt} + \frac{V + 2nS}{2n}(1 - e^{-2nt}) + \frac{p - r}{2n}\left[t + \frac{1}{2n}(e^{-2nt} - 1)\right]$$

$$+ \frac{a}{\varphi^2 + 4n^2}\left(e^{-2nt} + \frac{2n}{\varphi}\sin\varphi t - \cos\varphi t\right) \qquad (11.4.6)$$

Rearranging equation (11.4.6), we have

$$x = S + \frac{V}{2n} - \frac{p-r}{4n^2} + \frac{(p-r)t}{2n} - e^{-2nt}\left(\frac{V}{2n} - \frac{p-r}{4n^2} - \frac{a}{\varphi^2 + 4n^2}\right)$$

$$+ \frac{a}{\varphi^2 + 4n^2}\left(\frac{2n}{\varphi}\sin\varphi t - \cos\varphi t\right) \tag{11.4.7}$$

Transforming equation (11.4.7) by combining it with expression (10.3.9), we obtain

$$x = S + \frac{V}{2n} - \frac{p-r}{4n^2} + \frac{(p-r)t}{2n} - e^{-2nt}\left(\frac{V}{2n} - \frac{p-r}{4n^2} - \frac{a}{\varphi^2 + 4n^2}\right)$$

$$- \frac{a\cos(\varphi t + \lambda)}{\varphi\sqrt{\varphi^2 + 4n^2}} \tag{11.4.8}$$

The first derivative from equation (11.4.8) yields the velocity of the system:

$$\frac{dx}{dt} = \frac{p-r}{2n} + 2ne^{-2nt}\left(\frac{V}{2n} - \frac{p-r}{4n^2} - \frac{a}{\varphi^2 + 4n^2}\right) + \frac{a\sin(\varphi t + \lambda)}{\sqrt{\varphi^2 + 4n^2}} \tag{11.4.9}$$

Assuming that in equations (11.4.8) and (11.4.9), $t = 0$, we obtain $x = S$ and $\frac{dx}{dt} = V$, which is expected according to the initial conditions of motion presented in expression (2.1.2).

Taking the first derivative from equation (11.4.9), we determine the acceleration of the system:

$$\frac{d^2x}{dt^2} = -4n^2e^{-2nt}\left(\frac{V}{2n} - \frac{p-r}{4n^2} - \frac{a}{\varphi^2 + 4n^2}\right) + \frac{a\varphi\cos(\varphi t + \lambda)}{\sqrt{\varphi^2 + 4n^2}} \tag{11.4.10}$$

Hence, equations (11.4.8)–(11.4.10) describe the basic parameters of motion of the system.

11.4.1. *Numerical solution*

The following is a Python program to plot equation (11.4.7) as well as its output (Figure 11.4.2).

```
from matplotlib.pyplot import plot, show
from numpy import linspace, sin, cos, exp


S=0.05
V=0.5
n=1
p=2
r=1
a=20
phi=30


t = linspace(0, 2, 1000)
x = S + V/(2*n) - (p - r)/(4*n**2) + (p - r)*t/(2*n) \
    - exp(-2*n*t)*(V/(2*n) - (p - r)/(4*n**2) - a/(phi**2 + 4*n**2)) \
    + a/(phi**2 + 4*n**2)*((2*n/phi)*sin(phi*t) - cos(phi*t))


plot(t,x)
show()
```

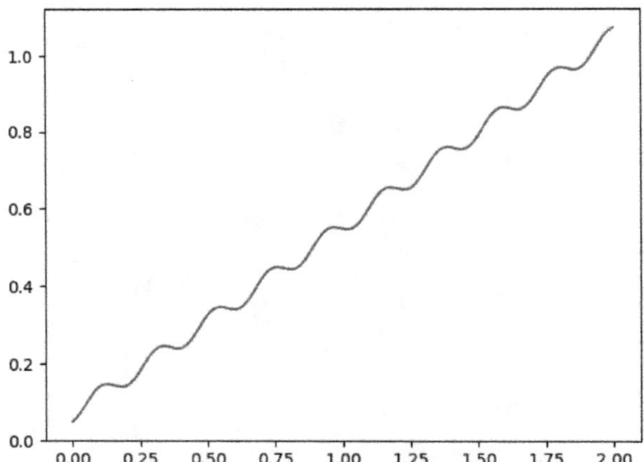

Fig. 11.4.2. Graph of equation (11.4.7).

MOTION OF A SYSTEM RESTRICTED BY A FLUID LINK AND SUBJECTED TO A DRY FRICTION FORCE

12.1. Motion of a System on a Horizontal Surface While Being Restricted by a Fluid Link and Subjected to a Dry Friction Force

Consider an operational process associated with a system restricted by a fluid link and moving on a horizontal surface under the action of its initial velocity, while the system is subjected to a constant dry friction force.

Figure 12.1.1 shows a schematic diagram of the system described above. The notations in the figure are self-explanatory.

Based on the considerations mentioned above and the schematic diagram shown in Figure 12.1.1, we compose the following differential equation of motion:

$$m\frac{d^2x}{dt^2} + C\frac{dx}{dt} + F = 0 \tag{12.1.1}$$

The initial conditions of motion are taken according to expression (2.1.2).

Dividing equation (12.1.1) by m, we have

$$\frac{d^2x}{dt^2} + 2n\frac{dx}{dt} + f = 0 \tag{12.1.2}$$

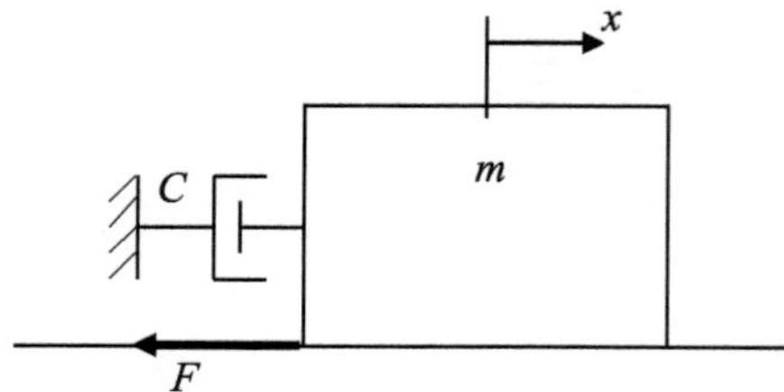

Fig. 12.1.1. Schematic diagram of a system moving on a horizontal surface while being restricted by a fluid link and subjected to a dry friction force.

With the help of Laplace Transform pairs 5, 2, 4, and 2, we convert equation (12.1.2) with initial conditions of motion according to expression (2.1.2) from the time domain into the Laplace domain and obtain a corresponding algebraic equation in the Laplace domain:

$$s^2 x(s) - sV - s^2 S + 2ns x(s) - 2nsS + f = 0 \qquad (12.1.3)$$

Rearranging equation (12.1.3), we have

$$x(s)s(s + 2n) = sV + s^2 S + 2nsS - f \qquad (12.1.4)$$

Solving equation (12.1.4) for the displacement $x(s)$ in the Laplace domain, we may write

$$x(s) = \frac{V}{s + 2n} + \frac{sS}{s + 2n} + \frac{2nS}{s + 2n} - \frac{f}{s(s + 2n)} \qquad (12.1.5)$$

Using Laplace Transform pairs 1, 8, 14, 8, and 16, we invert equation (12.1.5) from the Laplace domain into the time domain and obtain the solution of differential equation (12.1.1) with the initial conditions of motion according to expression (2.1.2):

$$x = \frac{V}{2n}(1 - e^{-2nt}) + Se^{-2nt} + S(1 - e^{-2nt}) - \frac{f}{2n}\left[t + \frac{1}{2n}(e^{-2nt} - 1)\right] \qquad (12.1.6)$$

Rearranging equation (12.1.6), we obtain

$$x = S + \frac{V}{2n} + \frac{f}{4n^2} - \frac{f}{2n}t - e^{-2nt}\left(\frac{V}{2n} + \frac{f}{4n^2}\right) \qquad (12.1.7)$$

Taking the first derivative from equation (12.1.7), we determine the velocity of the system:

$$\frac{dx}{dt} = 2ne^{-2nt}\left(\frac{V}{2n} + \frac{f}{4n^2}\right) - \frac{f}{2n} \qquad (12.1.8)$$

Supposing that in equations (12.1.7) and (12.1.8), $t = 0$, we determine that $x = S$ and $\frac{dx}{dt} = V$, as it should be according to initial conditions of motion (2.1.2).

The first derivative from equation (12.1.8) yields the acceleration of the system:

$$\frac{d^2x}{dt^2} = -4n^2 e^{-2nt}\left(\frac{V}{2n} + \frac{f}{4n^2}\right) \qquad (12.1.9)$$

Hence, equations (12.1.7)–(12.1.9) represent the basic parameters of motion of the system.

12.2. Motion of a System Restricted by a Fluid Link and Subjected to Dry Friction and Constant Active Forces

Consider the motion of a system restricted by a fluid link, moving on a horizontal surface, and being subjected to dry friction and constant active forces. Figure (12.2.1) shows a schematic diagram of the system. The notations in this figure are self-explanatory.

Based on the considerations mentioned above and the schematic diagram shown in Figure (12.2.1), we compose the differential

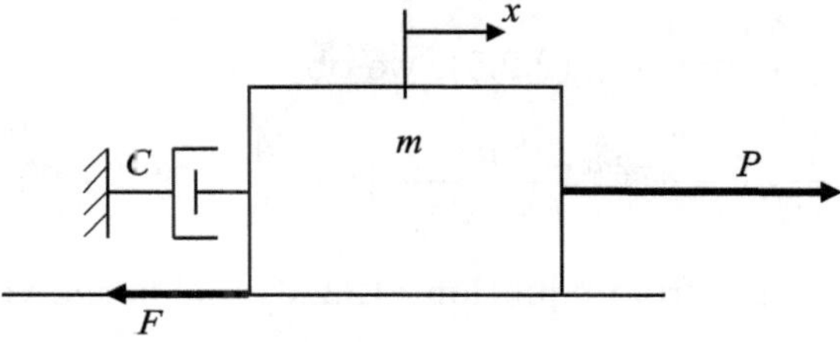

Fig. 12.2.1. Schematic diagram of the system moving on a horizontal surface, while being restricted by a fluid link and subjected to dry friction and constant active forces.

equation of motion of the system:

$$m\frac{d^2x}{dt^2} + C\frac{dx}{dt} + F = P \qquad (12.2.1)$$

The initial conditions of motion are taken according to expression (2.1.2).

Dividing equation (12.2.1) by m, we have

$$\frac{d^2x}{dt^2} + 2n\frac{dx}{dt} + f = p \qquad (12.2.2)$$

Applying Laplace Transform pairs 5, 2, 4, 2, and 2, we convert differential equation (12.2.2) with the initial conditions of motion according to expression (2.1.2) from the time domain into an algebraic equation in the Laplace domain:

$$s^2x(s) - sV - s^2S + 2nsx(s) - 2nsS + f = p \qquad (12.2.3)$$

The solution of equation (12.2.3) for the displacement $x(s)$ in the Laplace domain reads:

$$x(s) = \frac{V + 2nS}{s + 2n} + \frac{sS}{s + 2n} + \frac{p - f}{s(s + 2n)} \qquad (12.2.4)$$

Using the Laplace Transform pairs 1, 8, 14, and 16, we invert equation (12.2.4) from the Laplace domain into the time domain and obtain the solution of differential equation(12.2.1) with the initial conditions of motion according to expression (2.1.2):

$$x = \frac{V + 2nS}{2n}(1 - e^{-2nt}) + Se^{-2nt} + \frac{p - f}{2n}\left[t + \frac{1}{2n}(e^{-2nt} - 1)\right] \qquad (12.2.5)$$

Simplifying equation (12.2.5), we obtain

$$x = S + \frac{V}{2n}(1 - e^{-2nt}) + \frac{p - f}{2n}\left[t + \frac{1}{2n}(e^{-2nt} - 1)\right] \qquad (12.2.6)$$

The first derivative from equation (12.2.6) yields the velocity of the system:

$$\frac{dx}{dt} = Ve^{-2nt} + \frac{p - f}{2n}(1 - e^{-2nt}) \qquad (12.2.7)$$

Assuming that in equations (12.2.6) and (12.2.7), $t = 0$, we determine that $x = S$ and $\frac{dx}{dt} = V$, as expected according to the initial conditions of motion presented in expression (2.1.2).

Taking the first derivative from equation (12.2.7), we determine the acceleration of the system:

$$\frac{d^2 x}{dt^2} = e^{-2nt}(p - f - 2nV) \tag{12.2.8}$$

Hence, equations (12.2.6)–(12.2.8) are the basic parameters of motion of the system.

12.3. Motion of a System Restricted by a Fluid Link and Subjected to Dry Friction and Harmonic Forces

We consider an operational process of a system that is moving on a horizontal frictional surface. The system is restricted by a fluid link, and it is subjected to dry friction and harmonic forces.

Figure (12.3.1) shows a schematic diagram that characterizes the system described above. The notations in this figure are self-explanatory.

Based on the schematic diagram shown in Figure (12.3.1) and the considerations mentioned above, we compose the differential equation of motion of the system:

$$m\frac{d^2 x}{dt^2} + C\frac{dx}{dt} + F = A \cos \varphi t \tag{12.3.1}$$

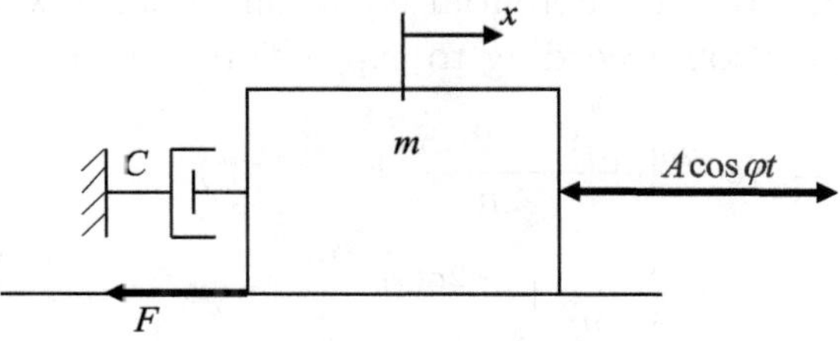

Fig. 12.3.1. Schematic diagram of a system moving on a horizontal surface, while being restricted by a fluid link and subjected to dry friction and harmonic forces.

The initial conditions of motion are taken according to expression (2.1.2).

Dividing equation (12.3.1) by m, we may write

$$\frac{d^2x}{dt^2} + 2n\frac{dx}{dt} + f = a\cos\varphi t \tag{12.3.2}$$

Applying Laplace Transform pairs 5, 2, 4, 2, and 29 to equation (12.3.2) with the initial conditions of motion according to expression (2.1.2), we convert this equation from the time domain into an algebraic equation in the Laplace domain:

$$s^2x(s) - Vs - s^2S + 2nsx(s) - 2nsS + f = \frac{as^2}{s^2 + \varphi^2} \tag{12.3.3}$$

Applying to equation (12.3.3) the conventional algebraic procedures, we may write

$$x(s)s(s + 2n) = s^2S + sV + \frac{as^2}{s^2 + \varphi^2} + 2nsS - f \tag{12.3.4}$$

The solution of equation (12.3.4) for the displacement $x(s)$ in the Laplace domain has the following form:

$$x(s) = \frac{sS}{s + 2n} + \frac{V + 2nS}{s + 2n} + \frac{as}{(s + 2n)(s^2 + \varphi^2)} - \frac{f}{s(s + 2n)} \tag{12.3.5}$$

Applying Laplace Transform pairs 1, 14, 8, 44, and 16, we invert equation (12.3.5) from the Laplace domain into the time domain and obtain the solution of differential equation (12.3.1) with the initial conditions of motion according to expression (2.1.2):

$$\begin{aligned}
x = {} & Se^{-2nt} + \frac{V + 2nS}{2n}\left(1 - e^{-2nt}\right) \\
& + \frac{a}{\varphi^2 + 4n^2}\left(e^{-2nt} + \frac{2n}{\varphi}\sin\varphi t - \cos\varphi t\right) \\
& - \frac{f}{2n}\left[t + \frac{1}{2n}\left(e^{-2nt} - 1\right)\right]
\end{aligned} \tag{12.3.6}$$

Rearranging equation (12.3.6), we may write

$$x = S + \frac{V}{2n} + e^{-2nt}\left(\frac{a}{\varphi^2 + 4n^2} - \frac{V}{2n}\right)$$

$$+ \frac{a}{\varphi^2 + 4n^2}\left(\frac{2n}{\varphi}\sin\varphi t - \cos\varphi t\right)$$

$$- \frac{f}{2n}\left[t + \frac{1}{2n}\left(e^{-2nt} - 1\right)\right] \qquad (12.3.7)$$

Combining equation (12.3.7) with expression (10.3.9), we have

$$x = S + \frac{V}{2n}(1 - e^{-2nt}) + \frac{a}{\varphi^2 + 4n^2}e^{-2nt}$$

$$- \frac{a\cos(\varphi t + \lambda)}{\varphi\sqrt{\varphi^2 + 4n^2}} - \frac{f}{2n}\left[t + \frac{1}{2n}\left(e^{-2nt} - 1\right)\right] \qquad (12.3.8)$$

Taking the first derivative from equation (12.3.8), we determine the velocity of the system:

$$\frac{dx}{dt} = Ve^{-2nt} - \frac{2nae^{-2nt}}{\varphi^2 + 4n^2} + \frac{a\sin(\varphi t + \lambda)}{\sqrt{\varphi^2 + 4n^2}} - \frac{f}{2n}(1 - e^{-2nt})$$

$$(12.3.9)$$

Supposing that in equations (12.3.8) and (12.3.9), we have $t = 0$, we obtain $x = S$ and $\frac{dx}{dt} = V$, which is expected according to the initial conditions of motion given by expression (2.1.2).

The first derivative from equation (12.3.9) yields the acceleration of the system:

$$\frac{d^2x}{dt^2} = -2nVe^{-2nt} + \frac{4n^2ae^{-2nt}}{\varphi^2 + 4n^2} + \frac{a\varphi\cos(\varphi t + \lambda)}{\sqrt{\varphi^2 + 4n^2}} - re^{-2nt}$$

$$(12.3.10)$$

Hence, equations (12.3.8)– (12.3.10) describe the basic parameters of motion of the system.

12.4. Motion of a System Restricted by a Fluid Link and Subjected to Dry Friction, Constant Active, and Harmonic Forces

This operational process is characterized by the motion of a system on a horizontal frictional surface. The system is restricted by a fluid link and is subjected to the actions of constant resisting, constant active, and harmonic forces. Figure 12.4.1 shows a schematic diagram of the system described above. The notations in the figure are self-explanatory.

Based on the schematic diagram shown in Figure 12.4.1 and the above-mentioned considerations, we compose the differential equation of motion of the system:

$$m\frac{d^2x}{dt^2} + C\frac{dx}{dt} + F = P + A\cos\varphi t \qquad (12.4.1)$$

The initial conditions of motion are taken according to expression (2.1.2).

Dividing equation (12.4.1) by m, we have

$$\frac{d^2x}{dt^2} + 2n\frac{dx}{dt} + f = p + a\cos\varphi t \qquad (12.4.2)$$

Applying Laplace Transform pairs 5, 2, 4, 2, 2, and 29, we convert equation (12.4.2) with the initial conditions of motion according to expression (2.1.2) from the time domain into an algebraic equation

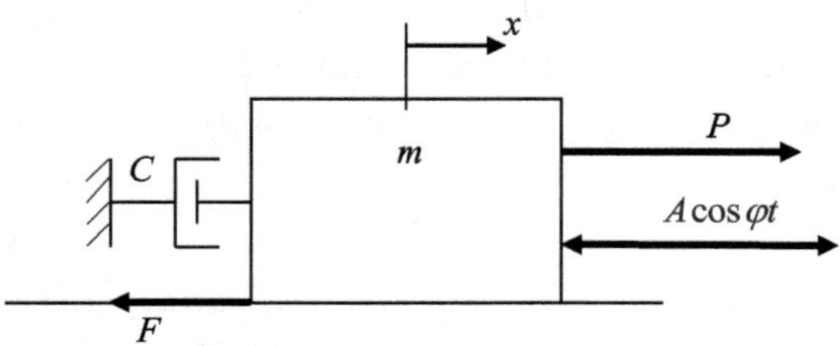

Fig. 12.4.1. Schematic diagram of a system moving on a horizontal frictional surface while being restricted by a fluid link and subjected to dry friction, constant active, and harmonic forces.

in the Laplace domain:

$$s^2 x(s) - Vs - s^2 S + 2nsx(s) - 2nsS + f = p + \frac{as^2}{s^2 + \varphi^2} \qquad (12.4.3)$$

Applying to equation (12.4.3) the conventional algebraic procedures, we have

$$x(s)s(s + 2n) = s^2 S + sV + 2nsS + p - f + \frac{as^2}{s^2 + \varphi^2} \qquad (12.4.4)$$

Solving equation (12.4.4) for the displacement $x(s)$ in the Laplace domain, we obtain

$$x(s) = \frac{sS}{s + 2n} + \frac{V + 2nS}{s + 2n} + \frac{p - f}{s(s + 2n)} + \frac{as}{(s + 2n)(s^2 + \varphi^2)}$$
$$(12.4.5)$$

Applying Laplace Transform pairs 1, 14, 8, 16, and 44, we invert equation (12.4.5) from the Laplace domain into the time domain and obtain the solution of differential equation (12.4.1) with the initial conditions of motion according to expression (2.1.2):

$$x = Se^{-2nt} + \frac{V + 2nS}{2n}(1 - e^{-2nt}) + \frac{p - f}{2n}\left[t + \frac{1}{2n}(e^{-2nt} - 1)\right]$$
$$+ \frac{a}{\varphi^2 + 4n^2}\left(e^{-2nt} + \frac{2n}{\varphi}\sin\varphi t - \cos\varphi t\right) \qquad (12.4.6)$$

Rearranging equation (12.4.6), we have

$$x = S + \frac{V}{2n}(1 - e^{-2nt}) + \frac{(p - f)t}{2n} + \frac{p - f}{4n^2}e^{-2nt} - \frac{p - f}{4n^2}$$
$$+ \frac{ae^{-2nt}}{\varphi^2 + 4n^2} + \frac{a}{\varphi^2 + 4n^2}\left(\frac{2n}{\varphi}\sin\varphi t - \cos\varphi t\right) \qquad (12.4.7)$$

Transforming equation (12.4.7) by combining it with expression (10.3.9), we obtain

$$x = S + \frac{V}{2n} - \frac{p-f}{4n^2} + \frac{(p-f)t}{2n} - e^{-2nt}\left(\frac{V}{2n} - \frac{p-f}{4n^2} - \frac{a}{\varphi^2 + 4n^2}\right)$$

$$- \frac{a\cos(\varphi t + \lambda)}{\varphi\sqrt{\varphi^2 + 4n^2}} \tag{12.4.8}$$

The first derivative from equation (12.4.8) yields the velocity of the system:

$$\frac{dx}{dt} = \frac{p-f}{2n} + 2ne^{-2nt}\left(\frac{V}{2n} - \frac{p-f}{4n^2} - \frac{a}{\varphi^2 + 4n^2}\right)$$

$$+ \frac{a\sin(\varphi t + \lambda)}{\sqrt{\varphi^2 + 4n^2}} \tag{12.4.9}$$

Assuming that in equations (12.4.8) and (12.4.9), $t = 0$, we obtain $x = S$ and $\frac{dx}{dt} = V$, which is expected according to the initial conditions of motion presented in expression (2.1.2).

Taking the first derivative from equation (12.4.9), we determine the acceleration of the system:

$$\frac{d^2x}{dt^2} = -4n^2e^{-2nt}\left(\frac{V}{2n} - \frac{p-f}{4n^2} - \frac{a}{\varphi^2 + 4n^2}\right)$$

$$+ \frac{a\varphi\cos(\varphi t + \lambda)}{\sqrt{\varphi^2 + 4n^2}} \tag{12.4.10}$$

Hence, equations (12.4.8)–(12.4.10) describe the basic parameters of motion of the system.

12.4.1. *Numerical solution*

Combining equation (12.4.9) with the expressions given in (10.3.9), we obtain

$$\frac{dx}{dt} = \frac{p-f}{2n} + 2ne^{-2nt}\left(\frac{V}{2n} - \frac{p-f}{4n^2} - \frac{a}{\varphi^2 + 4n^2}\right)$$

$$+ \frac{a}{\varphi^2 + 4n^2}(\varphi\sin\varphi t + 2n\cos\varphi t) \tag{12.4.11}$$

The following is a Python program to plot the velocity of the system represented by equation (12.4.11) and determine when the velocity first reaches zero.

```python
from matplotlib.pyplot import plot, show, text
from numpy import linspace, sin, cos, exp, sqrt
from scipy.optimize import fsolve

S = 0.1
V = 3
a = 20
f = 6
phi = 30
n = 0.02
p = 0.5

t = linspace(0, 2, 1000)

V1 = lambda t: \
    (p - f)/(2*n) \
    + 2*n*exp(-2*n*t)*(V/(2*n) - (p - f)/(4*n**2) \
            - a/(phi**2 + 4*n**2)) \
    + (a/(phi**2 + 4*n**2)) \
    *(phi*sin(phi*t) + 2*n*cos(phi*t))

[root] = fsolve(V1, 1.4)
root = round(root, 3)

plot(t, V1(t))
plot(root, 0, 'bo')
text(1.1, -0.5, f't={root}')
show()
```

As shown in Figure 12.4.2, the velocity first reaches zero at $t = 1.373$ s. The following is a Python program to plot equation (12.4.7) from $t = 0$ s to $t = 1.373$ s.

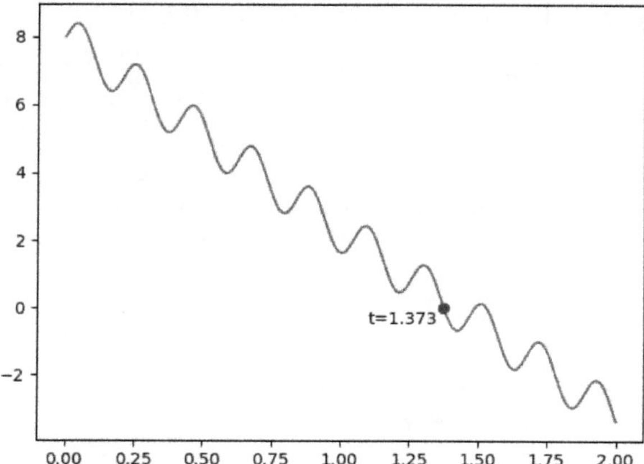

Fig. 12.4.2. Graph of equation (12.4.11).

```python
from matplotlib.pyplot import plot, show, text
from numpy import linspace, sin, cos, exp


S = 0.1
V = 8
a = 20
f = 6
phi = 30
n = 0.02
p = 0.5

x = lambda t: \
  S + (V/(2*n))*(1 - exp(-2*n*t)) + ((p - f)*t)/(2*n) \
  + ((p - f)/(4*n**2))*exp(-2*n*t) - (p - f)/(4*n**2) \
  + (a*exp(-2*n*t))/(phi**2 + 4*n**2) \
  + (a/(phi**2 + 4*n**2))*(((2*n)/phi)*sin(phi*t) - cos(phi*t))

stopTime = 1.373
t = linspace(0, stopTime, 1000)
plot(t, x(t))

stopPosition = round(x(stopTime), 3)
plot(stopTime, stopPosition, 'bo')
text(1.22, 5.2, f'x={stopPosition}')

show()
```

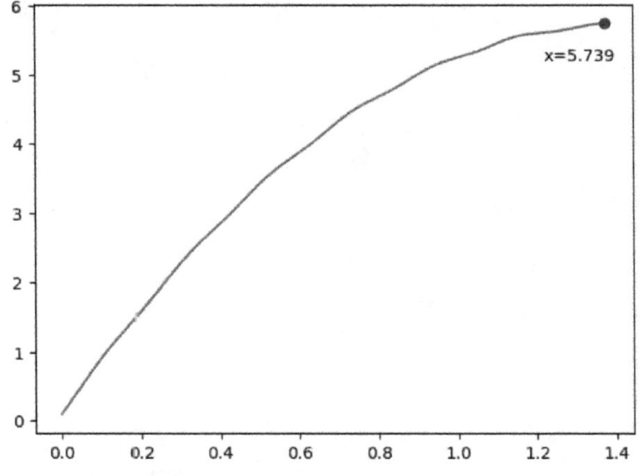

Fig. 12.4.3. Graph of equation (12.4.7).

For the next interaction, apply equation (12.4.1) with the following initial conditions of motion Figure 12.4.3

$$\text{for} \quad t = 0 \quad x = 5.739; \quad \frac{dx}{dt} = 0$$

For more details, see Section 1.5.

MOTION OF A SYSTEM RESTRICTED BY A FLUID LINK AND SUBJECTED TO A CONSTANT RESISTING FORCE AND A DRY FRICTION FORCE

13.1. Motion of a System on a Horizontal Surface Due to Its Initial Velocity While Being Restricted by a Fluid Link and Subjected to a Constant Resisting Force and a Dry Friction Force

Consider an operational process associated with a system restricted by a fluid link and moving on a horizontal surface, while the system is subjected to a constant resisting force and a dry friction force. Motion occurs due to an initial velocity. Figure 13.1.1 shows a schematic diagram of the system described above. The notations in the figure are self-explanatory.

Based on the considerations mentioned above and the schematic diagram shown in Figure 13.1.1, we compose the following differential equation of motion:

$$m\frac{d^2x}{dt^2} + C\frac{dx}{dt} + R + F = 0 \qquad (13.1.1)$$

The initial conditions of motion are taken according to expression (2.1.2).

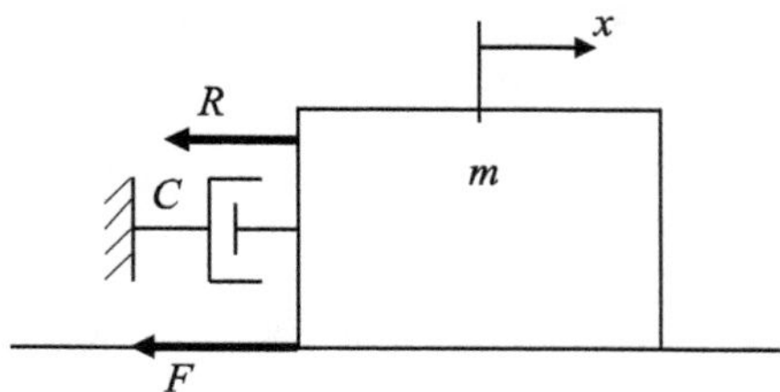

Fig. 13.1.1. Schematic diagram of a system moving on a horizontal surface while being restricted by a fluid link and subjected to a constant resisting force and a dry friction force.

Dividing equation (13.1.1) by m, we have

$$\frac{d^2x}{dt^2} + 2n\frac{dx}{dt} + r + f = 0 \tag{13.1.2}$$

With the help of Laplace Transform pairs 5, 2, 4, 2, and 2, we convert equation (13.1.2) with the initial conditions of motion according to expression (2.1.2) from the time domain into the Laplace domain and obtain the corresponding algebraic equation in the Laplace domain:

$$s^2 x(s) - sV - s^2 S + 2nsx(s) - 2nsS + r + f = 0 \tag{13.1.3}$$

Rearranging equation (13.1.3), we have

$$x(s)s(s + 2n) = sV + s^2 S + 2nsS - r - f \tag{13.1.4}$$

Solving equation (13.1.4) for the displacement $x(s)$ in the Laplace domain, we may write

$$x(s) = \frac{V}{s + 2n} + \frac{sS}{s + 2n} + \frac{2nS}{s + 2n} - \frac{r + f}{s(s + 2n)} \tag{13.1.5}$$

Using Laplace Transform pairs 1, 8, 14, 8, and 16, we invert equation (13.1.5) from the Laplace domain into the time domain and obtain the solution of differential equation (13.1.1) with the initial

conditions of motion according to expression (2.1.2):

$$x = \frac{V}{2n}(1 - e^{-2nt}) + Se^{-2nt} + S(1 - e^{-2nt})$$

$$-\frac{r+f}{2n}\left[t + \frac{1}{2n}(e^{-2nt} - 1)\right] \qquad (13.1.6)$$

Rearranging equation (13.1.6), we obtain

$$x = S + \frac{V}{2n} + \frac{r+f}{4n^2} - \frac{r+f}{2n}t - e^{-2nt}\left(\frac{V}{2n} + \frac{r+f}{4n^2}\right) \qquad (13.1.7)$$

Taking the first derivative from equation (13.1.7), we determine the velocity of the system:

$$\frac{dx}{dt} = 2ne^{-2nt}\left(\frac{V}{2n} + \frac{r+f}{4n^2}\right) - \frac{r+f}{2n} \qquad (13.1.8)$$

Supposing that in equation (13.1.7) and (13.1.8), $t = 0$, we determine that $x = S$ and $\frac{dx}{dt} = V$, as it should be according to the initial conditions of motion given by expression (2.1.2).

The first derivative from equation (13.1.8) yields the acceleration of the system:

$$\frac{d^2x}{dt^2} = -4n^2e^{-2nt}\left(\frac{V}{2n} + \frac{r+f}{4n^2}\right) \qquad (13.1.9)$$

Hence, equations (13.1.7)– (13.1.9) represent the basic parameters of motion of the system.

13.2. Motion of a System Restricted by a Fluid Link and Subjected to Constant Resisting, Dry Friction, and Constant Active Forces

Consider the motion of a system restricted by a fluid link, moving on a horizontal frictional surface, while being subjected to constant resisting, friction, and constant active forces. Figure 13.2.1 shows a schematic diagram of the system. The notations in this figure are self-explanatory.

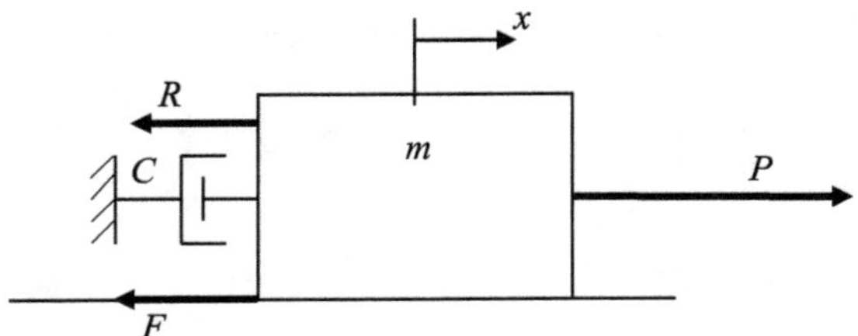

Fig. 13.2.1. Schematic diagram of a system moving on a horizontal frictional surface while being restricted by a fluid link and subjected to a constant resisting force, a dry friction force, and a constant active force.

Based on the considerations mentioned above and the schematic diagram shown in Figure 13.2.1, we compose the differential equation of motion of the system:

$$m\frac{d^2x}{dt^2} + C\frac{dx}{dt} + R + F = P \tag{13.2.1}$$

The initial conditions of motion are taken according to expression (2.1.2). Dividing equation (13.2.1) by m, we have

$$\frac{d^2x}{dt^2} + 2n\frac{dx}{dt} + r + f = p \tag{13.2.2}$$

Applying Laplace Transform pairs 5, 2, 4, 2, 2, and 2, we convert differential equation (13.2.2) with the initial conditions of motion according to expression (2.1.2) from the time domain into the algebraic equation in the Laplace domain:

$$s^2x(s) - sV - s^2S + 2nsx(s) - 2nsS + r + f = p \tag{13.2.3}$$

The solution of equation (13.2.3) for the displacement $x(s)$ in the Laplace domain reads:

$$x(s) = \frac{V + 2nS}{s + 2n} + \frac{sS}{s + 2n} + \frac{p - r - f}{s(s + 2n)} \tag{13.2.4}$$

Using Laplace Transform pairs 1, 8, 14, and 16, we invert equation (13.2.4) from the Laplace domain into the time domain and obtain the solution of differential equation (13.2.1) with the initial

conditions of motion according to expression (2.1.2):

$$x = \frac{V + 2nS}{2n}(1 - e^{-2nt}) + Se^{-2nt}$$

$$+ \frac{p - r - f}{2n}\left[t + \frac{1}{2n}(e^{-2nt} - 1)\right] \qquad (13.2.5)$$

Simplifying equation (13.2.5), we obtain

$$x = S + \frac{V}{2n}(1 - e^{-2nt}) + \frac{p - r - f}{2n}\left[t + \frac{1}{2n}(e^{-2nt} - 1)\right] \qquad (13.2.6)$$

The first derivative from equation (13.2.6) yields the velocity of the system:

$$\frac{dx}{dt} = Ve^{-2nt} + \frac{p - r - f}{2n}(1 - e^{-2nt}) \qquad (13.2.7)$$

Assuming that in equations (13.2.6) and (13.2.7), $t = 0$, we determine that $x = S$ and $\frac{dx}{dt} = V$, as expected according to the initial conditions of motion presented in expression (2.1.2).

Taking the first derivative from equation (13.2.7), we determine the acceleration of the system:

$$\frac{d^2x}{dt^2} = e^{-2nt}(p - r - f - 2nV) \qquad (13.2.8)$$

Hence, equations (13.2.6)–(13.2.8) are the basic parameters of motion of the system.

13.3. Motion of a System Restricted by a Fluid Link and Subjected to Constant Resisting, Dry Friction, and Harmonic Forces

We consider the operational process of a system that is moving on a horizontal surface. The system is restricted by a fluid link, and it is subjected to constant resisting, dry friction, and harmonic forces.

Figure 13.3.1 shows a schematic diagram that characterizes the system described above. The notations in this figure are self-explanatory.

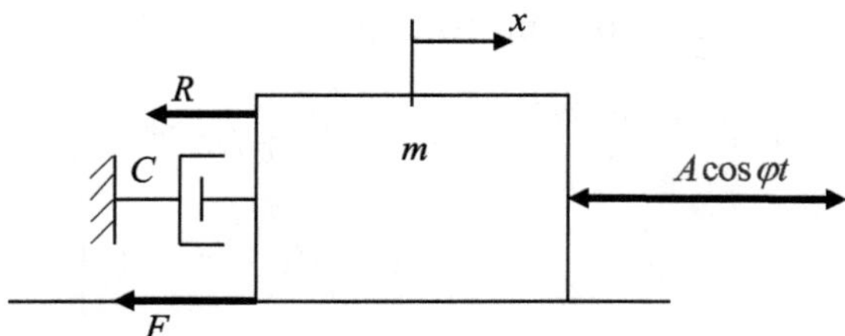

Fig. 13.3.1. Schematic diagram of a system moving on a horizontal surface while being restricted by a fluid link and subjected to constant resisting, dry friction, and harmonic forces.

Based on the schematic diagram shown in Figure 13.3.1 and the considerations mentioned above, we compose the differential equation of motion of the system:

$$m\frac{d^2x}{dt^2} + C\frac{dx}{dt} + R + F = A\cos\varphi t \qquad (13.3.1)$$

The initial conditions of motion are taken according to expression (2.1.2).

Dividing equation (13.3.1) by m, we may write

$$\frac{d^2x}{dt^2} + 2n\frac{dx}{dt} + r + f = a\cos\varphi t \qquad (13.3.2)$$

Applying Laplace Transform pairs 5, 2, 4, 2, 2, and 29 to equation (13.3.2) with the initial conditions of motion according to expression (2.1.2), we convert this equation from the time domain into an algebraic equation in the Laplace domain:

$$s^2x(s) - Vs - s^2S + 2nsx(s) - 2nsS + r + f = \frac{as^2}{s^2 + \varphi^2} \qquad (13.3.3)$$

Applying to equation (13.3.3) the conventional algebraic procedures, we may write

$$x(s)s(s + 2n) = s^2S + sV + \frac{as^2}{s^2 + \varphi^2} + 2nsS - r - f \qquad (13.3.4)$$

The solution of equation (13.3.4) for the displacement $x(s)$ in the Laplace domain has the following expression:

$$x(s) = \frac{sS}{s+2n} + \frac{V+2nS}{s+2n} + \frac{as}{(s+2n)(s^2+\varphi^2)} - \frac{r+f}{s(s+2n)} \tag{13.3.5}$$

Applying Laplace Transform pairs 1, 14, 8, 44, and 16, we invert equation (13.3.5) from the Laplace domain into the time domain and obtain the solution of differential equation (13.3.1) with the initial conditions of motion according to expression (2.1.2):

$$x = Se^{-2nt} + \frac{V+2nS}{2n}(1 - e^{-2nt}) + \frac{a}{\varphi^2+4n^2}$$
$$\times \left(e^{-2nt} + \frac{2n}{\varphi}\sin\varphi t - \cos\varphi t \right) - \frac{r+f}{2n}\left[t + \frac{1}{2n}(e^{-2nt} - 1) \right] \tag{13.3.6}$$

Rearranging equation (13.3.6), we may write

$$x = S + \frac{V}{2n} + e^{-2nt}\left(\frac{a}{\varphi^2+4n^2} - \frac{V}{2n} \right) + \frac{a}{\varphi^2+4n^2}$$
$$\times \left(\frac{2n}{\varphi}\sin\varphi t - \cos\varphi t \right) - \frac{r+f}{2n}\left[t + \frac{1}{2n}(e^{-2nt} - 1) \right] \tag{13.3.7}$$

Combining equation (13.3.7) with expression (10.3.9), we have

$$x = S + \frac{V}{2n}(1 - e^{-2nt}) + \frac{a}{\varphi^2+4n^2}e^{-2nt}$$
$$- \frac{a\cos(\varphi t + \lambda)}{\varphi\sqrt{\varphi^2+4n^2}} - \frac{r+f}{2n}\left[t + \frac{1}{2n}(e^{-2nt} - 1) \right] \tag{13.3.8}$$

Taking the first derivative from equation (13.3.8), we determine the velocity of the system:

$$\frac{dx}{dt} = Ve^{-2nt} - \frac{2nae^{-2nt}}{\varphi^2+4n^2} + \frac{a\sin(\varphi t + \lambda)}{\sqrt{\varphi^2+4n^2}} - \frac{r+f}{2n}(1 - e^{-2nt}) \tag{13.3.9}$$

Supposing that in equation (13.3.8) and (13.3.9), we have $t = 0$, then we obtain $x = S$ and $\frac{dx}{dt} = V$, which is expected according to the initial conditions of motion given by expression (2.1.2).

The first derivative from equation (13.3.9) yields the acceleration of the system:

$$\frac{d^2x}{dt^2} = -2nVe^{-2nt} + \frac{4n^2ae^{-2nt}}{\varphi^2 + 4n^2} + \frac{a\varphi\cos(\varphi t + \lambda)}{\sqrt{\varphi^2 + 4n^2}} - (r + f)e^{-2nt}$$

$$(13.3.10)$$

Hence, equations (13.3.8)–(13.3.10) describe the basic parameters of motion of the system.

13.4. Motion of a System Restricted by a Fluid Link and Subjected to Constant Resisting, Dry Friction, Constant Active, and Harmonic Forces

This operational process is characterized by the motion of a system on a horizontal frictional surface. The system is restricted by a fluid link and is subjected to the actions of constant resisting, dry friction, constant active, and harmonic forces. Figure 13.4.1 shows a schematic diagram of the system described above. The notations in the figure are self-explanatory.

Based on the schematic diagram shown in Figure 13.4.1 and the above-mentioned considerations, we compose the differential equation of motion of the system:

$$m\frac{d^2x}{dt^2} + C\frac{dx}{dt} + R + F = P + A\cos\varphi t \qquad (13.4.1)$$

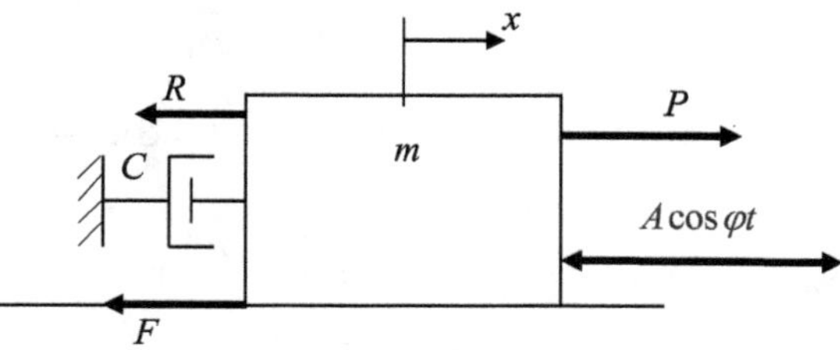

Fig. 13.4.1. Schematic diagram of a system moving on a horizontal frictional surface while being restricted by a fluid link and subjected to constant resisting, dry friction, constant active, and harmonic forces.

The initial conditions of motion are taken according to expression (2.1.2).

Dividing equation (13.4.1) by m, we have

$$\frac{d^2x}{dt^2} + 2n\frac{dx}{dt} + r + f = p + a\cos\varphi t \qquad (13.4.2)$$

Applying Laplace Transform pairs 5, 2, 4, 2, 2, 2, 2, and 29, we convert equation (13.4.2) with the initial conditions of motion according to expression (2.1.2) from the time domain into an algebraic equation in the Laplace domain:

$$s^2 x(s) - Vs - s^2 S + 2nsx(s) - 2nsS + r + f = p + \frac{as^2}{s^2 + \varphi^2}$$
$$(13.4.3)$$

Applying to equation (13.4.3) the conventional algebraic procedures, we have

$$x(s)s(s + 2n) = s^2 S + sV + 2nsS + p - r - f + \frac{as^2}{s^2 + \varphi^2}$$
$$(13.4.4)$$

Solving equation (13.4.4) for the displacement $x(s)$ in the Laplace domain, we obtain

$$x(s) = \frac{sS}{s + 2n} + \frac{V + 2nS}{s + 2n} + \frac{p - r - f}{s(s + 2n)} + \frac{as}{(s + 2n)(s^2 + \varphi^2)}$$
$$(13.4.5)$$

Applying Laplace Transform pairs 1, 14, 8, 16, and 44, we invert equation (13.4.5) from the Laplace domain into the time domain and obtain the solution of differential equation (13.4.1) with the initial conditions of motion according to expression (2.1.2):

$$x = Se^{-2nt} + \frac{V + 2nS}{2n}(1 - e^{-2nt}) + \frac{p - r - f}{2n}\left[t + \frac{1}{2n}(e^{-2nt} - 1)\right]$$

$$+ \frac{a}{\varphi^2 + 4n^2}\left(e^{-2nt} + \frac{2n}{\varphi}\sin\varphi t - \cos\varphi t\right) \qquad (13.4.6)$$

Rearranging equation (13.4.6), we have

$$x = S + \frac{V}{2n}(1 - e^{-2nt}) + \frac{(p - r - f)t}{2n} + \frac{p - r - f}{4n^2}e^{-2nt}$$

$$-\frac{p - r - f}{4n^2} + \frac{ae^{-2nt}}{\varphi^2 + 4n^2} + \frac{a}{\varphi^2 + 4n^2}\left(\frac{2n}{\varphi}\sin\varphi t - \cos\varphi t\right)$$

$$(13.4.7)$$

Transforming equation (13.4.7) by combining it with expression (10.3.9), we obtain

$$x = S + \frac{V}{2n} - \frac{p - r - f}{4n^2} - e^{-2nt}\frac{(p - r - f)t}{2n}$$

$$\times\left(\frac{V}{2n} - \frac{p - r - f}{4n^2} - \frac{a}{\varphi^2 + 4n^2}\right) - \frac{a\cos(\varphi t + \lambda)}{\varphi\sqrt{\varphi^2 + 4n^2}}$$

$$(13.4.8)$$

The first derivative from equation (13.4.8) yields the velocity of the system:

$$\frac{dx}{dt} = \frac{p - r - f}{2n} + 2ne^{-2nt}\left(\frac{V}{2n} - \frac{p - r - f}{4n^2} - \frac{a}{\varphi^2 + 4n^2}\right)$$

$$+\frac{a\sin(\varphi t + \lambda)}{\sqrt{\varphi^2 + 4n^2}}$$

$$(13.4.9)$$

Assuming that in equations (13.4.8) and (13.4.9), $t = 0$, we obtain $x = S$ and $\frac{dx}{dt} = V$, which is expected according to the initial conditions of motion given by expression (2.1.2).

Taking the first derivative from equation (13.4.9), we determine the acceleration of the system:

$$\frac{d^2x}{dt^2} = -4n^2e^{-2nt}\left(\frac{V}{2n} - \frac{p - r - f}{4n^2} - \frac{a}{\varphi^2 + 4n^2}\right)$$

$$+\frac{a\varphi\cos(\varphi t + \lambda)}{\sqrt{\varphi^2 + 4n^2}}$$

$$(13.4.10)$$

Hence, equations (13.4.8)–(13.4.10) describe the basic parameters of motion of the system.

13.4.1. *Numerical solution*

Combining equation (13.4.9) with the expressions presented in (10.3.9), we obtain

$$\frac{dx}{dt} = \frac{p-r-f}{2n} + 2ne^{-2nt}\left(\frac{V}{2n} - \frac{p-r-f}{4n^2} - \frac{a}{\varphi^2 + 4n^2}\right)$$

$$+ \frac{a}{\varphi^2 + 4n^2}(\varphi\sin\varphi t + 2n\cos\varphi t) \tag{13.4.11}$$

The following is a Python program to plot the velocity represented by equation (13.4.11) and to solve it for when it first reaches zero.

```python
from matplotlib.pyplot import plot, show, text
from numpy import linspace, sin, cos, exp, sqrt
from scipy.optimize import fsolve


S = 0.1
V = 8
a = 20
f = 6
r = 1
phi = 30
n = 0.02
p = 0.5

t = linspace(0, 2, 1000)

V1 = lambda t: \
    (p - r - f)/(2*n) \
    + 2*n*exp(-2*n*t)*(V/(2*n) - (p - r - f)/(4*n**2) \
            - a/(phi**2 + 4*n**2)) \
    + (a/(phi**2 + 4*n**2)) \
    *(phi*sin(phi*t) + 2*n*cos(phi*t))

[root] = fsolve(V1, 1.2)
root = round(root, 3)

plot(t, V1(t))
plot(root, 0, 'bo')
text(0.85, -0.5, f't={root}')
show()
```

As shown in Figure 13.4.2, the velocity first reaches zero at $t = 1.164$ s. The following is a Python program to plot equation (13.4.7) from $t = 0$ s to $t = 1.164$ s.

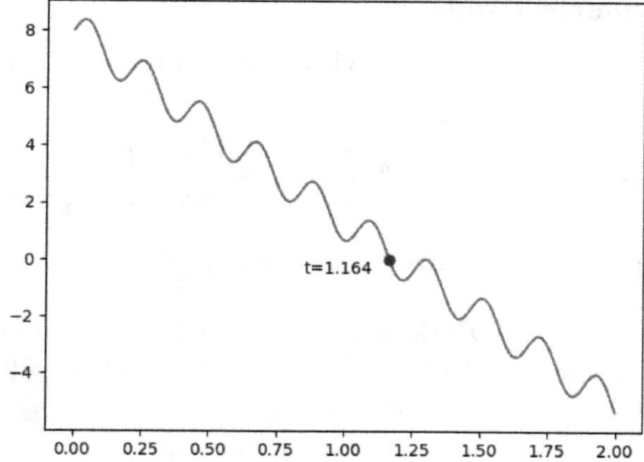

Fig. 13.4.2. Graph of equation (13.4.11).

```python
from matplotlib.pyplot import plot, show, text
from numpy import linspace, sin, cos, exp


S = 0.1
V = 8
a = 20
f = 6
r = 1
phi = 30
n = 0.02
p = 0.5

x1 = lambda t: \
    S + (V/(2*n))*(1 - exp(-2*n*t)) + (p - r - f)*t/(2*n) \
    + ((p - r - f)/(4*n**2))*exp(-2*n*t) - (p - r - f)/(4*n**2) \
    + a*exp(-2*n*t)/(phi**2 + 4*n**2) \
    + (a/(phi**2 + 4*n**2))*((2*n/phi)*sin(phi*t) - cos(phi*t))

stopTime = 1.164
t = linspace(0, stopTime, 1000)
plot(t, x1(t))

stopPosition = round(x1(stopTime), 3)
plot(stopTime, stopPosition, 'bo')
text(1, 4.4, f'x1={stopPosition}')

show()
```

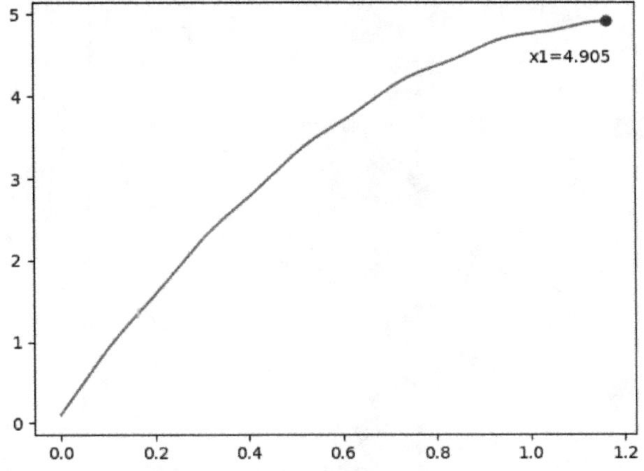

Fig. 13.4.3. Graph of equation (13.4.7).

For the next interaction, apply equation (13.4.1) with the following initial conditions of motion:

$$\text{for } t = 0 \quad x = 4.905; \quad \frac{dx}{dt} = 0$$

For more details, see Section 1.5.

MOTION OF A SYSTEM ON A HORIZONTAL FRICTIONLESS SURFACE WHILE BEING RESTRICTED BY A FLEXIBLE LINK AND A FLUID LINK

This chapter deals with systems, the motions of which occur on a horizontal frictionless surface and are caused by initial conditions of motion or by typical combinations of initial conditions of motion and external active forces.

14.1. Motion of a System Due to Its Initial Displacement and Initial Velocity

The operational process described in the following is characterized by the motion of a system on a horizontal frictionless surface, while no active forces are applied to the system. In this case, the motion occurs due to the system's initial displacement and velocity. The motion of the system is restricted by a flexible link and a fluid link that connect the mass of the system to a non-movable support.

Figure 14.1.1 shows a schematic diagram of the system, the notations in which are self-explanatory.

Based on the schematic diagram shown in Figure 14.1.1 and the considerations mentioned above, we compose the differential equation

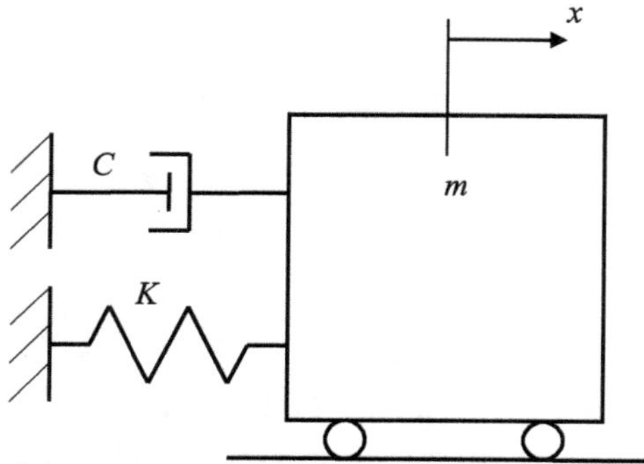

Fig. 14.1.1. Schematic diagram of a system moving on a horizontal frictionless surface while being restricted by a flexible link and a fluid link.

of motion of the system:

$$m\frac{d^2x}{dt^2} + C\frac{dx}{dt} + Kx = 0 \qquad (14.1.1)$$

The differential equation of motion (14.1.1) has different solutions depending on the combinations of the values of the damping coefficient C and the stiffness coefficient K. These solutions and their analyses are presented in the following.

Dividing equation (14.1.1) by m, we have

$$\frac{d^2x}{dt^2} + 2n\frac{dx}{dt} + \omega_0^2 x = 0 \qquad (14.1.2)$$

where

$$\omega_0^2 = \frac{K}{m} \qquad (14.1.3)$$

while ω_0 is the natural frequency.

It should be mentioned that in Chapters 14–17, the notations of the frequency ω and damping factor n may differ from similar notations used in Chapters 6–9 and 10–13.

The initial conditions of motion are taken according to expression (2.1.2).

Applying Laplace Transform pairs 5, 2, 4, 2, and 1, we convert equation (14.1.2) with the mentioned initial conditions of motion from the time domain into the Laplace domain and obtain the corresponding algebraic equation in the Laplace domain:

$$s^2 x(s) - sV - \varepsilon^2 S + 2nsx(s) - 2nsS + \omega_0^2 x(s) = 0 \qquad (14.1.4)$$

Rearranging equation (14.1.4), we may write

$$x(s)(s^2 + 2ns + \omega_0^2) = s(V + 2nS) + s^2 S \qquad (14.1.5)$$

Solving equation (14.1.5) for the displacement $x(s)$ in the Laplace domain, we obtain

$$x(s) = \frac{s(V + 2nS)}{s^2 + 2ns + \omega_0^2} + \frac{s^2 S}{s^2 + 2ns + \omega_0^2} \qquad (14.1.6)$$

Applying the conventional algebraic procedures to the denominators of the fractions in equation (14.1.6) and using different combinations of the values of the natural frequency and the values of the damping factors, we obtain different types of motion of the mass. Hence, rearranging the expression of the denominators in equation (14.1.6), we may write

$$s^2 + 2ns + \omega_0^2 + n^2 - n^2 = (s + n)^2 + \omega^2 \qquad (14.1.7)$$

where

$$\omega^2 = \omega_0^2 - n^2 \qquad (14.1.8)$$

while ω^2 could be positive, equal to zero, or negative. Hence, if $\omega^2 > 0$, then the system performs underdamped vibratory motion; in the case where $\omega^2 = 0$, we have critical damping, which is a non-vibratory motion; and in the case where $\omega^2 < 0$, the system performs overdamped non-vibratory motion. Rearranging equation (14.1.6) by combining it with equation (14.1.7), we arrive at an equation that allows us to obtain the solutions for the above-mentioned three cases:

$$x(s) = \frac{s(V + 2nS)}{(s + n)^2 + \omega^2} + \frac{s^2 S}{(s + n)^2 + \omega^2} \qquad (14.1.9)$$

In the following, we describe each of the three cases.

14.1.1. *Underdamped vibration, $\omega^2 > 0$*

If, in equation (14.1.9), it is assumed that ω^2 is positive, then by applying to this equation Laplace Transform pairs 1, 27, and 35, we invert this equation from the Laplace domain into the time domain and obtain the solution of differential equation of motion (14.1.1) with the initial conditions of motion according to expression (2.1.2):

$$x = \frac{V + 2nS}{\omega}e^{-nt}\sin\omega t + Se^{-nt}\left(\cos\omega t - \frac{n}{\omega}\sin\omega t\right) \qquad (14.1.10)$$

Taking the first derivative from equation (14.1.10), we obtain the equation describing the velocity of the system:

$$\frac{dx}{dt} = (V + 2nS)e^{-nt}\left(\cos\omega t - \frac{n}{\omega}\sin\omega t\right)$$

$$+ \frac{S}{\omega}e^{-nt}[(n^2 - \omega^2)\sin\omega t - 2n\omega\cos\omega t] \qquad (14.1.11)$$

Assuming that in equations (14.1.10) and (14.1.11), $t = 0$, we get that $x = S$ and $\frac{dx}{dt} = V$, as expected according to initial conditions of motion (2.1.2).

The second derivative from equation (14.1.10) yields the acceleration of the system:

$$\frac{d^2x}{dt^2} = \frac{V + 2nS}{\omega}e^{-nt}[(n^2 - \omega^2)\sin\omega t - 2n\omega\cos\omega t]$$

$$+ Se^{-nt}\left[\frac{n}{\omega}(3\omega^2 - n^2)\sin\omega t - (\omega^2 - 3n^2)\cos\omega t\right]$$

$$(14.1.12)$$

Equations (14.1.10)–(14.1.12) represent the basic parameters of motion of the system.

14.1.2. *Critical damping, $\omega^2 = 0$*

In equation (14.1.9), equating the parameter ω^2 to zero, we determine the displacement of the system in the Laplace domain for the current case:

$$x(s) = \frac{s(V + 2nS)}{(s + n)^2} + \frac{s^2S}{(s + n)^2} \qquad (14.1.13)$$

Applying Laplace Transform pairs 1, 25, and 31 to equation (14.1.3), we invert this equation from the Laplace domain into the time domain and obtain the solution of differential equation (14.1.1) with the initial conditions of motion according to expression (2.1.1) for the case of critical damping:

$$x = (V + 2nS)te^{-nt} + S(1 - nt)e^{-nt} \qquad (14.1.14)$$

Taking the first derivative from equation (14.1.14), we obtain the equation for calculating the velocity of the system:

$$\frac{dx}{dt} = (V + 2nS)e^{-nt}(1 - nt) + nSe^{-nt}(nt - 2) \qquad (14.1.15)$$

Supposing that in equation (14.1.14), $t = 0$, we obtain that $x = S$ and $\frac{dx}{dt} = V$, as it should be according to initial conditions of motion (2.1.2).

14.1.3. *Overdamped motion, $\omega^2 < 0$*

In this case, we change in equation (14.1.9) the sign of the parameter ω^2 from positive to negative, and we obtain

$$x(s) = \frac{s(V + 2nS)}{(s + n)^2 - \omega^2} + \frac{s^2 S}{(s + n)^2 - \omega^2} \qquad (14.1.16)$$

Applying Laplace Transform pairs 2, 28, and 36, we invert equation (14.1.16) from the Laplace domain into the time domain and obtain for this case the solution of differential equation of motion (14.1.1) with the initial conditions of motion according to expression (2.1.2):

$$x = \frac{V + 2nS}{\omega}e^{-nt}\sinh \omega t + Se^{-nt}\left(\cosh \omega t - \frac{n}{\omega}\sinh \omega t\right)$$

$$(14.1.17)$$

Taking the first derivative from equation (14.1.17), we determine the velocity of the system:

$$\frac{dx}{dt} = (V + 2nS)e^{-nt}\left(\cosh \omega t - \frac{n}{\omega}\sinh \omega t\right)$$

$$+ \frac{S}{\omega}e^{-nt}[(n^2 + \omega^2)\sinh \omega t - 2n\omega \cosh \omega t] \qquad (14.1.18)$$

Supposing that in equation (14.1.17), $t = 0$, we get that $x = S$ and $\frac{dx}{dt} = V$, as it should be according to the initial conditions of motion expressed by equation (2.1.2).

14.2. Motion of a System Due to Its Initial Displacement and Velocity and a Constant Active Force

In this section, we describe the operational process of a system moving on a horizontal frictionless surface due to its initial displacement and velocity and to a constant active force, while the motion of the system is restricted by a flexible link and a fluid link.

Figure 14.2.1 shows a schematic diagram of the system, the notations in which are self-explanatory.

Based on the schematic diagram shown in Figure 14.2.1 and the considerations mentioned above, we compose the differential equations of motion of the system:

$$m\frac{d^2x}{dt^2} + C\frac{dx}{dt} + Kx = P \tag{14.2.1}$$

Dividing equation (14.2.1) by m, we have

$$\frac{d^2x}{dt^2} + 2n\frac{dx}{dt} + \omega_0^2 x = p \tag{14.2.2}$$

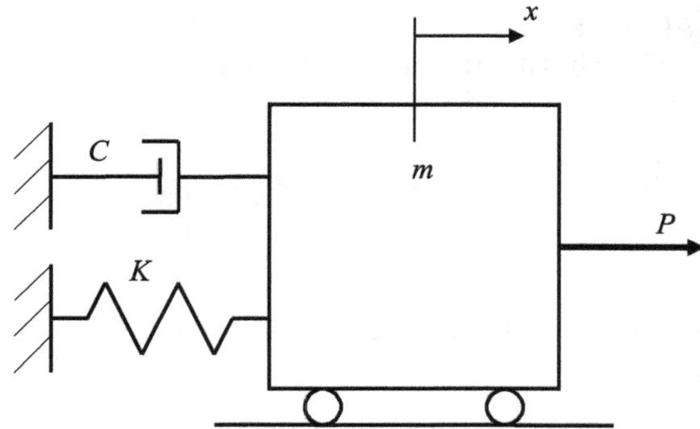

Fig. 14.2.1. Schematic diagram of a system moving on a horizontal frictionless surface while being restricted by a flexible link and a fluid link and subjected to a constant active force.

The initial conditions of motion are assigned according to expression (2.1.2).

Applying Laplace Transform pairs 5, 2, 4, 2, 1, and 2, we convert equation (14.2.2) with the mentioned initial conditions of motion from the time domain into the Laplace domain and obtain the corresponding algebraic equation in the Laplace domain:

Rearranging equation (14.2.3), we may write

$$x(s)(s^2 + 2ns + \omega_0^2) = s(V + 2nS) + s^2 S + p \tag{14.2.4}$$

Solving equation (14.2.4) for the displacement $x(s)$ in the Laplace domain, we obtain

$$x(s) = \frac{s(V + 2nS)}{s^2 + 2ns + \omega_0^2} + \frac{s^2 S}{s^2 + 2ns + \omega_0^2} + \frac{p}{s^2 + 2ns + \omega_0^2} \tag{14.2.5}$$

Combining equation (14.2.5) with equations (14.1.7) and (14.1.8), we may write

$$x(s) = \frac{s(V + 2nS)}{(s + n)^2 + \omega^2} + \frac{s^2 S}{(s + n)^2 + \omega^2} + \frac{p}{(s + n)^2 + \omega^2} \tag{14.2.6}$$

As mentioned above, if $\omega^2 > 0$, then the system performs underdamped vibratory motion; in the case where $\omega^2 = 0$, we have critical damping; and in the case where $\omega^2 < 0$, the system performs overdamped motion.

14.2.1. *Underdamped vibration, $\omega^2 > 0$*

In equation (14.2.6), the parameter ω^2 is supposed to be positive, which allows us to apply to this equation Laplace Transform pairs 1, 27, 35, and 21, and we invert this equation from the Laplace domain into the time domain and obtain the solution of differential equation of motion (14.2.1) at the initial conditions of motion according to expression (2.1.2):

$$x = \frac{V + 2nS}{\omega} e^{-nt} \sin \omega t + S e^{-nt} \left(\cos \omega t - \frac{n}{\omega} \sin \omega t \right)$$

$$+ \frac{p}{\omega^2 + n^2} \left[1 - e^{-nt} \left(\cos \omega t + \frac{n}{\omega} \sin \omega t \right) \right] \tag{14.2.7}$$

Taking the first derivative from equation (14.2.7), we obtain the equation describing the velocity of the system:

$$\frac{dx}{dt} = (V + 2nS)e^{-nt}\left(\cos\omega t - \frac{n}{\omega}\sin\omega t\right) + \frac{S}{\omega}e^{-nt}[(n^2 - \omega^2)\sin\omega t$$

$$-2n\omega\cos\omega t] + \frac{p}{\omega}e^{-nt}\sin\omega t \qquad (14.2.8)$$

Supposing that in equations (14.2.7) and (14.2.8), $t = 0$, we obtain that $x = S$ and $\frac{dx}{dt} = V$, as expected according to initial conditions of motion (2.1.2).

The second derivative from equation (14.2.7) yields the acceleration of the system:

$$\frac{d^2x}{dt^2} = \frac{V + 2nS}{\omega}e^{-nt}[(n^2 - \omega^2)\sin\omega t - 2n\omega\cos\omega t]$$

$$+Se^{-nt}\left[\frac{n}{\omega}(3\omega^2 - n^2)\sin\omega t - (\omega^2 - 3n^2)\cos\omega t\right]$$

$$+pe^{-nt}\left(\cos\omega t - \frac{n}{\omega}\sin\omega t\right) \qquad (14.2.9)$$

Equations (14.2.7)–(14.2.9) represent the basic parameters of motion of the system.

14.2.1.1. *Numerical solution, $\omega^2 > 0$*

The following is a Python program and the associated plot of equation (14.2.7). The graph is plotted in Figure 14.2.2.

```python
from matplotlib.pyplot import plot, show
from numpy import linspace, sin, cos, exp


S = 0.1
V = 0.15
n = 15
p = 10
omega = 60

t = linspace(0, 0.6, 1000)
x = ((V + 2*n*S)/omega)*exp(-n*t)*sin(omega*t) \
    + S*exp(-n*t)*(cos(omega*t) - (n/omega)*sin(omega*t)) \
    + (p/(omega**2 + n**2))*(1 - exp(-n*t)*(cos(omega*t) \
                + (n/omega)*sin(omega*t)))

plot(t, x)
show()
```

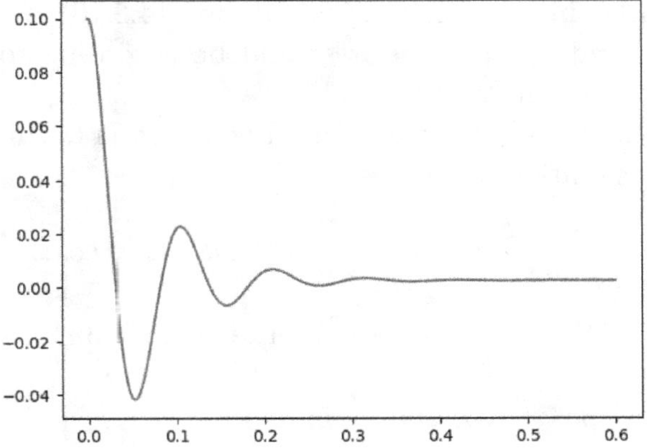

Fig. 14.2.2. Graph of equation (14.2.7) characterizing underdamped vibrations.

14.2.2. *Critical damping, $\omega^2 = 0$*

Taking in equation (14.2.6) that $\omega^2 = 0$, we obtain the expression for the displacement of the system in the Laplace domain for the case of critical damping:

$$x(s) = \frac{s(V + 2nS)}{(s+n)^2} + \frac{s^2 S}{(s+n)^2} + \frac{p}{(s+n)^2}$$

$$(14.2.14)$$

Applying Laplace Transform pairs 1, 25, 31, and 19 to equation (14.2.12), we invert this equation from the Laplace domain into the time domain and obtain the solution of differential equation (14.2.1) with the initial conditions of motion according to expression (2.1.2) for the case of critical damping:

$$x = (V + 2nS)te^{-nt} + S(1 - nt)e^{-nt} + \frac{p}{n^2}[1 - e^{-nt}(1 + nt)]$$

$$(14.2.15)$$

Taking the first derivative from equation (14.2.15), we obtain the equation for calculating the velocity of the system:

$$\frac{dx}{dt} = (V + 2nS)e^{-nt}(1 - nt) + nSe^{-nt}(nt - 2) + pte^{-nt}$$

$$(14.2.16)$$

Supposing that in equations (14.2.15) and (14.2.16), $t = 0$, we get that $x = S$ and $\frac{dx}{dt} = V$, as it should be according to the initial conditions of motion (2.1.2).

Taking the second derivative from equation (14.2.15), we determine the acceleration of the system:

$$\frac{d^2x}{dt^2} = \frac{V + 2nS}{\omega} e^{-nt}[(n^2 - \omega^2)\sin \omega t - 2n\omega \cos \omega t]$$

$$+ Se^{-nt}\left[\frac{n}{\omega}(3\omega^2 - n^2)\sin \omega t - (\omega^2 - 3n^2)\cos \omega t\right]$$

$$+ pe^{-nt}\left(\cos \omega t - \frac{n}{\omega}\sin \omega t\right) \qquad (14.2.17)$$

Hence, equations (14.2.15)–(14.2.17) represent the basic parameters of motion in the case of critical damping.

14.2.2.1. *Numerical solution, $\omega^2 = 0$*

The following is a Python program and the associated plot of equation (14.2.15). The graph is plotted in Figure 14.2.3.

```python
from matplotlib.pyplot import plot, show
from numpy import linspace, sin, cos, exp

S = 0
V = 0.15
n = 15
p = 10

t = linspace(0, 0.6, 1000)
x = (V + 2*n*S)*t*exp(-n*t) + S*(1 - n*t)*exp(-n*t) \
    + (p/(n**2))*(1 - exp(-n*t)*(1+n*t))

plot(t, x)
show()
```

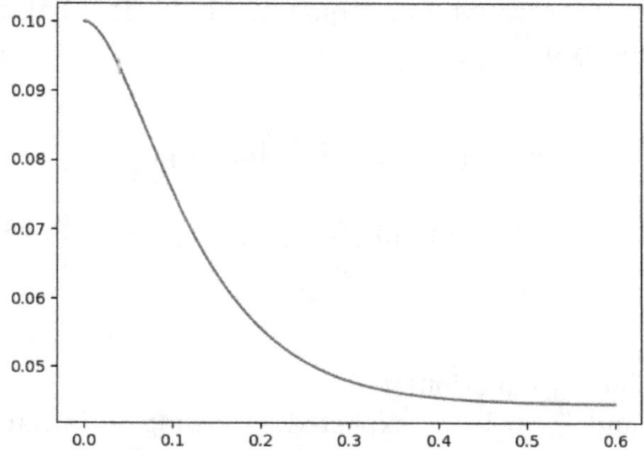

Fig. 14.2.3. Graph of equation (14.2.15) characterizing motion at critical damping.

14.2.3. *Overdamped motion, $\omega^2 < 0$*

In this case, we change in equation (14.2.6) the sign of the parameter ω^2 from positive to negative, and we obtain

$$x(s) = \frac{s(V + 2nS)}{(s+n)^2 - \omega^2} + \frac{s^2 S}{(s+n)^2 - \omega^2} + \frac{p}{(s+n)^2 - \omega^2} \qquad (14.2.22)$$

Applying Laplace Transform pairs 1, 28, 36, and 22, we invert equation (14.2.22) from the Laplace domain into the time domain and obtain for this case the solution of differential equation of motion (14.2.1) with the initial conditions of motion according to expression (2.1.2):

$$x = \frac{V + 2nS}{\omega} e^{-nt} \sinh \omega t + S e^{-nt} \left(\cosh \omega t - \frac{n}{\omega} \sinh \omega t \right)$$
$$+ \frac{p}{n^2 - \omega^2} \left[1 - e^{-nt} \left(\cosh \omega t + \frac{n}{\omega} \sinh \omega t \right) \right] \qquad (14.2.23)$$

Taking the first derivative from equation (14.2.23), we determine the velocity of the system:

$$\frac{dx}{dt} = (V + 2nS)e^{-nt}\left(\cosh \omega t - \frac{n}{\omega}\sinh \omega t\right)$$

$$+ \frac{S}{\omega}e^{-nt}[(n^2 + \omega^2)\sinh \omega t - 2n\omega \cosh \omega t] + \frac{p}{\omega}e^{-nt}\sinh \omega t$$

$$(14.2.24)$$

Supposing that in equations (14.2.23) and (14.2.24), $t = 0$, we get that $x = S$ and $\frac{dx}{dt} = V$, as expected according to initial conditions of motion (2.1.2).

14.2.3.1. *Numerical solution, $\omega^2 < 0$*

The following is a Python program and the associated plot of equation (14.2.23). The graph is plotted in Figure 14.2.4.

```python
from matplotlib.pyplot import plot, show
from numpy import linspace, sinh, cosh, exp

S = 0.1
V = 0.15
n = 80
p = 10
omega = 60

t = linspace(0, 0.6, 1000)
x = ((V + 2*n*S)/omega)*exp(-n*t)*sinh(omega*t) \
    + S*exp(-n*t)*(cosh(omega*t) - (n/omega)*sinh(omega*t)) \
    + (p/(n**2 - omega**2))*(1 - exp(-n*t)*(cosh(omega*t) \
                + (n/omega)*sinh(omega*t)))

plot(t, x)
show()
```

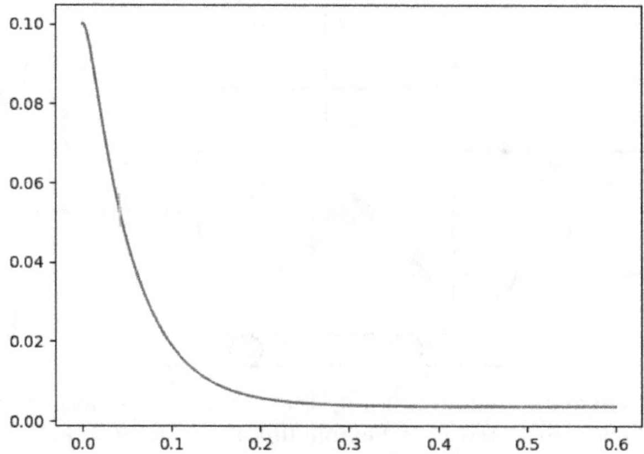

Fig. 14.2.4. Graph of equation (14.2.23) characterizing overdamped motion.

14.3. Motion of a System Due to Its Initial Displacement and Velocity and a Harmonic Force

The operational process of the system moving on a horizontal frictionless surface due to its initial displacement and initial velocity and subjected to a harmonic force is addressed in this section. The motion of the system is restricted by a flexible link and a fluid link.

Figure 14.3.1 shows a schematic diagram of the system that is described above. The notations in the figure are self-explanatory.

Based on the schematic diagram shown in Figure 14.3.1 and the considerations mentioned above, we compose the differential equations of motion of the system:

$$m\frac{d^2x}{dt^2} + C\frac{dx}{dt} + Kx = A\cos\varphi t \tag{14.3.1}$$

Dividing equation (14.3.1) by m, we may write

$$\frac{d^2x}{dt^2} + 2n\frac{dx}{dt} + \omega_0^2 x = a\cos\varphi t \tag{14.3.2}$$

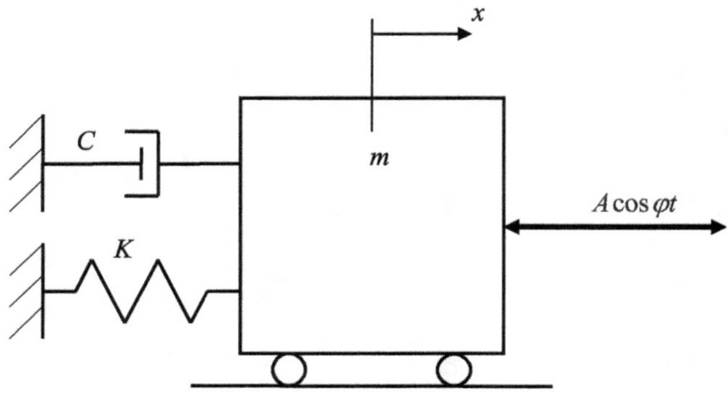

Fig. 14.3.1. Schematic diagram of a system moving on a horizontal frictionless surface while being restricted by a flexible link and a fluid link and subjected to a harmonic force.

The initial conditions of motion are taken according to expression (2.1.2).

Applying Laplace Transform pairs 5, 2, 4, 2, 1, and 29, we convert equation (14.3.2) with the above-mentioned initial conditions of motion from the time domain into the Laplace domain and obtain the corresponding algebraic equation in the Laplace domain:

$$s^2 x(s) - sV - s^2 S + 2nsx(s) - 2nsS + \omega_0^2 x(s) = \frac{as^2}{s^2 + \varphi^2} \qquad (14.3.3)$$

Rearranging equation (14.3.3), we have

$$x(s)(s^2 + 2ns + \omega_0^2) = s(V + 2nS) + s^2 S + \frac{as^2}{s^2 + \varphi^2} \qquad (14.3.4)$$

Solving equation (14.3.4) for the displacement $x(s)$ in the Laplace domain, we obtain

$$x(s) = \frac{s(V + 2ns)}{s^2 + 2ns + \omega_0^2} + \frac{s^2 S}{s^2 + 2ns + \omega_0^2} + \frac{as^2}{(s^2 + 2ns + \omega_0^2)(s^2 + \varphi^2)}$$

$$(14.3.5)$$

Combining equation (14.3.5) with equations (14.1.7) and (14.1.8), we may write

$$x(s) = \frac{s(V + 2nS)}{(s+n)^2 + \omega^2} + \frac{s^2 S}{(s+n)^2 + \omega^2} + \frac{s^2 a}{[(s+n)^2 + \omega^2](s^2 + \varphi^2)}$$

$$(14.3.6)$$

14.3.1. *Underdamped vibration, $\omega^2 > 0$*

Assuming that in equation (14.3.6) the parameter ω^2 is positive, we may apply to this equation Laplace Transform pairs 1, 27, 35, and 68 in order to invert it from the Laplace domain into the time domain and obtain the solution of differential equation of motion (14.3.1) with the initial conditions of motion according to expression (2.1.2):

$$x = \frac{V + 2nS}{\omega} e^{-nt} \sin \omega t + S e^{-nt} \left(\cos \omega t - \frac{n}{\omega} \sin \omega t \right)$$

$$+ \frac{a}{4n^2 \varphi^2 + (\omega^2 + n^2 - \varphi^2)^2} [(\omega^2 + n^2 - \varphi^2)(\cos \varphi t - e^{-nt} \cos \omega t)$$

$$+ 2n\varphi \sin \varphi t - \frac{n}{\omega}(\omega^2 + n^2 + \varphi^2) e^{-nt} \sin \omega t] \qquad (14.3.7)$$

The first derivative from equation (14.3.7) yields the velocity of the system:

$$\frac{dx}{dt} = (V + 2nS)e^{-nt} \left(\cos \omega t - \frac{n}{\omega} \sin \omega t \right)$$

$$+ \frac{S}{\omega} e^{-nt} [(n^2 - \omega^2) \sin \omega t - 2n\omega \cos \omega t]$$

$$+ \frac{a}{4n^2 \varphi^2 + (\omega^2 + n^2 - \varphi^2)^2} \left\{ 2n\varphi^2 (\cos \varphi t - e^{-nt} \cos \omega t) \right.$$

$$- \varphi(\omega^2 + n^2 - \varphi^2) \sin \varphi t$$

$$\left. + \frac{1}{\omega} e^{-nt} [(\omega^2 + n^2)^2 - \varphi^2(\omega^2 - n^2)] \sin \omega t \right\} \qquad (14.3.8)$$

Assuming that in equations (14.3.7) and (14.3.8), $t = 0$, we obtain that $x = S$ and $\frac{dx}{dt} = V$, as it is expected according to initial conditions of motion (2.1.2).

The first derivative from equation (14.3.8) allows us to determine the acceleration of the system:

$$\frac{d^2x}{dt^2} = \frac{V + 2nS}{\omega}e^{-nt}[(n^2 - \omega^2)\sin\omega t - 2n\omega\cos\omega t]$$

$$+ Se^{-nt}\left[\frac{n}{\omega}(3\omega^2 - n^2)\sin\omega t - (\omega^2 - 3n^2)\cos\omega t\right]$$

$$+ \frac{a}{4n^2\varphi^2 + (\omega^2 + n^2 - \varphi^2)^2}\left\{[(n^2 + \omega^2)^2 + \varphi^2(3n^2 - \omega^2)]\right.$$

$$\times e^{-nt}\cos\omega t - \varphi^2(\omega^2 + n^2 - \varphi^2)\cos\varphi t$$

$$\left. -2\varphi^3 n\sin\varphi t - \frac{n}{\omega}[(\omega^2 + n^2)^2 - \varphi^2(n^2 - 3\omega^2)]e^{-nt}\sin\omega t\right\}$$

$$(14.3.9)$$

Equations (14.3.7)–(14.3.9) represent the basic parameters of motion of the system.

14.3.2. *Critical damping, $\omega^2 = 0$*

Taking in equation (14.3.6) that $\omega^2 = 0$, we obtain the expression for the displacement of the system in the Laplace domain:

$$x(s) = \frac{s(V + 2nS)}{(s + n)^2} + \frac{s^2 S}{(s + n)^2} + \frac{s^2 a}{(s + n)^2(s^2 + \varphi^2)} \qquad (14.3.10)$$

Applying Laplace Transform pairs 1, 25, 31, and 71 to equation (14.3.10), we invert this equation from the Laplace domain into the time domain and obtain the solution of differential equation (14.3.1) with the initial conditions of motion according to expression (2.1.2) for the case of critical damping:

$$x = (V + 2nS)te^{-nt} + S(1 - nt)e^{-nt} + \frac{a}{(\varphi^2 + n^2)^2}$$

$$\times\{2n\varphi\sin\varphi t + e^{-nt}[\varphi^2 - n^2 - nt(\varphi^2 + n^2)] - (\varphi^2 - n^2)\cos\varphi t\}$$

$$(14.3.11)$$

Taking the first derivative from equation (14.3.11), we obtain the equation for calculating the velocity of the system:

$$\frac{dx}{dt} = (V - 2nS)e^{-nt}(1 - nt) + nSe^{-nt}(nt - 2)$$

$$+ \frac{a}{(\varphi^2 + n^2)^2}[(\varphi^2 - n^2)(\varphi \sin \varphi t + n^2 t e^{-nt})$$

$$- 2\varphi^2 n(\cos \varphi t - e^{-nt})] \tag{14.3.12}$$

Assuming that in equations (14.3.11) and (14.3.12), $t = 0$, we obtain that $x = S$ and $\frac{dx}{dt} = V$, as it should be according to initial conditions of motion (2.1.2).

14.3.3.   *Overdamped motion, $\omega^2 < 0$*

For this case, we change in equation (14.3.6) the sign of the parameter ω^2 from positive to negative, and we obtain

$$x(s) = \frac{s(V + 2nS)}{(s + n)^2 - \omega^2} + \frac{s^2 S}{(s + n)^2 - \omega^2} + \frac{s^2 a}{[(s + n)^2 - \omega^2](s^2 + \varphi^2)} \tag{14.3.13}$$

Applying Laplace Transform pairs 1, 28, 36, and 69, we invert equation (14.3.13) from the Laplace domain into the time domain and obtain for this case the solution of differential equation of motion (14.3.1) with the initial conditions of motion according to expression (2.1.2):

$$x = \frac{V + 2nS}{\omega}e^{-nt}\sinh \omega t + Se^{-nt}\left(\cosh \omega t - \frac{n}{\omega}\sinh \omega t\right)$$

$$+ \frac{a}{4n^2\varphi^2 + (n^2 + \omega^2 - \varphi^2)^2}[(n^2 + \omega^2 - \varphi^2)$$

$$\times (\cos \varphi t - e^{-nt}\cosh \omega t)$$

$$+ 2n\varphi \sin \varphi t - \frac{n}{\omega}(n^2 + \omega^2 + \varphi^2)e^{-nt}\sinh \omega t] \tag{14.3.14}$$

The first derivative from equation (14.3.14) yields the velocity of the system:

$$\frac{dx}{dt} = (V + 2nS)e^{-nt}\left(\cosh \omega t - \frac{n}{\omega}\sinh \omega t\right) + \frac{S}{\omega}e^{-nt}$$

$$\times \left[(n^2 + \omega^2)\sinh \omega t - 2n\omega \cosh \omega t\right]$$

$$+ \frac{a}{4n^2\varphi^2 + (n^2 - \omega^2 - \varphi^2)^2}$$

$$\times \left\{ 2n\varphi^2(\cos \varphi t - e^{-nt}\cosh \omega t) - \varphi(n^2 - \omega^2 - \varphi^2)\sin \varphi t \right.$$

$$\left. + \frac{1}{\omega}e^{-nt}[(n^2 - \omega^2)^2 + \varphi^2(\omega^2 + n^2)]\sinh \omega t \right\} \qquad (14.3.15)$$

Assuming that in equations (14.3.14) and (14.3.15), $t = 0$, we obtain that $x = S$ and $\frac{dx}{dt} = V$, as expected according to initial conditions of motion (2.1.2).

14.4. Motion of a System Due to Its Initial Displacement and Initial Velocity as Well as to a Constant Active Force and a Harmonic Force

Consider the operational process of a system moving on a horizontal frictionless surface due to its initial displacement and velocity and to a constant active force and a harmonic force. The motion of the system is restricted by a flexible link and a fluid link.

Figure 14.4.1 shows a schematic diagram of the system. The notations in the figure are self-explanatory.

Based on the schematic diagram shown in Figure 14.4.1 and the considerations mentioned above, we compose the differential equation of motion of the system:

$$m\frac{d^2x}{dt^2} + C\frac{dx}{dt} + Kx = P + A\cos \varphi t \qquad (14.4.1)$$

Dividing equation (14.4.1) by m, we may write

$$\frac{d^2x}{dt^2} + 2n\frac{dx}{dt} + \omega_0^2 x = p + a\cos \varphi t \qquad (14.4.2)$$

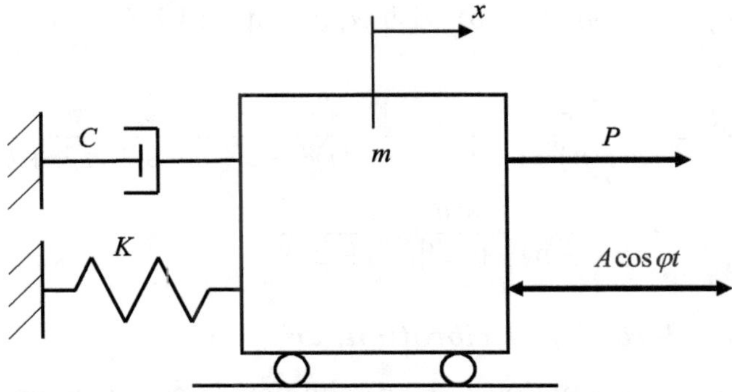

Fig. 14.4.1. Schematic diagram of a system moving on a horizontal frictionless surface while being restricted by a flexible link and a fluid link and subjected to a constant active force and a harmonic force.

The initial conditions of motion are taken according to expression (2.1.2).

Applying Laplace Transform pairs 5, 2, 4, 2, 1, 2, and 29, we convert equation (14.4.2) with the initial conditions of motion according to expression (2.1.2) from the time domain into the Laplace domain and obtain the corresponding algebraic equation in the Laplace domain:

$$s^2 x(s) - sV - s^2 S + 2nsx(s) - 2nsS + \omega_0^2 x(s) = p + \frac{as^2}{s^2 + \varphi^2} \qquad (14.4.3)$$

Rearranging equation (8.2.3), we have

$$x(s)(s^2 + 2ns + \omega_0^2) = s(V + 2nS) + s^2 S + p + \frac{as^2}{s^2 + \varphi^2} \qquad (14.4.4)$$

Solving equation (14.4.4) for the displacement $x(s)$ in the Laplace domain, we obtain

$$x(s) = \frac{s(V + 2nS)}{s^2 + 2ns + \omega_0^2} + \frac{s^2 S}{s^2 + 2ns + \omega_0^2} + \frac{p}{s^2 + 2ns + \omega_0^2}$$

$$+ \frac{as^2}{(s^2 + 2ns + \omega_0^2)(s^2 + \varphi^2)} \qquad (14.4.5)$$

Combining equation (14.4.5) with equations (14.1.7) and (14.1.8), we obtain

$$x(s) = \frac{s(V + 2nS)}{(s + n)^2 + \omega^2} + \frac{s^2 S}{(s + n)^2 + \omega^2} + \frac{p}{(s + n)^2 + \omega^2}$$

$$+ \frac{s^2 a}{[(s + n)^2 + \omega^2](s^2 + \varphi^2)} \tag{14.4.6}$$

14.4.1. *Underdamped vibration, $\omega^2 > 0$*

Realizing that in equation (14.4.6) the parameter ω^2 is positive, we may apply to this equation Laplace Transform pairs 1, 27, 35, 21, and 68 in order to invert it from the Laplace domain into the time domain and obtain the solution of differential equation of motion (14.4.1) with the initial conditions of motion according to expression (2.1.2):

$$x = \frac{V + 2nS}{\omega} e^{-nt} \sin \omega t + S e^{-nt} \left(\cos \omega t - \frac{n}{\omega} \sin \omega t \right)$$

$$+ \frac{p}{\omega^2 + n^2} \left[1 - e^{-nt} \left(\cos \omega t + \frac{n}{\omega} \sin \omega t \right) \right]$$

$$+ \frac{a}{4n^2 \varphi^2 + (\omega^2 + n^2 - \varphi^2)^2} \left[(\omega^2 + n^2 - \varphi^2)(\cos \varphi t - e^{-nt} \cos \omega t) \right.$$

$$\left. + 2n\varphi \sin \varphi t - \frac{n}{\omega}(\omega^2 + n^2 + \varphi^2) e^{-nt} \sin \omega t \right] \tag{14.4.7}$$

The first derivative from equation (14.4.7) yields the velocity of the system:

$$\frac{dx}{dt} = (V + 2nS) e^{-nt} \left(\cos \omega t - \frac{n}{\omega} \sin \omega t \right)$$

$$+ \frac{S}{\omega} e^{-nt} [(n^2 - \omega^2) \sin \omega t - 2n\omega \cos \omega t] + \frac{p}{\omega} e^{-nt} \sin \omega t$$

$$+ \frac{a}{4n^2 \varphi^2 + (\omega^2 + n^2 - \varphi^2)^2} \left\{ 2n\varphi^2 (\cos \varphi t - e^{-nt} \cos \omega t) \right.$$

$$- \varphi(\omega^2 + n^2 - \varphi^2) \sin \varphi t + \frac{1}{\omega} e^{-nt} [(\omega^2 + n^2)^2$$

$$\left. - \varphi^2(\omega^2 - n^2)] \sin \omega t \right\} \tag{14.4.8}$$

Assuming that in equations (14.4.7) and (14.4.8), $t = 0$, we obtain that $x = S$ and $\frac{dx}{dt} = V$, as expected according to initial conditions of motion (2.1.2).

The first derivative from equation (14.4.8) yields the acceleration of the system:

$$
\begin{aligned}
\frac{d^2 x}{dt^2} &= \frac{V + 2nS}{\omega} e^{-nt}[(n^2 - \omega^2)\sin\omega t - 2n\omega\cos\omega t] \\[2mm]
&\quad + Se^{-nt}\left[\frac{n}{\omega}(3\omega^2 - n^2)\sin\omega t - (\omega^2 - 3n^2)\cos\omega t\right] \\[2mm]
&\quad + pe^{-nt}\left(\cos\omega t - \frac{n}{\omega}\sin\omega t\right) + \frac{a}{4n^2\varphi^2 + (\omega^2 + n^2 - \varphi^2)^2} \\[2mm]
&\quad \times \Bigg\{ [(n^2 + \omega^2)^2 + \varphi^2(3n^2 - \omega^2)]e^{-nt}\cos\omega t \\[2mm]
&\quad - \varphi^2(\omega^2 + n^2 - \varphi^2)\cos\varphi t - 2\varphi^3 n\sin\varphi t - \frac{n}{\omega}[(\omega^2 + n^2)^2 \\[2mm]
&\quad - \varphi^2(n^2 - 3\omega^2)]e^{-nt}\sin\omega t \Bigg\}
\end{aligned}
\tag{14.4.9}
$$

Equations (14.4.7)–(14.4.9) represent the basic parameters of motion of the system.

14.4.2. *Critical damping, $\omega^2 = 0$*

In equation (14.4.6), equating the parameter ω^2 to zero, we obtain the expression for computing the displacement of the system in the Laplace domain:

$$
x(s) = \frac{s(V + 2nS)}{(s + n)^2} + \frac{s^2 S}{(s + n)^2} + \frac{p}{(s + n)^2} + \frac{s^2 a}{(s + n)^2(s^2 + \varphi^2)}
\tag{14.4.10}
$$

Applying Laplace Transform pairs 1, 25, 31, 19, and 71 to equation (14.4.10), we invert this equation from the Laplace domain into the time domain and obtain the solution of differential equation (14.4.1) with the initial conditions of motion according to expression (2.1.2) for the case of critical damping:

$$x = (V + 2nS)te^{-nt} + S(1 - nt)e^{-nt} + \frac{p}{n^2}[1 - e^{-nt}(1 + nt)]$$

$$+ \frac{a}{(\varphi^2 + n^2)^2}\{2n\varphi\sin\varphi t + e^{-nt}[\varphi^2 - n^2 - nt(\varphi^2 + n^2)]$$

$$- (\varphi^2 - n^2)\cos\varphi t\} \tag{14.4.11}$$

Taking the first derivative from equation (14.4.11), we determine the velocity of the system:

$$\frac{dx}{dt} = (V + 2nS)e^{-nt}(1 - nt) + nSe^{-nt}(nt - 2) + pte^{-nt}$$

$$+ \frac{a}{(\varphi^2 + n^2)^2}[(\varphi^2 - n^2)(\varphi\sin\varphi t + n^2te^{-nt})$$

$$- 2\varphi^2 n(\cos\varphi t - e^{-nt})] \tag{14.4.12}$$

Supposing that in equations (14.4.11) and (14.4.12), $t = 0$, we get that $x = S$ and $\frac{dx}{dt} = V$, as it should be according to initial conditions of motion (2.1.2)

14.4.3. *Overdamped motion, $\omega^2 < 0$*

For this case, we change in equation (14.4.6) the sign of the parameter ω^2 from positive to negative, and we obtain

$$x(s) = \frac{s(V + 2nS)}{(s + n)^2 - \omega^2} + \frac{s^2 S}{(s + n)^2 - \omega^2} + \frac{p}{(s + n)^2 - \omega^2}$$

$$+ \frac{s^2 a}{[(s + n)^2 - \omega^2](s^2 + \varphi^2)} \tag{14.4.13}$$

Applying Laplace Transform pairs 1, 28, 36, and 69, we invert equation (14.4.13) from the Laplace domain into the time domain and obtain for this case the solution of differential equation of motion (14.4.1) with the initial conditions of motion according to expression (2.1.2):

$$x = \frac{V + 2nS}{\omega} e^{-nt} \sinh \omega t + S e^{-nt} \left(\cosh \omega t - \frac{n}{\omega} \sinh \omega t \right)$$

$$+ \frac{p}{n^2 - \omega^2} \left[1 - e^{-nt} \left(\cosh \omega t - \frac{n}{\omega} \sinh \omega t \right) \right]$$

$$+ \frac{a}{4n^2\varphi^2 + (n^2 - \omega^2 - \varphi^2)^2} \left[(n^2 - \omega^2 - \varphi^2) \right.$$

$$\times (\cos \varphi t - e^{-nt} \cosh \omega t) + 2n\varphi \sin \varphi t$$

$$\left. - \frac{n}{\omega}(n^2 - \omega^2 + \varphi^2) e^{-nt} \sinh \omega t \right] \tag{14.4.14}$$

The first derivative from equation (14.4.14) yields the velocity of the system:

$$\frac{dx}{dt} = (V + 2nS) e^{-nt} \left(\cosh \omega t - \frac{n}{\omega} \sinh \omega t \right)$$

$$+ \frac{S}{\omega} e^{-nt} [(n^2 - \omega^2) \sinh \omega t - 2n\omega \cosh \omega t] + \frac{p}{\omega} e^{-nt} \sinh \omega t$$

$$+ \frac{a}{4n^2\varphi^2 + (n^2 - \omega^2 - \varphi^2)^2} \left\{ 2n\varphi^2 (\cos \varphi t - e^{-nt} \cosh \omega t) \right.$$

$$- \varphi(n^2 - \omega^2 - \varphi^2) \sin \varphi t + \frac{1}{\omega} e^{-nt} [(n^2 - \omega^2)^2$$

$$\left. + \varphi^2 (\omega^2 + n^2)] \sinh \omega t \right\} \tag{14.4.15}$$

If in equations (14.4.14) and (14.4.15), $t = 0$, we obtain that $x = S$ and $\frac{dx}{dt} = V$, as expected according to initial conditions of motion (2.1.2).

MOTION OF A SYSTEM ON A HORIZONTAL FRICTIONLESS SURFACE WHILE BEING RESTRICTED BY FLEXIBLE AND FLUID LINKS AND SUBJECTED TO A CONSTANT RESISTING FORCE

This chapter deals with a system that is subjected to a resisting force and is moving due to its initial conditions of motion.

15.1. Motion of a System Due to Its Initial Displacement and Velocity

The operational process described in the following is characterized by the motion of a system on a horizontal frictionless surface due to the initial conditions of motion. The system is subjected to a constant resisting force and is restricted by flexible and fluid links that connect the mass of the system to a non-movable support.

Figure 15.1.1 shows a schematic diagram of the system, the notations in which are self-explanatory.

Based on the schematic diagram shown in Figure 15.1.1 and the considerations mentioned above, we compose the differential

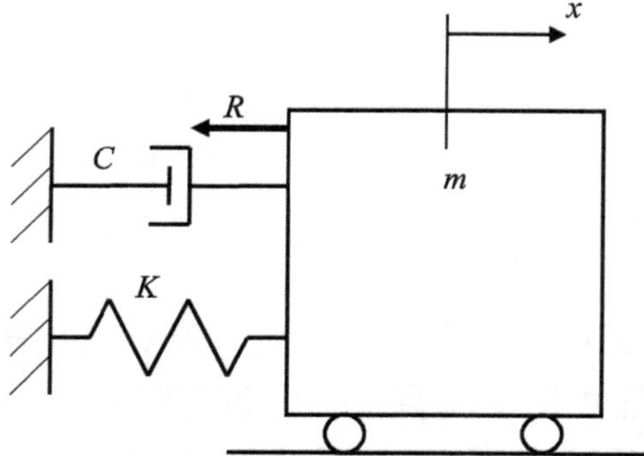

Fig. 15.1.1. Schematic diagram of a system moving on a horizontal frictionless surface while being restricted by flexible and fluid links and subjected to a constant resisting force.

equations of motion of the system:

$$m\frac{d^2x}{dt^2} + C\frac{dx}{dt} + Kx + R = 0 \tag{15.1.1}$$

Dividing equation (15.1.1) by m, we have

$$\frac{d^2x}{dt^2} + 2n\frac{dx}{dt} + \omega_0^2 x + r = 0 \tag{15.1.2}$$

The initial conditions of motion are assigned according to expression (2.1.2).

Applying Laplace Transform pairs 5, 2, 4, 2, 1, and 2, we convert equation (15.1.2) with the mentioned initial conditions of motion from the time domain into the Laplace domain and obtain the corresponding algebraic equation in the Laplace domain:

$$s^2x(s) - sV - s^2S + 2nsx(s) - 2nsS + \omega_0^2 x(s) + r = 0 \tag{15.1.3}$$

Rearranging equation (15.1.3), we have

$$x(s)(s^2 + 2ns + \omega_0^2) = s(V + 2nS) + s^2S - r \tag{15.1.4}$$

Solving equation (15.1.4) for the displacement $x(s)$ in the Laplace domain, we obtain

$$x(s) = \frac{s(V + 2nS)}{s^2 + 2ns + \omega_0^2} + \frac{s^2S}{s^2 + 2ns + \omega_0^2} - \frac{r}{s^2 + 2ns + \omega_0^2} \tag{15.1.5}$$

Combining equations (15.1.4) with equations (15.1.7) and (15.1.8), we obtain

$$x(s) = \frac{s(V + 2nS)}{(s+n)^2 + \omega^2} + \frac{s^2 S}{(s+n)^2 + \omega^2} - \frac{r}{(s+n)^2 + \omega^2} \qquad (15.1.6)$$

15.1.1. *Underdamped vibration, $\omega^2 > 0$*

It is assumed that in equation (15.1.6), ω^2 is positive; therefore, applying to this equation Laplace Transform pairs 1, 27, 35, and 21, we invert this equation from the Laplace domain into the time domain and obtain the solution of differential equation of motion (15.1.1) at the initial conditions of motion according to expression (2.1.2):

$$x = \frac{V + 2nS}{\omega} e^{-nt} \sin \omega t + S e^{-nt} \left(\cos \omega t - \frac{n}{\omega} \sin \omega t \right)$$
$$- \frac{r}{\omega^2 + n^2} \left[1 - e^{-nt} \left(\cos \omega t + \frac{n}{\omega} \sin \omega t \right) \right] \qquad (15.1.7)$$

The first derivative from equation (15.1.7) yields the velocity of the system:

$$\frac{dx}{dt} = (V + 2nS) e^{-nt} \left(\cos \omega t - \frac{n}{\omega} e^{-nt} \sin \omega t \right)$$
$$+ \frac{S}{\omega} e^{-nt} [(n^2 - \omega^2) \sin \omega t - 2n\omega \cos \omega t] - \frac{r}{\omega} e^{-nt} \sin \omega t$$
$$(15.1.8)$$

Assuming that in equations (15.1.7) and (15.1.8), $t = 0$, then we obtain that $x = S$ and $\frac{dx}{dt} = V$, as expected according to initial conditions of motion (2.1.2).

The second derivative from equation (15.1.7) represents the expression for calculating the acceleration of the system:

$$\frac{d^2 x}{dt^2} = \frac{V + 2nS}{\omega} e^{-nt} [(n^2 - \omega^2) \sin \omega t - 2n\omega \cos \omega t]$$
$$+ S e^{-nt} \left[\frac{n}{\omega} (3\omega^2 - n^2) \sin \omega t - (\omega^2 - 3n^2) \cos \omega t \right]$$
$$- r e^{-nt} \left(\cos \omega t - \frac{n}{\omega} \sin \omega t \right) \qquad (15.1.9)$$

Equations (15.1.7)–(15.1.9) represent the basic parameters of motion of the system.

15.1.2. *Critical damping, $\omega^2 = 0$*

In equation (15.1.6), equating the parameter ω^2 to zero, we determine the displacement of the system in the Laplace domain for the current case:

$$x(s) = \frac{s(V + 2nS)}{(s + n)^2} + \frac{s^2 S}{(s + n)^2} - \frac{r}{(s + n)^2} \qquad (15.1.10)$$

Applying Laplace Transform pairs 1, 25, 31, and 19 to equation (15.1.10), we invert this equation from the Laplace domain into the time domain and obtain the solution of differential equation (15.1.1) at the initial conditions of motion according to expression (2.1.2) for the case of critical damping:

$$x = (V + 2nS)te^{-nt} + S(1 - nt)e^{-nt} - \frac{r}{n^2}[1 - (1 + nt)e^{-nt}]$$

$$(15.1.11)$$

Taking the first derivative from equation (15.1.11), we obtain the equation for calculating the velocity of the system:

$$\frac{dx}{dt} = (V + 2nS)e^{-nt}(1 - nt) + nSe^{-nt}(nt - 2) - rte^{-nt} \qquad (15.1.12)$$

Assuming that in equations (15.1.11) and (15.1.12), $t = 0$, we obtain that $x = S$ and $\frac{dx}{dt} = V$, as it should be according to the initial conditions of motion (2.1.2).

15.1.3. *Overdamped motion, $\omega^2 < 0$*

For this case, we change in equation (15.1.6) the sign of parameter ω^2 from positive to negative, and we obtain

$$x(s) = \frac{s(V + 2nS)}{(s + n)^2 - \omega^2} + \frac{s^2 S}{(s + n)^2 - \omega^2} - \frac{r}{(s + n)^2 - \omega^2} \qquad (15.1.13)$$

Applying Laplace Transform pairs 1, 28, 36, and 22, we invert equation (15.1.13) from the Laplace domain into the time domain

and obtain for this case the solution of differential equation of motion (15.1.1) with the initial conditions of motion according to expression (2.1.2):

$$x = \frac{V + 2nS}{\omega} e^{-nt} \sinh \omega t + Se^{-nt} \left(\cosh \omega t - \frac{n}{\omega} \sinh \omega t \right)$$

$$- \frac{r}{n^2 - \omega^2} \left[1 - e^{-nt} \left(\cosh \omega t + \frac{n}{\omega} \sinh \omega t \right) \right] \qquad (15.1.14)$$

Taking the first derivative from equation (15.1.14), we determine the velocity of the system:

$$\frac{dx}{dt} = (V + 2nS)e^{-nt} \left(\cosh \omega t - \frac{n}{\omega} \sinh \omega t \right)$$

$$+ \frac{S}{\omega} e^{-nt} [(n^2 + \omega^2) \sinh \omega t - 2n\omega \cosh \omega t]$$

$$- \frac{r}{\omega} e^{-nt} \sinh \omega t \qquad (15.1.15)$$

Supposing that in equations (15.1.14) and (15.1.15), $t = 0$, we obtain that $x = S$ and $\frac{dx}{dt} = V$, as it should be according to the initial conditions of motion expressed by (2.1.2).

15.2. Motion of a System Due to Its Initial Displacement and Velocity and to a Constant Active Force

The operational process describes a system moving on a horizontal frictionless surface due to its initial displacement and velocity and to a constant active force while being subjected to the action of a constant resisting force.

Figure 15.2.1 shows a schematic diagram of the system, the notations in which are self-explanatory.

Based on the schematic diagram shown in Figure 15.2.1 and the considerations mentioned above, we compose the differential equations of motion of the system:

$$m\frac{d^2 x}{dt^2} + C\frac{dx}{dt} + Kx + R = P \qquad (15.2.1)$$

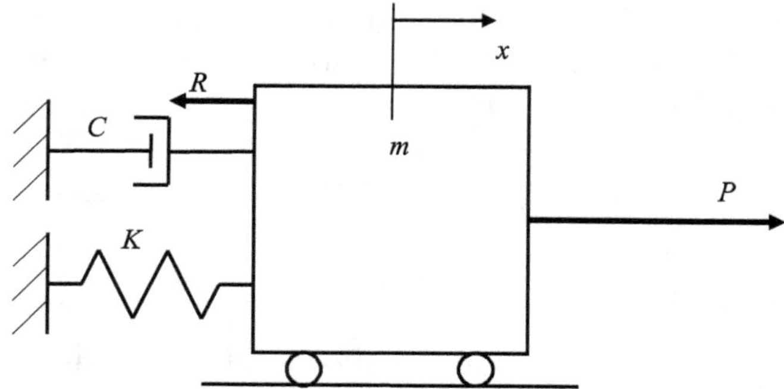

Fig. 15.2.1. Schematic diagram of a system moving on a horizontal frictionless surface while being restricted by a flexible link and by a fluid link and subjected to a constant resisting force and a constant active force.

Dividing equation (15.2.1) by m, we have

$$\frac{d^2x}{dt^2} + 2n\frac{dx}{dt} + \omega_0^2 x + r = p \tag{15.2.2}$$

The initial conditions of motion are assigned according to expression (2.1.2).

Applying Laplace Transform pairs 5, 2, 4, 2, 1, and 2, we convert equation (15.2.2) with the mentioned initial conditions of motion from the time domain into the Laplace domain and obtain the corresponding algebraic equation in the Laplace domain:

$$s^2 x(s) - sV - s^2 S + 2nsx(s) - 2nsS + \omega_0^2 x(s) = p - r \tag{15.2.3}$$

Rearranging equation (15.2.3), we may write

$$x(s)(s^2 + 2ns + \omega_0^2) = s(V + 2ns) + s^2 S + p - r \tag{15.2.4}$$

Solving equation (15.2.4) for the displacement $x(s)$ in the Laplace domain, we obtain

$$x(s) = \frac{s(V + 2ns)}{s^2 + 2ns + \omega_0^2} + \frac{s^2 S}{s^2 + 2ns + \omega_0^2} + \frac{p - r}{s^2 + 2ns + \omega_0^2} \tag{15.2.5}$$

Combining equation (15.2.5) with equation (14.1.7) and (14.1.8), we obtain

$$x(s) = \frac{s(V + 2nS)}{(s + n)^2 + \omega^2} + \frac{s^2 S}{(s + n)^2 + \omega^2} + \frac{p - r}{(s + n)^2 + \omega^2} \tag{15.2.6}$$

15.2.1. *Underdamped vibration, $\omega^2 > 0$*

We assume that in equation (15.2.6), the parameter ω^2 is positive; therefore, it is allowable to apply to this equation Laplace Transform pairs 1, 27, 35, and 21 and invert it from the Laplace domain into the time domain and obtain the solution of differential equation of motion (15.2.1) at the initial conditions of motion according to expression (2.1.2):

$$x = \frac{V + 2nS}{\omega}e^{-nt}\sin\omega t + Se^{-nt}\left(\cos\omega t - \frac{n}{\omega}\sin\omega t\right)$$
$$+\frac{p - r}{\omega^2 + n^2}\left[1 - e^{-nt}\left(\cos\omega t + \frac{n}{\omega}\sin\omega t\right)\right] \tag{15.2.7}$$

Taking the first derivative from equation (15.2.7), we obtain the equation describing the velocity of the system:

$$\frac{dx}{dt} = (V + 2nS)e^{-nt}\left(\cos\omega t - \frac{n}{\omega}\sin\omega t\right)$$
$$+\frac{S}{\omega}e^{-nt}[(n^2 - \omega^2)\sin\omega t - 2n\omega\cos\omega t]$$
$$+\frac{p - r}{\omega}e^{-nt}\sin\omega t \tag{15.2.8}$$

Supposing that in equations (15.2.7) and (15.2.8), $t = 0$, we obtain that $x = S$ and $\frac{dx}{dt} = V$, as expected according to initial conditions of motion (2.1.2).

The second derivative from equation (15.2.7) yields the acceleration of the system:

$$\frac{d^2x}{dt^2} = \frac{V + 2nS}{\omega}e^{-nt}[(n^2 - \omega^2)\sin\omega t - 2n\omega\cos\omega t]$$
$$+Se^{-nt}\left[\frac{n}{\omega}(3\omega^2 - n^2)\sin\omega t - (\omega^2 - 3n^2)\cos\omega t\right]$$
$$+(p - r)e^{-nt}\left(\cos\omega t - \frac{n}{\omega}\sin\omega t\right) \tag{15.2.9}$$

Equations (15.2.7)–(15.2.9) represent the basic parameters of motion of the system.

15.2.1.1. *Numerical solution, $\omega^2 > 0$*

The following is a Python program and the associated graph of equation (15.2.7). The graph is plotted in Figure 15.2.2.

```python
from matplotlib.pyplot import plot, show
from numpy import linspace, sin, cos, exp


S=0.1
V=0.15
n=8
p=10
r=5
omega=60


t = linspace(0, 0.6, 1000)
x = ((V + 2*n*S)/omega)*exp(-n*t)*sin(omega*t) \
    + S*exp(-n*t)*(cos(omega*t) - (n/omega)*sin(omega*t)) \
    + ((p - r)/(omega**2 + n**2))*(1 - exp(-n*t)*(cos(omega*t \
                        + (n/omega)*sin(omega*t))))


plot(t, x)
show()
```

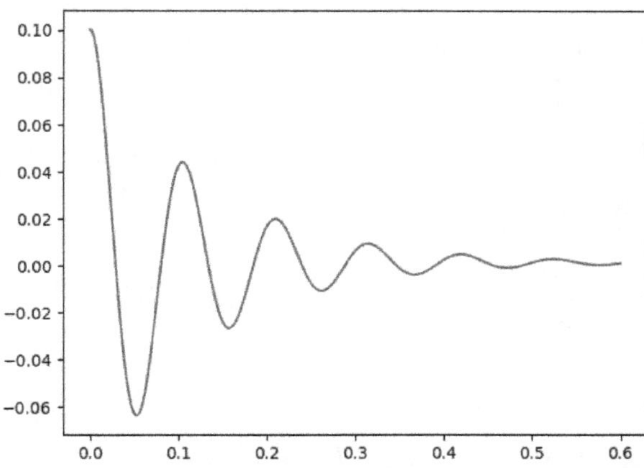

Fig. 15.2.2. Graph of equation (15.2.7) characterizing underdamped motion.

15.2.2. *Critical damping, $\omega^2 = 0$*

Taking in equation (15.2.6) that $\omega^2 = 0$, we obtain the expression for the displacement of the system in the Laplace domain for the case of critical damping:

$$x(s) = \frac{s(V + 2nS)}{(s+n)^2} + \frac{s^2 S}{(s+n)^2} + \frac{p-r}{(s+n)^2} \tag{15.2.11}$$

Applying Laplace Transform pairs 1, 25, 31, and 19 to equation (15.2.11), we invert this equation from the Laplace domain into the time domain and obtain the solution of differential equation (15.2.1) with the initial conditions of motion according to expression (2.1.2) for the case of critical damping:

$$x = (V + 2nS)te^{-nt} + S(1 - nt)e^{-nt} + \frac{p-r}{n^2}[1 - e^{-nt}(1 + nt)] \tag{15.2.12}$$

Taking the first derivative from equation (15.2.12), we obtain the equation for calculating the velocity of the system:

$$\frac{dx}{dt} = (V + 2nS)\epsilon^{-nt}(1 - nt) + nSe^{-nt}(nt - 2) + (p - r)te^{-nt} \tag{15.2.13}$$

Supposing that in equations (15.2.12) and (15.2.13), $t = 0$, we obtain that $x = S$ and $\frac{dx}{dt} = V$, as it should be according to the initial conditions of motion (2.1.2).

15.2.3. *Overdamped motion, $\omega^2 < 0$*

In this case, we change in equation (15.2.6) the sign of the parameter ω^2 from positive to negative, and we obtain

$$x(s) = \frac{s(V + 2nS)}{(s+n)^2 - \omega^2} + \frac{s^2 S}{(s+n)^2 - \omega^2} + \frac{p-r}{(s+n)^2 - \omega^2} \tag{15.2.14}$$

Applying Laplace Transform pairs 1, 28, 36, and 22, we invert equation (15.2.14) from the Laplace domain into the time domain

and obtain for this case the solution of differential equation of motion (15.2.1) with the initial conditions of motion according to expression (2.1.2):

$$x = \frac{V + 2nS}{\omega} e^{-nt} \sinh \omega t + S e^{-nt} \left(\cosh \omega t - \frac{n}{\omega} \sinh \omega t \right)$$
$$+ \frac{p - r}{n^2 - \omega^2} \left[1 - e^{-nt} \left(\cosh \omega t - \frac{n}{\omega} \sinh \omega t \right) \right] \qquad (15.2.15)$$

Taking the first derivative from equation (15.2.15), we determine the velocity of the system:

$$\frac{dx}{dt} = (V + 2nS) e^{-nt} \left(\cosh \omega t - \frac{n}{\omega} \sinh \omega t \right) + \frac{S}{\omega} e^{-nt} [(n^2 + \omega^2)$$
$$\times \sinh \omega t - 2n\omega \cosh \omega t] + \frac{p - r}{\omega} e^{-nt} \sinh \omega t \qquad (15.2.16)$$

Supposing that in equations (15.2.15) and (15.2.16), $t = 0$, we obtain that $x = S$ and $\frac{dx}{dt} = V$, as expected according to initial conditions of motion (2.1.2).

15.3. Motion of a System Due to Its Initial Displacement and Velocity and a Harmonic Force

The operational process of the system subjected to a constant resisting force while moving on a horizontal frictionless surface due to its initial displacement and velocity and to a harmonic force is addressed in this section. The motion of the system is restricted by a flexible link and a fluid link.

Figure 15.3.1 shows a schematic diagram of the system, the notations in which are self-explanatory.

Based on the schematic diagram shown in Figure 15.3.1 and the considerations mentioned above, we compose the differential equations of motion of the system:

$$m \frac{d^2 x}{dt^2} + C \frac{dx}{dt} + K x + R = A \cos \varphi t \qquad (15.3.1)$$

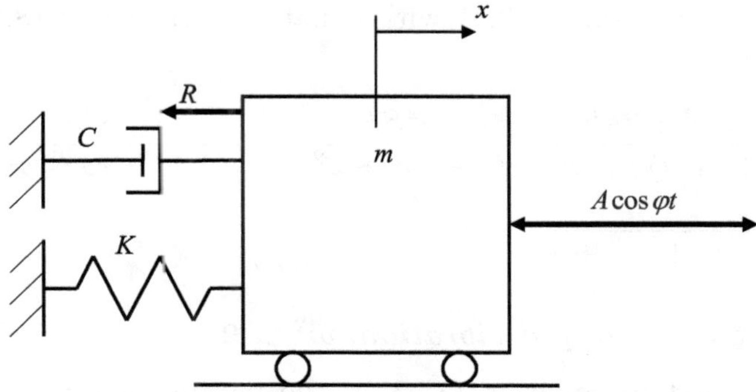

Fig. 15.3.1. Motion of system restricted by a flexible link and a fluid link and subjected to a constant resisting force and a harmonic force.

Dividing equation (15.3.1) by m, we may write

$$\frac{d^2x}{dt^2} + 2n\frac{dx}{dt} + \omega_0^2 x + r = a \cos \varphi t \qquad (15.3.2)$$

The initial conditions of motion are taken according to expression (2.1.2).

Applying Laplace Transform pairs 5, 2, 4, 2, 1, 2, and 29, we convert equation (15.3.2) with the above-mentioned initial conditions of motion from the time domain into the Laplace domain and obtain the corresponding algebraic equation in the Laplace domain:

$$s^2 x(s) - sV - s^2 S + 2ns x(s) - 2ns S + \omega_0^2 x(s) + r = \frac{as^2}{s^2 + \varphi^2}$$

$$(15.3.3)$$

Rearranging equation (15.3.3), we have

$$x(s)(s^2 + 2ns + \omega_0^2) = s(V + 2nS) + s^2 S + \frac{as^2}{s^2 + \varphi^2} - r \qquad (15.3.4)$$

Solving equation (15.3.4) for the displacement $x(s)$ in the Laplace domain, we obtain

$$x(s) = \frac{s(V + 2nS)}{s^2 + 2ns + \omega_0^2} + \frac{s^2 S}{s^2 + 2ns + \omega_0^2} + \frac{as^2}{(s^2 + 2ns + \omega_0^2)(s^2 + \varphi^2)}$$

$$- \frac{r}{s^2 + 2ns + \omega_0^2} \qquad (15.3.5)$$

Combining equation (15.3.5) with equation (8.1.7) and (8.1.8), we may write

$$x(s) = \frac{s(V + 2nS)}{(s+n)^2 + \omega^2} + \frac{s^2 S}{(s+n)^2 + \omega^2} + \frac{s^2 a}{[(s+n)^2 + \omega^2](s^2 + \varphi^2)}$$

$$- \frac{r}{(s+n)^2 + \omega^2} \tag{15.3.6}$$

15.3.1. *Underdamped vibration, $\omega^2 > 0$*

Assuming that in equation (15.3.6), the parameter ω^2 is positive, we may apply to this equation Laplace Transform pairs 1, 27, 35, 68, and 21 in order to invert this equation from the Laplace domain into the time domain and obtain the solution of differential equation of motion (15.3.1) at the initial conditions of motion according to expression (2.1.2):

$$x = \frac{V + 2nS}{\omega} e^{-nt} \sin \omega t + S e^{-nt} \left(\cos \omega t - \frac{n}{\omega} \sin \omega t \right)$$

$$+ \frac{a}{4n^2 \varphi^2 + (\omega^2 + n^2 - \varphi^2)^2} \Big[(\omega^2 + n^2 - \varphi^2)(\cos \varphi t$$

$$- e^{-nt} \cos \omega t) + 2n\varphi \sin \varphi t - \frac{n}{\omega}(\omega^2 + n^2 + \varphi^2) e^{-nt} \sin \omega t \Big]$$

$$- \frac{r}{\omega^2 + n^2} \left[1 - e^{-nt} \left(\cos \omega t + \frac{n}{\omega} \sin \omega t \right) \right] \tag{15.3.7}$$

The first derivative from equation (15.3.7) yields the velocity of the system:

$$\frac{dx}{dt} = (V + 2nS) e^{-nt} \left(\cos \omega t - \frac{n}{\omega} \sin \omega t \right) + \frac{S}{\omega} e^{-nt} [(n^2 - \omega^2) \sin \omega t$$

$$- 2n\omega \cos \omega t] + \frac{a}{4n^2 \varphi^2 + (\omega^2 + n^2 - \varphi^2)^2}$$

$$\times \left\{ 2n\varphi^2 (\cos \varphi t - e^{-nt} \cos \omega t) - \varphi(\omega^2 + n^2 - \varphi^2) \sin \varphi t \right.$$

$$\left. + \frac{1}{\omega} e^{-nt} [(\omega^2 + n^2)^2 - \varphi^2(\omega^2 - n^2)] \sin \omega t \right\} - \frac{r}{\omega} e^{-nt} \sin \omega t$$

$$\tag{15.3.8}$$

Assuming that in equations (15.3.7) and (15.3.8), $t = 0$, we obtain that $x = S$ and $\frac{dx}{dt} = V$, as expected according to initial conditions of motion (2.1.2).

The first derivative from equation (15.3.8) allows us to determine the acceleration of the system:

$$\frac{d^2x}{dt^2} = \frac{V + 2nS}{\omega} e^{-nt}[(n^2 - \omega^2)\sin \omega t - 2n\omega \cos \omega t]$$

$$+ Se^{-nt}\left[\frac{n}{\omega}(3\omega^2 - n^2)\sin \omega t - (\omega^2 - 3n^2)\cos \omega t\right]$$

$$+ \frac{a}{4n^2\varphi^2 + (\omega^2 + n^2 - \varphi^2)^2}\left\{[(n^2 + \omega^2)^2 + \varphi^2(3n^2 - \omega^2)]e^{-nt}\right.$$

$$\times \cos \omega t - \varphi^2(\omega^2 + n^2 - \varphi^2)\cos \varphi t - 2\varphi^3 n \sin \varphi t$$

$$\left. - \frac{n}{\omega}[(\omega^2 + n^2)^2 - \varphi^2(n^2 - 3\omega^2)]e^{-nt}\sin \omega t\right\} \qquad (15.3.9)$$

Equations (15.3.7)–(15.3.9) represent the basic parameters of motion of the system.

15.3.2. *Critical damping, $\omega^2 = 0$*

Taking in equation (15.3.6) that $\omega^2 = 0$, we obtain the expression for the displacement of the system in the Laplace domain:

$$x(s) = \frac{s(V + 2nS)}{(s + n)^2} + \frac{s^2 S}{(s + n)^2} + \frac{s^2 a}{(s + n)^2(s^2 + \varphi^2)} - \frac{r}{(s + n)^2}$$

$$(15.3.10)$$

Applying Laplace Transform pairs 1, 25, 31, 71, and 19 to equation (15.3.10), we invert this equation from the Laplace domain into the time domain and obtain the solution of differential equation (15.3.1) with the initial conditions of motion according to expression (2.1.2) for the case of critical damping:

$$x = (V + 2nS)te^{-nt} + S(1 - nt)e^{-nt} + \frac{a}{(\varphi^2 + n^2)^2}\{2n\varphi \sin \varphi t$$

$$+ e^{-nt}[\varphi^2 - n^2 - nt(\varphi^2 + n^2)] - (\varphi^2 - n^2)\cos \varphi t\}$$

$$- \frac{r}{n^2}[1 - e^{-nt}(1 + nt)] \qquad (15.3.11)$$

Taking the first derivative from equation (15.3.11), we obtain the equation for calculating the velocity of the system:

$$\frac{dx}{dt} = (V + 2nS)e^{-nt}(1 - nt) + nSe^{-nt}(nt - 2)$$

$$+ \frac{a}{(\varphi^2 + n^2)^2}[(\varphi^2 - n^2)(\varphi \sin \varphi t + n^2 t e^{-nt})$$

$$- 2\varphi^2 n(\cos \varphi t - e^{-nt})] - rte^{-nt} \tag{15.3.12}$$

Assuming that in equations (15.3.11) and (15.3.12), $t = 0$, we obtain that $x = S$ and $\frac{dx}{dt} = V$, as it should be according to initial conditions of motion (2.1.2).

15.3.3. *Overdamped motion, $\omega^2 < 0$*

For this case, we replace in equation (15.3.6) the sign of the parameter ω^2 from positive to negative, and we obtain

$$x(s) = \frac{s(V + 2nS)}{(s + n)^2 - \omega^2} + \frac{s^2 S}{(s + n)^2 - \omega^2} + \frac{s^2 a}{[(s + n)^2 - \omega^2](s^2 + \varphi^2)}$$

$$- \frac{r}{(s + n)^2 - \omega^2} \tag{15.3.13}$$

Applying Laplace Transform pairs 1, 28, 36, 69, and 22, we invert equation (15.3.13) from the Laplace domain into the time domain and obtain for this case the solution of differential equation of motion (15.3.1) with the initial conditions of motion according to expression (2.1.2):

$$x = \frac{V + 2nS}{\omega}e^{-nt}\sinh \omega t + Se^{-nt}\left(\cosh \omega t - \frac{n}{\omega}\sinh \omega t\right)$$

$$+ \frac{a}{4n^2\varphi^2 + (n^2 - \omega^2 - \varphi^2)^2}\Big[(n^2 - \omega^2 - \varphi^2)(\cos \varphi t$$

$$- e^{-nt}\cosh \omega t) + 2n\varphi \sin \varphi t - \frac{n}{\omega}(n^2 - \omega^2 + \varphi^2)e^{-nt}\sinh \omega t\Big]$$

$$- \frac{r}{n^2 - \omega^2}\left[1 - e^{-nt}\left(\cosh \omega t - \frac{n}{\omega}\sinh \omega t\right)\right] \tag{15.3.14}$$

The first derivative from equation (15.3.13) yields the velocity of the system:

$$\frac{dx}{dt} = (V + 2nS)e^{-nt}\left(\cosh \omega t - \frac{n}{\omega}\sinh \omega t\right)$$

$$+ \frac{S}{\omega}e^{-nt}[(n^2 + \omega^2)\sinh \omega t - 2n\omega \cosh \omega t]$$

$$+ \frac{a}{4n^2\varphi^2 + (n^2 - \omega^2 - \varphi^2)^2}\left\{2n\varphi^2(\cos \varphi t - e^{-nt}\cosh \omega t)\right.$$

$$-\varphi(n^2 - \omega^2 - \varphi^2)\sin \varphi t + \frac{1}{\omega}e^{-nt}[(n^2 - \omega^2)^2$$

$$\left.+\varphi^2(\omega^2 + n^2)]\sinh \omega t\right\} - \frac{r}{\omega}e^{-nt}\sinh \omega t \qquad (15.3.15)$$

Assuming that in equations (15.3.14) and (15.3.15), $t = 0$, we obtain that $x = S$ and $\frac{dx}{dt} = V$, as expected according to the initial conditions of motion (2.1.2).

15.4. Motion of a System Due to Its Initial Displacement and Velocity as Well as Constant Active and Harmonic Forces

This operational process is associated with a system moving on a horizontal frictionless surface due to its initial displacement and velocity and to constant active and harmonic forces.

Figure 15.4.1 shows a schematic diagram of the system, the notations in which are self-explanatory.

Based on the schematic diagram shown in Figure 15.4.1 and the considerations mentioned above, we compose the differential equations of motion of the system:

$$m\frac{d^2x}{dt^2} + C\frac{dx}{dt} + Kx + R = P + A\cos \varphi t \qquad (15.4.1)$$

Dividing equation (15.4.1) by m, we may write

$$\frac{d^2x}{dt^2} + 2n\frac{dx}{dt} + \omega_0^2 x + r = p + a\cos \varphi t \qquad (15.4.2)$$

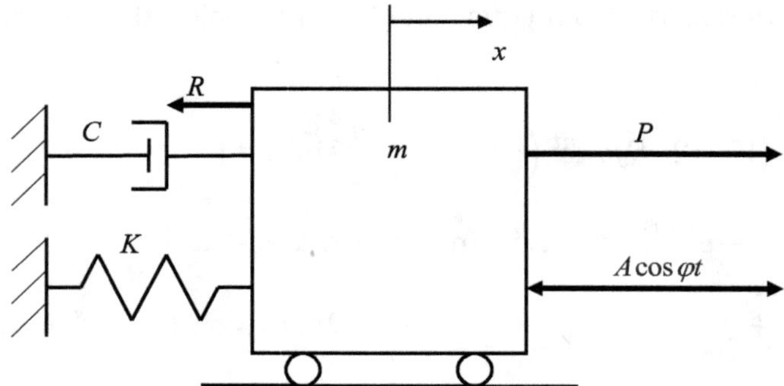

Fig. 15.4.1. Schematic diagram of a system moving on a horizontal frictionless surface while being restricted by a flexible link and a fluid link and subjected to a constant resisting force, a constant active force, and a harmonic force.

The initial conditions of motion are taken according to expression (2.1.2).

Applying Laplace Transform pairs 5, 2, 4, 2, 1, 2, 2, and 29, we convert equation (15.4.2) with the initial conditions of motion according to expression (2.1.2) from the time domain into the Laplace domain and obtain the corresponding algebraic equation in the Laplace domain:

$$s^2 x(s) - sV - s^2 S + 2nsx(s) - 2nsS + \omega_0^2 x(s) + r = p + \frac{as^2}{s^2 + \varphi^2}$$

$$(15.4.3)$$

Rearranging equation (15.4.3), we have

$$x(s)(s^2 + 2ns + \omega_0^2) = s(V + 2nS) + s^2 S + p - r + \frac{as^2}{s^2 + \varphi^2}$$

$$(15.4.4)$$

Solving equation (15.4.4) for the displacement $x(s)$ in the Laplace domain, we obtain

$$x(s) = \frac{s(V + 2nS)}{s^2 + 2ns + \omega_0^2} + \frac{s^2 S}{s^2 + 2ns + \omega_0^2} + \frac{p - r}{s^2 + 2ns + \omega_0^2}$$

$$+ \frac{as^2}{(s^2 + 2ns + \omega_0^2)(s^2 + \varphi^2)} \qquad (15.4.5)$$

Combining equation (15.4.5) with equations (14.1.7) and (14.1.8), we obtain

$$x(s) = \frac{s(V + 2nS)}{(s+n)^2 + \omega^2} + \frac{s^2 S}{(s+n)^2 + \omega^2} + \frac{p-r}{(s+n)^2 + \omega^2}$$

$$+ \frac{s^2 a}{[(s+n)^2 + \omega^2](s^2 + \varphi^2)} \tag{15.4.6}$$

15.4.1. *Underdamped vibration, $\omega^2 > 0$*

Since in equation (15.4.6), the parameter ω^2 is positive, we may therefore apply to this equation Laplace Transform pairs 1, 27, 35, 21, and 68 in order to invert this equation from the Laplace domain into the time domain and obtain the solution of differential equation of motion (15.4.1) with the initial conditions of motion according to expression (2.1.2):

$$x = \frac{V + 2nS}{\omega} e^{-nt} \sin \omega t + S e^{-nt} \left(\cos \omega t - \frac{n}{\omega} \sin \omega t \right) + \frac{p-r}{\omega^2 + n^2}$$

$$\times \left[1 - e^{-nt} \left(\cos \omega t + \frac{n}{\omega} \sin \omega t \right) \right] + \frac{a}{4n^2 \varphi^2 + (\omega^2 + n^2 - \varphi^2)^2}$$

$$\times \left[(\omega^2 + n^2 - \varphi^2)(\cos \varphi t - e^{-nt} \cos \omega t) \right.$$

$$\left. + 2n\varphi \sin \varphi t - \frac{n}{\omega}(\omega^2 + n^2 + \varphi^2) e^{-nt} \sin \omega t \right] \tag{15.4.7}$$

The first derivative from equation (15.4.7) yields the velocity of the system:

$$\frac{dx}{dt} = (V + 2nS) e^{-nt} \left(\cos \omega t - \frac{n}{\omega} \sin \omega t \right)$$

$$+ \frac{S}{\omega} e^{-nt} [(n^2 - \omega^2) \sin \omega t - 2n\omega \cos \omega t] + \frac{p-r}{\omega} e^{-nt} \sin \omega t$$

$$+ \frac{a}{4n^2 \varphi^2 + (\omega^2 + n^2 - \varphi^2)^2} \left\{ 2n\varphi^2 (\cos \varphi t - e^{-nt} \cos \omega t) \right.$$

$$- \varphi(\omega^2 + n^2 - \varphi^2) \sin \varphi t + \frac{1}{\omega} e^{-nt} [(\omega^2 + n^2)^2$$

$$\left. - \varphi^2(\omega^2 - n^2)] \sin \omega t \right\} \tag{15.4.8}$$

Assuming that in equations (15.4.7) and (15.4.8), $t = 0$, we obtain that $x = S$ and $\frac{dx}{dt} = V$, as expected according to the initial conditions of motion (2.1.2).

The first derivative from equation (15.4.8) allows us to obtain the equation describing the acceleration of the system:

$$
\begin{aligned}
\frac{d^2 x}{dt^2} &= \frac{V + 2nS}{\omega} e^{-nt}[(n^2 - \omega^2)\sin \omega t - 2n\omega \cos \omega t] \\
&\quad + Se^{-nt}\left[\frac{n}{\omega}(3\omega^2 - n^2)\sin \omega t - (\omega^2 - 3n^2)\cos \omega t\right] \\
&\quad + (p - r)e^{-nt}\left(\cos \omega t - \frac{n}{\omega}\sin \omega t\right) + \frac{a}{4n^2\varphi^2 + (\omega^2 + n^2 - \varphi^2)^2} \\
&\quad \times \left\{[(n^2 + \omega^2)^2 + \varphi^2(3n^2 - \omega^2)]e^{-nt}\cos \omega t - \varphi^2(\omega^2 + n^2 - \varphi^2)\right. \\
&\quad \times \cos \varphi t - 2\varphi^3 n \sin \varphi t \\
&\quad \left. - \frac{n}{\omega}[(\omega^2 + n^2)^2 - \varphi^2(n^2 - 3\omega^2)]e^{-nt}\sin \omega t\right\}
\end{aligned}
\tag{15.4.9}
$$

Equations (15.4.7)– (15.4.9) represent the basic parameters of motion of the system.

15.4.2. *Critical damping, $\omega^2 = 0$*

In equation (15.4.6), equating the parameter ω^2 to zero, we obtain the expression for determining the displacement of the system in the Laplace domain:

$$
x(s) = \frac{s(V + 2nS)}{(s + n)^2} + \frac{s^2 S}{(s + n)^2} + \frac{p - r}{(s + n)^2} + \frac{s^2 a}{(s + n)^2(s^2 + \varphi^2)}
\tag{15.4.10}
$$

Applying Laplace Transform pairs 1, 25, 31, 19, and 71 to equation (15.4.10), we invert this equation from the Laplace domain into the time domain and obtain the solution of differential equation (15.4.1) with the initial conditions of motion according to expression (2.1.2) for the case of critical damping:

$$x = (V + 2nS)te^{-nt} + S(1 - nt)e^{-nt}$$

$$+ \frac{p - r}{n^2}[1 - e^{-nt}(1 + nt)] + \frac{a}{(\varphi^2 + n^2)^2}\{2n\varphi \sin \varphi t$$

$$+ e^{-nt}[\varphi^2 - n^2 - nt(\varphi^2 + n^2)] - (\varphi^2 - n^2)\cos \varphi t\}$$

$$\tag{15.4.11}$$

Taking the first derivative from equation (15.4.11), we determine the velocity of the system:

$$\frac{dx}{dt} = (V + 2nS)e^{-nt}(1 - nt) + nSe^{-nt}(nt - 2) + (p - r)te^{-nt}$$

$$+ \frac{a}{(\varphi^2 + n^2)^2}[(\varphi^2 - n^2)(\varphi \sin \varphi t + n^2te^{-nt})$$

$$- 2\varphi^2 n(\cos \varphi t - e^{-nt})]\tag{15.4.12}$$

Supposing that in equations (15.4.11) and (15.4.12), $t = 0$, we obtain that $x = S$ and $\frac{dx}{dt} = V$, as it should be according to the initial conditions of motion (2.1.2).

15.4.3. *Overdamped motion, $\omega^2 < 0$*

For this case, we change in equation (15.4.6) the sign of the parameter ω^2 from positive to negative, and we obtain

$$x(s) = \frac{s(V + 2nS)}{(s + n)^2 - \omega^2} + \frac{s^2 S}{(s + n)^2 - \omega^2} + \frac{p - r}{(s + n)^2 - \omega^2}$$

$$+ \frac{s^2 a}{[(s + n)^2 - \omega^2](s^2 + \varphi^2)}\tag{15.4.13}$$

Applying Laplace Transform pairs 1, 28, 36, and 69, we invert equation (15.4.13) from the Laplace domain into the time domain

and obtain for this case the solution of differential equation of motion (15.4.1) with the initial conditions of motion according to expression (2.1.2):

$$x = \frac{V + 2nS}{\omega} e^{-nt} \sinh \omega t + S e^{-nt} \left(\cosh \omega t - \frac{n}{\omega} \sinh \omega t \right)$$

$$+ \frac{p - r}{n^2 - \omega^2} \left[1 - e^{-nt} \left(\cosh \omega t - \frac{n}{\omega} \sinh \omega t \right) \right]$$

$$+ \frac{a}{4n^2\varphi^2 + (n^2 - \omega^2 - \varphi^2)^2} [(n^2 - \omega^2 - \varphi^2)(\cos \varphi t - e^{-nt} \cosh \omega t)$$

$$+ 2n\varphi \sin \varphi t - \frac{n}{\omega}(n^2 - \omega^2 + \varphi^2) e^{-nt} \sinh \omega t] \tag{15.4.14}$$

The first derivative from equation (15.4.14) yields the velocity of the system:

$$\frac{dx}{dt} = (V + 2nS) e^{-nt} \left(\cosh \omega t - \frac{n}{\omega} \sinh \omega t \right)$$

$$+ \frac{S}{\omega} e^{-nt} [(n^2 - \omega^2) \sinh \omega t - 2n\omega \cosh \omega t] + \frac{p - r}{\omega} e^{-nt} \sinh \omega t$$

$$+ \frac{a}{4n^2\varphi^2 + (n^2 - \omega^2 - \varphi^2)^2} \left\{ 2n\varphi^2 (\cos \varphi t - e^{-nt} \cosh \omega t) \right.$$

$$- \varphi(n^2 - \omega^2 - \varphi^2) \sin \varphi t + \frac{1}{\omega} e^{-nt} [(n^2 - \omega^2)^2$$

$$\left. + \varphi^2(\omega^2 + n^2)] \sinh \omega t \right\} \tag{15.4.15}$$

Assuming that in equations (15.4.14) and (15.4.15), $t = 0$, we obtain that $x = S$ and $\frac{dx}{dt} = V$, as expected according to initial conditions of motion (2.1.2).

CHAPTER 16

MOTION OF A SYSTEM ON A HORIZONTAL SURFACE WHILE BEING SUBJECTED TO A DRY FRICTION FORCE AND RESTRICTED BY A FLEXIBLE LINK AND A FLUID LINK

This chapter deals with systems, the motions of which are associated with their initial conditions of motion and with appropriate loading factors.

16.1. Motion of a System Due to Its Initial Displacement and Velocity

The operational process described in the following is characterized by the motion of a system on a horizontal surface due to the initial conditions of motion. The system is subjected to a dry fiction force. The motion of the system is restricted by a flexible link and a fluid link that connect the mass of the system to a non-movable support.

Figure 16.1.1 shows a schematic diagram of the system described above. The notations in the figure are self-explanatory.

Based on the schematic diagram shown in Figure 16.1.1 and the considerations mentioned above, we compose the differential equation

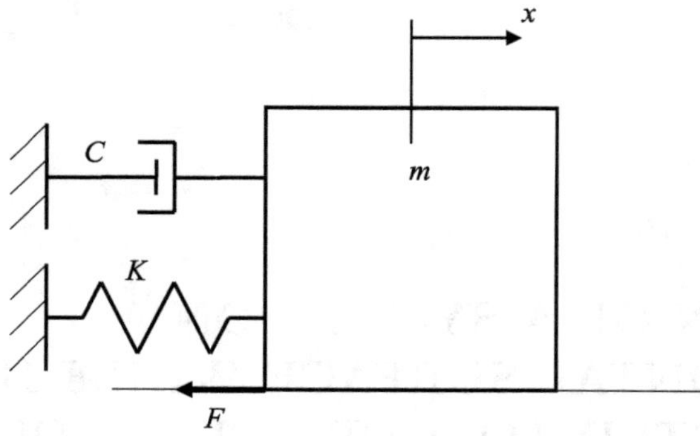

Fig. 16.1.1. Schematic diagram of a system moving on a horizontal surface being restricted by a flexible link and a fluid link while subjected to a dry friction force.

of motion of the system:

$$m\frac{d^2x}{dt^2} + C\frac{dx}{dt} + Kx + F = 0 \qquad (16.1.1)$$

Dividing equation (16.1.1) by m, we have

$$\frac{d^2x}{dt^2} + 2n\frac{dx}{dt} + \omega_0^2 x + f = 0 \qquad (16.1.2)$$

The initial conditions of motion are assigned according to expression (2.1.2).

Applying Laplace Transform Pairs 5, 2, 4, 2, 1, and 2, we convert equation (16.1.2) with the mentioned initial conditions of motion from the time domain into the Laplace domain and obtain the corresponding algebraic equation in the Laplace domain:

$$s^2 x(s) - sV - s^2 S + 2nsx(s) - 2nsS + \omega_0^2 x(s) + f = 0 \quad (16.1.3)$$

Rearranging equation (16.1.3), we have

$$x(s)(s^2 + 2ns + \omega_0^2) = s(V + 2ns) + s^2 S - f \qquad (16.1.4)$$

Solving equation (16.1.4) for the displacement $x(s)$ in the Laplace domain, we obtain

$$x(s) = \frac{s(V + 2ns)}{s^2 + 2ns + \omega_0^2} + \frac{s^2 S}{s^2 + 2ns + \omega_0^2} - \frac{f}{s^2 + 2ns + \omega_0^2} \qquad (16.1.5)$$

Combining equations (16.1.4) with equations (14.1.7) and (14.1.8), we may write

$$x(s) = \frac{s(V + 2nS)}{(s + n)^2 + \omega^2} + \frac{s^2 S}{(s + n)^2 + \omega^2} - \frac{f}{(s + n)^2 + \omega^2} \qquad (16.1.6)$$

16.1.1. *Underdamped motion, $\omega^2 > 0$*

Applying to equation (16.1.6) Laplace Transform pairs 1, 27, 35, and 21, we invert this equation from the Laplace domain into the time domain and obtain the solution of differential equation of motion (16.1.1) at the initial conditions of motion according to expression (2.1.2):

$$x = \frac{V + 2nS}{\omega} e^{-nt} \sin \omega t + S e^{-nt} \left(\cos \omega t - \frac{n}{\omega} \sin \omega t \right)$$

$$- \frac{f}{\omega^2 + n^2} \left[1 - e^{-nt} \left(\cos \omega t + \frac{n}{\omega} \sin \omega t \right) \right] \qquad (16.1.7)$$

The first derivative from equation (16.1.7) yields the velocity of the system:

$$\frac{dx}{dt} = (V + 2nS) e^{-nt} \left(\cos \omega t - \frac{n}{\omega} e^{-nt} \sin \omega t \right) + \frac{S}{\omega} e^{-nt} [(n^2 - \omega^2)$$

$$\times \sin \omega t - 2n\omega \cos \omega t] - \frac{f}{\omega^2 + n^2} e^{-nt} \sin \omega t \qquad (16.1.8)$$

Assuming that in equations (16.1.7) and (16.1.8), $t = 0$, then we obtain that $x = S$ and $\frac{dx}{dt} = V$, as expected according to the initial conditions of motion (2.1.2).

The second derivative from equation (16.1.7) represents the expression for calculating the acceleration of the system:

$$\frac{d^2 x}{dt^2} = \frac{V - 2nS}{\omega} e^{-nt} [(n^2 - \omega^2) \sin \omega t - 2n\omega \cos \omega t]$$

$$+ S e^{-nt} \left[\frac{n}{\omega} (3\omega^2 - n^2) \sin \omega t - (\omega^2 - 3n^2) \cos \omega t \right]$$

$$- f e^{-nt} \left(\cos \omega t - \frac{n}{\omega} \sin \omega t \right) \qquad (16.1.9)$$

Equations (16.1.7)–(16.1.9) represent the basic parameters of motion of the system.

16.1.2.　*Critical damping, $\omega^2 = 0$*

In equation (16.1.6), equating the parameter ω^2 to zero, we determine the displacement of the system in the Laplace domain for the current case:

$$x(s) = \frac{s(V + 2nS)}{(s + n)^2} + \frac{s^2 S}{(s + n)^2} - \frac{f}{(s + n)^2} \qquad (16.1.10)$$

Applying Laplace Transform pairs 1, 25, 31, and 19 to equation (16.1.10), we invert this equation from the Laplace domain into the time domain and obtain the solution of differential equation (16.1.1) at the initial conditions of motion according to expression (2.1.1) for the case of critical damping:

$$x = (V + 2nS)te^{-nt} + S(1 - nt)e^{-nt} - \frac{f}{n^2}[1 - (1 + nt)e^{-nt}] \qquad (16.1.11)$$

Taking the first derivative from equation (16.1.11), we obtain the equation for calculating the velocity of the system:

$$\frac{dx}{dt} = (V + 2nS)e^{-nt}(1 - nt) + nSe^{-nt}(nt - 2) - rte^{-nt} \qquad (16.1.12)$$

Assuming that in equations (16.1.11) and (16.1.12), $t = 0$, we obtain that $x = S$ and $\frac{dx}{dt} = V$, as it should be according to the initial conditions of motion (2.1.2).

16.1.3.　*Overdamped motion, $\omega^2 < 0$*

For this case, we change in equation (16.1.6) the sign of parameter ω^2 from positive to negative, and we obtain

$$x(s) = \frac{s(V + 2nS)}{(s + n)^2 - \omega^2} + \frac{s^2 S}{(s + n)^2 - \omega^2} - \frac{f}{(s + n)^2 - \omega^2} \qquad (16.1.13)$$

Applying Laplace Transform Pairs 1, 28, 36, and 22, we invert equation (16.1.13) from the Laplace domain into the time domain and obtain for this case the solution of differential equation of motion

(16.1.1) with the initial conditions of motion according to expression (2.1.2):

$$x = \frac{V + 2nS}{\omega} e^{-nt} \sinh \omega t + Se^{-nt}\left(\cosh \omega t - \frac{n}{\omega}\sinh \omega t\right)$$

$$- \frac{f}{n^2 - \omega^2}\left[1 - e^{-nt}\left(\cosh \omega t + \frac{n}{\omega}\sinh \omega t\right)\right] \qquad (16.1.14)$$

Taking the first derivative from equation (16.1.14), we determine the velocity of the system:

$$\frac{dx}{dt} = (V + 2nS)e^{-nt}\left(\cosh \omega t - \frac{n}{\omega}\sinh \omega t\right) + \frac{S}{\omega}e^{-nt}[(n^2 + \omega^2)$$

$$\times \sinh \omega t - 2n\omega \cosh \omega t] - \frac{f}{\omega}e^{-nt}\sinh \omega t \qquad (16.1.15)$$

Supposing that in equations (16.1.14) and (16.1.15), $t = 0$, we obtain that $x = S$ and $\frac{dx}{dt} = V$, as it should be according to the initial conditions of motion expressed by (2.1.2).

16.2. Motion of a System Due to Its Initial Displacement and Velocity and to a Constant Active Force

The operational process represents a system moving on a horizontal surface due to its initial displacement and velocity and to a constant active force while being subjected to the action of a dry friction force.

Figure 16.2.1 shows a schematic diagram of the system, the notations in which are self-explanatory.

Based on the schematic diagram shown in Figure 16.2.1 and the considerations mentioned above, we compose the differential equation of motion of the system:

$$m\frac{d^2x}{dt^2} + C\frac{dx}{dt} + Kx + F = P \qquad (16.2.1)$$

Dividing equation (16.2.1) by m, we have

$$\frac{d^2x}{dt^2} + 2n\frac{dx}{dt} + \omega_0^2 x + f = p \qquad (16.2.2)$$

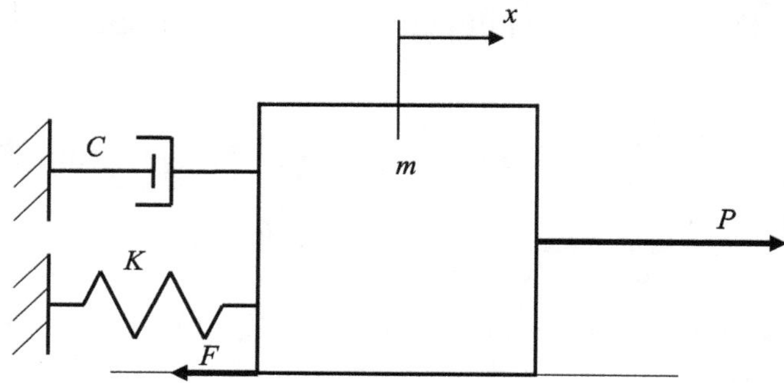

Fig. 16.2.1. Schematic diagram of a system moving on a horizontal surface while being restricted by flexible and fluid links and subjected to dry friction and constant active forces.

The initial conditions of motion are assigned according to expression (2.1.2).

Applying Laplace Transform pairs 5, 2, 4, 2, 1, 2, and 2, we convert equation (16.2.2) with the mentioned initial conditions of motion from the time domain into the Laplace domain and obtain the corresponding algebraic equation in the Laplace domain:

$$s^2 x(s) - sV - s^2 S + 2nsx(s) - 2nsS + \omega_0^2 x(s) = p - f \quad (16.2.3)$$

Rearranging equation (16.2.3), we may write

$$x(s)(s^2 + 2ns + \omega_0^2) = s(V + 2ns) + s^2 S + p - f \quad (16.2.4)$$

Solving equation (16.2.4) for the displacement $x(s)$ in the Laplace domain, we obtain

$$x(s) = \frac{s(V + 2ns)}{s^2 + 2ns + \omega_0^2} + \frac{s^2 S}{s^2 + 2ns + \omega_0^2} + \frac{p - f}{s^2 + 2ns + \omega_0^2} \quad (16.2.5)$$

Combining equation (16.2.5) with equations (14.1.7) and (14.1.8), we obtain

$$x(s) = \frac{s(V + 2nS)}{(s + n)^2 + \omega^2} + \frac{s^2 S}{(s + n)^2 + \omega^2} + \frac{p - f}{(s + n)^2 + \omega^2} \quad (16.2.6)$$

16.2.1. *Underdamped motion, $\omega^2 > 0$*

In equation (16.2.6), the parameter ω^2 is positive; therefore, it is allowable to apply Laplace Transform pairs 1, 27, 35, and 21 to this equation and invert it from the Laplace domain into the time domain and obtain the solution of differential equation of motion (16.2.1) at the initial conditions of motion according to expression (2.1.2):

$$x = \frac{V + 2nS}{\omega} e^{-nt} \sin \omega t + Se^{-nt} \left(\cos \omega t - \frac{n}{\omega} \sin \omega t \right)$$

$$+ \frac{p - f}{\omega^2 + n^2} \left[1 - e^{-nt} \left(\cos \omega t + \frac{n}{\omega} \sin \omega t \right) \right] \qquad (16.2.7)$$

Taking the first derivative from equation (16.2.7), we obtain the equation describing the velocity of the system:

$$\frac{dx}{dt} = (V + 2nS)e^{-nt} \left(\cos \omega t - \frac{n}{\omega} \sin \omega t \right) + \frac{S}{\omega} e^{-nt} [(n^2 - \omega^2)$$

$$\times \sin \omega t - 2n\omega \cos \omega t] + \frac{p - f}{\omega} e^{-nt} \sin \omega t \qquad (16.2.8)$$

Supposing that in equations (16.2.7) and (16.2.8), $t = 0$, we obtain that $x = S$ and $\frac{dx}{dt} = V$, as expected according to the initial conditions of motion (2.1.2).

The second derivative from equation (16.2.7) yields the acceleration of the system:

$$\frac{d^2x}{dt^2} = \frac{V + 2nS}{\omega} e^{-nt} [(n^2 - \omega^2) \sin \omega t - 2n\omega \cos \omega t]$$

$$+ Se^{-nt} \left[\frac{n}{\omega}(3\omega^2 - n^2) \sin \omega t - (\omega^2 - 3n^2) \cos \omega t \right] \qquad (16.2.9)$$

$$+ (p - f)e^{-nt} \left(\cos \omega t - \frac{n}{\omega} \sin \omega t \right)$$

Equations (16.2.7)–(16.2.9) represent the basic parameters of motion of the system.

16.2.1.1. *Numerical solution, $\omega^2 > 0$*

```python
from matplotlib.pyplot import plot, show
from numpy import linspace, sin, cos, exp


S = 0.1
V = 0.15
n = 15
w = 60
p = 10
f = 2


t = linspace(0,0.6,1000)


x = ((V + 2*n*S)/omega)*exp(-n*t)*sin(omega*t) \
    + S*exp(-n*t)*(cos(omega*t) - (n/omega)*sin(omega*t)) \
    + ((p-f)/(omega**2 + n**2))*(1 - exp(-n*t)*(cos(omega*t) \
                    + (n/omega)*sin(omega*t)))

plot(t, x)
show()
```

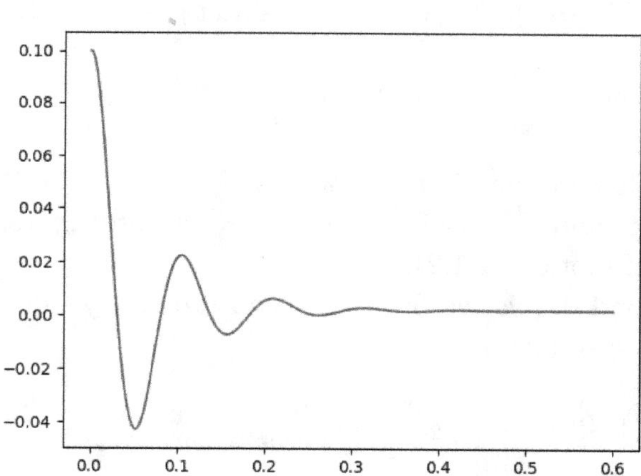

Fig. 16.2.2. Graph of equation (16.2.7) characterizing underdamped motion.

16.2.2. *Critical damping, $\omega^2 = 0$*

Taking in equation (16.2.6) that $\omega^2 = 0$, we obtain the expression for the displacement of the system in the Laplace domain:

$$x(s) = \frac{s(V + 2nS)}{(s + n)^2} + \frac{s^2 S}{(s + n)^2} + \frac{p - f}{(s + n)^2} \qquad (16.2.11)$$

Applying Laplace Transform pairs 1, 25, 31, and 19 to equation (16.2.11), we invert this equation from the Laplace domain into the time domain and obtain the solution of differential equation (16.2.1) at the initial conditions of motion according to expression (2.1.2) for the case of critical damping:

$$x = (V + 2nS)te^{-nt} + S(1 - nt)e^{-nt} + \frac{p - f}{n^2}[1 - e^{-nt}(1 + nt)]$$

$$(16.2.12)$$

Taking the first derivative from equation (16.2.12), we obtain the equation for calculating the velocity of the system:

$$\frac{dx}{dt} = (V + 2nS)e^{-nt}(1 - nt) + nSe^{-nt}(nt - 2) + (p - f)te^{-nt}$$

$$(16.2.13)$$

Supposing that in equations (16.2.12) and (16.2.13), $t = 0$, we obtain that $x = S$ and $\frac{dx}{dt} = V$, as it should be according to the initial conditions of motion (2.1.2).

16.2.3. *Overdamped motion, $\omega^2 < 0$*

In this case, we change in equation (16.2.6) the sign of the parameter ω^2 from positive to negative, and we obtain

$$x(s) = \frac{s(V + 2nS)}{(s + n)^2 - \omega^2} + \frac{s^2 S}{(s + n)^2 - \omega^2} + \frac{p - f}{(s + n)^2 - \omega^2} \quad (16.2.14)$$

Applying Laplace Transform pairs 1, 28, 36, and 22, we invert equation (16.2.14) from the Laplace domain into the time domain and obtain for this case the solution of differential equation of motion (16.2.1) with the initial conditions of motion according to expression (2.1.2):

$$x = \frac{V + 2nS}{\omega}e^{-nt}\sinh\omega t + Se^{-nt}\left(\cosh\omega t - \frac{n}{\omega}\sinh\omega t\right)$$

$$+ \frac{p - f}{n^2 - \omega^2}\left[1 - e^{-nt}\left(\cosh\omega t - \frac{n}{\omega}\sinh\omega t\right)\right]$$

$$(16.2.15)$$

Taking the first derivative from equation (16.2.15), we determine the velocity of the system:

$$\frac{dx}{dt} = (V + 2nS)e^{-nt}\left(\cosh \omega t - \frac{n}{\omega}\sinh \omega t\right) + \frac{S}{\omega}e^{-nt}[(n^2 + \omega^2)$$

$$\times \sinh \omega t - 2n\omega \cosh \omega t] + \frac{p-f}{\omega}e^{-nt}\sinh \omega t \qquad (16.2.16)$$

Supposing that in equations (16.2.15) and (16.2.16), $t = 0$, we obtain that $x = S$ and $\frac{dx}{dt} = V$, as it is expected according to initial conditions of motion (2.1.2).

16.3. Motion of a System Due to Its Initial Displacement and Velocity and to a Harmonic Force

The operational process of the system subjected to a dry friction force while moving on a horizontal frictional surface due to its initial displacement and velocity and to a harmonic force is addressed in this section. The motion of the system is restricted by a flexible link and a fluid link.

Figure 16.3.1 shows a schematic diagram of the system, the notations in which are self-explanatory.

Based on the schematic diagram shown in Figure 16.3.1 and the considerations mentioned above, we compose the differential equation

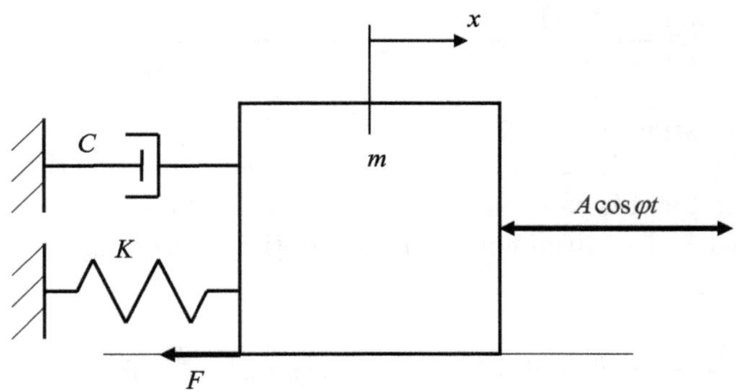

Fig. 16.3.1. Schematic diagram of a system moving on a horizontal frictional surface while being restricted by flexible and fluid links and subjected to dry friction and harmonic forces.

of motion of the system:

$$m\frac{d^2x}{dt^2} + C\frac{dx}{dt} + Kx + F = A\cos\varphi t \qquad (16.3.1)$$

Dividing equation (16.3.1) by m, we may write

$$\frac{d^2x}{dt^2} + 2n\frac{dx}{dt} + \omega_0^2 x + f = a\cos\varphi t \qquad (16.3.2)$$

The initial conditions of motion are taken according to expression (2.1.2).

Applying Laplace Transform pairs 5, 2, 4, 2, 1, 2, and 29, we convert equation (16.3.2) with the above-mentioned initial conditions of motion from the time domain into the Laplace domain and obtain the corresponding algebraic equation in the Laplace domain:

$$s^2 x(s) - sV - s^2 S + 2nsx(s) - 2nsS + \omega_0^2 x(s) + f = \frac{as^2}{s^2 + \varphi^2}$$

$$(16.3.3)$$

Rearranging equation (16.3.3), we have

$$x(s)(s^2 + 2ns + \omega_0^2) = s(V + 2ns) + s^2 S + \frac{as^2}{s^2 + \varphi^2} - f \qquad (16.3.4)$$

Solving equation (16.3.4) for the displacement $x(s)$ in the Laplace domain, we obtain

$$x(s) = \frac{s(V + 2ns)}{s^2 + 2ns + \omega_0^2} + \frac{s^2 S}{s^2 + 2ns + \omega_0^2} + \frac{as^2}{(s^2 + 2ns + \omega_0^2)(s^2 + \varphi^2)}$$

$$- \frac{f}{s^2 + 2ns + \omega_0^2} \qquad (16.3.5)$$

Combining equation (16.3.5) with equations (14.1.7) and (14.1.8), we may write

$$x(s) = \frac{s(V + 2nS)}{(s + n)^2 + \omega^2} + \frac{s^2 S}{(s + n)^2 + \omega^2} + \frac{s^2 a}{[(s + n)^2 + \omega^2](s^2 + \varphi^2)}$$

$$- \frac{f}{(s + n)^2 + \omega^2} \qquad (16.3.6)$$

16.3.1. *Underdamped vibration, $\omega^2 > 0$*

It is seen that in equation (16.3.6) the parameter ω^2 is positive; therefore, we may apply to this equation Laplace Transform pairs 1, 27, 35, 68, and 21 in order to invert this equation from the Laplace domain into the time domain and obtain the solution of differential equation of motion (16.3.1) at the initial conditions of motion according to expression (2.1.2):

$$
\begin{aligned}
x = {} & \frac{V + 2nS}{\omega} e^{-nt} \sin \omega t + S e^{-nt} \left(\cos \omega t - \frac{n}{\omega} \sin \omega t \right) \\
& + \frac{a}{4n^2 \varphi^2 + (\omega^2 + n^2 - \varphi^2)^2} \Big[(\omega^2 + n^2 - \varphi^2)(\cos \varphi t - e^{-nt} \cos \omega t) \\
& \quad + 2n\varphi \sin \varphi t - \frac{n}{\omega}(\omega^2 + n^2 + \varphi^2) e^{-nt} \sin \omega t \Big] \\
& - \frac{f}{\omega^2 + n^2} \left[1 - e^{-nt} \left(\cos \omega t + \frac{n}{\omega} \sin \omega t \right) \right]
\end{aligned}
\tag{16.3.7}
$$

The first derivative from equation (16.3.7) yields the velocity of the system:

$$
\begin{aligned}
\frac{dx}{dt} = {} & (V + 2nS) e^{-nt} \left(\cos \omega t - \frac{n}{\omega} \sin \omega t \right) + \frac{S}{\omega} e^{-nt} [(n^2 - \omega^2) \\
& \times \sin \omega t - 2n\omega \cos \omega t] + \frac{a}{4n^2 \varphi^2 + (\omega^2 + n^2 - \varphi^2)^2} \\
& \times \Big\{ 2n\varphi^2 (\cos \varphi t - e^{-nt} \cos \omega t) - \varphi(\omega^2 + n^2 - \varphi^2) \sin \varphi t \\
& + \frac{1}{\omega} e^{-nt} [(\omega^2 + n^2)^2 - \varphi^2 (\omega^2 - n^2)] \sin \omega t \Big\} - \frac{f}{\omega} e^{-nt} \sin \omega t
\end{aligned}
\tag{16.3.8}
$$

Assuming that in equations (16.3.7) and (16.3.8) $t = 0$, we obtain that $x = S$ and $\frac{dx}{dt} = V$, as expected according to initial conditions of motion (2.1.2).

The first derivative from equation (16.1.8) allows for determining the acceleration of the system:

$$\frac{d^2x}{dt^2} = \frac{V+2nS}{\omega}e^{-nt}[(n^2-\omega^2)\sin\omega t - 2n\omega\cos\omega t]$$

$$+ Se^{-nt}\left[\frac{n}{\omega}(3\omega^2-n^2)\sin\omega t - (\omega^2-3n^2)\cos\omega t\right]$$

$$+ \frac{a}{4n^2\varphi^2+(\omega^2+n^2-\varphi^2)^2}\Big\{[(n^2+\omega^2)^2+\varphi^2(3n^2-\omega^2)]e^{-nt}$$

$$\times \cos\omega t - \varphi^2(\omega^2+n^2-\varphi^2)\cos\varphi t - 2\varphi^3 n\sin\varphi t$$

$$- \frac{n}{\omega}[(\omega^2+n^2)^2-\varphi^2(n^2-3\omega^2)]e^{-nt}\sin\omega t\Big\} \tag{16.3.9}$$

Equations (16.3.7)–(16.3.9) represent the basic parameters of motion of the system.

16.3.2. *Critical damping, $\omega^2 = 0$*

Taking in equation (16.3.6) that $\omega^2 = 0$, we obtain the expression for the displacement of the system in the Laplace domain:

$$x(s) = \frac{s(V+2nS)}{(s+n)^2} + \frac{s^2S}{(s+n)^2} + \frac{s^2a}{(s+n)^2(s^2+\varphi^2)} - \frac{f}{(s+n)^2} \tag{16.3.10}$$

Applying Laplace Transform pairs 1, 25, 31, 71, and 19 to equation (16.3.10), we invert this equation from the Laplace domain into the time domain and obtain the solution of differential equation (16.3.1) at the initial conditions of motion according to expression (2.1.2) for the case of critical damping:

$$x = (V+2nS)te^{-nt} + S(1-nt)e^{-nt} + \frac{a}{(\varphi^2+n^2)^2}\{2n\varphi\sin\varphi t$$

$$+ e^{-nt}[\varphi^2-n^2-nt(\varphi^2+n^2)] - (\varphi^2-n^2)\cos\varphi t\}$$

$$- \frac{f}{n^2}[1-e^{-nt}(1+nt)] \tag{16.3.11}$$

Taking the first derivative from equation (16.3.11), we obtain the equation for calculating the velocity of the system:

$$\frac{dx}{dt} = (V + 2nS)e^{-nt}(1 - nt) + nSe^{-nt}(nt - 2)$$

$$+ \frac{a}{(\varphi^2 + n^2)^2}[(\varphi^2 - n^2)(\varphi \sin \varphi t + n^2 t e^{-nt}) \qquad (16.3.12)$$

$$- 2\varphi^2 n(\cos \varphi t - e^{-nt})] - rte^{-nt}$$

Assuming that in equations (16.3.11) and (16.3.12), $t = 0$, we obtain that $x = S$ and $\frac{dx}{dt} = V$, as it should be according to the initial conditions of motion (2.1.2).

16.3.3. *Overdamped motion, $\omega^2 < 0$*

For this case, we replace in equation (16.3.6) the sign of the parameter ω^2 from positive to negative, and we obtain

$$x(s) = \frac{s(V + 2nS)}{(s + n)^2 - \omega^2} + \frac{s^2 S}{(s + n)^2 - \omega^2} + \frac{s^2 a}{[(s + n)^2 - \omega^2](s^2 + \varphi^2)}$$

$$- \frac{f}{(s + n)^2 - \omega^2} \qquad (16.3.13)$$

Applying Laplace Transform pairs 1, 28, 36, 69, and 22, we invert the equation (16.3.13) from the Laplace domain into the time domain and obtain for this case the solution of differential equation of motion (16.3.1) with the initial conditions of motion according to expression (2.1.2).

$$x = \frac{V + 2nS}{\omega}e^{-nt}\sinh \omega t + Se^{-nt}\left(\cosh \omega t - \frac{n}{\omega}\sinh \omega t\right)$$

$$+ \frac{a}{4n^2\varphi^2 + (n^2 - \omega^2 - \varphi^2)^2}\Big[(n^2 - \omega^2 - \varphi^2)(\cos \varphi t$$

$$- e^{-nt}\cosh \omega t) + 2n\varphi \sin \varphi t - \frac{n}{\omega}(n^2 - \omega^2 + \varphi^2)e^{-nt}\sinh \omega t\Big]$$

$$- \frac{f}{n^2 - \omega^2}\left[1 - e^{-nt}\left(\cosh \omega t - \frac{n}{\omega}\sinh \omega t\right)\right] \qquad (16.3.14)$$

The first derivative from equation (16.3.13) yields the velocity of the system:

$$
\frac{dx}{dt} = (V + 2nS)e^{-nt}\left(\cosh \omega t - \frac{n}{\omega}\sinh \omega t\right) + \frac{S}{\omega}e^{-nt}[(n^2 + \omega^2)
$$

$$
\times \sinh \omega t - 2n\omega \cosh \omega t] + \frac{a}{4n^2\varphi^2 + (n^2 - \omega^2 - \varphi^2)^2}
$$

$$
\times \left\{ 2n\varphi^2(\cos \varphi t - e^{-nt}\cosh \omega t) - \varphi(n^2 - \omega^2 - \varphi^2)\sin \varphi t \right.
$$

$$
\left. + \frac{1}{\omega}e^{-nt}[(n^2 - \omega^2)^2 + \varphi^2(\omega^2 + n^2)]\sinh \omega t \right\} - \frac{f}{\omega}e^{-nt}\sinh \omega t
$$

$$
\tag{16.3.15}
$$

Assuming that in equations (16.3.14) and (16.3.15), $t = 0$, we obtain that $x = S$ and $\frac{dx}{dt} = V$, as expected according to initial conditions of motion (2.1.2).

16.4. Motion of a System Due to Its Initial Displacement and Velocity as Well as to Constant Active and Harmonic Forces

This operational process is associated with a system moving on a horizontal frictional surface due to its initial displacement and velocity and to constant active and harmonic forces.

Figure 16.4.1 shows a schematic diagram of the system, the notations in which are self-explanatory.

Based on the schematic diagram shown in Figure 16.4.1 and the considerations mentioned above, we compose the differential equation of motion of the system:

$$
m\frac{d^2x}{dt^2} + C\frac{dx}{dt} + Kx + F = P + A\cos \varphi t
\tag{16.4.1}
$$

Dividing equation (16.4.1) by m, we may write

$$
\frac{d^2x}{dt^2} + 2n\frac{dx}{dt} + \omega_0^2 x + f = p + a\cos \varphi t
\tag{16.4.2}
$$

The initial conditions of motion are taken according to expression (2.1.2).

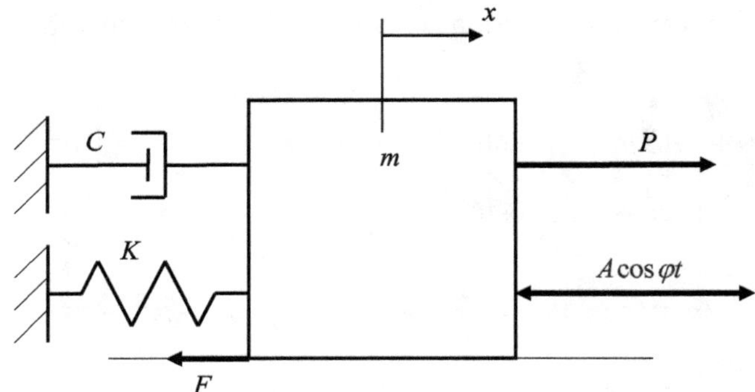

Fig. 16.4.1. Schematic diagram of a system moving on a horizontal frictional surface while being restricted by a flexible link and a fluid link and subjected to a dry friction force, a constant active force, and a harmonic force.

Applying Laplace Transform pairs 5, 2, 4, 2, 1, 2, 2, and 29, we convert equation (16.4.2) with the initial conditions of motion according to expression (2.1.2) from the time domain into the Laplace domain and obtain the corresponding algebraic equation in the Laplace domain:

$$s^2 x(s) - sV - s^2 S + 2nsx(s) - 2nsS + \omega_0^2 x(s) + f = p + \frac{as^2}{s^2 + \varphi^2}$$

$$(16.4.3)$$

Rearranging equation (16.4.3), we have

$$x(s)(s^2 + 2ns + \omega_0^2) = s(V + 2ns) + s^2 S + p - f + \frac{as^2}{s^2 + \varphi^2}$$

$$(16.4.4)$$

Solving equation (16.4.4) for the displacement $x(s)$ in the Laplace domain, we obtain

$$x(s) = \frac{s(V + 2ns)}{s^2 + 2ns + \omega_0^2} + \frac{s^2 S}{s^2 + 2ns + \omega_0^2} + \frac{p - f}{s^2 + 2ns + \omega_0^2}$$

$$+ \frac{as^2}{(s^2 + 2ns + \omega_0^2)(s^2 + \varphi^2)} \qquad (16.4.5)$$

Combining equation (16.4.5) with equations (14.1.7) and (14.1.8), we obtain

$$x(s) = \frac{s(V + 2nS)}{(s+n)^2 + \omega^2} + \frac{s^2 S}{(s+n)^2 + \omega^2} + \frac{p - f}{(s+n)^2 + \omega^2}$$

$$+ \frac{s^2 a}{[(s+n)^2 + \omega^2](s^2 + \varphi^2)} \tag{16.4.6}$$

16.4.1. *Underdamped vibration, $\omega^2 > 0$*

Since in equation (16.4.6), the parameter ω^2 is positive; therefore, we may apply to this equation Laplace Transform pairs 1, 27, 35, 21, and 68 in order to invert this equation from the Laplace domain into the time domain and obtain the solution of differential equation of motion (16.4.1) at the initial conditions of motion according to expression (2.1.2):

$$x = \frac{V + 2nS}{\omega} e^{-nt} \sin \omega t + S e^{-nt} \left(\cos \omega t - \frac{n}{\omega} \sin \omega t \right)$$

$$+ \frac{p - f}{\omega^2 + n^2} \left[1 - e^{-nt} \left(\cos \omega t + \frac{n}{\omega} \sin \omega t \right) \right]$$

$$+ \frac{a}{4n^2 \varphi^2 + (\omega^2 + n^2 - \varphi^2)^2} \tag{16.4.7}$$

$$\times \left[(\omega^2 + n^2 - \varphi^2)(\cos \varphi t - e^{-nt} \cos \omega t) \right.$$

$$\left. + 2n\varphi \sin \varphi t - \frac{n}{\omega} (\omega^2 + n^2 + \varphi^2) e^{-nt} \sin \omega t \right]$$

The first derivative from equation (16.4.7) yields the velocity of the system:

$$\frac{dx}{dt} = (V + 2nS) e^{-nt} \left(\cos \omega t - \frac{n}{\omega} \sin \omega t \right) + \frac{S}{\omega} e^{-nt} [(n^2 - \omega^2) \sin \omega t$$

$$- 2n\omega \cos \omega t] + \frac{p - f}{\omega} e^{-nt} \sin \omega t + \frac{a}{4n^2 \varphi^2 + (\omega^2 + n^2 - \varphi^2)^2}$$

$$\times \left\{ 2n\varphi^2 (\cos \varphi t - e^{-nt} \cos \omega t) - \varphi(\omega^2 + n^2 - \varphi^2) \sin \varphi t \right.$$

$$\left. + \frac{1}{\omega} e^{-nt} [(\omega^2 + n^2)^2 - \varphi^2(\omega^2 - n^2)] \sin \omega t \right\} \tag{16.4.8}$$

Assuming that in equations (16.4.7) and (16.4.8), $t = 0$, we obtain that $x = S$ and $\frac{dx}{dt} = V$, as expected according to initial conditions of motion (2.1.2).

The first derivative from equation (16.4.8) allows for obtaining the equation describing the acceleration of the system:

$$\frac{d^2x}{dt^2} = \frac{V + 2nS}{\omega}e^{-nt}[(n^2 - \omega^2)\sin\omega t - 2n\omega\cos\omega t]$$

$$+ Se^{-nt}\left[\frac{n}{\omega}(3\omega^2 - n^2)\sin\omega t - (\omega^2 - 3n^2)\cos\omega t\right]$$

$$+ (p - f)e^{-nt}\left(\cos\omega t - \frac{n}{\omega}\sin\omega t\right) + \frac{a}{4n^2\varphi^2 + (\omega^2 + n^2 - \varphi^2)^2}$$

$$\times \left\{[(n^2 + \omega^2)^2 + \varphi^2(3n^2 - \omega^2)]e^{-nt}\cos\omega t\right.$$

$$- \varphi^2(\omega^2 + n^2 - \varphi^2)\cos\varphi t - 2\varphi^3 n\sin\varphi t$$

$$\left.- \frac{n}{\omega}[(\omega^2 + n^2)^2 - \varphi^2(n^2 - 3\omega^2)]e^{-nt}\sin\omega t\right\} \tag{16.4.9}$$

Equations (16.4.7)–(16.4.9) represent the basic parameters of motion of the system.

16.4.2. *Critical damping, $\omega^2 = 0$*

In equation (16.4.6), equating the parameter ω^2 to zero, we obtain the expression for determining the displacement of the system in the Laplace domain:

$$x(s) = \frac{s(V + 2nS)}{(s + n)^2} + \frac{s^2 S}{(s + n)^2} + \frac{p - f}{(s + n)^2} + \frac{s^2 a}{(s + n)^2(s^2 + \varphi^2)} \tag{16.4.10}$$

Applying Laplace Transform pairs 1, 25, 31, 19, and 71 to equation (16.4.10), we invert this equation from the Laplace domain into the time domain and obtain the solution of differential equation (16.4.1) at the initial conditions of motion according to expression (2.1.2) for the case of critical damping:

$$x = (V + 2nS)te^{-nt} + S(1 - nt)e^{-nt}$$

$$+ \frac{p - f}{n^2}[1 - e^{-nt}(1 + nt)] + \frac{a}{(\varphi^2 + n^2)^2}$$

$$\times \{2n\varphi \sin \varphi t + e^{-nt}[\varphi^2 - n^2 - nt(\varphi^2 + n^2)]$$

$$- (\varphi^2 - n^2) \cos \varphi t\} \tag{16.4.11}$$

Taking the first derivative from equation (16.4.11), we determine the velocity of the system:

$$\frac{dx}{dt} = (V + 2nS)e^{-nt}(1 - nt) + nSe^{-nt}(nt - 2) + (p - f)te^{-nt}$$

$$+ \frac{a}{(\varphi^2 + n^2)^2}[(\varphi^2 - n^2)(\varphi \sin \varphi t + n^2 te^{-nt})$$

$$- 2\varphi^2 n(\cos \varphi t - e^{-nt})] \tag{16.4.12}$$

Supposing that in equations (16.4.11) and (16.4.12), $t = 0$, we obtain that $x = S$ and $\frac{dx}{dt} = V$, as it should be according to the initial conditions of motion (2.1.2).

16.4.3. *Overdamped motion, $\omega^2 < 0$*

For this case, we change in equation (16.4.6) the sign of the parameter ω^2 from positive to negative, and we obtain

$$x(s) = \frac{s(V + 2nS)}{(s + n)^2 - \omega^2} + \frac{s^2 S}{(s + n)^2 - \omega^2} + \frac{p - f}{(s + n)^2 - \omega^2}$$

$$+ \frac{s^2 a}{[(s + n)^2 - \omega^2](s^2 + \varphi^2)} \tag{16.4.13}$$

Applying Laplace Transform pairs 1, 28, 36, and 69, we invert equation (16.4.13) from the Laplace domain into the time domain and obtain for this case the solution of differential equation of motion

(16.4.1) with the initial conditions of motion according to expression (2.1.2):

$$x = \frac{V + 2nS}{\omega} e^{-nt} \sinh \omega t + S e^{-nt} \left(\cosh \omega t - \frac{n}{\omega} \sinh \omega t \right)$$

$$+ \frac{p - f}{n^2 - \omega^2} \left[1 - e^{-nt} \left(\cosh \omega t - \frac{n}{\omega} \sinh \omega t \right) \right]$$

$$+ \frac{a}{4n^2 \varphi^2 + (n^2 - \omega^2 - \varphi^2)^2} \left[(n^2 - \omega^2 - \varphi^2)(\cos \varphi t \right.$$

$$\left. - e^{-nt} \cosh \omega t) + 2n\varphi \sin \varphi t - \frac{n}{\omega}(n^2 - \omega^2 + \varphi^2) e^{-nt} \sinh \omega t \right]$$

$$(16.4.14)$$

The first derivative from equation (16.4.14) yields the velocity of the system:

$$\frac{dx}{dt} = (V + 2nS)e^{-nt} \left(\cosh \omega t - \frac{n}{\omega} \sinh \omega t \right) + \frac{S}{\omega} e^{-nt}$$

$$\times [(n^2 - \omega^2) \sinh \omega t - 2n\omega \cosh \omega t] + \frac{p - f}{\omega} e^{-nt} \sinh \omega t$$

$$+ \frac{a}{4n^2 \varphi^2 + (n^2 - \omega^2 - \varphi^2)^2} \left\{ 2n\varphi^2 (\cos \varphi t - e^{-nt} \cosh \omega t) \right.$$

$$- \varphi(n^2 - \omega^2 - \varphi^2) \sin \varphi t + \frac{1}{\omega} e^{-nt} [(n^2 - \omega^2)^2$$

$$\left. + \varphi^2(\omega^2 + n^2)] \sinh \omega t \right\} \tag{16.4.15}$$

If in equations (16.4.14) and (16.4.15), $t = 0$, we obtain that $x = S$ and $\frac{dx}{dt} = V$, as expected according to the initial conditions of motion (2.1.2).

MOTION OF A SYSTEM ON A HORIZONTAL SURFACE WHILE BEING SUBJECTED TO CONSTANT RESISTING AND DRY FRICTION FORCES AND RESTRICTED BY FLEXIBLE AND FLUID LINKS

This chapter deals with systems, the motions of which are associated with their initial conditions of motion, constant resisting and dry friction forces, and constant active and harmonic forces.

17.1. Motion of a System Due to Its Initial Displacement and Velocity

The operational process described in the following is characterized by the motion of a system on a horizontal surface due to the initial conditions of motion. The system is subjected to a constant resisting force and a dry friction force. The motion of the system is restricted by a flexible link and a fluid link that connect the mass of the system to a non-movable support. Figure 17.1.1 shows a schematic diagram of the system, the notations in which are self-explanatory.

Based on the schematic diagram shown in Figure 17.1.1 and the considerations mentioned above, we compose the differential

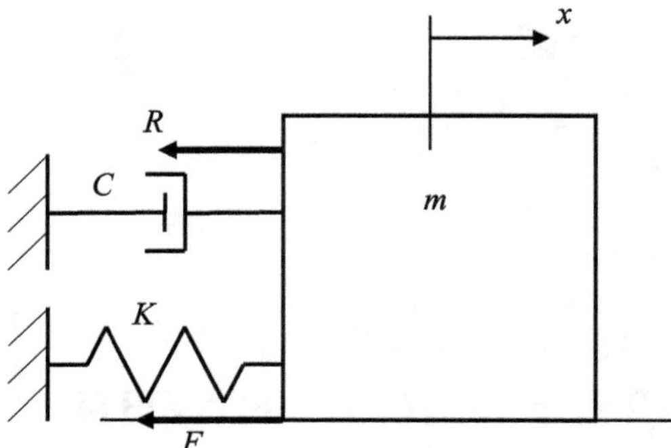

Fig. 17.1.1. Schematic diagram of a system moving on a horizontal surface while being restricted by flexible and fluid links and subjected to constant resisting and dry friction forces.

equations of motion of the system:

$$m\frac{d^2x}{dt^2} + C\frac{dx}{dt} + Kx + F + R = 0 \qquad (17.1.1)$$

Dividing equation (17.1.1) by m, we have

$$\frac{d^2x}{dt^2} + 2n\frac{dx}{dt} + \omega_0^2 x + f + r = 0 \qquad (17.1.2)$$

The initial conditions of motion are assigned according to expression (2.1.2).

Applying Laplace Transform pairs 5, 2, 4, 2, 1, 2, and 2, we convert equation (17.1.2) with the mentioned initial conditions of motion from the time domain into the Laplace domain and obtain the corresponding algebraic equation in the Laplace domain:

$$s^2x(s) - sV - s^2S + 2nsx(s) - 2nsS + \omega_0^2 x(s) + f + r = 0$$
$$(17.1.3)$$

Rearranging equation (17.1.3), we have

$$x(s)(s^2 + 2ns + \omega_0^2) = s(V + 2nS) + s^2S - f - r \qquad (17.1.4)$$

Solving equation (17.1.4) for the displacement $x(s)$ in the Laplace domain, we obtain

$$x(s) = \frac{s(V + 2nS)}{s^2 + 2ns + \omega_0^2} + \frac{s^2S}{s^2 + 2ns + \omega_0^2} - \frac{r + f}{s^2 + 2ns + \omega_0^2} \qquad (17.1.5)$$

Combining equations (17.1.4) with equations (14.1.7) and (14.1.8), we obtain

$$x(s) = \frac{s(V + 2nS)}{(s+n)^2 + \omega^2} + \frac{s^2 S}{(s+n)^2 + \omega^2} - \frac{f+r}{(s+n)^2 + \omega^2} \qquad (17.1.6)$$

17.1.1. *Underdamped motion, $\omega^2 > 0$*

Equation (17.1.6) implies that ω^2 is positive; therefore, applying to this equation Laplace Transform pairs 1, 27, 35, and 21, we invert this equation from the Laplace domain into the time domain and obtain the solution of differential equation of motion (17.1.1) with the initial conditions of motion according to expression (2.1.2):

$$x = \frac{V + 2nS}{\omega} e^{-nt} \sin \omega t + Se^{-nt} \left(\cos \omega t - \frac{n}{\omega} \sin \omega t \right)$$
$$- \frac{f+r}{\omega^2 + n^2} \left[1 - e^{-nt} \left(\cos \omega t + \frac{n}{\omega} \sin \omega t \right) \right] \qquad (17.1.7)$$

The first derivative from equation (17.1.7) yields the velocity of the system:

$$\frac{dx}{dt} = (V + 2nS)e^{-nt} \left(\cos \omega t - \frac{n}{\omega} e^{-nt} \sin \omega t \right)$$
$$+ \frac{S}{\omega} e^{-nt} [(n^2 - \omega^2) \sin \omega t - 2n\omega \cos \omega t]$$
$$- \frac{f+r}{\omega^2 + n^2} e^{-nt} \sin \omega t \qquad (17.1.8)$$

Assuming that in equations (17.1.7) and (17.1.8), $t = 0$, then we obtain that $x = S$ and $\frac{dx}{dt} = V$, as expected according to the initial conditions of motion (2.1.2).

The second derivative from equation (17.1.7) represents the expression for calculating the acceleration of the system:

$$\frac{d^2 x}{dt^2} = \frac{V + 2nS}{\omega} e^{-nt} [(n^2 - \omega^2) \sin \omega t - 2n\omega \cos \omega t]$$
$$+ Se^{-nt} \left[\frac{n}{\omega} (3\omega^2 - n^2) \sin \omega t - (\omega^2 - 3n^2) \cos \omega t \right]$$
$$- (f+r)e^{-nt} \left(\cos \omega t - \frac{n}{\omega} \sin \omega t \right) \qquad (17.1.9)$$

Equations (17.1.7)–(17.1.9) represent the basic parameters of motion of the system.

17.1.2. *Critical damping, $\omega^2 = 0$*

In equation (17.1.6), equating the parameter ω^2 to zero, we determine the displacement of the system in the Laplace domain for the case of critical damping:

$$x(s) = \frac{s(V + 2nS)}{(s + n)^2} + \frac{s^2 S}{(s + n)^2} - \frac{f + r}{(s + n)^2} \qquad (17.1.10)$$

Applying Laplace Transform pairs 1, 25, 31, and 19 to equation (17.1.10), we invert this equation from the Laplace domain into the time domain and obtain the solution of differential equation (17.1.1) at the initial conditions of motion according to expression (2.1.2) for the case of critical damping:

$$x = (V + 2nS)te^{-nt} + S(1 - nt)e^{-nt} - \frac{f + r}{n^2}[1 - (1 + nt)e^{-nt}]$$

$$(17.1.11)$$

Taking the first derivative from equation (17.1.11), we obtain the equation for calculating the velocity of the system:

$$\frac{dx}{dt} = (V + 2nS)e^{-nt}(1 - nt) + nSe^{-nt}(nt - 2) - (f + r)te^{-nt}$$

$$(17.1.12)$$

Assuming that in equations (17.1.11) and (17.1.12), $t = 0$, we obtain that $x = S$ and $\frac{dx}{dt} = V$, as it should be according to initial conditions of motion (2.1.2)

17.1.3. *Overdamped motion, $\omega^2 < 0$*

For this case, we change in equation (17.1.6) the sign of parameter ω^2 from positive to negative, and we obtain

$$x(s) = \frac{s(V + 2nS)}{(s + n)^2 - \omega^2} + \frac{s^2 S}{(s + n)^2 - \omega^2} - \frac{f + r}{(s + n)^2 - \omega^2} \qquad (17.1.13)$$

Applying Laplace Transform pairs 1, 28, 36, and 22, we invert equation (17.1.13) from the Laplace domain into the time domain and obtain for this case the solution of differential equation of motion (17.1.1) with the initial conditions of motion according to expression (2.1.2):

$$x = \frac{V + 2nS}{\omega} e^{-nt} \sinh \omega t + S e^{-nt} \left(\cosh \omega t - \frac{n}{\omega} \sinh \omega t \right)$$

$$- \frac{f + r}{n^2 - \omega^2} \left[1 - e^{-nt} \left(\cosh \omega t + \frac{n}{\omega} \sinh \omega t \right) \right] \qquad (17.1.14)$$

Taking the first derivative from equation (17.1.14), we determine the velocity of the system:

$$\frac{dx}{dt} = (V + 2nS) e^{-nt} \left(\cosh \omega t - \frac{n}{\omega} \sinh \omega t \right)$$

$$+ \frac{S}{\omega} e^{-nt} [(n^2 + \omega^2) \sinh \omega t - 2n\omega \cosh \omega t]$$

$$- \frac{f + r}{\omega} e^{-nt} \sinh \omega t \qquad (17.1.15)$$

Supposing that in equations (17.1.14) and (17.1.15), $t = 0$, we obtain that $x = S$ and $\frac{dx}{dt} = V$, as it should be according to the initial conditions of motion expressed by (2.1.2).

17.2. Motion of a System Due to Its Initial Displacement and Velocity and to a Constant Active Force

This operational process describes a system moving on a horizontal surface due to its initial displacement and velocity and to a constant active force while being subjected to the actions of constant resisting and dry friction forces.

Figure 17.2.1 shows a schematic diagram of the system, the notations in which are self-explanatory.

Based on the schematic diagram shown in Figure 17.2.1 and the considerations mentioned above, we compose the differential equation

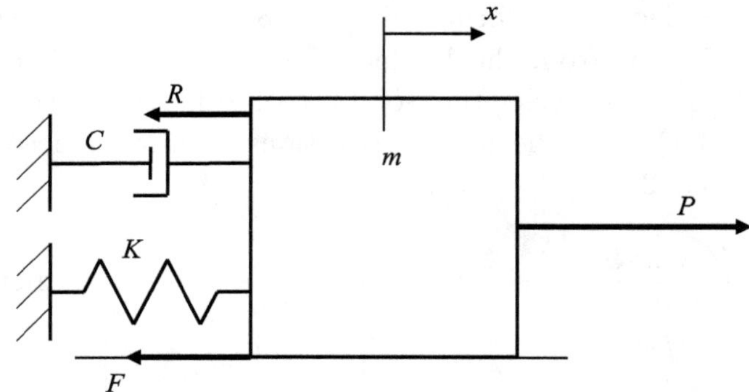

Fig. 17.2.1. Schematic diagram of a system moving on a horizontal surface being restricted by flexible and fluid links and subjected to constant resisting, dry friction, and constant active forces.

of motion of the system:

$$m\frac{d^2x}{dt^2} + C\frac{dx}{dt} + Kx + F + R = P \qquad (17.2.1)$$

Dividing equation (17.2.1) by m, we have

$$\frac{d^2x}{dt^2} + 2n\frac{dx}{dt} + \omega_0^2 x + f + r = p \qquad (17.2.2)$$

The initial conditions of motion are assigned according to expression (2.1.2).

Applying Laplace Transform pairs 5, 2, 4, 2, 1, 2, 2, and 2, we convert equation (17.2.2) with the mentioned initial conditions of motion from the time domain into the Laplace domain and obtain the corresponding algebraic equation in the Laplace domain:

$$s^2 x(s) - sV - s^2 S + 2nsx(s) - 2nsS + \omega_0^2 x(s) = p - f - r \quad (17.2.3)$$

Rearranging equation (17.2.3), we may write

$$x(s)(s^2 + 2ns + \omega_0^2) = s(V + 2nS) + s^2 S + p - f - r \qquad (17.2.4)$$

Solving equation (17.2.4) for the displacement $x(s)$ in the Laplace domain, we obtain

$$x(s) = \frac{s(V + 2ns)}{s^2 + 2ns + \omega_0^2} + \frac{s^2 S}{s^2 + 2ns + \omega_0^2} + \frac{p - f - r}{s^2 + 2ns + \omega_0^2} \qquad (17.2.5)$$

Combining equation (17.2.5) with equations (14.1.7) and (14.1.8), we obtain

$$x(s) = \frac{s(V + 2nS)}{(s + n)^2 + \omega^2} + \frac{s^2 S}{(s + n)^2 + \omega^2} + \frac{p - f - r}{(s + n)^2 + \omega^2} \qquad (17.2.6)$$

17.2.1. *Underdamped motion, $\omega^2 > 0$*

We assume in equation (17.2.6) that the parameter ω^2 is positive; therefore, it is allowable to apply Laplace Transform pairs 1, 27, 35, and 21 to this equation and invert it from the Laplace domain into the time domain and obtain the solution of differential equation of motion (17.2.1) with the initial conditions of motion according to expression (2.1.2):

$$x = \frac{V + 2nS}{\omega} e^{-nt} \sin \omega t + S e^{-nt} \left(\cos \omega t - \frac{n}{\omega} \sin \omega t \right)$$

$$+ \frac{p - f - r}{\omega^2 + n^2} \left[1 - e^{-nt} \left(\cos \omega t + \frac{n}{\omega} \sin \omega t \right) \right] \qquad (17.2.7)$$

Taking the first derivative from equation (17.2.7), we obtain the equation describing the velocity of the system:

$$\frac{dx}{dt} = (V + 2nS) e^{-nt} \left(\cos \omega t - \frac{n}{\omega} \sin \omega t \right) + \frac{S}{\omega} e^{-nt} [(n^2 - \omega^2) \sin \omega t$$

$$- 2n\omega \cos \omega t] + \frac{p - f - r}{\omega} e^{-nt} \sin \omega t \qquad (17.2.8)$$

Supposing that in equations (17.2.7) and (17.2.8), $t = 0$, we obtain that $x = S$ and $\frac{dx}{dt} = V$, as expected according to initial conditions of motion (2.1.2).

The second derivative from equation (17.2.7) yields the acceleration of the system:

$$\frac{d^2 x}{dt^2} = \frac{V + 2nS}{\omega} e^{-nt} [(n^2 - \omega^2) \sin \omega t - 2n\omega \cos \omega t]$$

$$+ S e^{-nt} \left[\frac{n}{\omega} (3\omega^2 - n^2) \sin \omega t - (\omega^2 - 3n^2) \cos \omega t \right]$$

$$+ (p - f - r) e^{-nt} \left(\cos \omega t - \frac{n}{\omega} \sin \omega t \right) \qquad (17.2.9)$$

Equations (17.2.7)–(17.2.9) represent the basic parameters of motion of the system.

17.2.1.1. *Numerical solution*

The following is a Python program and the associated graph (Figure 17.2.2) of equation (17.2.7).

```python
from matplotlib.pyplot import plot, show
from numpy import linspace, sin, cos, exp


S = 0.1
V = 0.15
n = 15
omega = 60
p = 10
f = 2
r = 1
t = linspace(0,0.6,1000)

x = ((V + 2*n*S)/omega)*exp(-n*t)*sin(omega*t) \
  + S*exp(-n*t)*(cos(omega*t) - (n/omega)*sin(omega*t)) \
  + ((p - f - r)/(omega**2 + n**2))*(1 - exp(-n*t)*(cos(omega*t) \
                    + (n/omega)*sin(omega*t)))

plot(t, x)
show()
```

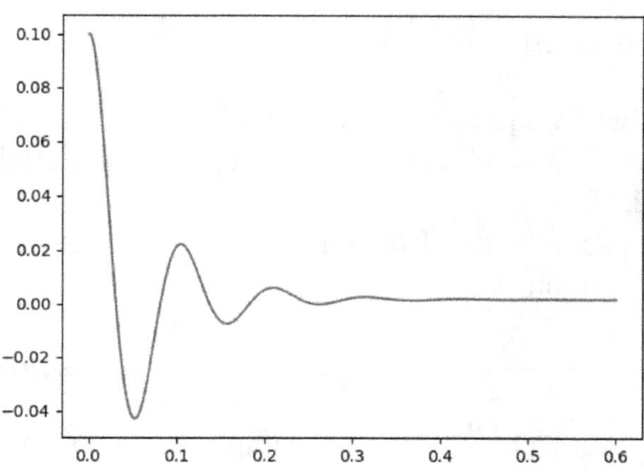

Fig. 17.2.2. Graph of equation (17.2.7) characterizing underdamped motion.

17.2.2. *Critical damping, $\omega^2 = 0$*

Taking in equation (17.2.6) that $\omega^2 = 0$, we obtain the expression for the displacement of the system in the Laplace domain:

$$x(s) = \frac{s(V + 2nS)}{(s + n)^2} + \frac{s^2 S}{(s + n)^2} + \frac{p - f - r}{(s + n)^2} \qquad (17.2.13)$$

Applying Laplace Transform pairs 1, 25, 31, and 19 to equation (17.2.13), we invert this equation from the Laplace domain into the time domain and obtain the solution of differential equation (17.2.1) at the initial conditions of motion according to expression (2.1.2) for the case of critical damping:

$$x = (V + 2nS)te^{-nt} + S(1 - nt)e^{-nt} + \frac{p - f - r}{n^2}[1 - e^{-nt}(1 + nt)] \qquad (17.2.14)$$

Taking the first derivative from equation (17.2.14), we obtain the equation for calculating the velocity of the system:

$$\frac{dx}{dt} = (V + 2nS)e^{-nt}(1 - nt) + nSe^{-nt}(nt - 2) + (p - f - r)te^{-nt} \qquad (17.2.15)$$

Supposing that in equations (17.2.14) and (17.2.15), $t = 0$, we obtain that $x = S$ and $\frac{dx}{dt} = V$, as it should be according to initial conditions of motion (2.1.2).

17.2.3. *Overdamped motion, $\omega^2 < 0$*

In this case, we change in equation (17.2.6) the sign of the parameter ω^2 from positive to negative, and we obtain

$$x(s) = \frac{s(V + 2nS)}{(s + n)^2 - \omega^2} + \frac{s^2 S}{(s + n)^2 - \omega^2} + \frac{p - f - r}{(s + n)^2 - \omega^2} \qquad (17.2.16)$$

Applying Laplace Transform Pairs 1, 28, 36, and 22, we invert the equation (17.2.16) from the Laplace domain into the time domain and obtain for this case the solution of differential equation of motion (17.2.1) with the initial conditions of motion according to expression (2.1.2):

$$x = \frac{V + 2nS}{\omega} e^{-nt} \sinh \omega t + S e^{-nt} \left(\cosh \omega t - \frac{n}{\omega} \sinh \omega t \right)$$

$$+ \frac{p - f - r}{n^2 - \omega^2} \left[1 - e^{-nt} \left(\cosh \omega t - \frac{n}{\omega} \sinh \omega t \right) \right] \qquad (17.2.17)$$

Taking the first derivative from equation (17.2.17), we determine the velocity of the system:

$$\frac{dx}{dt} = (V + 2nS) e^{-nt} \left(\cosh \omega t - \frac{n}{\omega} \sinh \omega t \right)$$

$$+ \frac{S}{\omega} e^{-nt} [(n^2 + \omega^2) \sinh \omega t - 2n\omega \cosh \omega t]$$

$$+ \frac{p - f - r}{\omega} e^{-nt} \sinh \omega t \qquad (17.2.18)$$

Supposing that in equations (17.2.17) and (17.2.18), $t = 0$, we obtain that $x = S$ and $\frac{dx}{dt} = V$, as expected according to the initial conditions of motion (2.1.2).

17.3. Motion of a System Due to Its Initial Displacement and Velocity and to a Harmonic Force

The operational process of a system subjected to a constant resisting force and a dry friction force while moving on a horizontal surface due to its initial displacement and velocity and to a harmonic force is addressed in this section. The motion of the system is restricted by flexible and fluid links.

Figure 17.3.1 shows a schematic diagram of the system described above. The notations in the figure are self-explanatory.

Based on the schematic diagram shown in Figure 17.3.1 and the considerations mentioned above, we compose the differential equations of motion of the system:

$$m \frac{d^2 x}{dt^2} + C \frac{dx}{dt} + Kx + F + R = A \cos \varphi t \qquad (17.3.1)$$

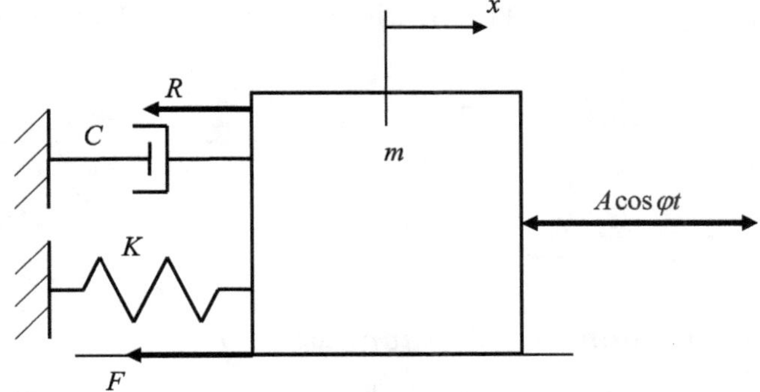

Fig. 17.3.1. Schematic diagram of a system moving on a horizontal surface while being restricted by flexible and fluid links and subjected to constant resisting, dry friction, and harmonic forces.

Dividing equation (17.3.1) by m, we may write

$$\frac{d^2x}{dt^2} + 2n\frac{dx}{dt} + \omega_0^2 x + f + r = a\cos\varphi t \qquad (17.3.2)$$

The initial conditions of motion are taken according to expression (2.1.2).

Applying Laplace Transform pairs 5, 2, 4, 2, 1, 2, 2, and 29, we convert equation (17.3.2) with the above-mentioned initial conditions of motion from the time domain into the Laplace domain and obtain the corresponding algebraic equation in the Laplace domain:

$$s^2 x(s) - sV - s^2 S + 2ns x(s) - 2ns S + \omega_0^2 x(s) + f + r = \frac{as^2}{s^2 + \varphi^2}$$
$$(17.3.3)$$

Rearranging equation (17.3.3), we have

$$x(s)(s^2 + 2ns + \omega_0^2) = s(V + 2ns) + s^2 S + \frac{as^2}{s^2 + \varphi^2} - f - r$$
$$(17.3.4)$$

Solving equation (17.3.4) for the displacement $x(s)$ in the Laplace domain, we obtain

$$x(s) = \frac{s(V + 2ns)}{s^2 + 2ns + \omega_0^2} + \frac{s^2 S}{s^2 + 2ns + \omega_0^2} + \frac{as^2}{(s^2 + 2ns + \omega_0^2)(s^2 + \varphi^2)}$$

$$- \frac{f + r}{s^2 + 2ns + \omega_0^2} \qquad (17.3.5)$$

Combining equation (17.3.5) with equations (14.1.7) and (14.1.8), we may write

$$x(s) = \frac{s(V + 2nS)}{(s+n)^2 + \omega^2} + \frac{s^2 S}{(s+n)^2 + \omega^2} + \frac{s^2 a}{[(s+n)^2 + \omega^2](s^2 + \varphi^2)}$$

$$- \frac{f + r}{(s+n)^2 + \omega^2} \tag{17.3.6}$$

17.3.1. *Underdamped vibration, $\omega^2 > 0$*

It is assumed that in equation (17.3.6) the parameter ω^2 is positive; therefore, we may apply to this equation Laplace Transform pairs 1, 27, 35, 68, and 21 in order to invert this equation from the Laplace domain into the time domain and obtain the solution of the differential equation of motion (17.3.1) at the initial conditions of motion according to expression (2.1.2):

$$x = \frac{V + 2nS}{\omega} e^{-nt} \sin \omega t + S e^{-nt} \left(\cos \omega t - \frac{n}{\omega} \sin \omega t \right)$$

$$+ \frac{a}{4n^2 \varphi^2 + (\omega^2 + n^2 - \varphi^2)^2} \Big[(\omega^2 + n^2 - \varphi^2)$$

$$\times (\cos \varphi t - e^{-nt} \cos \omega t) + 2n\varphi \sin \varphi t$$

$$- \frac{n}{\omega}(\omega^2 + n^2 + \varphi^2) e^{-nt} \sin \omega t \Big]$$

$$- \frac{f + r}{\omega^2 + n^2} \Big[1 - e^{-nt} \left(\cos \omega t + \frac{n}{\omega} \sin \omega \right) \Big] \tag{17.3.7}$$

The first derivative from equation (17.3.7) yields the velocity of the system:

$$\frac{dx}{dt} = (V + 2nS)e^{-nt} \left(\cos \omega t - \frac{n}{\omega} \sin \omega t \right) + \frac{S}{\omega} e^{-nt}[(n^2 - \omega^2) \sin \omega t$$

$$- 2n\omega \cos \omega t] + \frac{a}{4n^2\varphi^2 + (\omega^2 + n^2 - \varphi^2)^2} \Big\{ 2n\varphi^2 (\cos \varphi t$$

$$- e^{-nt} \cos \omega t) - \varphi(\omega^2 + n^2 - \varphi^2) \sin \varphi t + \frac{1}{\omega} e^{-nt}[(\omega^2 + n^2)^2$$

$$- \varphi^2 (\omega^2 - n^2)] \sin \omega t \Big\} - \frac{f + r}{\omega} e^{-nt} \sin \omega t \tag{17.3.8}$$

Assuming that in equations (17.3.7) and (17.3.8), $t = 0$, we obtain that $x = S$ and $\frac{dx}{dt} = V$, as expected according to the initial conditions of motion (2.1.2).

The first derivative from equation (17.1.8) allows for determining the acceleration of the system:

$$\frac{d^2 x}{dt^2} = \frac{V + 2nS}{\omega} e^{-nt}[(n^2 - \omega^2)\sin \omega t - 2n\omega \cos \omega t]$$

$$+ S e^{-nt}\left[\frac{n}{\omega}(3\omega^2 - n^2)\sin \omega t - (\omega^2 - 3n^2)\cos \omega t\right]$$

$$+ \frac{a}{4n^2\varphi^2 + (\omega^2 + n^2 - \varphi^2)^2}\Big\{[(n^2 + \omega^2)^2$$

$$+ \varphi^2(3n^2 - \omega^2)]e^{-nt}\cos \omega t - \varphi^2(\omega^2 + n^2 - \varphi^2)\cos \varphi t$$

$$- 2\varphi^3 n \sin \varphi t - \frac{n}{\omega}[(\omega^2 + n^2)^2 - \varphi^2(n^2 - 3\omega^2)]e^{-nt}\sin \omega t\Big\}$$

$$(17.3.9)$$

Equations (17.3.7)–(17.3.9) represent the basic parameters of motion of the system.

17.3.2. *Critical damping, $\omega^2 = 0$*

Taking in equation (17.3.6) that $\omega^2 = 0$, we obtain the expression for the displacement of the system in the Laplace domain:

$$x(s) = \frac{s(V + 2nS)}{(s + n)^2} + \frac{s^2 S}{(s + n)^2} + \frac{s^2 a}{(s + n)^2(s^2 + \varphi^2)} - \frac{f + r}{(s + n)^2}$$

$$(17.3.10)$$

Applying Laplace Transform pairs 1, 25, 31, 71, and 19 to equation (17.3.10), we invert this equation from the Laplace domain into the time domain and obtain the solution of differential equation (17.3.1) with the initial conditions of motion according to expression (2.1.2) for the case of critical damping:

$$x = (V + 2nS)te^{-nt} + S(1 - nt)e^{-nt} + \frac{a}{(\varphi^2 + n^2)^2}\{2n\varphi \sin \varphi t$$

$$+ e^{-nt}[\varphi^2 - n^2 - nt(\varphi^2 + n^2)] - (\varphi^2 - n^2)\cos \varphi t\}$$

$$- \frac{f + r}{n^2}[1 - e^{-nt}(1 + nt)] \qquad (17.3.11)$$

Taking the first derivative from equation (17.3.11), we obtain the equation for calculating the velocity of the system:

$$\frac{dx}{dt} = (V + 2nS)e^{-nt}(1 - nt) + nSe^{-nt}(nt - 2)$$

$$+ \frac{a}{(\varphi^2 + n^2)^2}[(\varphi^2 - n^2)(\varphi \sin \varphi t + n^2 t e^{-nt})$$

$$- 2\varphi^2 n(\cos \varphi t - e^{-nt})] - (f + r)t e^{-nt} \qquad (17.3.12)$$

Assuming that in equations (17.3.11) and (17.3.12), $t = 0$, we obtain that $x = S$ and $\frac{dx}{dt} = V$, as it should be according to the initial conditions of motion (2.1.2).

17.3.3. *Overdamped motion, $\omega^2 < 0$*

For this case, we replace in equation (17.3.6) the sign of the parameter ω^2 from positive to negative, and we obtain

$$x(s) = \frac{s(V + 2nS)}{(s + n)^2 - \omega^2} + \frac{s^2 S}{(s + n)^2 - \omega^2} + \frac{s^2 a}{[(s + n)^2 - \omega^2](s^2 + \varphi^2)}$$

$$- \frac{f + r}{(s + n)^2 - \omega^2} \qquad (17.3.13)$$

Applying Laplace Transform pairs 1, 28, 36, 69, and 22, we invert equation (17.3.13) from the Laplace domain into the time domain and obtain for this case the solution of differential equation of motion (17.3.1) with the initial conditions of motion according to expression (2.1.2):

$$x = \frac{V + 2nS}{\omega}e^{-nt}\sinh \omega t + Se^{-nt}\left(\cosh \omega t - \frac{n}{\omega}\sinh \omega t\right)$$

$$+ \frac{a}{4n^2\varphi^2 + (n^2 - \omega^2 - \varphi^2)^2}\Big[(n^2 - \omega^2 - \varphi^2)$$

$$\times (\cos \varphi t - e^{-nt}\cosh \omega t) + 2n\varphi \sin \varphi t$$

$$- \frac{n}{\omega}(n^2 - \omega^2 + \varphi^2)e^{-nt}\sinh \omega t\Big]$$

$$- \frac{f + r}{n^2 - \omega^2}\left[1 - e^{-nt}\left(\cosh \omega t - \frac{n}{\omega}\sinh \omega t\right)\right] \qquad (17.3.14)$$

The first derivative from equation (17.3.14) yields the velocity of the system:

$$\frac{dx}{dt} = (V + 2nS)e^{-nt}\left(\cosh \omega t - \frac{n}{\omega}\sinh \omega t\right)$$

$$+ \frac{S}{\omega}e^{-nt}[(n^2 + \omega^2)\sinh \omega t - 2n\omega \cosh \omega t]$$

$$+ \frac{a}{4n^2\varphi^2 + (n^2 - \omega^2 - \varphi^2)^2}\left\{2n\varphi^2(\cos \varphi t - e^{-nt}\cosh \omega t)\right.$$

$$- \varphi(n^2 - \omega^2 - \varphi^2)\sin \varphi t + \frac{1}{\omega}e^{-nt}[(n^2 - \omega^2)^2$$

$$\left. + \varphi^2(\omega^2 + n^2)]\sinh \omega t\right\} - \frac{f + r}{\omega}e^{-nt}\sinh \omega t \qquad (17.3.15)$$

Assuming that in equations (17.3.14) and (17.3.15), $t = 0$, we obtain that $x = S$ and $\frac{dx}{dt} = V$, as expected according to the initial conditions of motion (2.1.2).

17.4. Motion of a System Due to Its Initial Displacement and Velocity as Well as to a Constant Active Force and a Harmonic Force

This operational process is associated with a system subjected to a constant resisting force and a dry friction force while moving on a horizontal surface due to its initial displacement and velocity and to a constant active force and harmonic force.

Figure 17.4.1 shows a schematic diagram of the system described above. The notations in the figure are self-explanatory.

Based on the schematic diagram shown in Figure 17.4.1 and the considerations mentioned above, we compose the differential equations of motion of the system:

$$m\frac{d^2x}{dt^2} + C\frac{dx}{dt} + Kx + F + R = P + A\cos \varphi t \qquad (17.4.1)$$

Dividing equation (17.4.1) by m, we may write

$$\frac{d^2x}{dt^2} + 2n\frac{dx}{dt} + \omega_0^2 x + f + r = p + a\cos \varphi t \qquad (17.4.2)$$

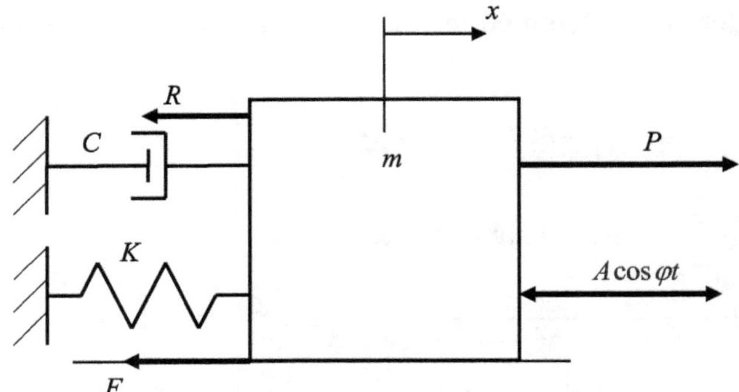

Fig. 17.4.1. Schematic diagram of a system moving on a horizontal surface while being restricted by flexible and fluid links and subjected to constant resisting, dry friction, constant active, and harmonic forces.

The initial conditions of motion are taken according to expression (2.1.2).

Applying Laplace Transform pairs 5, 2, 4, 2, 1, 2, 2, and 29, we convert equation (17.4.2) with the initial conditions of motion according to expression (2.1.2) from the time domain into the Laplace domain and obtain the corresponding algebraic equation in the Laplace domain:

$$s^2 x(s) - sV - s^2 S + 2nsx(s) - 2nsS + \omega_0^2 x(s) + f + r = p + \frac{as^2}{s^2 + \varphi^2}$$
$$(17.4.3)$$

Rearranging equation (17.4.3), we have

$$x(s)(s^2 + 2ns + \omega_0^2) = s(V + 2nS) + s^2 S + p - f - r + \frac{as^2}{s^2 + \varphi^2}$$
$$(17.4.4)$$

Solving equation (17.4.4) for the displacement $x(s)$ in the Laplace domain, we obtain

$$x(s) = \frac{s(V + 2nS)}{s^2 + 2ns + \omega_0^2} + \frac{s^2 S}{s^2 + 2ns + \omega_0^2} + \frac{p - f - r}{s^2 + 2ns + \omega_0^2}$$

$$+ \frac{as^2}{(s^2 + 2ns + \omega_0^2)(s^2 + \varphi^2)} \qquad (17.4.5)$$

Combining equation (17.4.5) with equations (14.1.7) and (14.1.8), we obtain

$$x(s) = \frac{s(V + 2nS)}{(s+n)^2 + \omega^2} + \frac{s^2 S}{(s+n)^2 + \omega^2} + \frac{p - f - r}{(s+n)^2 + \omega^2}$$

$$+ \frac{s^2 a}{[(s+n)^2 + \omega^2](s^2 + \varphi^2)} \tag{17.4.6}$$

17.4.1. *Underdamped motion, $\omega^2 > 0$*

Since in equation (17.4.6), the parameter ω^2 is assumed to be positive, we may apply to this equation Laplace Transform pairs 1, 27, 35, 21, and 68 in order to invert this equation from the Laplace domain into the time domain and obtain the solution of differential equation of motion (17.4.1) with the initial conditions of motion according to expression (2.1.2):

$$x = \frac{V + 2nS}{\omega} e^{-nt} \sin \omega t + S e^{-nt} \left(\cos \omega t - \frac{n}{\omega} \sin \omega t \right) + \frac{p - f - r}{\omega^2 + n^2}$$

$$\times \left[1 - e^{-nt} \left(\cos \omega t + \frac{n}{\omega} \sin \omega t \right) \right] + \frac{a}{4n^2\varphi^2 + (\omega^2 + n^2 - \varphi^2)^2}$$

$$\times \left[(\omega^2 + n^2 - \varphi^2)(\cos \varphi t - e^{-nt} \cos \omega t) + 2n\varphi \sin \varphi t \right.$$

$$\left. - \frac{n}{\omega}(\omega^2 + n^2 + \varphi^2) e^{-nt} \sin \omega t \right] \tag{17.4.7}$$

The first derivative from equation (17.4.7) yields the velocity of the system:

$$\frac{dx}{dt} = (V + 2nS) e^{-nt} \left(\cos \omega t - \frac{n}{\omega} \sin \omega t \right) + \frac{S}{\omega} e^{-nt} [(n^2 - \omega^2)$$

$$\times \sin \omega t - 2n\omega \cos \omega t] + \frac{p - f - r}{\omega} e^{-nt} \sin \omega t$$

$$+ \frac{a}{4n^2\varphi^2 + (\omega^2 + n^2 - \varphi^2)^2} \left\{ 2n\varphi^2 (\cos \varphi t - e^{-nt} \cos \omega t) \right.$$

$$- \varphi(\omega^2 + n^2 - \varphi^2) \sin \varphi t + \frac{1}{\omega} e^{-nt} [(\omega^2 + n^2)^2$$

$$\left. - \varphi^2(\omega^2 - n^2)] \sin \omega t \right\} \tag{17.4.8}$$

Assuming that in equations (17.4.7) and (17.4.8), $t = 0$, we obtain that $x = S$ and $\frac{dx}{dt} = V$, as it is expected according to the initial conditions of motion (2.1.2).

The first derivative from equation (17.4.8) allows for obtaining the equation describing the acceleration of the system:

$$
\begin{aligned}
\frac{d^2 x}{dt^2} &= \frac{V + 2nS}{\omega} e^{-nt} [(n^2 - \omega^2) \sin \omega t - 2n\omega \cos \omega t] \\
&\quad + S e^{-nt} \left[\frac{n}{\omega} (3\omega^2 - n^2) \sin \omega t - (\omega^2 - 3n^2) \cos \omega t \right] \\
&\quad + (p - f - r) e^{-nt} \left(\cos \omega t - \frac{n}{\omega} \sin \omega t \right) \\
&\quad + \frac{a}{4n^2 \varphi^2 + (\omega^2 + n^2 - \varphi^2)^2} \Big\{ [(n^2 + \omega^2)^2 + \varphi^2 (3n^2 - \omega^2)] e^{-nt} \\
&\quad \times \cos \omega t - \varphi^2 (\omega^2 + n^2 - \varphi^2) \cos \varphi t - 2\varphi^3 n \sin \varphi t \\
&\quad - \frac{n}{\omega} [(\omega^2 + n^2)^2 - \varphi^2 (n^2 - 3\omega^2)] e^{-nt} \sin \omega t \Big\}
\end{aligned}
\tag{17.4.9}
$$

Equations (17.4.7)–(17.4.9) represent the basic parameters of motion of the system.

17.4.2. *Critical damping, $\omega^2 = 0$*

In equation (17.4.6), equating the parameter ω^2 to zero, we obtain the expression for determining the displacement of the system in the Laplace domain:

$$
x(s) = \frac{s(V + 2nS)}{(s + n)^2} + \frac{s^2 S}{(s + n)^2} + \frac{p - f - r}{(s + n)^2} + \frac{s^2 a}{(s + n)^2 (s^2 + \varphi^2)}
\tag{17.4.10}
$$

Applying Laplace Transform pairs 1, 25, 31, 19, and 71 to equation (17.4.10), we invert this equation from the Laplace domain into the time domain and obtain the solution of differential equation (17.4.1) with the initial conditions of motion according to expression (2.1.2) for the case of critical damping:

$$x = (V + 2nS)te^{-nt} + S(1 - nt)e^{-nt} + \frac{p - f - r}{n^2}[1 - e^{-nt}(1 + nt)]$$

$$+ \frac{a}{(\varphi^2 + n^2)^2}\{2n\varphi \sin \varphi t + e^{-nt}[\varphi^2 - n^2 - nt(\varphi^2 + n^2)]$$

$$- (\varphi^2 - n^2)\cos \varphi t\} \tag{17.4.11}$$

Taking the first derivative from equation (17.4.11), we determine the velocity of the system:

$$\frac{dx}{dt} = (V + 2nS)e^{-nt}(1 - nt) + nSe^{-nt}(nt - 2) + (p - f - r)te^{-nt}$$

$$+ \frac{a}{(\varphi^2 + n^2)^2}[(\varphi^2 - n^2)(\varphi \sin \varphi t + n^2 te^{-nt})$$

$$- 2\varphi^2 n(\cos \varphi t - e^{-nt})] \tag{17.4.12}$$

Supposing that in equations (17.4.11) and (17.4.12), $t = 0$, we get that $x = S$ and $\frac{dx}{dt} = V$, as it should be according to the initial conditions of motion (2.1.2).

17.4.3. *Overdamped motion, $\omega^2 < 0$*

For this case, we change in equation (17.4.6) the sign of the parameter ω^2 from positive to negative, and we obtain

$$x(s) = \frac{s(V + 2nS)}{(s + n)^2 - \omega^2} + \frac{s^2 S}{(s + n)^2 - \omega^2} + \frac{p - f - r}{(s + n)^2 - \omega^2}$$

$$+ \frac{s^2 a}{[(s + n)^2 - \omega^2](s^2 + \varphi^2)} \tag{17.4.13}$$

Applying Laplace Transform pairs 1, 28, 36, and 69, we invert the equation (17.4.13) from the Laplace domain into the time domain and obtain for this case the solution of differential equation of motion

(17.4.1) with the initial conditions of motion according to expression (2.1.2):

$$x = \frac{V + 2nS}{\omega} e^{-nt} \sinh \omega t + S e^{-nt} \left(\cosh \omega t - \frac{n}{\omega} \sinh \omega t \right) + \frac{p - f - r}{n^2 - \omega^2}$$

$$\times \left[1 - e^{-nt} \left(\cosh \omega t - \frac{n}{\omega} \sinh \omega t \right) \right] + \frac{a}{4n^2 \varphi^2 + (n^2 - \omega^2 - \varphi^2)^2}$$

$$\times \left[(n^2 - \omega^2 - \varphi^2)(\cos \varphi t - e^{-nt} \cosh \omega t) + 2n\varphi \sin \varphi t \right.$$

$$\left. - \frac{n}{\omega}(n^2 - \omega^2 + \varphi^2) e^{-nt} \sinh \omega t \right] \tag{17.4.14}$$

The first derivative from equation (17.4.14) yields the velocity of the system:

$$\frac{dx}{dt} = (V + 2nS)e^{-nt} \left(\cosh \omega t - \frac{n}{\omega} \sinh \omega t \right) + \frac{S}{\omega} e^{-nt}[(n^2 - \omega^2)$$

$$\times \sinh \omega t - 2n\omega \cosh \omega t] + \frac{p - f - p}{\omega} e^{-nt} \sinh \omega t$$

$$+ \frac{a}{4n^2 \varphi^2 + (n^2 - \omega^2 - \varphi^2)^2} \left\{ 2n\varphi^2 (\cos \varphi t - e^{-nt} \cosh \omega t) \right.$$

$$- \varphi(n^2 - \omega^2 - \varphi^2) \sin \varphi t + \frac{1}{\omega} e^{-nt}[(n^2 - \omega^2)^2$$

$$\left. + \varphi^2(\omega^2 + n^2)] \sinh \omega t \right\} \tag{17.4.15}$$

Assuming that in equations (17.4.14) and (17.4.15), $t = 0$, we obtain that $x = S$ and $\frac{dx}{dt} = V$, as expected according to the initial conditions of motion (2.1.2).

PART 2

Two-Degree-of-Freedom Systems

FUNDAMENTALS OF THE STUDY OF TWO-DEGREE-OF-FREEDOM SYSTEMS

Movable mechanical engineering systems predominantly have one- or two-degree-of-freedom structures, while most of them possess one degree of freedom. The descriptions of theoretical fundamentals that allow us to analyze the operational processes of one-degree-of-freedom systems were presented in Part 1 of this text.

In Part 2, we consider the analyses of two-degree-of-freedom systems, the masses of which are moving along the line crossing the centers of gravity of these two masses.

The existence of two- or multi-degree-of-freedom systems is based on the specific features of the kinematic links connecting the two or multiple masses to each other as well as to the non-movable supports. These links allow relative motion between connected masses of the system. In the case when two masses are in a state of relative motion, the forces exerted by the links on these masses are proportional to the differences between the absolute values of the parameters of motion of these two masses. The mentioned parameters of motion of these masses represent their displacements and velocities.

However, it should be stressed that connecting the two masses to each other by the above-mentioned kinematic link or links in parallel does not represent a sufficient condition for establishing a state of relative motion between the two masses. In addition, it is necessary that these two masses have different laws of motion, according to which the absolute values of the velocities of these masses are different, while it may happen that their instantaneous velocities are equal to zero. All these means that the two masses have to possess different laws of motion that could be obtained due to the corresponding initial conditions of motion of each mass, the corresponding loading factors applied to each mass, or a corresponding combination of the initial conditions of motion and loading factors applied to the masses.

18.1. Composing and Solving Second-Order Linear Differential Equations of Motion of Two-Degree-of-Freedom Systems

The analysis of operational processes is based on the investigation of the basic parameters of motion, the mathematical expressions of which are obtained from the solutions of corresponding second-order differential equations of motion. Therefore, it is necessary first to get familiar with the structural compositions of the differential equations of the motion of two-degree-of-freedom mechanical systems. These systems comprise two masses connected to each other by corresponding kinematic links. As is well known, the masses must be connected to each other just by the two types of kinematic links, namely flexible links or fluid links, or by flexible and fluid links in parallel. A flexible link symbolizes an elastic medium, such as a spring, while a fluid link symbolizes a liquid or gaseous medium, such as a dashpot. In this text, the system that consists of two masses connected to each other by any of the kinematic links, even if neither of the masses is attached to a non-movable support, is considered an unrestricted system.

By attaching one of the masses of an unrestricted system by a spring or dashpot or by a spring and a dashpot in parallel to a non-movable support, we obtain a two-degree-of-freedom system comprising a sequence of two connecting arrangements, while each of these arrangements may have its own structural composition of kinematic links. Since at least one of the masses of this system is attached to a non-movable support, we define in this text this structural composition as a restricted two-degree-of-freedom system with two connecting arrangements.

Finally, by attaching both masses to each other and each mass to a non-movable support by the above-mentioned three connecting arrangements, we obtain a corresponding two-degree-of-freedom system with a sequence of three connecting arrangements. It is accepted in this text to define this structural composition as a restricted two-degree-of-freedom system with three connecting arrangements. All these allow us to conclude that two-degree-of-freedom structures consist of unrestricted systems characterized by an arrangement of connecting links, while restricted systems are characterized by a sequence of two or three connecting arrangements of kinematic links.

A survey of two-degree-of-freedom systems shows that most publications deal with restricted systems.

18.2. Composing Pairs of Simultaneous Differential Equations of Motion of Two-Degree-of-Freedom Systems

As we already know from Part 1 of this book, to perform an analysis of the operational process of a one-degree-of-freedom mechanical system, it is necessary to investigate the law of motion of the system and other basic parameters of motion of the system. The mathematical expression of the law of motion of a mechanical system can be determined by composing and solving a corresponding second-order differential equation of motion of the system. Therefore, since a two-degree-of-freedom system is defined by two differential equations describing the motion of each mass of the system, for the analysis of its operational process, it is necessary to compose and solve a corresponding pair of simultaneous differential equations of motion (for each mass, its differential equation of motion). Since a two-degree-of-freedom system contains two masses connected to each other by a kinematic link or links, which, by definition, allow relative motion between these two masses, it is obvious that these two masses move according to different laws of motion. It becomes clear that in the case where these two masses are not in a state of relative motion, they must move according to one (same) law of motion.

In this case, the two masses move as one mass, while the system is characterized by one degree of freedom. In a one-degree-of-freedom system, the resisting forces exerted by the kinematic links (the flexible and fluid links) on the masses that they interact with are proportional to the absolute values of the displacement or velocity of the masses. However, since in a two-degree-of-freedom system, the masses due to these kinematic links permanently perform a relative motion to each other, these kinematic links exert resisting forces on the masses that are proportional to the differences between the corresponding parameters of motion. All these are reflected in the following mathematical expressions related to the differential equation of motion of two-degree-of-freedom systems:

$$m_1\frac{d^2x_1}{dt^2} + \Sigma C_i\left(\frac{dx_1}{dt} - \frac{dx_2}{dt}\right) + \Sigma K_j(x_1 - x_2) = 0 \qquad (18.2.1)$$

$$m_2\frac{d^2x_2}{dt^2} + \Sigma C_i\left(\frac{dx_2}{dt} - \frac{dx_1}{dt}\right) + \Sigma K_j(x_2 - x_1) = 0 \qquad (18.2.2)$$

In addition, it should be noted that all other loading factors should be accounted for in the differential equation of motion of two-degree-of-freedom systems the same way they are in the equations of motion of one-degree-of-freedom systems. This is because the flexible or fluid links alone or both of these links in parallel can facilitate the relative motion of the masses of the system. All these should be reflected in the structures of the analytical expressions of the terms related to these two links in the corresponding differential equations of motion.

It should be emphasized that the surveyed published sources issued during the past 75 years present the corresponding schematic diagrams of two and three sequences of flexible links in two-degree-of-freedom systems in a consistent way, and there are no contradictions between publications in using them.

All surveyed publications suggest that, for each schematic diagram, certain pairs of simultaneous differential equations of motion supposedly describe the motion of the systems. Thus, according to my survey, in the book by J. Den Hartog, *Mechanical Vibrations*, Fourth Edition, McGraw-Hill Book Company, published in 1956, the correct composition of a pair of simultaneous differential equations of motion of the two masses of an unrestricted two-degree-of-freedom system was first published. (It is possible that this pair of equations was introduced in the earlier editions of this book.) This pair of equations has the following form:

$$m_1\ddot{x}_1 + k(x_1 - x_2) = P_0 \sin \omega t$$
$$m_2\ddot{x}_2 + k(x_2 - x_1) = 0$$

where k is the stiffness coefficient of the flexible link, P_0 is the amplitude of the harmonic force, and ω is the frequency of the harmonic function. However, there are no solutions provided for these differential equations in the book mentioned above.

According to my survey, each of the published pairs of differential equations of motion related to two-degree-of-freedom systems contains a mix of mathematical terms of one- and two-degree-of-freedom systems, while for none of these differential equations was ever presented a solution that resembles expressions as functions of time. Figure 18.2.1 shows a schematic diagram of a restricted two-degree-of-freedom system that is often discussed in published sources.

Many publications suggest a pair of simultaneous differential equations that supposedly describe the motion of the system characterized by the schematic diagram shown in Figure 18.2.1.

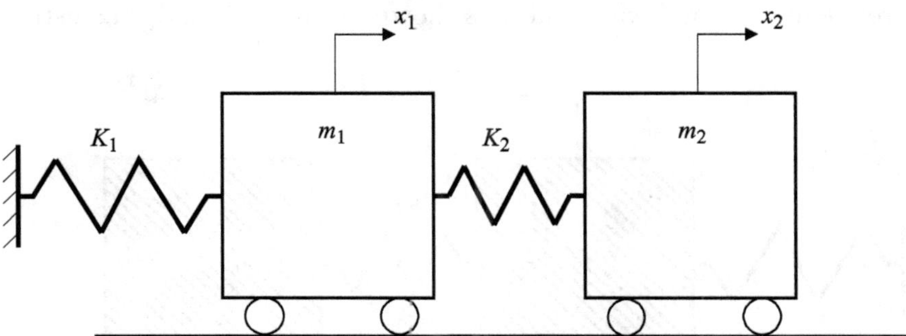

Fig. 18.2.1. Schematic diagram of a restricted two-degree-of-freedom system with two sequences of flexible links.

This pair of differential equations is presented as follows:

$$m_1 \frac{d^2 x_1}{dt^2} + K_1 x_1 + K_2(x_1 - x_2) = 0 \qquad (18.2.3)$$

$$m_2 \frac{d^2 x_2}{dt^2} + (K_1 + K_2)(x_2 - x_1) = 0 \qquad (18.2.4)$$

In the book *Vibration Problems in Engineering* by S. Timoshenko, Third Edition, published in 1955 by D. Van Nostrand Company, Inc., a schematic diagram of a two-degree-of-freedom system is presented, which is similar to the schematic diagram shown in Figure 18.2.1. In the above-mentioned book, a pair of differential equations related to the description of motion of the system shown in Figure 18.2.2 is also presented.

This pair of equations reads:

$$m_1 \ddot{x}_1 = -k_1 x_1 + k_2(x_2 - x_1) \qquad (18.2.5)$$

$$m_2 \ddot{x}_2 = -k_2(x_2 - x_1) \qquad (18.2.6)$$

The terms $K_1 x_1$ and $k_1 x_1$, respectively, in equations (18.2.3) and (18.2.5) have a structure that belongs to a differential equation describing the motion of a one-degree-of-freedom system, while the terms $K_2(x_1 - x_2)$ in equation (18.2.3) and $k_2(x_2 - x_1)$ in equation (18.2.5) belong to a two-degree-of-freedom system.

Many publications discuss two-degree-of-freedom systems with three sequences of flexible links, as shown in Figure 18.2.3.

Many publications reported two-degree-of-freedom systems with three sequences of flexible links, as shown in Figure 18.2.3, suggesting

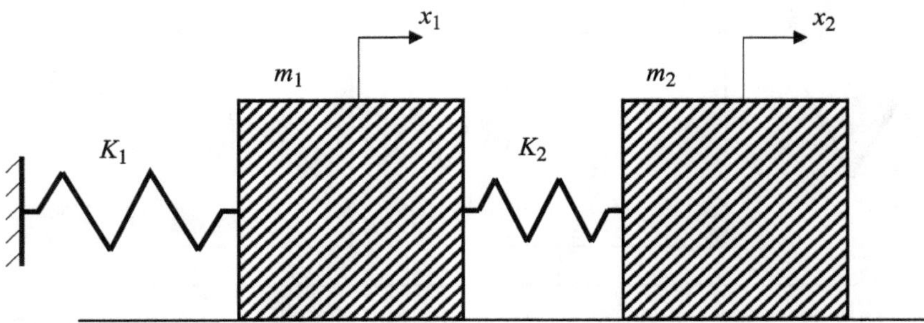

Fig. 18.2.2. Schematic diagram of a restricted two-degree-of-freedom system presented in the book by S. Timoshenko.

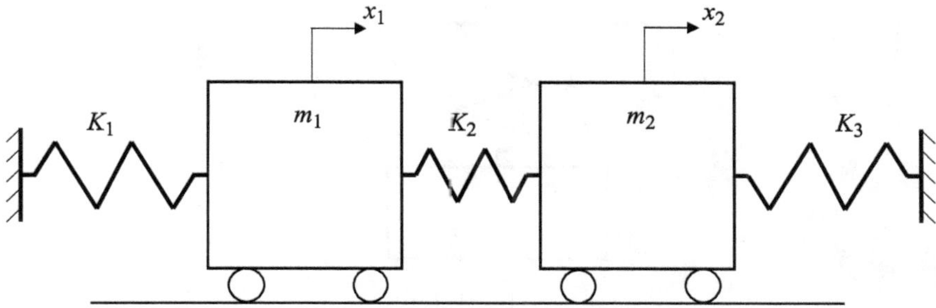

Fig. 18.2.3. Schematic diagram of a restricted two-degree-of-freedom system with three sequences of flexible links.

that the motion of the masses can be described using the following pair of differential equations:

$$m_1 \frac{d^2 x_1}{dt^2} + K_1 x_1 + K_2(x_1 - x_2) = 0 \qquad (18.2.7)$$

$$m_2 \frac{d^2 x_2}{dt^2} + K_2(x_2 - x_1) + K_3 x_2 = 0 \qquad (18.2.8)$$

However, the terms $K_1 x_1$ in equation (18.2.7) and $K_3 x_2$ in equation (18.2.8) do not fit into the description of the motion of two-degree-of-freedom systems. Hence, the term $K_1 x_1$ in equation (18.2.7) and the term $K_3 x_2$ in equation (18.2.8) relate to one-degree-of-freedom systems; therefore, the pair of equations (18.2.5) and (18.2.6) cannot describe the motion of a two-degree-of-freedom system.

My survey of published sources did not reveal even one publication that disclosed a pair of differential equations of motion that comply with the fundamentals of relative motion.

In the book *Mechanical Vibrations*, Fourth Edition, by J. Den Hartog, published in 1956, a schematic diagram of a two-degree-of-freedom system oriented in the vertical direction was presented. This schematic diagram is shown in Figure 18.2.4.

Below are Den Hartog's differential equations that supposedly describe the motion of the masses shown in Figure 18.2.4.

$$m_1 \ddot{x}_1 + (k_1 + k_3)x_1 - k_3 x_2 = 0 \qquad (18.2.9)$$

$$m_2 \ddot{x}_2 + (k_2 + k_3)x_2 - k_3 x_1 = 0 \qquad (18.2.10)$$

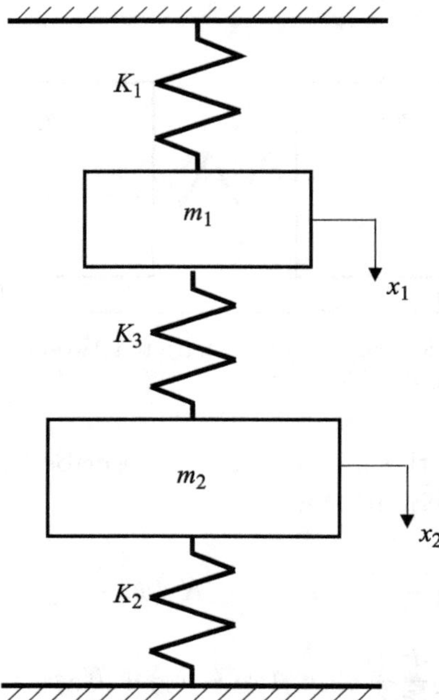

Fig. 18.2.4. Schematic diagram of a restricted two-degree-of-freedom system presented in the book by J. Den Hartog.

First, it should be noted that on the right-hand side of equation (18.2.9), the term representing the gravity force of the first mass $(m_1 g)$ is missing, while on the right-hand side of equation (18.2.10), the term representing the gravity force of the second mass $(m_2 g)$ is missing. However, in these equations, the terms $k_3 x_2$ and $k_3 x_1$ have the structure of terms that describe one-degree-of-freedom systems and should not be used for the description of two-degree-of-freedom systems.

Several published sources presented attempts to solve the pair of differential equations (18.2.3) and (18.2.4) and the pair of equations (18.2.7) and (18.2.8) that supposedly describe the motion of restricted two-degree-of-freedom systems. However, all these attempts were unsuccessful. It should be noted that my survey did not reveal even one complete, mathematically rigorous solution of any two-degree-of-freedom differential equation of motion.

It should be stressed that following the fundamentals of relative motion, I revised the structural compositions of all the surveyed differential equations of motion and obtained for all of them rigorous mathematical solutions based on the Laplace Transform methodology. In addition, I composed and solved numerous pairs of simultaneous differential equations describing the motion of two-degree-of-freedom systems. Additionally, it should be mentioned that in my book *Solving Engineering Problems in Dynamics*, published in 2014 by Industrial Press, several examples of mathematically rigorous solutions of unrestricted two-degree-of-freedom systems were presented for the first time. Among these examples, the solutions of the above-mentioned Den Hartog's pair of differential equations, describing the motion of the two masses of an unrestricted two-degree-of-freedom system, are also presented.

Part 2 of this book presents the entire spectrum of possible pairs of simultaneous differential equations of motion of two-degree-of-freedom systems. All these differential equations are solved using the methodology of Laplace transforms. It should be noted that the structural compositions and solutions of all these differential equations of motion of two-degree-of-freedom systems have never been published before. Also, the Laplace transform methodology has never been used for solving the differential equations of motion of two-degree-of-freedom systems.

Several examples demonstrating the use of Laplace Transform methodology for solving the differential equations of motion of two-degree-of-freedom systems are presented in the following.

18.3. Structures and Solutions of Differential Equations of Motion of Two-Degree-of-Freedom Systems

18.3.1. *Structure and solution of the pairs of differential equations describing motion of a restricted two-degree-of-freedom system with two sequences of flexible links*

Equations (18.2.3) and (18.2.4) represent the above-mentioned pair of equations that are often discussed in published sources; however, my survey did not reveal any published rigorous mathematical solutions of these equations.

Based on the schematic diagram shown in Figure 18.2.1 and the fundamentals of relative motion, I revised the structural composition of equation (18.2.3) to attain the form shown in equation (18.3.1). Equation (18.2.4) does not require any revision, and just for convenience, it is rewritten here as equation (18.3.2).

$$m_1 \frac{d^2 x_1}{dt^2} + (K_1 + K_2)(x_1 - x_2) = 0 \qquad (18.3.1)$$

$$m_2 \frac{d^2 x_2}{dt^2} + (K_1 + K_2)(x_2 - x_1) = 0 \qquad (18.3.2)$$

Dividing equation (18.3.1) by m_1 and equation (18.3.2) by m_2, we may write

$$\frac{d^2 x_1}{dt^2} + (\omega_{11}^2 + \omega_{21}^2)(x_1 - x_2) = 0 \qquad (18.3.3)$$

$$\frac{d^2 x_2}{dt^2} + (\omega_{12}^2 + \omega_{22}^2)(x_2 - x_1) = 0 \qquad (18.3.4)$$

where

$$\omega_{11}^2 = \frac{K_1}{m_1}; \quad \omega_{21}^2 = \frac{K_2}{m_1} \quad \omega_{12}^2 = \frac{K_1}{m_2}; \quad \omega_{22}^2 = \frac{K_2}{m_2} \qquad (18.3.5)$$

The initial conditions of motion are

$$x_1 = -S_1; \quad \frac{dx_1}{dt} = 0;$$

for $t = 0$

$$x_2 = 0; \quad \frac{dx_2}{dt} = 0 \qquad (18.3.6)$$

Applying Laplace Transform pairs 5, 2, and 1 to equations (18.3.3) and (18.3.4), we convert these differential equations with the initial conditions of motion according to expression (18.3.6) from the time domain into a corresponding system of simultaneous algebraic equations in the Laplace domain:

$$s^2 x_1(s) - s^2 S_1 + \omega_{11}^2 x_1(s) - \omega_{11}^2 x_2(s) + \omega_{21}^2 x_1(s) - \omega_{21}^2 x_2(s) = 0$$
$$(18.3.7)$$

$$s^2 x_2(s) + \omega_{12}^2 x_2(s) - \omega_{12}^2 x_1(s) + \omega_{22}^2 x_2(s) - \omega_{22}^2 x_1(s) = 0$$
$$(18.3.8)$$

Each of the algebraic equations (18.3.7) and (18.3.8) contains two unknowns $x_1(s)$ and $x_2(s)$ that, respectively, represent the displacements of the masses m_1 and m_2 in the Laplace domain. Rearranging equation (18.3.7), we may write

$$x_1(s)(s^2 + \omega_1^2) = s^2 S_1 + x_2(s)\omega_1^2 \tag{18.3.9}$$

where

$$\omega_1^2 = \omega_{11}^2 + \omega_{21}^2 \tag{18.3.10}$$

Solving equation (18.3.8) for $x_2(s)$, we have

$$x_2(s) = x_1(s)\frac{\omega_2^2}{s^2 + \omega_2^2} \tag{18.3.11}$$

where

$$\omega_2^2 = \omega_{12}^2 + \omega_{22}^2 \tag{18.3.12}$$

Combining equations (18.3.9) and (18.3.11), we eliminate the unknown $x_2(s)$ and obtain

$$x_1(s) = \frac{s^2 S_1}{s^2 + \omega^2} + \frac{\omega_2^2 S_1}{s^2 + \omega^2} \tag{18.3.13}$$

where

$$\omega^2 = \omega_1^2 + \omega_2^2 \tag{18.3.14}$$

Applying Laplace Transform pairs 1, 29, 2. 2, and 17 to equation (18.3.13), we invert this equation from the Laplace domain into the time domain and obtain the expression of the displacement of the first mass as a function of time:

$$x_1 = S_1 \cos \omega t + \frac{S_1 \omega_2^2}{\omega^2}(1 - \cos \omega t) \tag{18.3.15}$$

Rearranging equations (18.3.8) and (18.3.9), we may, respectively, write

$$x_2(s)(s^2 + \omega_2^2) = x_1(s)\omega_2^2 \tag{18.3.16}$$

and

$$x_1(s) = \frac{s^2 S_1}{s^2 + \omega_1^2} + \frac{x_2(s)\omega_1^2}{s^2 + \omega_1^2} \tag{18.3.17}$$

Combining equation (18.3.16) with equation (18.3.17), we have

$$x_2(s) = \frac{S_1\omega_2^2}{s^2 + \omega^2} \tag{18.3.18}$$

Applying Laplace Transform pairs 1, 2, and 17 to equation (18.3.18), we invert this equation from the Laplace domain into the time domain and obtain the solution of differential equation (18.3.4) with the initial conditions of motion according to expression (18.3.6):

$$x_2 = \frac{S_1\omega_2^2}{\omega^2}(1 - \cos\omega t) \tag{18.3.19}$$

Taking the first derivatives from equations (18.3.15) and (18.3.19), we, respectively, determine the velocities of the masses:

$$\frac{dx_1}{dt} = -\frac{S_1\omega_1^2}{\omega}\sin\omega t \tag{18.3.20}$$

$$\frac{dx_2}{dt} = \frac{S_1\omega_2^2}{\omega}\sin\omega t \tag{18.3.21}$$

Assuming that in equations (18.3.15), (18.3.19), (18.3.20), and (18.3.21), $t = 0$, we determine that $x_1 = 0$; $\frac{dx_2}{dt} = 0$; $x_2 = 0$; $\frac{dx_2}{dt} = 0$, as expected according to the initial conditions of motion given by expression (18.3.6).

Taking the first derivatives from equations (18.3.19) and (18.3.20), we, respectively, determine the accelerations of the masses of the system:

$$\frac{d^2x_1}{dt^2} = -S_1\omega_1^2 \tag{18.3.22}$$

$$\frac{d^2x_2}{dt^2} = S_1\omega_2^2 \tag{18.3.23}$$

Hence, equations (18.3.15), (18.3.19), (18.3.20), (18.3.21), (18.3.22), and (18.3.23) describe the basic parameters of motion and allow us to carry out the analysis of the related operational process.

18.3.2. *Structure and solution of a pair of simultaneous differential equations describing motion of a restricted two-degree-of-freedom system containing two sequences of fluid links*

Consider a restricted two-degree-of-freedom system containing two sequences of fluid links and moving on a horizontal frictionless surface, while the first mass is subjected to a constant resisting force and a constant active force. The schematic diagram of the system is shown in Figure 18.3.1.

Based on the schematic diagram shown in Figure 18.3.1 and the considerations mentioned above, we compose the following pair of simultaneous differential equations of motion:

$$m_1 \frac{d^2 x_1}{dt^2} + (C_1 + C_2)\left(\frac{dx_1}{dt} - \frac{dx_2}{dt}\right) + R_1 = P_1 \quad (18.3.24)$$

$$m_2 \frac{d^2 x_2}{dt^2} + (C_1 + C_2)\left(\frac{dx_2}{dt} - \frac{dx_1}{dt}\right) = 0 \quad (18.3.25)$$

$$x_1 = 0; \quad \frac{dx_1}{dt} = 0;$$

for $t = 0$

$$x_2 = 0; \quad \frac{dx_2}{dt} = 0; \quad (18.3.26)$$

It is assumed that $P_1 > R_1$.

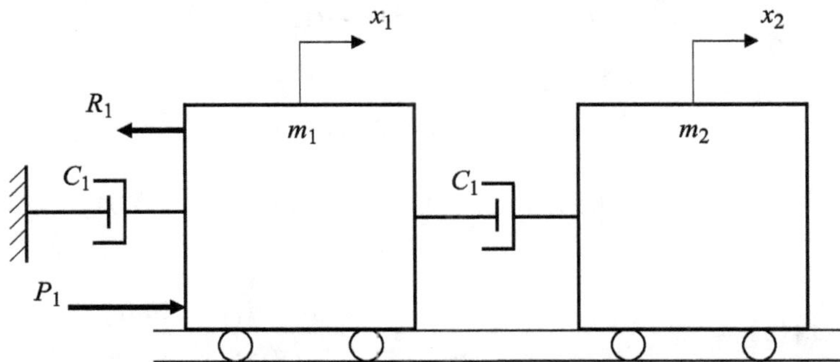

Fig. 18.3.1. Schematic diagram of a restricted two-degree-of-freedom system containing two sequences of fluid links, while the first mass is subjected to a constant resisting force and a constant active force.

Dividing equation (18.3.24) by m_1 and equation (18.3.25) by m_2, we have

$$\frac{d^2 x_1}{dt^2} + 2n_{11}\frac{dx_1}{dt} - 2n_{11}\frac{dx_2}{dt} + 2n_{21}\frac{dx_1}{dt}$$

$$-2n_{21}\frac{dx_2}{dt} + r_1 = p_1 \tag{18.3.27}$$

$$\frac{d^2 x_2}{dt^2} + 2n_{12}\frac{dx_2}{dt} - 2n_{12}\frac{dx_1}{dt} + 2n_{22}\frac{dx_2}{dt}$$

$$-2n_{22}\frac{dx_1}{dt} = 0 \tag{18.3.28}$$

where

$$2n_{11} = \frac{C_1}{m_1}; \quad 2n_{21} = \frac{C_2}{m_1}; \quad 2n_{12} = \frac{C_1}{m_2};$$

$$2n_{22} = \frac{C_2}{m_2}; \quad r_1 = C; \quad p_1 = \frac{P_1}{m_1} \tag{18.3.29}$$

Applying Laplace Transform pairs 5, 2, 4, 2, 4, 2, and 2 to equation (18.3.27) and Laplace Transform pairs 5, 2, 4, 2, 4, and 2 to equation (18.3.28), we, respectively, convert differential equations (18.3.27) and (18.3.28) from the time domain into the Laplace domain and obtain two simultaneous algebraic equations in the Laplace domain:

$$x_1(s)(s^2 + 2n_{11}s + 2n_{21}s)$$

$$= p_1 - r_1 + x_2(s)s(2n_{11} + 2n_{21}) \tag{18.3.30}$$

$$x_2(s)(s^2 + 2n_{12}s + 2n_{22}s)$$

$$= x_1(s)s(2n_{12} + 2n_{22}) \tag{18.3.31}$$

Solving equation (18.3.30) for $x_1(s)$ and equation (18.3.31) for $x_2(s)$, we may write

$$x_1(s) = \frac{p_1 - r_1}{s^2 + 2n_{11}s + 2n_{21}s} + \frac{x_2(s)s(2n_{11} + 2n_{21})}{s^2 + 2n_{11}s + 2n_{21}s} \tag{18.3.32}$$

$$x_2(s) = \frac{x_1(s)s(2n_{12} + 2n_{22})}{s^2 + 2n_{12}s + 2n_{22}s} \tag{18.3.33}$$

Combining equation (18.3.30) with equation (18.3.33), we eliminate from equation (18.3.30) one unknown, and we may write

$$x_1(s)[(s^2 + 2n_{11}s + 2n_{21}s)(s^2 + 2n_{12}s + 2n_{22}s)$$
$$-s^2(2n_{12} + 2n_{22})(2n_{11} + 2n_{21})]$$
$$= s^2(p_1 - r_1) + 2s(p_1 - r_1)(n_{12} + n_{22}) \qquad (18.3.34)$$

Solving equation (18.3.34) for $x_1(s)$, we determine the displacement of the first mass as a function of Laplace variable s:

$$x_1(s) = \frac{p_1 - r_1}{s(s + 2n)} + \frac{2(p_1 - r_1)(n_{12} + n_{22})}{s^2(s + 2n)} \qquad (18.3.35)$$

where

$$n = n_{11} + n_{12} + n_{21} + n_{22} \qquad (18.3.36)$$

Combining equations (18.3.31) and (18.3.32), we eliminate from equation (18.3.31) one unknown, and we have

$$x_2(s)[(s^2 + 2n_{12}s + 2n_{22}s)(s^2 + 2n_{11}s + 2n_{21}s)$$
$$-s^2(2n_{11} + 2n_{21})(2n_{12} + 2n_{22})]$$
$$= 2s(p_1 - r_1)(n_{12} + n_{22}) \qquad (18.3.37)$$

Solving equation (18.3.37) for $x_1(s)$, we determine the displacement of the second mass as a function of the Laplace variable s:

$$x_2 = \frac{2(p_1 - r_1)(n_{12} + n_{22})}{s^2(s + 2n)} \qquad (18.3.38)$$

Applying to equation (18.3.35) Laplace Transform pairs 16 and 48, while to equation (18.3.38) Laplace Transform pair 48, we invert these equations from the Laplace domain into the time domain and obtain the solutions of differential equations (18.3.24) and (18.3.25) with initial conditions of motion (18.3.26):

$$x_1 = \frac{p_1 - r_1}{2n}\left\{\left[t + \frac{1}{2n}(e^{-2nt} - 1)\right]\right.$$

$$\left. + (n_{12} + n_{22})\left[t^2 - \frac{1}{n}t - \frac{1}{2n^2}(e^{-2nt} - 1)\right]\right\} \qquad (18.3.39)$$

$$x_2 = \frac{(p_1 - r_1)(n_{12} + n_{22})}{2n}\left[t^2 - \frac{1}{n}t - \frac{1}{2n^2}(e^{-2nt} - 1)\right]$$

$$(18.3.40)$$

Taking the first derivative from equations (18.3.39) and (18.3.40), we determine the velocities of the corresponding two masses:

$$\frac{dx_1}{dt} = \frac{p_1 - r_1}{2n}\left\{1 - e^{-2nt} + (n_{12} + n_{22})\left[2t - \frac{1}{n}(1 - e^{-2nt})\right]\right\}$$

$$(18.3.41)$$

$$\frac{dx_2}{dt} = \frac{(p_1 - r_1)}{2n}(n_{12} + n_{22})\left[2t - \frac{1}{n}(1 - e^{-2nt})\right] \qquad (18.3.42)$$

Assuming that in equations (18.3.41) and (18.3.42), $t = 0$, we determine that $x_1 = 0$ and $x_2 = 0$, as expected according to initial conditions of motion (18.3.26).

Taking the first derivatives from equations (18.3.41) and (18.3.42), we, respectively, determine the accelerations of the masses of the system.

Hence, the equations describing the displacements, velocities, and accelerations of the masses are the basic parameters of motion of the system and allow us to carry out the analysis of the operational process of the system.

18.3.3. *Structure and solution of a pair of simultaneous differential equations describing flexible and fluid links*

The system is moving on a horizontal frictionless surface, while the first mass is subjected to constant resisting and constant active forces, as shown in Figure 18.3.2.

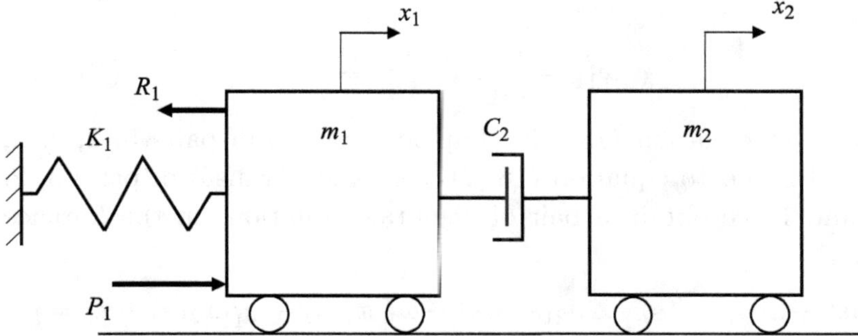

Fig. 18.3.2. Schematic diagram of a restricted two-degree-of-freedom system containing a sequence of flexible and fluid links, while the first mass is subjected to constant resisting and constant active forces.

Based on the schematic diagram shown in Figure 18.3.2 and the above-mentioned considerations, we compose the following pair of simultaneous differential equations of motion:

$$m_1 \frac{d^2 x_1}{dt^2} + C_2 \left(\frac{dx_1}{dt} - \frac{dx_2}{dt} \right) + K_1(x_1 - x_2) + R_1 = P_1 \qquad (18.3.43)$$

$$m_2 \frac{d^2 x_2}{dt^2} + C_2 \left(\frac{dx_2}{dt} - \frac{dx_1}{dt} \right) + K_1(x_2 - x_1) = 0 \qquad (18.3.44)$$

It is assumed that $P_1 > R_1$.

The initial conditions of motion are

$$x_1 = 0; \quad \frac{dx_1}{dt} = 0;$$

for $t = 0$

$$x_2 = 0; \quad \frac{dx_2}{dt} = 0 \qquad (18.3.45)$$

Dividing equations (18.3.43) and (18.3.44) by m_1 and m_2, we, respectively, obtain

$$\frac{d^2 x_1}{dt^2} + 2n_{21} \left(\frac{dx_1}{dt} - \frac{dx_2}{dt} \right) + \omega_{11}^2 (x_1 - x_2) + r_1 = p_1 \qquad (18.3.46)$$

$$\frac{d^2 x_2}{dt^2} + 2n_{22} \left(\frac{dx_2}{dt} - \frac{dx_1}{dt} \right) + \omega_{12}^2 (x_2 - x_1) = 0 \qquad (18.3.47)$$

where

$$\omega_{11}^2 = \frac{K_1}{m_1}; \quad \omega_2^2 = \frac{K_1}{m_2} \tag{18.3.48}$$

Applying to equation (18.3.46) Laplace Transform pairs 5, 2, 4, 2, 1, 2, and 2 and to equation (18.3.47) Laplace Transform pairs 5, 2, 4, 2, and 1, we obtain a pair of algebraic equations in the Laplace domain:

$$x_1(s)s^2 + 2n_{21}sx_1(s) - 2n_{21}sx_2(s) + \omega_{11}^2 x_1(s) - \omega_{11}^2 x_2(s) + r_1 = p_1 \tag{18.3.49}$$

$$x_2(s)s^2 + 2n_{22}sx_2(s) - 2n_{22}sx_1(s) + \omega_{12}^2 x_2(s) - \omega_{12}^2 x_1(s) = 0 \tag{18.3.50}$$

Rearranging equations (18.3.49) and (18.3.50), we may, respectively, write

$$x_1(s)(s^2 + 2n_{21}s + \omega_{11}^2) = p_1 - r_1 + x_2(s)(2n_{21}s + \omega_{11}^2) \tag{18.3.51}$$

$$x_2(s)(s^2 + 2n_{22}s + \omega_{12}^2) = x_1(s)(2n_{22}s + \omega_{12}^2) \tag{18.3.52}$$

Solving equation (18.3.52) for $x_2(s)$, we have

$$x_2(s) = \frac{x_1(s)(2n_{22}s + \omega_{12}^2)}{s^2 + 2n_{22}s + \omega_{12}^2} \tag{18.3.53}$$

Combining equations (18.3.51) and (18.3.53), we eliminate in equation (18.3.51) one unknown and obtain an equation with one unknown characterizing the displacement of the first mass in the Laplace domain:

$$x_1(s)s^2(s^2 + 2n_{21}s + 2n_{22}s + \omega_{11}^2 + \omega_{12}^2)$$
$$= (p_1 - r_1)(s^2 + 2n_{21}s + 2n_{22}s + \omega_{11}^2 + \omega_{12}^2) \tag{18.3.54}$$

Denoting

$$n = n_{21} + n_{22}; \quad \omega_0^2 = \omega_{11}^2 + \omega_{12}^2 \tag{18.3.55}$$

we may write

$$x_1(s) = \frac{p_1 - r_1}{s^2 + 2ns + \omega_0^2} + \frac{2n(p_1 - r_1)}{s(s^2 + 2ns + \omega_0^2)} + \frac{\omega_0^2(p_1 - r_1)}{s^2(s^2 + 2ns + \omega_0^2)} \tag{18.3.56}$$

Solving equation (18.3.51) for $x_1(s)$, we have

$$x_1(s) = \frac{p_1 - r_1}{s^2 + 2n_{21}s + \omega_{11}^2} + \frac{x_2(s)(2n_{21}s + \omega_{11}^2)}{s^2 + 2n_{21}s + \omega_{11}^2} \qquad (18.3.57)$$

Combining equation (18.3.52) with equation (18.3.57), we eliminate one unknown and obtain an algebraic equation with one unknown representing the displacement of the second mass in the Laplace domain.

$$x_2(s)[(s^2 + 2n_{22}s + \omega_{12}^2)(s^2 + 2n_{21}s + \omega_{11}^2)$$
$$-(2n_{21}s + \omega_{11}^2)(2n_{22}s + \omega_{12}^2)]$$
$$= (p_1 - r_1)(2n_{22}s + \omega_{12}^2) \qquad (18.3.58)$$

Solving equation (18.3.58) for the displacement in the Laplace domain, we may write

$$x_2(s) = \frac{2n_{22}(p_1 - r_1)}{s(s^2 + 2ns + \omega_0^2)} + \frac{\omega_{12}^2(p_1 - r_1)}{s^2(s^2 + 2ns + \omega_0^2)} \qquad (18.3.59)$$

Rearranging the denominators of the fractions in equations (18.3.56) and (18.3.59), we may write

$$s^2 + 2ns + \omega_0^2 = s^2 + 2ns + n^2 - n^2 + \omega_0^2 = (s + n)^2 + \omega_0^2 - n^2$$
$$= (s + n)^2 + \omega^2 \qquad (18.3.60)$$

while

$$\omega^2 = \omega_0^2 - n^2 \qquad (18.3.61)$$

Adjusting the denominators of the fractions in equations (18.3.56) and (18.3.59) according to equation (18.3.60), we obtain

$$x_1(s) = \frac{p_1 - r_1}{(s + n)^2 + \omega^2} + \frac{2n(p_1 - r_1)}{s[(s + n)^2 + \omega^2]} + \frac{\omega_0^2(p_1 - r_1)}{s^2[(s + n)^2 + \omega^2]}$$
$$(18.3.62)$$

$$x_2(s) = \frac{2n_{22}(p_1 - r_1)}{s[(s + n)^2 + \omega^2]} + \frac{\omega_{12}^2(p_1 - r_1)}{s^2[(s + n)^2 + \omega^2]} \qquad (18.3.63)$$

The absolute value of the parameter ω^2 has a significant influence on the type of motion of the system. Thus, in the case when $\omega^2 > 0$, we have underdamped vibrations; in the case when $\omega^2 = 0$, we have critical damping; and, finally, in the case when $\omega^2 < 0$, we have overdamped motion.

Underdamped Motion, $\omega^2 > 0$

Assuming that $\omega^2 > 0$, we apply to equation (18.3.62) Laplace Transform pairs 21, 45, and 56, and to equation (18.3.63) Laplace Transform pairs 45, and 56, we invert these equations from the Laplace domain into the time domain and obtain the solutions of differential equations (18.3.43) and (18.3.44) with initial conditions (18.3.45):

$$
\begin{aligned}
x_1 = {} & \frac{p_1 - r_1}{\omega^2 + n^2}\left[1 - e^{-nt}(\cos\omega t + \frac{n}{\omega}\sin\omega t)\right] \\
& + \frac{2n(p_1 - r_1)}{\omega(\omega^2 + n^2)^2}\{(\omega^2 + n^2)\omega t - 2n\omega \\
& - e^{-nt}[(\omega^2 - n^2)\sin\omega t - 2n\omega\cos\omega t]\} \\
& + \frac{\omega_0^2(p_1 - r_1)}{\omega(\omega^2 + n^2)^2}\left\{\frac{\omega t^2}{2}(\omega^2 + n^2) - 2n\omega t\right. \\
& - \frac{\omega(\omega^2 - n^2)}{\omega^2 + n^2}\left[1 - e^{-nt}(\cos\omega t + \frac{n}{\omega}\sin\omega t)\right] \\
& \left. - \frac{2n^2\omega}{\omega^2 + n^2}\left[1 - e^{-nt}\left(\frac{\omega}{n}\sin\omega t - \cos\omega t\right)\right]\right\} \qquad (18.3.64)
\end{aligned}
$$

$$
\begin{aligned}
x_2 = {} & \frac{2n_{22}(p_1 - r_1)}{\omega(\omega^2 + n^2)^2}\{(\omega^2 + n^2)\omega t - 2n\omega \\
& - e^{-nt}[(\omega^2 - n^2)\sin\omega t - 2n\omega\cos\omega t]\} \\
& + \frac{\omega_{12}^2(p_1 - r_1)}{\omega(\omega^2 + n^2)^2}\left\{\frac{\omega t^2}{2}(\omega^2 + n^2) - 2n\omega t\right. \\
& - \frac{\omega(\omega^2 - n^2)}{\omega^2 + n^2}\left[1 - e^{-nt}(\cos\omega t + \frac{n}{\omega}\sin\omega t)\right] \\
& \left. - \frac{2n^2\omega}{\omega^2 + n^2}\left[1 - e^{-nt}\left(\frac{\omega}{n}\sin\omega t - \cos\omega t\right)\right]\right\} \qquad (18.3.65)
\end{aligned}
$$

Taking the first derivatives from equations (18.3.64) and (18.3.65), we determine the velocities of the masses:

$$\frac{dx_1}{dt} = \frac{p_1 - r_1}{\omega} e^{-nt} \sin \omega t + \frac{n(p_1 - r_1)}{\omega^2 + n^2}$$

$$\times \left[1 - e^{-nt} \left(\cos \omega t + \frac{n}{\omega} \sin \omega t \right) \right]$$

$$+ \frac{\omega_0^2 (p_1 - r_1)}{\omega(\omega^2 + n^2)^2} \{ (\omega^2 + n^2)\omega t - 2n\omega$$

$$- e^{-nt}[(\omega^2 - n^2) \sin \omega t - 2n\omega \cos \omega t] \} \qquad (18.3.66)$$

$$\frac{dx_2}{dt} = \frac{2n_{22}(p_1 - r_1)}{\omega^2 + n^2} \left[1 - e^{-nt} \left(\cos \omega t + \frac{n}{\omega} \sin \omega t \right) \right]$$

$$+ \frac{\omega_{12}^2 (p_1 - r_1)}{\omega(\omega^2 + n^2)^2} \{ (\omega^2 + n^2)\omega t - 2n\omega$$

$$- e^{-nt}[(\omega^2 - n^2) \sin \omega t - 2n\omega \cos \omega t] \} \qquad (18.3.67)$$

Taking in equations (18.3.64)–(18.3.67) that $t = 0$, we determine that $x_1 = 0$; $x_2 = 0$; $\frac{dx_1}{dt} = 0$; $\frac{dx_2}{dt} = 0$ as expected according to initial conditions of motion (18.3.45).

The derivations and expressions presented above allow us to carry out the analysis of the corresponding operational processes.

Critical damping, $\omega^2 = 0$

In the denominators of equations (18.3.62) and (18.3.63), equating the parameter ω^2 to zero, we have

$$x_1(s) = \frac{p_1 - r_1}{(s + n)^2} + \frac{2n(p_1 - r_1)}{s(s + n)^2} + \frac{\omega_0^2 (p_1 - r_1)}{s^2(s + n)^2} \qquad (18.3.68)$$

$$x_2(s) = \frac{2n_{22}(p_1 - r_1)}{s(s + n)^2} + \frac{\omega_{12}^2 (p_1 - r_1)}{s^2(s + n)^2} \qquad (18.3.69)$$

Applying to equation (18.3.68) Laplace Transform pairs 19, 37, and 47 and to equation (18.3.69) Laplace Transform pairs 37 and 47, we obtain

$$x_1 = \frac{p_1 - r_1}{n^2}[1 - e^{-nt}(1 + nt)]$$

$$+ \frac{2n(p_1 - r_1)}{n^2}\left[t - \frac{2}{n} + e^{-nt}\left(\frac{2}{n} + t\right)\right]$$

$$+ \frac{\omega_0^2(p_1 - r_1)}{2n}\left[t^2 - \frac{2}{n}t - \frac{2}{n^2}(e^{-nt} - 1)\right] \qquad (18.3.70)$$

$$x_2 = \frac{2n_{22}(p_1 - r_1)}{n^2}\left[t - \frac{2}{n} + e^{-nt}\left(\frac{2}{n} + t\right)\right]$$

$$+ \frac{\omega_{12}^2(p_1 - r_1)}{2n}\left[t^2 - \frac{2}{n}t - \frac{2}{n^2}(e^{-nt} - 1)\right] \qquad (18.3.71)$$

Overdamped motion, $\omega^2 < 0$

In the denominators of equations (18.3.62) and (18.3.63), changing the sign of ω^2 from positive to negative, we obtain

$$x_1(s) = \frac{p_1 - r_1}{(s + n)^2 - \omega^2} + \frac{2n(p_1 - r_1)}{s[(s + n)^2 - \omega^2]} + \frac{\omega_0^2(p_1 - r_1)}{s^2[(s + n)^2 - \omega^2]}$$

$$(18.3.73)$$

$$x_2(s) = \frac{2n_{22}(p_1 - r_1)}{s[(s + n)^2 - \omega^2]} + \frac{\omega_{12}^2(p_1 - r_1)}{s^2[(s + n)^2 - \omega^2]} \qquad (18.3.74)$$

Applying to equation (18.3.73) Laplace Transform pairs 22, 46, and 57, while to equation (18.3.74) Laplace Transform pairs 46, and 57, we invert these equations from the Laplace domain into the time domain and obtain the solutions of differential equations (18.3.43) and (18.3.44) with initial conditions of motion (18.3.45):

$$x_1 = \frac{p_1 - r_1}{n^2 - \omega^2}\left[1 - e^{-nt}\left(\cosh \omega t + \frac{n}{\omega}\sinh \omega t\right)\right]$$

$$+ \frac{2n(p_1 - r_1)}{\omega(n^2 - \omega^2)^2}\{\omega t((n^2 - \omega^2)$$

$$- 2n\omega + e^{-nt}[(n^2 - \omega^2)\sinh \omega t + 2n\omega \cosh \omega t]\}$$

$$+ \frac{\omega_0^2(p_1 - r_1)}{\omega(n^2 - \omega^2)^2}\left\{\frac{\omega t^2}{2}(n^2 - \omega^2)\right.$$

$$-2n\omega t + \frac{\omega(\omega^2 + n^2)}{n^2 - \omega^2}\left[1 - e^{-nt}\left(\cosh\omega t + \frac{n}{\omega}\sinh\omega t\right)\right]$$

$$-\frac{2n^2\omega}{n^2 - \omega^2}\left[1 + e^{-nt}\left[1 + \left(\frac{\omega}{n}\sinh\omega t - \cosh\omega t\right)\right]\right]\Big\}$$

$$(18.3.75)$$

$$x_2 = \frac{2n_{22}(p_1 - r_1)}{\omega(n^2 - \omega^2)^2}\{\omega t((n^2 - \omega^2) - 2n\omega$$

$$+e^{-nt}[(n^2 - \omega^2)\sinh\omega t + 2n\omega\cosh\omega t]\}$$

$$+\frac{\omega_{12}^2(p_1 - r_1)}{\omega(n^2 - \omega^2)^2}\left\{\frac{\omega t^2}{2}(n^2 - \omega^2) - 2n\omega t\right.$$

$$+\frac{\omega(\omega^2 + n^2)}{n^2 - \omega^2}\left[1 - e^{-nt}(\cosh\omega t + \frac{n}{\omega}\sinh\omega t)\right]$$

$$-\frac{2n^2\omega}{n^2 - \omega^2}\left[1 + e^{-nt}\left[1 + \left(\frac{\omega}{n}\sinh\omega t - \cosh\omega t\right)\right]\right]\Big\}$$

$$(18.3.76)$$

The derivations and expressions presented above allow us to carry out the analysis of the corresponding operational processes.

18.3.4. *Solutions of the differential equations describing the motion of masses of a restricted two-degree-of-freedom system with three sequences of flexible links*

A preliminary discussion of this system is presented in Section 18.2, where it was mentioned that certain contradictions in the structural composition of related differential equations of motion are found in the surveyed published sources. In this section, we consider the revised versions of the related equations.

It should be noted that the notations of all the involved parameters are valid only for the current section.

We consider the operational process of a restricted two-degree-of-freedom system, the masses of which are connected to each other by a spring, while each mass of the system is attached by a spring to a non-movable support. The system is moving on a horizontal frictionless surface due to the initial velocity of the first mass. The air resistance to the motion of the system is negligible. Figure 18.2.3 shows a schematic diagram of the system, where K_1, K_2, and K_3 are the stiffness coefficients of the respective springs, while the rest of the notations in this figure are self-explanatory.

Based on the schematic diagram shown in Figure 18.2.3 and the considerations mentioned above, we compose the pair of simultaneous differential equations describing the motion of the system:

$$m_1 \frac{d^2 x_1}{dt^2} + (K_1 + K_2 + K_3)(x_1 - x_2) = 0 \qquad (18.3.77)$$

$$m_2 \frac{d^2 x_2}{dt^2} + (K_1 + K_2 + K_3)(x_2 - x_1) = 0 \qquad (18.3.78)$$

The initial conditions of motion are

$$x_1 = 0; \quad \frac{dx_1}{dt} = V_1;$$

for $t = 0$ \hfill (18.3.79)

$$x_2 = 0; \quad \frac{dx_2}{dt} = 0$$

Dividing equation (18.3.77) by m_1 and equation (18.3.78) by m_2 and applying to them

Laplace Transform pairs 5, 2, and 1, we convert these equations from the time domain into the Laplace domain and obtain a system of two simultaneous algebraic equations with two unknowns in each equation:

$$s^2 x_1(s) - sV_1 + \omega_{11}^2 x_1(s) - \omega_{11}^2 x_2(s) + \omega_{21}^2 x_1(s)$$

$$-\omega_{21}^2 x_2(s) + \omega_{31}^2 x_1(s) - \omega_{31}^2 x_2(s) = 0 \qquad (18.3.80)$$

$$s^2 x_2(s) + \omega_{12}^2 x_2(s) - \omega_{12}^2 x_1(s) + \omega_{22}^2 x_2(s)$$

$$-\omega_{22}^2 x_1(s) + \omega_{32}^2 x_2(s) - \omega_{32}^2 x_1(s) = 0 \qquad (18.3.81)$$

where, just for this section, we denote

$$\omega_{11}^2 = \frac{K_1}{m_1}; \quad \omega_{21}^2 = \frac{K_2}{m_1}; \quad \omega_{31}^2 = \frac{K_3}{m_1};$$

$$\omega_{12}^2 = \frac{K_1}{m_2}; \quad \omega_{22}^2 = \frac{K_2}{m_2}; \quad \omega_{32}^2 = \frac{K_3}{m_2} \qquad (18.3.82)$$

Rearranging equations (18.3.80) and (18.3.81), we may write

$$x_1(s)(s^2 + \omega_{11}^2 + \omega_{21}^2 + \omega_{31}^2) = sV_1 + x_2(s)(\omega_{11}^2 + \omega_{21}^2 + \omega_{31}^2)$$
(18.3.83)

$$x_2(s)(s^2 + \omega_{12}^2 + \omega_{22}^2 + \omega_{32}^2) = x_1(s)(\omega_{12}^2 + \omega_{22}^2 + \omega_{32}^2)$$
(18.3.84)

Solving, respectively, for $x_1(s)$ and $x_2(s)$ equations (18.3.83) and (18.3.84), we have

$$x_1(s) = \frac{sV_1}{s^2 + \omega_{11}^2 + \omega_{21}^2 + \omega_{31}^2} + \frac{x_2(s)(\omega_{11}^2 + \omega_{21}^2 + \omega_{31}^2)}{s^2 + \omega_{11}^2 + \omega_{21}^2 + \omega_{31}^2}$$
(18.3.85)

$$x_2(s) = \frac{x_1(s)(\omega_{12}^2 + \omega_{22}^2 + \omega_{32}^2)}{(s^2 + \omega_{12}^2 + \omega_{22}^2 + \omega_{32}^2)}$$
(18.3.86)

Combining equations (18.3.83) and (18.3.86), we eliminate one unknown from equation (18.3.83) and obtain an algebraic equation with one unknown in the Laplace domain:

$$x_1(s)(s^2 + \omega_{11}^2 + \omega_{21}^2 + \omega_{31}^2)$$
$$= sV_1 + \frac{x_1(s)(\omega_{11}^2 + \omega_{21}^2 + \omega_{31}^2)((\omega_{12}^2 + \omega_{22}^2 + \omega_{32}^2)}{(s^2 + \omega_{12}^2 + \omega_{22}^2 + \omega_{32}^2)}$$
(18.3.87)

Similarly, combining equations (18.3.84) and (18.3.85), we eliminate from equation (18.3.84) one unknown and obtain

$$x_2(s)(s^2 + \omega_{12}^2 + \omega_{22}^2 + \omega_{32}^2) = \frac{sV_1(\omega_{12}^2 + \omega_{22}^2 + \omega_{32}^2)}{s^2 + \omega_{11}^2 + \omega_{21}^2 + \omega_{31}^2}$$
$$+ \frac{x_2(s)(\omega_{11}^2 + \omega_{21}^2 + \omega_{31}^2)(\omega_{12}^2 + \omega_{22}^2 + \omega_{32}^2)}{s^2 + \omega_{11}^2 + \omega_{21}^2 + \omega_{31}^2}$$
(18.3.88)

Solving equation (18.3.87) for $x_1(s)$ and equation (18.3.88) for $x_2(s)$, we have

$$x_1(s) = \frac{sV_1}{s^2 + \omega^2} + \frac{V_1(\omega_{12}^2 + \omega_{22}^2 + \omega_{32}^2)}{s(s^2 + \omega^2)} \qquad (18.3.89)$$

$$x_2(s) = \frac{V_1(\omega_{12}^2 + \omega_{22}^2 + \omega_{32}^2)}{s(s^2 + \omega^2)} \qquad (18.3.90)$$

where

$$\omega^2 = \omega_{11}^2 + \omega_{21}^2 + \omega_{31}^2 + \omega_{11}^2 + \omega_{21}^2 + \omega_{31}^2 \qquad (18.3.91)$$

Applying to equation (18.3.89) Laplace Transform pairs 23 and 39 while to equation (18.3.90) Laplace Transform pair 39, we invert these equations into the time domain and obtain the solutions of differential equations (18.3.77) and (18.3.78) with initial conditions of motion according to (18.3.79):

$$x_1 = \frac{V_1}{\omega}\sin\omega t + \frac{V_1(\omega_{12}^2 + \omega_{22}^2 + \omega_{32}^2)}{\omega^2}\left(t - \frac{1}{\omega}\sin\omega t\right) \qquad (18.3.92)$$

$$x_2 = \frac{V_1(\omega_{12}^2 + \omega_{22}^2 + \omega_{32}^2)}{\omega^2}\left(t - \frac{1}{\omega}\sin\omega t\right) \qquad (18.3.93)$$

Taking from equations (18.3.92) and (18.3.93) the first derivatives, we determine the velocities of the masses:

$$\frac{dx_1}{dt} = V_1\cos\omega t + \frac{V_1(\omega_{12}^2 + \omega_{22}^2 + \omega_{32}^2)}{\omega^2}(1 - \cos\omega t) \qquad (18.3.94)$$

$$\frac{dx_2}{dt} = \frac{V_1(\omega_{12}^2 + \omega_{22}^2 + \omega_{32}^2)}{\omega^2}(1 - \cos\omega t) \qquad (18.3.95)$$

Assuming that in equations (18.3.92)–(18.3.95), $t = 0$, we determine that $x_1 = 0$; $x_2 = 0$; $\frac{dx_1}{dt} = V_1$; and $\frac{dx_2}{dt} = 0$, as it should be according to initial conditions of motion (18.3.79).

The derivations and expressions presented above allow the reader to carry out the analysis of the corresponding operational processes.

CHAPTER 19

UNRESTRICTED SYSTEMS WITH MASSES CONNECTED TO EACH OTHER BY A FLEXIBLE LINK

As mentioned in the previous chapter, an unrestricted system consists of two masses connected to each other by a kinematic link (a spring), a fluid kinematic link (a dashpot), or by both links in parallel. We consider three typical groups of unrestricted systems, namely the masses that are connected by one spring or dashpot or by a spring and a dashpot in parallel. The analyses are carried out for the cases where the motion is caused by the initial conditions of motion, the appropriate loading factors, or common combinations of both.

As emphasized above, the masses of any two-degree-of-freedom system are permanently in a state of relative motion, and as mentioned earlier, the terms of the related differential equations describing the motion have structures that correspond to relative motion.

It is extremely important to emphasize that the above-mentioned kinematic links, connecting (coupling) these two masses to each other allow motions of these two masses relative to each other. The system remains a two-degree-of-freedom system if its two masses remain in a state of relative motion.

The only alternative to the relative motion of these masses is that they move with the same velocity as one whole. In this case, the system is characterized by one degree of freedom.

19.1. Motion of an Unrestricted System on a Horizontal Frictionless Surface Due to Initial Displacements and Initial Velocities of the Masses

This section deals with an investigation of the operational processes of the above-mentioned unrestricted system.

An analysis of the operational process of an unrestricted system moving on a horizontal surface due to the initial conditions of motion reveals many important characteristics of the motion of the system.

The schematic diagram of the system, shown in Figure 19.1.1, indicates that the two masses of the system are connected by a spring, the stiffness coefficient of which is denoted by K.

The notations in this figure are self-explanatory. Based on the schematic diagram shown in Figure 19.1.1 and the considerations mentioned above, we compose the following pair of simultaneous

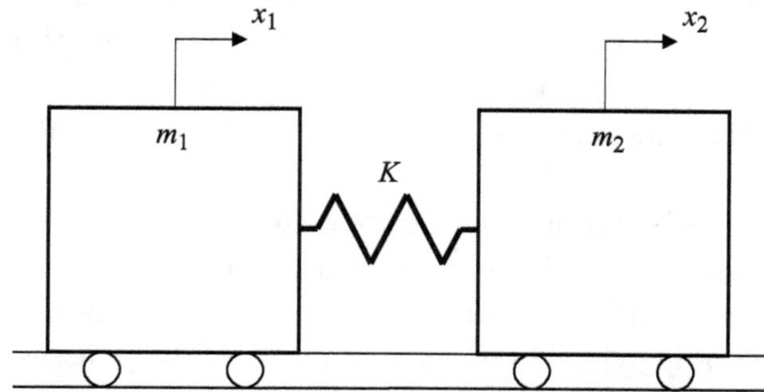

Fig. 19.1.1. Schematic diagram of an unrestricted system moving on a horizontal surface while the masses of the system are connected to each other by a flexible link.

differential equations of motion of the system:

$$m_1 \frac{d^2 x_1}{dt^2} + K(x_1 - x_2) = 0 \qquad (19.1.1)$$

$$m_2 \frac{d^2 x_2}{dt^2} + K(x_2 - x_1) = 0 \qquad (19.1.2)$$

The system of two simultaneous differential equations (19.1.1) and (19.1.2) contains two unknowns x_1 and x_2 in each of the two mentioned equations. This prevents the possibility of solving these equations using the methodologies based on the characteristic equations as well as on Laplace Transform. However, contrary to the characteristic equation methodology, the Laplace Transform methodology allows us to convert the above-mentioned two simultaneous differential equations into two algebraic equations. This is demonstrated in my book *Solving Engineering Problems in Dynamics* published by Industrial Press in 2014. Since these two simultaneous equations after the conversion become algebraic, it is possible to apply to them all conventional procedures that allow for eliminating one unknown from each of these equations. Therefore, for instance, applying to these equations the method of substitutions, we obtain two equations with one unknown in each of the equations that, respectively, represent the displacements of these two masses in the Laplace domain having the Laplace independent variable s. The following mathematical procedures clarify the above-mentioned approach:

The initial conditions of motion are

$$x_1 = S_1; \quad \frac{dx_1}{dt} = V_1;$$

for $t = 0$ \qquad (19.1.3)

$$x_2 = S_2; \quad \frac{dx_2}{dt} = V_2$$

Dividing, respectively, equations (19.1.1) and (19.1.2) by m_1 and m_2 while applying to them Laplace Transform pairs 5, 2, and 1, we convert differential equations (19.1.1) and (19.1.2) with initial conditions of motion (19.1.3) from the time domain into a system of two simultaneous algebraic equations with two unknowns in each equation in the Laplace domain:

$$s^2 x_1(s) - sV_1 - s^2 S_1 + \omega_1^2 x_1(s) - \omega_1^2 x_2(s) = 0 \qquad (19.1.4)$$

$$s^2 x_2(s) - sV_2 - s^2 S_2 + \omega_2^2 x_2(s) - \omega_2^2 x_1(s) = 0 \qquad (19.1.5)$$

where

$$\omega_1^2 = \frac{K}{m_1}; \quad \omega_2^2 = \frac{K}{m_2} \qquad (19.1.6)$$

The system of two simultaneous algebraic equations (19.1.4) and (19.1.5) contains two unknowns $x_1(s)$ and $x_2(s)$ in each of the two mentioned equations.

Rearranging equations (19.1.4) and (19.1.5), we may write.

$$x_1(s)(s^2 + \omega_1^2) = sV_1 + s^2 S_1 + \omega_1^2 x_2(s) \qquad (19.1.7)$$

$$x_2(s)(s^2 + \omega_2^2) = sV_2 + s^2 S_1 + \omega_2^2 x_1(s) \qquad (19.1.8)$$

Solving equation (19.1.8) for $x_2(s)$, we have

$$x_2(s) = \frac{sV_2}{s^2 + \omega_2^2} + \frac{s^2 S_2}{s^2 + \omega_2^2} + \frac{\omega_2^2 x_1(s)}{s^2 + \omega_2^2} \qquad (19.1.9)$$

Applying the method of substitutions, we combine equation (19.1.7) with equation (19.1.9) and eliminate one unknown from equation (19.1.7). Hence, we obtain an algebraic equation with one unknown $x_1(s)$ that represents the displacement of the first mass in the Laplace domain:

$$x_1(s)(s^2 + \omega_1^2) = sV_1 + s^2 S_1 + \frac{sV_2 \omega_1^2}{s^2 + \omega_2^2} + \frac{s^2 S_2 \omega_1^2}{s^2 + \omega_2^2} + \frac{\omega_1^2 \omega_2^2 x_1(s)}{s^2 + \omega_2^2}$$

$$(19.1.10)$$

Rearranging equation (19.1.10), we may write

$$x_1(s)[(s^2 + \omega_1^2)(s^2 + \omega_2^2) - \omega_1^2 \omega_2^2]$$

$$= sV_1(s^2 + \omega_2^2) + s^2 S_1(s^2 + \omega_2^2) + sV_2 \omega_1^2 + s^2 S_2 \omega_1^2$$

$$(19.1.11)$$

Simplifying equation (19.1.11), we have

$$x_1(s)(s^2 + \omega_1^2 + \omega_2^2) = sV_1 + \frac{1}{s}(V_1\omega_2^2 + V_2\omega_1^2) + s^2S_1 + S_1\omega_2^2 + S_2\omega_1^2$$

$$(19.1.12)$$

Similarly, we obtain

$$x_2(s)(s^2 + \omega_1^2 + \omega_2^2) = sV_2 + \frac{1}{s}(V_2\omega_1^2 + V_1\omega_2^2) + s^2S_2 + S_2\omega_1^2 + S_1\omega_2^2$$

$$(19.1.13)$$

Denote

$$\omega^2 = \omega_1^2 + \omega_2^2 \qquad (19.1.14)$$

where ω is the natural compound frequency of the system.

Combining equations (19.1.12)–(19.1.14), we obtain the equations for the displacements of the masses in the Laplace domain:

$$x_1(s) = \frac{sV_1}{s^2 + \omega^2} + \frac{V_1\omega_2^2 + V_2\omega_1^2}{s(s^2 + \omega^2)} + \frac{s^2S_1}{s^2 + \omega^2} + \frac{S_1\omega_2^2 + S_2\omega_1^2}{s^2 + \omega^2}$$

$$(19.1.15)$$

$$x_2(s) = \frac{sV_2}{s^2 + \omega^2} + \frac{V_2\omega_1^2 + V_1\omega_2^2}{s(s^2 + \omega^2)} + \frac{s^2S_2}{s^2 + \omega^2} + \frac{S_2\omega_1^2 + S_1\omega_2^2}{s^2 + \omega^2}$$

$$(19.1.16)$$

Applying to equations (19.1.15) and (19.1.16) Laplace Transform pairs 23, 39, 29, and 17, we invert these equations from the Laplace domain into the time domain and obtain the solutions of differential equations (19.1.1) and (19.1.2) for the initial conditions of motion (19.1.3):

$$x_1 = \frac{V_1}{\omega}\sin\omega t + \frac{V_1\omega_2^2 + V_2\omega_1^2}{\omega^2}\left(t - \frac{1}{\omega}\sin\omega t\right)$$

$$+ S_1\cos\omega t + \frac{S_1\omega_2^2 + S_2\omega_1^2}{\omega^2}(1 - \cos\omega t) \qquad (19.1.17)$$

$$x_2 = \frac{V_2}{\omega}\sin\omega t + \frac{V_2\omega_1^2 + V_1\omega_2^2}{\omega^2}\left(t - \frac{1}{\omega}\sin\omega t\right)$$

$$+ S_2\cos\omega t + \frac{S_2\omega_1^2 + S_1\omega_2^2}{\omega^2}(1 - \cos\omega t) \qquad (19.1.18)$$

Applying to equations (19.1.17) and (19.1.18) the corresponding algebraic procedures, we obtain

$$x_1 = \frac{\omega_1^2(S_1 - S_2)}{\omega^2}\cos\omega t + \frac{\omega_1^2(V_1 - V_2)}{\omega^3}\sin\omega t$$

$$+ \frac{S_1\omega_2^2 + S_2\omega_1^2}{\omega^2} + \frac{V_1\omega_2^2 + V_2\omega_1^2}{\omega^2}t \qquad (19.1.19)$$

$$x_2 = \frac{\omega_2^2(S_2 - S_1)}{\omega^2}\cos\omega t + \frac{\omega_2^2(V_2 - V_1)}{\omega^3}\sin\omega t$$

$$+ \frac{S_1\omega_2^2 + S_2\omega_1^2}{\omega^2} + \frac{V_1\omega_2^2 + V_2\omega_1^2}{\omega^2}t \qquad (19.1.20)$$

Equations (19.1.19) and (19.1.20) describe the free vibrational motion of the two masses of the unrestricted system. These equations lead to the following conclusions:

(1) The system possesses one compound frequency ω, and both masses, obviously, vibrate with the same compound frequency.

(2) In the cases when $S_1 \neq S_2$ and $V_1 \neq V_2$, $S_1 = S_2 = 0$ and $V_1 \neq V_2$, or $S_1 \neq S_2$ and $V_1 = V_2 = 0$, then these two masses are in the antiphase vibration mode.

(3) In the case when $S_1 = S_2 = 0$ while $V_1 = V_2 = V$, there is no relative motion between these two masses that move uniformly, representing a one-degree-of-freedom system, and differential equations (19.1.1) and (19.1.2) are not applicable to describe the motion of this system. In this case, the spring that connects the two masses remains undeformed. Therefore, these two masses behave as one mass in a one-degree-of-freedom system. Hence, to compose a differential equation of motion for this system, we should consider that these two masses without any spring between them and combined into one mass, namely $m_1 + m_2 = m$. Then, the differential equation describing the motion of these masses as one mass reads $m\frac{d^2x}{dt^2} = 0$; if the initial velocity of the mass is $\frac{dx}{dt} = V$, then we have $x = Vt$.

(4) In the case when $S_1 = S_2$ and $V_1 = V_2 = 0$, we have a static system with two bodies.

It should be emphasized that the analysis of free vibratory motion of an unrestricted system presented above is characterized by a single value of the compound frequency and an anti-phased mode of vibration.

Taking the first derivatives from equations (19.1.19) and (19.1.20), we obtain the equations describing the velocities of the masses:

$$\frac{dx_1}{dt} = \frac{\omega_1^2(V_1 - V_2)}{\omega^2}\cos\omega t - \frac{\omega_1^2(S_1 - S_2)}{\omega}\sin\omega t + \frac{V_1\omega_2^2 + V_2\omega_1^2}{\omega^2}$$

$$(19.1.21)$$

$$\frac{dx_2}{dt} = \frac{\omega_2^2(V_2 - V_1)}{\omega^2}\cos\omega t - \frac{\omega_2^2(S_2 - S_1)}{\omega}\sin\omega t + \frac{V_1\omega_2^2 + V_2\omega_1^2}{\omega^2}$$

$$(19.1.22)$$

Supposing for the equations (19.1.19)–(19.1.22) that $t = 0$, we obtain that the initial displacements and velocities of the masses, respectively, equal: $x_1 = S_1$, $x_2 = S_2$, $\frac{dx_1}{dt} = V_1$, and $\frac{dx_2}{dt} = V_2$, as it should be according to initial conditions of motion (19.1.3).

Taking from equations (19.1.21) and (19.1.22) the first derivatives, we obtain the equations of the accelerations of the masses, respectively:

$$\frac{d^2x_1}{dt^2} = -\left(V_1 - \frac{V_1\omega_2^2 + V_2\omega_1^2}{\omega^2}\right)\omega\sin\omega t - S_1\omega^2\cos\omega t$$
$$+ (S_1\omega_2^2 + S_2\omega_1^2)\cos\omega t \qquad (19.1.23)$$

$$\frac{d^2x_2}{dt^2} = -\left(V_2 - \frac{V_2\omega_1^2 + V_1\omega_2^2}{\omega^2}\right)\omega\sin\omega t - S_2\omega^2\cos\omega t$$
$$+ (S_2\omega_1^2 + S_1\omega_2^2)\cos\omega t \qquad (19.1.24)$$

Assuming in equations (19.1.23) and (19.1.24) that $t = 0$, we determine the initial accelerations of the masses at the beginning of the motion:

$$\frac{d^2x_1}{dt^2} = -\omega_1^2(S_1 - S_2) \qquad (19.1.25)$$

$$\frac{d^2x_2}{dt^2} = -\omega_2^2(S_2 - S_1) \qquad (19.1.26)$$

Hence, expressions (19.1.19)–(19.1.24) represent the basic parameters of motion of the two masses of a two-degree-of-freedom system.

19.2. Motion of an Unrestricted System on a Horizontal Surface While Being Subjected to Dry Friction, Constant Resisting, and Constant Active Forces

We consider an unrestricted system, the two masses of which are connected by a spring while these masses are moving on a horizontal surface and are subjected to dry friction, constant resisting, and constant active forces. The schematic diagram of the system is shown in Figure 19.2.1, where F_1 and F_2 are the respective dry friction forces, R_1 and R_2 represent constant resisting forces, and P_1 and P_2 are constant active forces. The rest of the notations in the figure are self-explanatory.

Based on the schematic diagram shown in Figure 19.2.1 and the considerations mentioned above, we compose the following pair of simultaneous differential equations of motion:

$$m_1 \frac{d^2 x_1}{dt^2} + K(x_1 - x_2) + F_1 + R_1 = P_1 \qquad (19.2.1)$$

$$m_2 \frac{d^2 x_2}{dt^2} + K(x_2 - x_1) + F_2 + R_2 = P_2 \qquad (19.2.2)$$

The general initial conditions of motion are taken according to expression (19.1.3).

Dividing, respectively, differential equations (19.2.1) and (19.2.2) by m_1 and m_2 and applying Laplace Transform pairs 2, 5, 2, 1, 2, and 2 to these equations with the initial conditions of motion

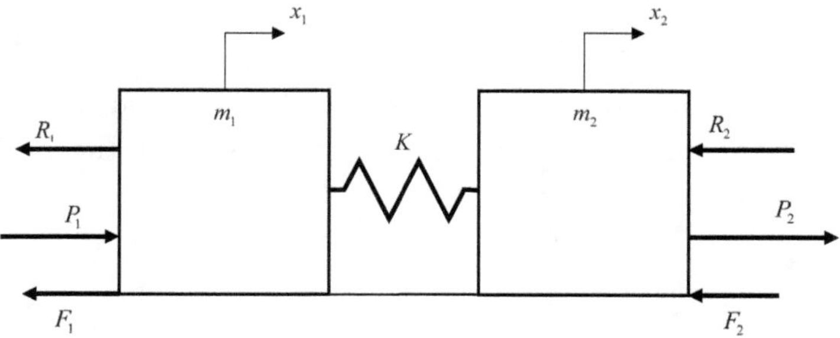

Fig. 19.2.1. Schematic diagram of an unrestricted system, the masses of which are connected by a flexible link and moving on a horizontal frictional surface, while the masses are subjected to dry friction, constant resisting, and constant active forces.

according to expression (19.1.3), we convert these equations from the time domain into a system of two simultaneous algebraic equations with two unknowns in the Laplace domain:

$$s^2 x_1(s) - sV_1 - s^2 S_1 + \omega_1^2 x_1(s) - \omega_1^2 x_2(s) + f_1 + r_1 = p_1$$

$$(19.2.3)$$

$$s^2 x_2(s) - sV_2 - s^2 S_2 + \omega_2^2 x_2(s) - \omega_2^2 x_1(s) + f_2 + r_2 = p_2$$

$$(19.2.4)$$

where ω_1^2 and ω_2^2 are determined using (19.1.6), while

$$f_1 = \frac{F_1}{m_1}; \quad f_2 = \frac{F_2}{m_2}; \quad r_1 = \frac{R_1}{m_1};$$

$$r_2 = \frac{R_2}{m_2}; \quad p_1 = \frac{P_1}{m_1}; \quad p_2 = \frac{P_2}{m_2}$$

$$(19.2.5)$$

Rearranging the pair of simultaneous algebraic equations (19.2.3) and (19.2.4) and applying to them the method of substitutions, we obtain the equations describing the displacements $x_1(s)$ and $x_2(s)$ in the Laplace domain:

$$x_1(s) = \frac{V_1 s}{s^2 + \omega^2} + \frac{V_1 \omega_2^2 + V_2 \omega_1^2}{s(s^2 + \omega^2)} + \frac{S_1 s^2}{s^2 + \omega^2} + \frac{S_1 \omega_2^2 + S_2 \omega_1^2}{s^2 + \omega^2}$$

$$+ \frac{p_1 - f_1 - r_1}{s^2 + \omega^2} + \frac{\omega_2^2(p_1 - f_1 - r_1) + \omega_1^2(p_2 - f_2 - r_2)}{s^2(s^2 + \omega^2)}$$

$$(19.2.6)$$

$$x_2(s) = \frac{V_2 s}{s^2 + \omega^2} + \frac{V_2 \omega_1^2 + V_1 \omega_2^2}{s(s^2 + \omega^2)} + \frac{S_2 s^2}{s^2 + \omega^2} + \frac{S_2 \omega_1^2 + S_1 \omega_2^2}{s^2 + \omega^2}$$

$$+ \frac{p_2 - f_2 - r_2}{s^2 + \omega^2} + \frac{\omega_1^2(p_2 - f_2 - r_2) + \omega_2^2(p_1 - f_1 - r_1)}{s^2(s^2 + \omega^2)}$$

$$(19.2.7)$$

where ω^2 is defined by equation (19.1.14).

Applying to these equations Laplace Transform pairs 23, 39, 29, 17, 17, and 51, we invert these equations from the Laplace domain into the time domain and obtain the solutions of differential equations of motion (19.2.1) and (19.2.2) with their initial conditions of motion:

$$x_1 = \frac{V_1}{\omega}\sin\omega t + \frac{V_1\omega_2^2 + V_2\omega_1^2}{\omega^2}\left(t - \frac{1}{\omega}\sin\omega t\right)$$

$$+ S_1\cos\omega t + \frac{S_1\omega_2^2 + S_2\omega_1^2}{\omega^2}(1 - \cos\omega t)$$

$$+ \frac{p_1 - f_1 - r_1}{\omega^2}(1 - \cos\omega t) + \left[\omega_2^2(p_1 - f_1 - r_1)\right.$$

$$\left. + \omega_1^2(p_2 - f_2 - r_2)\right]\left[\frac{1}{\omega^4}(\cos\omega t - 1) + \frac{1}{2\omega^2}t^2\right] \quad (19.2.8)$$

$$x_2 = \frac{V_2}{\omega}\sin\omega t + \frac{V_2\omega_1^2 + V_1\omega_2^2}{\omega^2}\left(t - \frac{1}{\omega}\sin\omega t\right)$$

$$+ S_2\cos\omega t + \frac{S_2\omega_1^2 + S_1\omega_2^2}{\omega^2}(1 - \cos\omega t)$$

$$+ \frac{p_2 - f_2 - r_2}{\omega^2}(1 - \cos\omega t) + \left[\omega_1^2(p_2 - f_2 - r_2)\right.$$

$$\left. + \omega_2^2(p_1 - f_1 - r_1)\right]\left[\frac{1}{\omega^4}(\cos\omega t - 1) + \frac{1}{2\omega^2}t^2\right] \quad (19.2.9)$$

Assuming that in equations (19.2.8) and (19.2.9), $t = 0$, we determine that $x_1 = S_1$ and $x_2 = S_2$, which correspond to the initial conditions of motion. Taking the first derivatives from equations (19.2.8) and (19.2.9), we determine the velocities of the two masses:

$$\frac{dx_1}{dt} = V_1\cos\omega t + \frac{V_1\omega_2^2 + V_2\omega_1^2}{\omega^2}(1 - \cos\omega t)$$

$$- S_1\omega\sin\omega t + \frac{S_1\omega_2^2 + S_2\omega_1^2}{\omega}\sin\omega t$$

$$+ \frac{p_1 - f_1 - r_1}{\omega}\sin\omega t + \left[\omega_2^2(p_1 - f_1 - r_1)\right.$$

$$\left. + \omega_1^2(p_2 - f_2 - r_2)\right]\left[\frac{1}{\omega^2}\left(t - \frac{1}{\omega}\sin\omega t\right)\right] \quad (19.2.10)$$

$$\frac{dx_2}{dt} = V_2\cos\omega t + \frac{V_2\omega_1^2 + V_1\omega_2^2}{\omega^2}(1 - \cos\omega t)$$

$$- S_2\omega\sin\omega t + \frac{S_2\omega_1^2 + S_1\omega_2^2}{\omega}\sin\omega t$$

$$+ \frac{p_2 - f_2 - r_2}{\omega} \sin \omega t + [\omega_1^2 (p_2 - f_2 - r_2)$$

$$+ \omega_2^2 (p_1 - f_1 - r_1)] \left[\frac{1}{\omega^2} \left(t - \frac{1}{\omega} \sin \omega t \right) \right] \qquad (19.2.11)$$

Supposing that in equations (19.2.10) and (eq19.2.11), $t = 0$, we calculate that $\frac{dx_1}{dt} = V_1$ and $\frac{dx_2}{dt} = V_2$, as it should be according to the initial conditions of motion.

Taking the first derivatives from equations (19.2.10) and (19.2.11), we determine the accelerations of the masses, respectively:

$$\frac{d^2 x_1}{dt^2} = -V_1 \omega \sin \omega t + \frac{V_1 \omega_2^2 + V_2 \omega_1^2}{\omega} \sin \omega t$$

$$- S_1 \omega^2 \cos \omega t + (S_1 \omega_2^2 + S_2 \omega_1^2) \cos \omega t$$

$$+ (p_1 - f_1 - r_1) \cos \omega t + [\omega_2^2 (p_1 - f_1 - r_1)$$

$$+ \omega_1^2 (p_2 - f_2 - r_2)] \frac{1}{\omega^2} (1 - \cos \omega t) \qquad (19.2.12)$$

$$\frac{d^2 x_2}{dt^2} = -V_2 \omega \sin \omega t + \frac{V_2 \omega_1^2 + V_1 \omega_2^2}{\omega} \sin \omega t$$

$$- S_2 \omega^2 \cos \omega t + (S_2 \omega_1^2 + S_1 \omega_2^2) \cos \omega t$$

$$+ (p_2 - f_2 - r_2) \cos \omega t + [\omega_1^2 (p_2 - f_2 - r_2) + \omega_2^2 (p_1 - f_1 - r_1)]$$

$$\times \frac{1}{\omega^2} (1 - \cos \omega t)] \qquad (19.2.13)$$

Equations (19.2.8)–(19.2.13) represent the basic parameters of motion and allow us to carry out the analysis of the corresponding operational process.

Taking in equations (19.2.12) and (19.2.13) that $t = 0$, we determine the initial accelerations of the masses, respectively, at the beginning of the motion:

$$\frac{d^2 x_1}{dt^2} = -\omega_1^2 (S_1 - S_2) + p_1 - f_1 - r_1 \qquad (19.2.14)$$

$$\frac{d^2 x_2}{dt^2} = -\omega_2^2 (S_2 - S_1) + p_2 - f_2 - r_2 \qquad (19.2.15)$$

It should be stressed that since this section is associated with the dry friction force F, we must apply all considerations presented in Section 1.5 of Chapter 01.

19.2.1. *Motion of an unrestricted system on a horizontal surface while being subjected to constant resisting and constant active forces*

The schematic diagram of the system has the shape shown in Figure 19.2.2.

Based on the schematic diagram shown in Figure 19.2.2 and the considerations mentioned above, we compose the following pair of simultaneous differential equations of motion:

$$m_1 \frac{d^2 x_1}{dt^2} + K(x_1 - x_2) + R_1 = P_1 \tag{19.2.16}$$

$$m_2 \frac{d^2 x_2}{dt^2} + K(x_2 - x_1) + R_2 = P_2 \tag{19.2.17}$$

The general initial conditions of motion are taken according to expression (19.1.3).

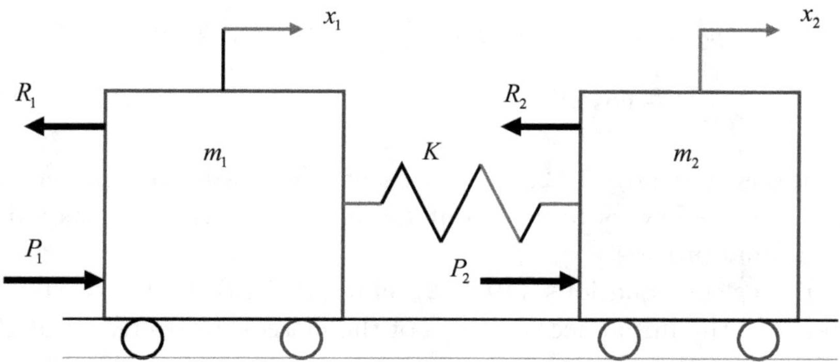

Fig. 19.2.2. Schematic diagram of an unrestricted system, the masses of which are connected by a flexible link and moving on a horizontal frictionless surface, while being subjected to a constant resisting force and a constant active force.

In equations (19.2.8) and (19.2.9), considering that $f_1 = f_2 = 0$, we obtain based on equations (19.2.8) and (19.2.9) the solutions of differential equations (19.2.16) and (19.2.17):

$$x_1 = \frac{V_1}{\omega} \sin \omega t + \frac{V_1 \omega_2^2 + V_2 \omega_1^2}{\omega^2} \left(t - \frac{1}{\omega} \sin \omega t \right)$$

$$+ S_1 \cos \omega t + \frac{S_1 \omega_2^2 + S_2 \omega_1^2}{\omega^2} (1 - \cos \omega t)$$

$$+ \frac{p_1 - r_1}{\omega^2} (1 - \cos \omega t) + [\omega_2^2 (p_1 - r_1) + \omega_1^2 (p_2 - r_2)]$$

$$\times \left[\frac{1}{\omega^4} (\cos \omega t - 1) + \frac{1}{2\omega^2} t^2 \right] \tag{19.2.18}$$

$$x_2 = \frac{V_2}{\omega} \sin \omega t + \frac{V_2 \omega_1^2 + V_1 \omega_2^2}{\omega^2} \left(t - \frac{1}{\omega} \sin \omega t \right)$$

$$+ S_2 \cos \omega t + \frac{S_2 \omega_1^2 - S_1 \omega_2^2}{\omega^2} (1 - \cos \omega t)$$

$$+ \frac{p_2 - r_2}{\omega^2} (1 - \cos \omega t) + [\omega_1^2 (p_2 - r_2) + \omega_2^2 (p_1 - r_1)]$$

$$\times \left[\frac{1}{\omega^4} (\cos \omega t - 1) + \frac{1}{2\omega^2} t^2 \right] \tag{19.2.19}$$

The first and second derivatives from equations (19.2.18) and (19.2.19), respectively, yield the velocity and acceleration of the masses.

19.2.2. *Numerical solution*

The following is a Python program to plot the graph (Figure 19.2.3) of the displacement of the first mass according to equation (19.2.18).

```python
from matplotlib.pyplot import plot, show
from numpy import linspace, sin, cos, sqrt

S1 = 0.1
V1 = 0.1
S2 = 0.2
V2 = 0.2
omega1 = 15
omega2 = 20
omega = sqrt(omega1**2 + omega2**2)
p1 = 10
r1 = 5
p2 = 15
r2 = 8

t = linspace(0, 1, 1000)

x1 = (V1/omega)*sin(omega*t) \
  + ((V1*omega2**2 + V2*omega1**2)/omega**2)*(t - (1/omega)*sin(omega*t)) \
  + S1*cos(omega*t) \
  + ((S1*omega2**2 + S2*omega1**2)/omega**2)*(1 - cos(omega*t)) \
  + ((p1 - r1)/omega**2)*(1 - cos(omega*t)) \
  + (omega2**2*(p1 - r1) + omega1**2*(p2 - r2)) \
  *((1/omega**4)*(cos(omega*t) - 1) + (1/(2*omega**2))*t**2)

plot(t, x1)
show()
```

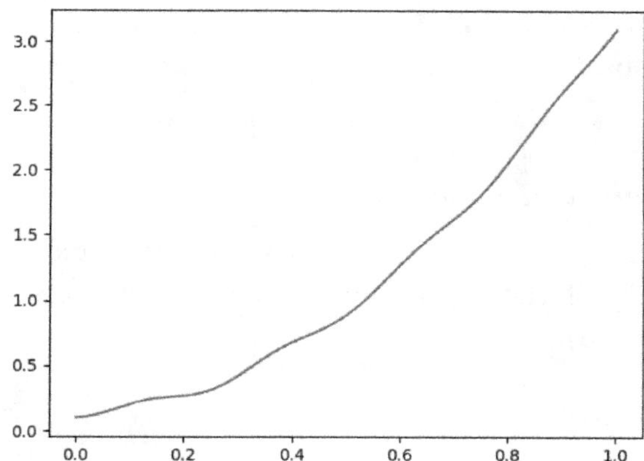

Fig. 19.2.3. Graph of equation (19.2.18).

19.3. Motion of an Unrestricted System on a Horizontal Frictionless Surface Being Subjected to Harmonic Forces

The schematic diagram of an unrestricted system subjected to the action of harmonic forces is shown in Figure 19.3.1, where A_1 and A_2 are the amplitudes of the harmonic forces, φ_1 and φ_2 are, respectively, the frequencies of the harmonic functions. The rest of the notations are self-explanatory.

The pair of simultaneous differential equations of motion of the system reads:

$$m_1 \frac{d^2 x_1}{dt^2} + K(x_1 - x_2) = A_1 \cos \varphi_1 t \qquad (19.3.1)$$

$$m_2 \frac{d^2 x_2}{dt^2} + K(x_2 - x_1) = A_2 \cos \varphi_2 t \qquad (19.3.2)$$

The initial conditions of motion are taken according to expression (3.1.3).

Dividing equations (19.3.1) and (19.3.2) by m_1 and m_2, respectively, and applying to them Laplace Transform pairs 2, 5, 2, 1, 2, and 29, we convert these equations with their initial conditions of motion from the time domain into a system of two simultaneous algebraic equations with two unknowns in each of the equations in

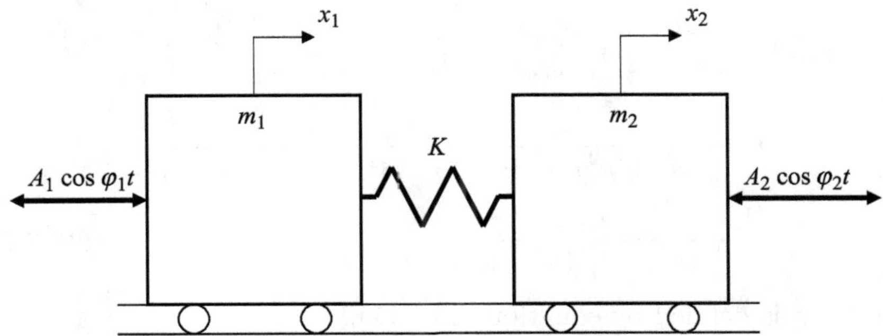

Fig. 19.3.1. Schematic diagram of an unrestricted system, the masses of which are connected by a flexible link, while moving on a horizontal frictionless surface and subjected to harmonic forces.

the Laplace domain:

$$s^2 x_1(s) - sV_1 - s^2 S_1 + \omega_1^2 x_1(s) - \omega_1^2 x_2(s) = a_1 \frac{s^2}{s^2 + \varphi_1^2}$$

$$(19.3.3)$$

$$s^2 x_2(s) - sV_2 - s^2 S_2 + \omega_2^2 x_2(s) - \omega_2^2 x_1(s) = a_2 \frac{s^2}{s^2 + \varphi_2^2}$$

$$(19.3.4)$$

where

$$a_1 = \frac{A_1}{m_1}; \quad a_2 = \frac{A_2}{m_2} \qquad (19.3.5)$$

while ω_1^2 and ω_2^2 are taken according to expression (19.1.6).

Rearranging equations (19.3.3) and (19.3.4) and applying to them the method of substitutions, we eliminate one unknown from each of the equations and obtain expressions that allow us to determine the displacements $x_1(s)$ and $x_2(s)$ in the Laplace domain:

$$x_1(s) = \frac{sV_1}{s^2 + \omega^2} + \frac{V_1 \omega_2^2 + V_2 \omega_1^2}{s(s^2 + \omega^2)} + \frac{s^2 S_1}{s^2 + \omega^2}$$

$$+ \frac{S_1 \omega_2^2 + S_2 \omega_1^2}{s^2 + \omega^2} + \frac{s^2 a_1}{(s^2 + \varphi_1^2)(s^2 + \omega^2)}$$

$$+ \frac{a_1 \omega_2^2}{(s^2 + \varphi_1^2)(s^2 + \omega^2)} + \frac{a_2 \omega_1^2}{(s^2 + \varphi_2^2)(s^2 + \omega^2)} \qquad (19.3.6)$$

$$x_2(s) = \frac{sV_2}{s^2 + \omega^2} + \frac{V_2 \omega_1^2 + V_1 \omega_2^2}{s(s^2 + \omega^2)} + \frac{s^2 S_2}{s^2 + \omega^2} + \frac{S_2 \omega_1^2 + S_1 \omega_2^2}{s^2 + \omega^2}$$

$$+ \frac{s^2 a_2}{(s^2 + \varphi_2^2)(s^2 + \omega^2)} + \frac{a_2 \omega_1^2}{(s^2 + \varphi_2^2)(s^2 + \omega^2)}$$

$$+ \frac{a_1 \omega_2^2}{(s^2 + \varphi_1^2)(s^2 + \omega^2)} \qquad (19.3.7)$$

while ω^2 is defined by equation (19.1.11).

Applying to equations (19.3.6) and (19.3.7) Laplace Transform pairs 23, 39, 29, 17, 70, 55, and 55, we invert these equations from the Laplace domain into the time domain and obtain the solutions of differential equations (19.3.1) and (19.3.2) with their initial conditions of motion:

$$x_1 = \frac{V_1}{\omega}\sin\omega t + \frac{V_1\omega_2^2 + V_2\omega_1^2}{\omega^2}\left(t - \frac{1}{\omega}\sin\omega t\right)$$

$$+ S_1\cos\omega t + \frac{S_1\omega_2^2 + S_2\omega_1^2}{\omega^2}(1 - \cos\omega t)$$

$$+ a_1\frac{\cos\omega t - \cos\varphi_1 t}{\varphi_1^2 - \omega^2} + \frac{a_1\omega_2^2}{\varphi_1^2 - \omega^2}$$

$$\times\left[\frac{1}{\omega^2}(1 - \cos\omega t) - \frac{1}{\varphi_1^2}(1 - \cos\varphi_1 t)\right]$$

$$+ \frac{a_2\omega_1^2}{\varphi_2^2 - \omega^2}\left[\frac{1}{\omega^2}(1 - \cos\omega t) - \frac{1}{\varphi_2^2}(1 - \cos\varphi_2 t)\right] \qquad (19.3.8)$$

$$x_2 = \frac{V_2}{\omega}\sin\omega t + \frac{V_2\omega_1^2 + V_1\omega_2^2}{\omega^2}\left(t - \frac{1}{\omega}\sin\omega t\right)$$

$$+ S_2\cos\omega t + \frac{S_2\omega_1^2 + S_1\omega_2^2}{\omega^2}(1 - \cos\omega t)$$

$$+ a_2\frac{\cos\omega t - \cos\varphi_2 t}{\varphi_2^2 - \omega^2} + \frac{a_2\omega_1^2}{\varphi_2^2 - \omega^2}$$

$$\times\left[\frac{1}{\omega^2}(1 - \cos\omega t) - \frac{1}{\varphi_2^2}(1 - \cos\varphi_2 t)\right]$$

$$+ \frac{a_1\omega_2^2}{\varphi_1^2 - \omega^2}\left[\frac{1}{\omega^2}(1 - \cos\omega t) - \frac{1}{\varphi_1^2}(1 - \cos\varphi_1 t)\right] \qquad (19.3.9)$$

Assuming that in equations (19.3.8) and (19.3.9), $t = 0$, we determine that $x_1 = S_1$ and $x_2 = S_2$, as it should be according to the initial conditions of motion.

Taking from equations (19.3.8) and (19.3.9) the first derivatives, we determine the velocities of the masses:

$$\frac{dx_1}{dt} = V_1\cos\omega t + \frac{V_1\omega_2^2 + V_2\omega_1^2}{\omega^2}(1 - \cos\omega t)$$

$$- S_1\omega\sin\omega t + \frac{1}{\omega}(S_1\omega_2^2 + S_2\omega_1^2)\sin\omega t$$

$$+ \frac{a_1}{\varphi_1^2 - \omega^2}(\varphi_1 \sin \varphi_1 t - \omega \sin \omega t)$$

$$+ \frac{a_1 \omega_2^2}{\varphi_1^2 - \omega^2}\left(\frac{1}{\omega}\sin \omega t - \frac{1}{\varphi_1}\sin \varphi_1 t\right)$$

$$+ \frac{a_2 \omega_1^2}{\varphi_2^2 - \omega^2}\left(\frac{1}{\omega^2}\sin \omega t - \frac{1}{\varphi_2}\sin \varphi_2 t\right) \qquad (19.3.10)$$

$$\frac{dx_2}{dt} = V_2 \cos \omega t + \frac{V_2 \omega_1^2 + V_1 \omega_2^2}{\omega^2}(1 - \cos \omega t)$$

$$- S_2 \omega \sin \omega t + \frac{1}{\omega}(S_2 \omega_1^2 + S_1 \omega_1^2)\sin \omega t$$

$$+ \frac{a_2}{\varphi_2^2 - \omega^2}(\varphi_2 \sin \varphi_2 t - \omega \sin \omega t)$$

$$+ \frac{a_2 \omega_1^2}{\varphi_2^2 - \omega^2}\left(\frac{1}{\omega}\sin \omega t - \frac{1}{\varphi_2}\sin \varphi_2 t\right)$$

$$+ \frac{a_1 \omega_2^2}{\varphi_1^2 - \omega^2}\left(\frac{1}{\omega}\sin \omega t - \frac{1}{\varphi_1}\sin \varphi_1 t\right) \qquad (19.3.11)$$

Assuming that in equations (19.3.10) and (19.3.11), $t = 0$, we determine that $\frac{dx_1}{dt} = V_1$ and $\frac{dx_2}{dt} = V_2$, as expected according to the initial conditions of motion.

19.3.1. *Resonance*

The phenomenon of resonance is possible if the compound natural frequency is equal to the frequency of the exciting force. Therefore, we accept that in equations (19.3.8) and (19.3.9), the compound natural frequency equals the frequency of the exiting force applied to one of the masses, for instance to the first mass:

$$\omega = \varphi_1 \qquad (19.3.12)$$

while $\omega^2 = \omega_1^2 + \omega_2^2 = \varphi_1^2$. In this case, according to equation (19.3.12), several denominators of the fractional members of equations (19.3.8) and (19.3.9) become equal to zero. These members have the potential to tend to infinity and cause the resonance of the system. The rest

of the members of these equations that do not influence resonance
are ignorable and, consequently, are removed from these equations.
Therefore, for the purpose of the investigation on resonance, we
restructure equations (19.3.8) and (19.3.9) by keeping in them just
the terms that may potentially tend to infinity:

$$x_1' = \frac{a_1(\cos \omega t - \cos \varphi_1 t)}{\varphi_1^2 - \omega^2} + \frac{a_1 \omega_2^2}{\varphi_1^2 - \omega^2}$$

$$\times \left[\frac{1}{\omega^2}(1 - \cos \omega t) - \frac{1}{\varphi_1^2}(1 - \cos \varphi_1 t) \right] \tag{19.3.13}$$

$$x_2' = \frac{a_1 \omega_2^2}{\varphi_1^2 - \omega^2} \left[\frac{1}{\omega^2}(1 - \cos \omega t) - \frac{1}{\varphi_1^2}(1 - \cos \varphi_1 t) \right]$$

$$\tag{19.3.14}$$

where x_1' and x_2' represent the potential displacements of the
respective masses.

Based on equation (19.3.12), we can see that both terms in
equations (19.3.13) and (19.3.14) represent indeterminates of type $\frac{0}{0}$.

Rearranging these equations, we may write

$$x_1' = \frac{a_1(\cos \omega t - \cos \varphi_1 t)}{\varphi_1^2 - \omega^2} + \frac{a_1 \omega_2^2(\varphi_1^2 - \omega^2 - \varphi_1^2 \cos \omega t + \omega^2 \cos \varphi_1 t)}{\varphi_1^2 \omega^2(\varphi_1^2 - \omega^2)}$$

$$\tag{19.3.15}$$

$$x_2' = \frac{a_1 \omega_2^2(\varphi_1^2 - \omega^2 - \varphi_1^2 \cos \omega t + \omega^2 \cos \varphi_1 t)}{\varphi_1^2 \omega^2(\varphi_1^2 - \omega^2)} \tag{19.3.16}$$

To evaluate these indeterminate forms, we apply L'Hopital's rule to
equations (19.3.15) and (19.3.16), and we obtain

$$x_1' = \lim_{\varphi_1 \to \omega} \frac{a_1(\cos \omega t - \cos \varphi_1 t)}{\varphi_1^2 - \omega^2}$$

$$+ \lim_{\varphi_1 \to \omega} \frac{a_1 \omega_2^2(\varphi_1^2 - \omega^2 - \varphi_1^2 \cos \omega t + \omega^2 \cos \varphi_1 t)}{\omega^2 \varphi_1^2(\varphi_1^2 - \omega^2)}$$

$$\tag{19.3.17}$$

$$x_2' = \lim_{\varphi_1 \to \omega} \frac{a_1 \omega_2^2(\varphi_1^2 - \omega^2 - \varphi_1^2 \cos \omega t + \omega^2 \cos \varphi_1 t)}{\omega^2 \varphi_1^2(\varphi_1^2 - \omega^2)} \tag{19.3.18}$$

Differentiating the numerators and denominators of expressions (19.3.17) and (19.3.18) with respect to ω, we obtain

$$x_1' = \lim_{\varphi_1 \to \omega} \frac{a_1(-t\sin\omega t)}{-2\omega} + \lim_{\varphi_1 \to \omega} \frac{a_1\omega_2^2(-2\omega + \varphi_1^2 t\sin\omega t + 2\omega\cos\varphi_1 t)}{2\omega\varphi_1^4 - 4\varphi_1\omega^4}$$

$$(19.3.19)$$

$$x_2' = \lim_{\varphi_1 \to \omega} \frac{a_1\omega_2^2(-2\omega + \varphi_1^2 t\sin\omega t + 2\omega\cos\varphi_1 t)}{2\omega\varphi_1^4 - 4\omega^3\varphi_1^2} \qquad (19.3.20)$$

Completing the actions with limits, we have

$$x_1' = \frac{a_1 t\sin\omega t}{2\omega} + \frac{a_1\omega_2^2(-2 + \omega t\sin\omega t + 2\cos\omega t)}{-2\omega^4} \qquad (19.3.21)$$

$$x_2' = \frac{a_1\omega_2^2(-2 + \omega t\sin\omega t + 2\cos\omega t)}{-2\omega^4} \qquad (19.3.22)$$

Eliminating from expressions (19.3.21) and (19.3.22) the terms that do not have influence on the resonance, we obtain

$$x_1' = \frac{a_1 t\sin\omega t}{2\omega} - \frac{a_1\omega_2^2 t\sin\omega t}{2\omega^3} \qquad (19.3.23)$$

$$x_2' = -\frac{a_1\omega_2^2 t\sin\omega t}{2\omega^3} \qquad (19.3.24)$$

Further analysis of equation (19.3.23) yields

$$x_1' = \frac{a_1\omega^2 t\sin\omega t}{2\omega^3} - \frac{a_1\omega_2^2 t\sin\omega t}{2\omega^3}$$

$$= \frac{a_1 t(\omega_1^2 + \omega_2^2)\sin\omega t}{2\omega^3} - \frac{a_1\omega_2^2 t\sin\omega t}{2\omega^3} \qquad (19.3.25)$$

Hence, equations (19.3.24) and (19.3.25) allow us to evaluate the above-mentioned indeterminates:

$$X_1 = \frac{a_1\omega_1^2 t\sin\omega t}{2\omega^3} \qquad (19.3.26)$$

$$X_2 = -\frac{a_1\omega_2^2 t\sin\omega t}{2\omega^3} \qquad (19.3.27)$$

where X_1 and X_2 represent the respective evaluations of the indeterminates.

It should be emphasized that the multiplier t in equations (19.3.26) and (19.3.27) causes the amplitudes of both masses to tend to infinity, bringing the system to the state of resonance.

19.3.2. *Numerical solution*

The following is a Python program to plot equation (19.3.26). The graph is plotted in Figure 19.3.2.

```python
from matplotlib.pyplot import plot, show
from numpy import linspace, sin

a1 = 5
omega1 = 15
omega = 25

t = linspace(0,2,1000)

x1 = a1*omega1**2*t*sin(omega*t)/(2*omega**3)

plot(t, x1)
show()
```

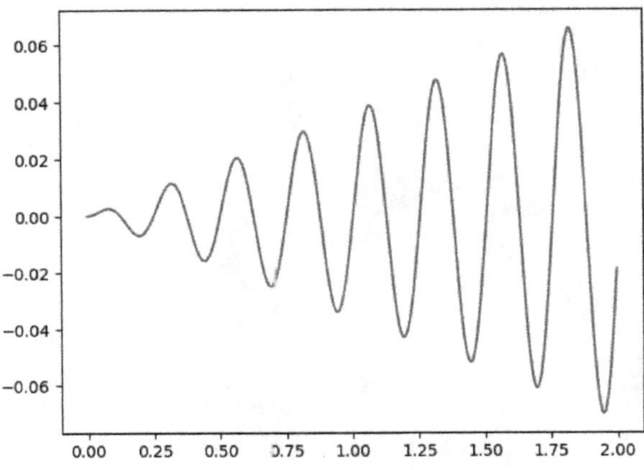

Fig. 19.3.2. Graph of equation (19.3.26).

UNRESTRICTED TWO-DEGREE-OF-FREEDOM SYSTEM WITH THE TWO MASSES CONNECTED TO EACH OTHER BY A FLUID LINK

In this chapter, we analyze the operational processes of unrestricted two-degree-of-freedom systems, the motions of which are caused by the initial conditions of motion, constant resisting, and harmonic forces. Unrestricted two-degree-of-freedom systems are widely used in pneumatically operated impact machinery, hydraulic impact tools, hydraulic dashpots, and other industrial applications.

20.1. Motion of an Unrestricted System on a Horizontal Frictionless Surface Due to Initial Displacements and Velocities

The schematic diagram representing an unrestricted system, the masses of which are connected to each other by a fluid kinematic link (dashpot) with the damping coefficient C, is presented in Figure 20.1.1, the notations in which are self-explanatory. The masses of the system are not subjected to any external forces, while the motion of the masses on the horizontal frictionless surface occurs due to their initial displacements and velocities.

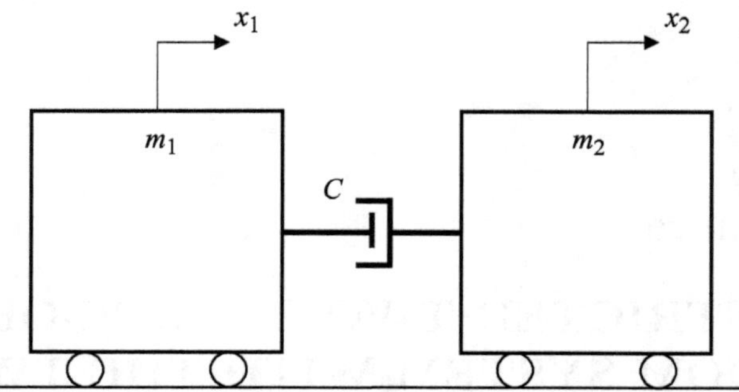

Fig. 20.1.1. Schematic diagram of an unrestricted system moving on a horizontal frictionless surface, the masses of which are connected by a fluid link, while the motions of these masses occur due to their initial conditions of motion.

Based on the schematic diagram shown in Figure 20.1.1 and the related considerations, we compose the following pair of simultaneous differential equations describing the motion of the masses of the system:

$$m_1 \frac{d^2 x_1}{dt^2} + C\left(\frac{dx_1}{dt} - \frac{dx_2}{dt}\right) = 0 \tag{20.1.1}$$

$$m_2 \frac{d^2 x_2}{dt^2} + C\left(\frac{dx_2}{dt} - \frac{dx_1}{dt}\right) = 0 \tag{20.1.2}$$

The initial conditions of motion are taken according to expression (19.1.3).

Dividing equations (20.1.1) and (20.1.2), respectively, by m_1 and m_2, and applying to these equations Laplace Transform pairs 5, 2, and 4, we convert them from the time domain into a system of two simultaneous algebraic equations with two unknowns in each equation in the Laplace domain:

$$s^2 x_1(s) - sV_1 - s^2 S_1 + 2n_1 s x_1(s) - 2n_1 s S_1$$
$$- 2n_1 s x_2(s) + 2n_1 s S_2 = 0 \tag{20.1.3}$$
$$s^2 x_2(s) - sV_2 - s^2 S_2 + 2n_2 s x_2(s)$$
$$- 2n_2 s S_2 - 2n_2 s x_1(s) + 2n s S_1 = 0 \tag{20.1.4}$$

where

$$2n_1 = \frac{C}{m_1}; \quad 2n_2 = \frac{C}{m_2} \tag{20.1.5}$$

Rearranging the terms in equations (20.1.3) and (20.1.4) and applying to them the method of substitutions, we obtain a pair of two simultaneous algebraic equations with one unknown in each equation, respectively, describing the displacements of the masses in the Laplace domain:

$$x_1(s) = \frac{V_1}{s+2n} + \frac{2(V_1 n_2 + V_2 n_1)}{s(s+2n)} + \frac{sS_1}{s+2n} + \frac{2nS_1}{s+2n} \tag{20.1.6}$$

$$x_2(s) = \frac{V_2}{s+2n} + \frac{2(V_2 n_1 + V_1 n_2)}{s(s+2n)} + \frac{sS_2}{s+2n} + \frac{2nS_2}{s+2n} \tag{20.1.7}$$

where

$$n = n_1 + n_2 \tag{20.1.8}$$

Applying to equations (20.1.6) and (20.1.7) Laplace Transform pairs 8, 16, 14, and 8, we invert these equations from the Laplace domain into the time domain and obtain the solutions of the mentioned differential equations (20.1.1) and (20.1.2) with their initial conditions according to expression (19.1.3):

$$x_1 = S_1 + \frac{V_1}{2n}(1 - e^{-2nt}) + \frac{V_1 n_2 + V_2 n_1}{n}\left[t + \frac{1}{2n}(e^{-2nt} - 1)\right] \tag{20.1.9}$$

$$x_2 = S_2 + \frac{V_2}{2n}(1 - e^{-2nt}) + \frac{V_2 n_1 + V_1 n_2}{n}\left[t + \frac{1}{2n}(e^{-2nt} - 1)\right] \tag{20.1.10}$$

Assuming in equations (20.1.9) and (20.1.10) that $t = 0$, we obtain $x_1 = S_1$ and $x_2 = S_2$, as it should be according to initial conditions of motion (19.1.3). Taking the first derivatives from equations (20.1.9) and (20.1.10), we determine the velocities of the masses:

$$\frac{dx_1}{dt} = V_1 e^{-2nt} + \frac{V_1 n_2 + V_2 n_1}{n}(1 - e^{-2nt}) \qquad (20.1.11)$$

$$\frac{dx_2}{dt} = V_2 e^{-2nt} + \frac{V_2 n_1 + V_1 n_2}{n}(1 - e^{-2nt}) \qquad (20.1.12)$$

Supposing in equations (20.1.11) and (20.1.12) that $t = 0$, we determine $\frac{dx_1}{dt} = V_1$ and $\frac{dx_2}{dt} = V_2$, as expected according to initial conditions of motion (19.1.3). Taking the first derivatives from equations (20.1.11) and (20.1.12), we determine the accelerations of the masses:

$$\frac{d^2 x_1}{dt^2} = -2nV_1 e^{-2nt} + 2(V_1 n_2 + V_2 n_1)e^{-2nt} \qquad (20.1.13)$$

$$\frac{d^2 x_2}{dt^2} = -2nV_2 e^{-2nt} + 2(V_2 n_1 + V_1 n_2)e^{-2nt} \qquad (20.1.14)$$

Hence, equations (20.1.9)–(20.1.14) represent the basic parameters of motion of the masses.

Taking in equations (20.1.13) and (20.1.14) that $t = 0$, we determine the accelerations of the masses at the beginning of the motion:

$$\frac{d^2 x_1}{dt^2} = -2n_1(V_1 - V_2) \qquad (20.1.15)$$

$$\frac{d^2 x_2}{dt^2} = -2n_2(V_2 - V_1) \qquad (20.1.16)$$

20.2. Motion of an Unrestricted System on a Horizontal Surface While the System Is Being Subjected to Constant Resisting, Dry Friction, and Constant Active Forces

We consider the operational process of a system, the masses m_1 and m_2 of which are, respectively, subjected to the action of constant resisting, dry friction, and constant active forces. The system is moving on a horizontal surface.

The schematic diagram of the described system is shown in Figure 20.2.1, the notations in which are self-explanatory.

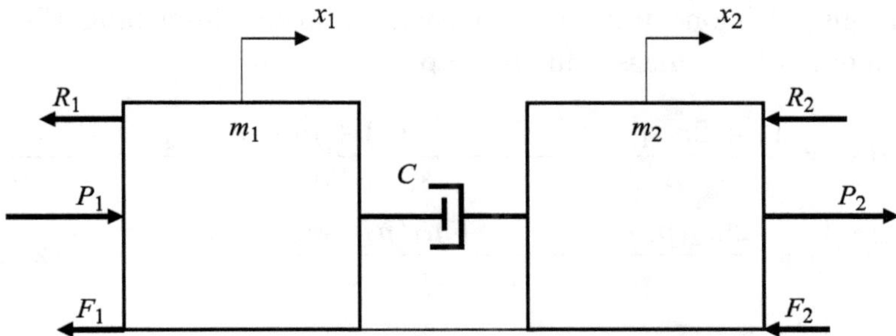

Fig. 20.2.1. Schematic diagram of an unrestricted system moving on a horizontal surface and subjected to constant resisting, dry friction, and constant active forces.

Based on the schematic diagram shown in Figure 20.2.1 and related considerations, we compose a pair of simultaneous differential equations describing the motion of the system:

$$m_1 \frac{d^2 x_1}{dt^2} + C\left(\frac{dx_1}{dt} - \frac{dx_2}{dt}\right) + R_1 + F_1 = P_1 \qquad (20.2.1)$$

$$m_2 \frac{d^2 x_2}{dt^2} + C\left(\frac{dx_2}{dt} - \frac{dx_1}{dt}\right) + R_2 + F_2 = P_2 \qquad (20.2.2)$$

The general initial conditions of motion are taken according to expression (19.1.3).

Dividing equations (20.2.1) and (20.2.2), respectively, by m_1 and m_2, while applying to them Laplace Transform pairs 5, 2, 4, 2, 2, and 2, we convert these equations with their initial conditions of motion from the time domain into a system of two simultaneous algebraic equations with two unknowns in each equation in the Laplace domain:

$$s^2 x_1(s) - sV_1 - s^2 S_1 + 2n_1 s x_1(s) - 2n_1 s x_1(s)$$

$$-2n_1 s x_2(s) + 2n_1 s S_2 + r_1 + f_1 = p_1 \qquad (20.2.3)$$

$$s^2 x_2(s) - sV_2 - s^2 S_2 + 2n_2 s x_2(s) - 2n_2 s x_2(s)$$

$$-2n_2 s x_1(s) + 2ns S_1 + r_2 + f_2 = p_2 \qquad (20.2.4)$$

Rearranging equations (20.2.3) and (20.2.4) and applying to them the method of substitutions, we obtain two simultaneous algebraic equations with one unknown in each equation describing the displacements of the masses in the Laplace domain:

$$x_1(s) = \frac{V_1 + 2nS_1}{s + 2n} + \frac{2(n_2V_1 + n_1V_2) + p_1 - r_1 - f_1}{s(s + 2n)} + \frac{sS_1}{s + 2n}$$

$$+ \frac{2[n_2(p_1 - r_1 - f_1) + n_1(p_2 - r_2 - f_2)]}{s^2(s + 2n)} \tag{20.2.5}$$

$$x_2(s) = \frac{V_2 + 2nS_2}{s + 2n} + \frac{2(n_1V_2 + n_2V_1) + p_2 - r_2 - f_2}{s(s + 2n)} + \frac{sS_2}{s + 2n}$$

$$+ \frac{2[n_1(p_2 - r_2 - f_2) + n_2(p_1 - r_1 - f_1)]}{s^2(s + 2n)} \tag{20.2.6}$$

Applying to equations (20.2.5) and (20.2.6) Laplace Transform pairs 8, 16, 14, and 48, we invert these equations from the Laplace domain into the time domain and obtain the solutions of differential equations (20.2.1) and (20.2.2) with initial conditions of motion (19.1.3):

$$x_1 = \frac{V_1 + 2nS_1}{2n}(1 - e^{-2nt}) + \frac{2(n_2V_1 + n_1V_2) + p_1 - r_1 - f_1}{2n}$$

$$\times \left[t + \frac{1}{2n}(e^{-2nt} - 1)\right] + S_1 e^{-2nt} + \frac{1}{4n}2[n_2(p_1 - r_1 - f_1)$$

$$+ n_1(p_2 - r_2 - f_2)]\left[t^2 - \frac{1}{n}t - \frac{1}{2n^2}(e^{-2nt} - 1)\right] \tag{20.2.7}$$

$$x_2 = \frac{V_2 + 2nS_2}{2n}(1 - e^{-2nt}) + \frac{2(n_1V_2 + n_2V_1) + p_2 - r_2 - f_2}{2n}$$

$$\times \left[t + \frac{1}{2n}(e^{-2nt} - 1)\right] + S_2 e^{-2nt} + \frac{1}{4n}2[n_1(p_2 - r_2 - f_2)$$

$$+ n_2(p_1 - r_1 - f_1)]\left[t^2 - \frac{1}{n}t - \frac{1}{2n^2}(e^{-2nt} - 1)\right] \tag{20.2.8}$$

Rearranging equations (20.2.7) and (20.2.8), we have

$$x_1 = S_1 + \frac{V_1}{2n}(1 - e^{-2nt}) + \frac{2(n_2 V_1 + n_1 V_2) + p_1 - r_1 - f_1}{2n}$$

$$\times \left[t + \frac{1}{2n}(e^{-2nt} - 1) \right] + \frac{n_2(p_1 - r_1 - f_1) + n_1(p_2 - r_2 - f_2)}{2n}$$

$$\times \left[t^2 - \frac{t}{n} - \frac{1}{2n^2}(e^{-2nt} - 1) \right] \tag{20.2.9}$$

$$x_2 = S_2 + \frac{V_2}{2n}(1 - e^{-2nt}) + \frac{2(n_1 V_2 + n_2 V_1) + p_2 - r_2 - f_2}{2n}$$

$$\times \left[t + \frac{1}{2n}(e^{-2nt} - 1) \right] + \frac{n_1(p_2 - r_2 - f_2) + n_2(p_1 - r_1 - f_1)}{2n}$$

$$\times \left[t^2 - \frac{t}{n} - \frac{1}{2n^2}(e^{-2nt} - 1) \right] \tag{20.2.10}$$

Assuming in equations (20.2.9) and (20.2.10) that $t = 0$, we determine that $x_1 = S_1$ and $x_2 = S_2$, as expected according to initial conditions of motion (19.1.3). Taking from equations (20.2.9) and (20.2.10) the first derivatives, we obtain the equations of the velocities of the masses:

$$\frac{dx_1}{dt} = V_1 e^{-2nt} + \frac{2(n_1 V_2 + n_2 V_1) + p_1 - r_1 - f_1}{2n}(1 - e^{-2nt})$$

$$+ \frac{n_2(p_1 - r_1 - f_1) + n_1(p_2 - r_2 - f_2)}{2n}\left(2t - \frac{1}{n} + \frac{1}{n}e^{-2nt} \right)$$

$$\tag{20.2.11}$$

$$\frac{dx_2}{dt} = V_2 e^{-2nt} + \frac{2(n_2 V_1 + n_1 V_2) + p_2 - r_2 - f_2}{2n}(1 - e^{-2nt})$$

$$+ \frac{n_1(p_2 - r_2 - f_2) + n_2(p_1 - r_1 - f_1)}{2n}\left(2t - \frac{1}{n} + \frac{1}{n}e^{-2nt} \right)$$

$$\tag{20.2.12}$$

Supposing for the equations (20.2.11) and (20.2.12) that $t = 0$, we determine that $\frac{dx_1}{dt} = V_1$ and $\frac{dx_2}{dt} = V_2$, as it should be according to the initial conditions of motion.

Taking the first derivatives from equations (20.2.11) and (20.2.12), we determine the accelerations of the masses:

$$\frac{d^2 x_1}{dt^2} = -2nV_1 e^{-2nt} + [2(n_1 V_2 + n_2 V_1) + p_1 - r_1 - f_1]e^{-2nt}$$

$$+ \frac{1}{n}[n_2(p_1 - r_1 - f_1) + n_1(p_2 - r_2 - f_2)](1 - e^{-2nt})$$

$$(20.2.13)$$

$$\frac{d^2 x_2}{dt^2} = -2nV_2 e^{-2nt} + [2(n_2 V_1 + n_1 V_2) + p_2 - r_2 - f_2]e^{-2nt}$$

$$+ \frac{1}{n}[n_1(p_2 - r_2 - f_2) + n_2(p_1 - r_1 - f_1)](1 - e^{-2nt})$$

$$(20.2.14)$$

Hence, equations (20.2.9)–(20.2.14) represent the basic parameters of motion of the masses.

Taking in equations (20.2.13) and (20.2.14) that $t = 0$, we obtain

$$\frac{d^2 x_1}{dt^2} = -2n_1(V_1 - V_2) + p_1 - r_1 - f_1 \qquad (20.2.15)$$

$$\frac{d^2 x_2}{dt^2} = -2n_2(V_2 - V_1) + p_2 - r_2 - f_2 \qquad (20.2.16)$$

20.3. Motion of an Unrestricted System on a Horizontal Frictionless Surface While Being Subjected to Harmonic Forces

Figure 20.3.1 shows a schematic diagram of an unrestricted system, the masses of which are connected to each other by a fluid link (dashpot) and are subjected to the actions of harmonic forces. The notations of this figure are self-explanatory. The masses are mowing on a horizontal frictionless surface.

Based on the schematic diagram shown in Figure 20.3.1 and the related considerations, we compose a pair of simultaneous differential

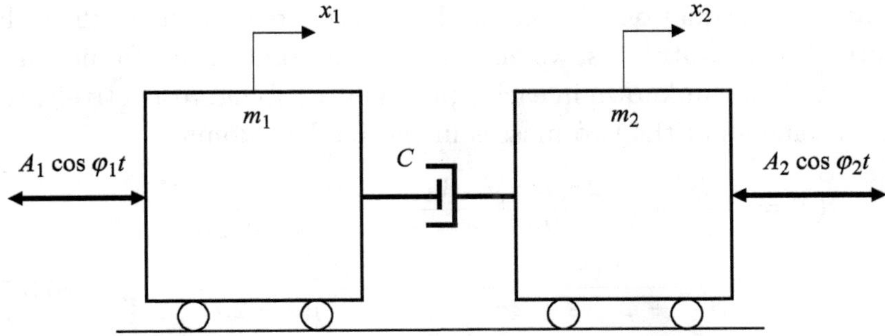

Fig. 20.3.1. Schematic diagram of an unrestricted system moving on a horizontal frictionless surface and subjected to harmonic forces.

equations describing the motion of the masses of this system:

$$m_1 \frac{d^2 x_1}{dt^2} + C\left(\frac{dx_1}{dt} - \frac{dx_2}{dt}\right) = A_1 \cos \varphi_1 t \qquad (20.3.1)$$

$$m_2 \frac{d^2 x_2}{dt^2} + C\left(\frac{dx_2}{dt} - \frac{dx_1}{dt}\right) = A_2 \cos \varphi_2 t \qquad (20.3.2)$$

The general initial conditions of motion are taken according to expression (19.1.3).

Dividing equations (20.3.1) and (20.3.2) by m_1 and m_2, respectively, and applying to them Laplace Transform pairs 2, 5, 2, 4, 2, and 29, we convert these equations with their initial conditions of motion from the time domain into the Laplace domain and obtain a corresponding system of two simultaneous algebraic equations with two unknowns in each equation in the Laplace domain:

$$s^2 x_1(s) - sV_1 - s^2 S_1 + 2n_1 s x_1(s) - 2n_1 s S_1$$

$$- 2n_1 s x_2(s) + 2n_1 s S_2 = \frac{a_1 s^2}{s^2 + \varphi_1^2} \qquad (20.3.3)$$

$$s^2 x_2(s) - sV_2 - s^2 S_2 + 2n_2 s x_2(s) - 2n_2 s S_2$$

$$- 2n_2 s x_1(s) + 2n_2 s S_1 = \frac{a_2 s^2}{s^2 + \varphi_2^2} \qquad (20.3.4)$$

Rearranging equations (20.3.3) and (20.3.4) and applying to them the method of substitutions, we obtain two simultaneous algebraic equations with one unknown in each equation describing, respectively, the displacements of the two masses in the Laplace domain:

$$x_1(s) = \frac{sS_1}{s+2n} + \frac{2S_1 n}{s+2n} + \frac{V_1}{s+2n} + \frac{2(V_1 n_2 + V_2 n_1)}{s(s+2n)}$$

$$+ \frac{a_1 s}{(s^2 + \varphi_1^2)(s+2n)} + \frac{2n_2 a_1}{(s^2 + \varphi_1^2)(s+2n)}$$

$$+ \frac{2n_1 a_2}{(s^2 + \varphi_2^2)(s+2n)} \tag{20.3.5}$$

$$x_2(s) = \frac{sS_2}{l+2n} + \frac{2S_2 n}{s+2n} + \frac{V_2}{s+2n} + \frac{2(V_2 n_1 + V_1 n_2)}{s(s+2n)}$$

$$+ \frac{a_2 s}{(s^2 + \varphi_2^2)(s+2n)} + \frac{2n_1 a_2}{(s^2 + \varphi_2^2)(s+2n)}$$

$$+ \frac{2n_2 a_1}{(s^2 + \varphi_1^2)(s+2n)} \tag{20.3.6}$$

Applying to equations (20.3.5) and (20.3.6) Laplace Transform pairs 14, 8, 8, 16, 44, 42, and 42, we invert these equations from the Laplace domain into the time domain and obtain the solutions of differential equations (20.3.1) and (20.3.2) with the initial conditions of motion according to expression (19.1.3):

$$x_1 = S_1 e^{-2nt} + S_1(1 - e^{-2nt}) + \frac{V_1}{2n}(1 - e^{-2nt}) + \frac{V_1 n_2 + V_2 n_1}{n}$$

$$\times \left[t + \frac{1}{2n}(e^{-2nt} - 1) \right] + \frac{a_1}{\varphi_1^2 + 4n^2}\left(e^{-2nt} + \frac{2n}{\varphi_1}\sin\varphi_1 t \right.$$

$$\left. - \cos\varphi_1 t \right) + \frac{2n_2 a_1}{\varphi_1^2 + 4n^2}\left[\frac{2n}{\varphi_1^2}(1 - \cos\varphi_1 t) - \frac{1}{\varphi_1}\sin\varphi_1 t \right.$$

$$\left. + \frac{1}{2n}(1 - e^{-2nt}) \right] + \frac{2n_1 a_2}{\varphi_2^2 + 4n^2}\left[\frac{2n}{\varphi_2^2}(1 - \cos\varphi_2 t) - \frac{1}{\varphi_2}\sin\varphi_2 t \right.$$

$$\left. + \frac{1}{2n}(1 - e^{-2nt}) \right] \tag{20.3.7}$$

$$x_2 = S_2 e^{-2nt} + S_2(1 - e^{-2nt}) + \frac{V_2}{2n}(1 - e^{-2nt}) + \frac{V_2 n_1 + V_1 n_2}{n}$$

$$\times \left[t + \frac{1}{2n}(e^{-2nt} - 1) \right] + \frac{a_2}{\varphi_2^2 + 4n^2}\left(e^{-2nt} + \frac{2n}{\varphi_2}\sin\varphi_2 t \right.$$

$$\left. - \cos\varphi_2 t \right) + \frac{2n_1 a_2}{\varphi_2^2 + 4n^2}\left[\frac{2n}{\varphi_2^2}(1 - \cos\varphi_2 t) - \frac{1}{\varphi_2}\sin\varphi_2 t \right.$$

$$\left. + \frac{1}{2n}(1 - e^{-2nt}) \right] + \frac{2n_2 a_1}{\varphi_1^2 + 4n^2}\left[\frac{2n}{\varphi_1^2}(1 - \cos\varphi_1 t) \right.$$

$$\left. - \frac{1}{\varphi_1}\sin\varphi_1 t + \frac{1}{2n}(1 - e^{-2nt}) \right] \tag{20.3.8}$$

Simplifying equations (20.3.7) and (20.3.8), we obtain

$$x_1 = S_1 + \frac{V_1}{2n}(1 - e^{-2nt}) + \frac{V_1 n_2 + V_2 n_1}{n}\left[t + \frac{1}{2n}(e^{-2nt} - 1) \right]$$

$$+ \frac{a_1}{\varphi_1^2 + 4n^2}\left(e^{-2nt} + \frac{2n}{\varphi_1}\sin\varphi_1 t - \cos\varphi_1 t \right)$$

$$+ \frac{2n_2 a_1}{\varphi_1^2 + 4n^2}\left[\frac{2n}{\varphi_1^2}(1 - \cos\varphi_1 t) - \frac{1}{\varphi_1}\sin\varphi_1 t + \frac{1}{2n}(1 - e^{-2nt}) \right]$$

$$+ \frac{2n_1 a_2}{\varphi_2^2 + 4n^2}\left[\frac{2n}{\varphi_2^2}(1 - \cos\varphi_2 t) - \frac{1}{\varphi_2}\sin\varphi_2 t + \frac{1}{2n}(1 - e^{-2nt}) \right]$$

$$\tag{20.3.9}$$

$$x_2 = S_2 + \frac{V_2}{2n}(1 - e^{-2nt}) + \frac{V_2 n_1 + V_1 n_2}{n}\left[t + \frac{1}{2n}(e^{-2nt} - 1) \right]$$

$$+ \frac{a_2}{\varphi_2^2 + 4n^2}\left(e^{-2nt} + \frac{2n}{\varphi_2}\sin\varphi_2 t - \cos\varphi_2 t \right)$$

$$+ \frac{2n_1 a_2}{\varphi_2^2 + 4n^2}\left[\frac{2n}{\varphi_2^2}(1 - \cos\varphi_2 t) - \frac{1}{\varphi_2}\sin\varphi_2 t + \frac{1}{2n}(1 - e^{-2nt}) \right]$$

$$+ \frac{2n_2 a_1}{\varphi_1^2 + 4n^2}\left[\frac{2n}{\varphi_1^2}(1 - \cos\varphi_1 t) - \frac{1}{\varphi_1}\sin\varphi_1 t + \frac{1}{2n}(1 - e^{-2nt}) \right]$$

$$\tag{20.3.10}$$

Supposing in equations (20.3.9) and (20.3.10) that $t = 0$, we obtain that $x_1 = S_1$ and $x_2 = S_2$, which conform with the initial conditions of motion according to (19.1.3). Taking from these equations the first derivatives, we determine the velocities of the first and second masses, respectively:

$$\frac{dx_1}{dt} = V_1 e^{-2nt} + \frac{1}{n}(V_1 n_2 + V_2 n_1)(1 - e^{-2nt})$$

$$+ \frac{a_1}{\varphi_1^2 + 4n^2}(\varphi_1 \sin \varphi_1 t + 2n \cos \varphi_1 t - 2n e^{-2nt})$$

$$+ \frac{2n_2 a_1}{\varphi_1^2 + 4n^2}\left(\frac{2n}{\varphi_1} \sin \varphi_1 t - \cos \varphi_1 t + e^{-2nt}\right)$$

$$+ \frac{2n_1 a_2}{\varphi_2^2 + 4n^2}\left(\frac{2n}{\varphi_2} \sin \varphi_2 t - \cos \varphi_2 t + e^{-2nt}\right) \qquad (20.3.11)$$

$$\frac{dx_2}{dt} = V_2 e^{-2nt} + \frac{1}{n}(V_2 n_1 + V_1 n_2)(1 - e^{-2nt})$$

$$+ \frac{a_2}{\varphi_2^2 + 4n^2}(\varphi_2 \sin \varphi_2 t + 2n \cos \varphi_2 t - 2n e^{-2nt})$$

$$+ \frac{2n_1 a_2}{\varphi_2^2 + 4n^2}\left(\frac{2n}{\varphi_2^2} \sin \varphi_2 t - \cos \varphi_2 t + e^{-2nt}\right)$$

$$+ \frac{2n_2 a_1}{\varphi_1^2 + 4n^2}\left(\frac{2n}{\varphi_1} \sin \varphi_1 t - \cos \varphi_1 t + e^{-2nt}\right) \qquad (20.3.12)$$

Assuming in equations (20.3.11) and (20.3.12) that $t = 0$, we obtain that $\frac{dx_1}{dt} = V_1$ and $\frac{dx_2}{dt} = V_2$, as it is supposed to be according to the initial conditions of motion.

Taking from equations (20.3.11) and (20.3.12) the first derivatives, we determine the accelerations of the masses:

$$\frac{d^2 x_1}{dt^2} = -2n V_1 e^{-2nt} + 2(V_1 n_2 + V_2 n_1)e^{-2nt}$$

$$+ \frac{a_1}{\varphi_1^2 + 4n^2}(\varphi_1^2 \cos \varphi_1 t - 2n\varphi_1 \sin \varphi_1 t + 4n^2 e^{-2nt})$$

$$+ \frac{2n_2 a_1}{\varphi_1^2 + 4n^2}(2n\cos\varphi_1 t + \varphi_1 \sin\varphi_1 t - 2ne^{-2nt})$$

$$+ \frac{2n_1 a_2}{\varphi_2^2 + 4n^2}(2n\cos\varphi_2 t + \varphi_2 \sin\varphi_2 t - 2ne^{-2nt})$$

$$(20.3.13)$$

$$\frac{d^2 x_2}{dt^2} = -2nV_1 e^{-2nt} + 2(V_1 n_2 + V_2 n_1)e^{-2nt}$$

$$+ \frac{a_2}{\varphi_2^2 + 4n^2}(\varphi_2^2 \cos\varphi_2 t - 2n\varphi_2 \sin\varphi_2 t + 4n^2 e^{-2nt})$$

$$+ \frac{2n_1 a_2}{\varphi_1^2 + 4n^2}(2n\cos\varphi_2 t + \varphi_2 \sin\varphi_2 t - 2ne^{-2nt})$$

$$+ \frac{2n_2 a_1}{\varphi_2^2 + 4n^2}(2n\cos\varphi_1 t + \varphi_1 \sin\varphi_1 t - 2ne^{-2nt})$$

$$(20.3.14)$$

Simplifying equations (20.3.13) and (20.3.14), we obtain

$$\frac{d^2 x_1}{dt^2} = -2n_1(V_1 - V_2)e^{-2nt} + \frac{a_1}{\varphi_1^2 + 4n^2}(\varphi_1^2 \cos\varphi_1 t - 2n\varphi_1 \sin\varphi_1 t$$

$$+ 4n^2 e^{-2nt}) + \frac{2(a_1 n_2 + a_2 n_1)}{\varphi_1^2 + 4n^2}(2n\cos\varphi_1 t + \varphi_1 \sin\varphi_1 t$$

$$- 2ne^{-2nt}) \tag{20.3.15}$$

$$\frac{d^2 x_2}{dt^2} = -2n_2(V_2 - V_1)e^{-2nt} + \frac{a_2}{\varphi_2^2 + 4n^2}(\varphi_2^2 \cos\varphi_2 t - 2n\varphi_2 \sin\varphi_2 t$$

$$+ 4n^2 e^{-2nt}) + \frac{2(a_2 n_1 + a_1 n_2)}{\varphi_2^2 + 4n^2}(2n\cos\varphi_2 t + \varphi_2 \sin\varphi_2 t$$

$$- 2ne^{-2nt}) \tag{20.3.16}$$

Hence, equations (20.3.9)–(20.3.12), (20.3.15), and (20.3.16) represent the basic parameters of motion of the system.

20.3.1. *Numerical solution*

Following is a Python program to plot a graph of equation (20.3.9). Figure 20.3.2 shows a graph that characterizes the motion that occurs during the first four seconds of the beginning of the motion.

```python
from matplotlib.pyplot import plot, show
from numpy import linspace, sin, cos, exp

S1 = 0.1
V1 = 0.15
V2 = 0.25
n1 = 3
n2 = 2
n = n1 + n2
phi1 = 20
phi2 = 15
a1 = 15
a2 = 25

t = linspace(0, 4, 1000)
x1 = S1 + (V1/(2*n))*(1 - exp(-2*n*t)) \
    + ((V2*n2 + V2*n1)/n)*(t + (1/(2*n))*(exp(-2*n*t) - 1)) \
    + (a1/(phi1**2 + 4*n**2)) \
      *(exp(-2*n*t) + ((2*n)/phi1)*sin(phi1*t) - cos(phi1*t)) \
    + ((2*n2*a1)/(phi1**2 + 4*n**2))*(((2*n)/phi1**2)*(1 - cos(phi1*t)) \
                    - (1/phi1)*sin(phi1*t) \
                    + (1/(2*n))*(1 - exp(-2*n*t))) \
    + ((2*n1*a2)/(phi2**2 + 4*n**2))*(((2*n)/phi2**2)*(1 - cos(phi2*t)) \
                    - (1/phi2)*sin(phi2*t) \
                    + (1/(2*n))*(1 - exp(-2*n*t)))

plot(t, x1)
show()
```

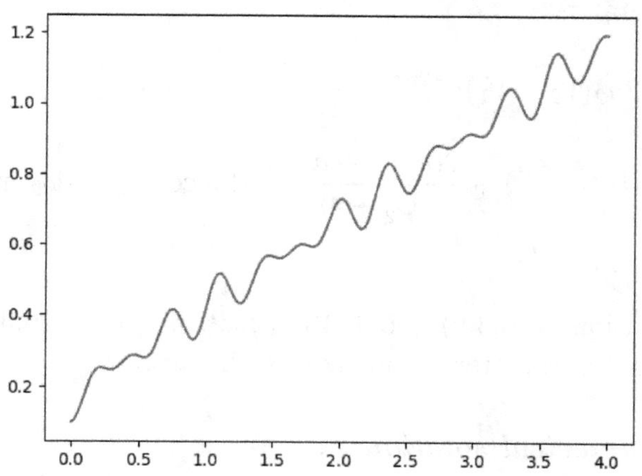

Fig. 20.3.2. Graph of equation (20.3.9).

UNRESTRICTED TWO-DEGREE-OF-FREEDOM SYSTEMS WITH MASSES CONNECTED TO EACH OTHER BY FLEXIBLE AND FLUID LINKS IN PARALLEL

A variety of shock-absorbing mechanisms represent unrestricted systems, the masses of which are connected by a flexible link and a fluid link in parallel. Due to this structural composition, the masses of the systems, depending on the values of the stiffness and damping coefficients, can perform underdamped motion, critically damped motion, or overdamped motion.

The notations of the involved parameters are valid only for the current chapter.

The analysis of the types of motion of the masses of the above-mentioned unrestricted systems is addressed in this chapter.

21.1. Motion of an Unrestricted System Due to the Initial Displacements and Velocities of the Masses Connected by a Flexible Link and a Fluid Link in Parallel

The masses m_1 and m_2 of the considered unrestricted system are connected by a spring having the stiffness coefficient K and a dashpot having the damping coefficient C. The schematic diagram of the

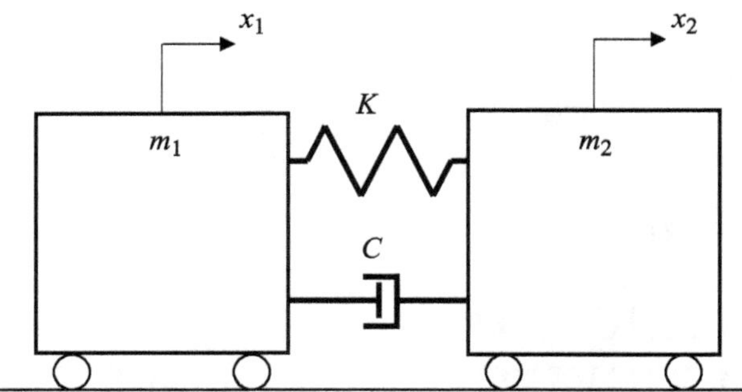

Fig. 21.1.1. Schematic diagram of an unrestricted system, the masses of which are connected by a flexible link and a fluid link in parallel.

system is shown in Figure 21.1.1. The rest of the notations in this figure are self-explanatory. The masses are moving on a horizontal frictionless surface.

The pair of simultaneous differential equations describing the motion of the masses of the system reads:

$$m_1 \frac{d^2 x_1}{dt^2} + C\left(\frac{dx_1}{dt} - \frac{dx_2}{dt}\right) + K\left(x_1 - x_2\right) = 0 \qquad (21.1.1)$$

$$m_2 \frac{d^2 x_2}{dt^2} + C\left(\frac{dx_2}{dt} - \frac{dx_1}{dt}\right) + K\left(x_2 - x_1\right) = 0 \qquad (21.1.2)$$

The initial conditions of motion are assigned according to expression (19.1.3).

Dividing, respectively, equations (21.1.1) and (21.1.2) by m_1 and m_2, while applying to them Laplace Transform pairs 5, 2, 4, 2, and 1, we convert these equations with their initial conditions of motion from the time domain into a system of two simultaneous algebraic equations with two unknowns in each equation in the Laplace domain, namely $x_1(s)$ and $x_2(s)$:

$$s^2 x_1(s) - sV_1 - s^2 S_1 + 2n_1 s x_1(s) - 2n_1 s S_1$$

$$- 2n_1 s x_2(s) + 2n_1 s S_2 + \omega_1^2 x_1(s) - \omega_1^2 x_2(s) = 0 \qquad (21.1.3)$$

$$s^2 x_2(s) - sV_2 - s^2 S_2 + 2n_2 s x_2(s) - 2n_2 s S_2$$

$$- 2n_2 s x_1(s) + 2n_2 s S_1 + \omega_2^2 x_2(s) - \omega_2^2 x_1(s) = 0 \qquad (21.1.4)$$

where ω_1^2 and ω_2^2 are the respective squares of stiffness coefficients, while n_1 and n_2 are the respective damping coefficients.

Rearranging the terms in equations (21.1.3) and (21.1.4) and applying to them the method of substitutions, we obtain two simultaneous algebraic equations with one unknown in each equation that describe in the Laplace domain the displacements of the first and second masses, respectively:

$$x_1(s) = \frac{S_1 s^2}{s^2 + 2ns + \omega_0^2} + \frac{2nS_1 s}{s^2 + 2ns + \omega_0^2} + \frac{S_1\omega_2^2 + S_2\omega_1^2}{s^2 + 2ns + \omega_0^2}$$

$$+ \frac{2(S_1 - S_2)(n_1\omega_2^2 - n_2\omega_1^2)}{s(s^2 + 2nl + \omega_0^2)} + \frac{V_1 s}{s^2 + 2ns + \omega_0^2}$$

$$+ \frac{2(V_1 n_2 + V_2 n_1)}{s^2 + 2ns + \omega_0^2} + \frac{V_1\omega_2^2 + V_2\omega_1^2}{s(s^2 + 2ns + \omega_0^2)} \tag{21.1.5}$$

$$x_2(s) = \frac{S_2 s^2}{s^2 + 2ns + \omega_0^2} + \frac{2nS_2 s}{s^2 + 2ns + \omega_0^2} + \frac{S_2\omega_1^2 + S_1\omega_2^2}{s^2 + 2ns + \omega_0^2}$$

$$+ \frac{2(S_2 - S_1)(n_2\omega_1^2 - n_1\omega_2^2)}{s(s^2 + 2ns + \omega_0^2)} + \frac{V_2 s}{s^2 + 2ns + \omega_0^2}$$

$$+ \frac{2(V_2 n_1 + V_1 n_2)}{s^2 + 2ns + \omega_0^2} + \frac{V_2\omega_1^2 + V_1\omega_2^2}{s(s^2 + 2ns + \omega_0^2)} \tag{21.1.6}$$

where

$$2n_1 = \frac{C}{m_1}; \quad 2n_2 = \frac{C}{m_2}; \quad n = n_1 + n_2 \tag{21.1.7}$$

while

$$\omega_1^2 = \frac{K}{m_1}; \quad \omega_2^2 = \frac{K}{m_2}; \quad \omega_0^2 = \omega_1^2 + \omega_2^2 \tag{21.1.8}$$

In order to determine the characteristics of the damped motion of the masses according to equations (21.1.5) and (21.1.6), we should rearrange the denominators of the fractions of these equations in the following way:

$$s^2 + 2ns + \omega_0^2 = s^2 + 2ns + n^2 - n^2 + \omega_0^2 = s^2 + 2ns + n^2 + \omega^2$$

$$= (s + n)^2 + \omega^2 \tag{21.1.9}$$

where

$$\omega^2 = \omega_0^2 - n^2 \qquad (21.1.10)$$

To continue solving these equations, we need to determine the value of the parameter ω^2. Hence, as was emphasized earlier, in the case when $\omega^2 > 0$, we have the an underdamped decaying vibratory motion of the masses; however, when $\omega^2 = 0$, we have critical damping, and the masses are in a state of a decelerated (non-vibrational) motion; and finally, in the case when $\omega^2 < 0$, we have an overdamped non-vibrational motion. These three cases are considered in the following.

21.1.1. *Underdamped decaying vibratory motion*

Assuming that $\omega^2 > 0$, we continue the solutions by combining equations (21.1.9) and (21.1.10) with equations (21.1.5) and (21.1.6), and as a result, we obtain

$$x_1(s) = \frac{S_1 s^2}{(s+n)^2 + \omega^2} + \frac{2nS_1 s}{(s+n)^2 + \omega^2} + \frac{S_1 \omega_2^2 + S_2 \omega_1^2}{(s+n)^2 + \omega^2}$$

$$+ \frac{2(S_1 - S_2)(n_1\omega_2^2 - n_2\omega_1^2)}{s[(s+n)^2 + \omega^2]} + \frac{V_1 s}{(s+n)^2 + \omega^2}$$

$$+ \frac{2(V_1 n_2 + V_2 n_1)}{(s+n)^2 + \omega^2} + \frac{V_1 \omega_2^2 + V_2 \omega_1^2}{s[(s+n)^2 + \omega^2]} \qquad (21.1.11)$$

$$x_2(s) = \frac{S_2 s^2}{(s+n)^2 + \omega^2} + \frac{2nS_2 s}{(s+n)^2 + \omega^2} + \frac{S_2 \omega_1^2 + S_1 \omega_2^2}{(s+n)^2 + \omega^2}$$

$$+ \frac{2(S_2 - S_1)(n_2\omega_1^2 - n_1\omega_2^2)}{s[(s+n)^2 + \omega^2]} + \frac{V_2 s}{(s+n)^2 + \omega^2}$$

$$+ \frac{2(V_2 n_1 + V_1 n_2)}{(s+n)^2 + \omega^2} + \frac{V_2 \omega_1^2 + V_1 \omega_2^2}{s[(s+n)^2 + \omega^2]} \qquad (21.1.12)$$

Applying to equations (21.1.11) and (21.1.12) Laplace Transform pairs 35, 27, 21, 45, 27, 21, and 45, we invert them from the Laplace domain into the time domain and obtain the solutions of differential equations (21.1.1) and (21.1.2) with initial conditions of motion (19.1.6):

$$x_1 = S_1 e^{-nt}(\cos\omega t - \frac{n}{\omega}\sin\omega t) + 2nS_1 e^{-nt}\sin\omega t + \frac{S_1\omega_2^2 + S_2\omega_1^2}{\omega^2 + n^2}$$

$$\times \left[1 - e^{-nt}\left(\cos\omega t + \frac{n}{\omega}\sin\omega t\right)\right] + \frac{2(S_1 - S_2)(n_2\omega_1^2 - n_1\omega_2^2)}{\omega(\omega^2 + n^2)^2}$$

$$\times \left\{(\omega^2 + n^2)\omega t - 2n\omega - e^{-nt}\left[(\omega^2 - n^2)\sin\omega t - 2n\omega\cos\omega t\right]\right\}$$

$$+ \frac{V_1}{\omega}e^{-nt}\sin\omega t + \frac{2(V_1 n_2 + V_2 n_1)}{\omega^2 + n^2}$$

$$\times \left[1 - e^{-nt}\left(\cos\omega t + \frac{n}{\omega}\sin\omega t\right)\right] + \frac{V_1\omega_2^2 + V_2\omega_1^2}{\omega(\omega^2 + n^2)^2}$$

$$\times \left\{(\omega^2 + n^2)\omega t - 2n\omega - \epsilon^{-nt}\left[(\omega^2 - n^2)\sin\omega t - 2n\omega\cos\omega t\right]\right\}$$

$$\tag{21.1.13}$$

$$x_2 = S_2 e^{-nt}\left(\cos\omega t - \frac{n}{\omega}\sin\omega t\right) + 2nS_2 e^{-nt}\sin\omega t$$

$$+ \frac{S_2\omega_1^2 + S_1\omega_2^2}{\omega^2 + n^2}\left[1 - e^{-nt}\left(\cos\omega t + \frac{n}{\omega}\sin\omega t\right)\right]$$

$$+ \frac{2(S_2 - S_1)(n_2\omega_1^2 - n_1\omega_2^2)}{\omega(\omega^2 + n^2)^2}\{(\omega^2 + n^2)\omega t - 2n\omega$$

$$- e^{-nt}[(\omega^2 - n^2)\sin\omega t - 2n\omega\cos\omega t]\} + \frac{V_2}{\omega}e^{-nt}\sin\omega t$$

$$+ \frac{2(V_2 n_1 + V_1 n_2)}{\omega^2 + n^2}\left[1 - e^{-nt}\left(\cos\omega t + \frac{n}{\omega}\sin\omega t\right)\right]$$

$$+ \frac{V_2\omega_1^2 + V_1\omega_2^2}{\omega(\omega^2 + n^2)^2}\{(\omega^2 + n^2)\omega t - 2n\omega$$

$$- e^{-nt}[(\omega^2 - n^2)\sin\omega t - 2n\omega\cos\omega t]\} \tag{21.1.14}$$

Assuming that in equations (21.1.13) and (21.1.14), $t = 0$, we have $x_1 = S_1$ and $x_2 = S_2$, as it should be according to the initial conditions of motion. Taking from these equations the first derivatives, we obtain the velocities of the two masses, respectively:

$$\frac{dx_1}{dt} = \frac{S_1}{\omega} e^{-nt} \left[(n^2 - \omega^2) \sin \omega t - 2n\omega \cos \omega t) \right] + 2nS_1 e^{-nt}$$

$$\times \left(\cos \omega t - \frac{n}{\omega} \sin \omega t \right) + \frac{S_1 \omega_2^2 + S_2 \omega_1^2}{\omega^2 + n^2} \left(\frac{1}{\omega} e^{-nt} \sin \omega t \right)$$

$$+ \frac{2\left(S_1 - S_2\right)\left(n_1 \omega_2^2 - n_2 \omega_1^2\right)}{\omega^2 + n^2} \left[1 - e^{-nt} \left(\cos \omega t + \frac{n}{\omega} \sin \omega t \right) \right]$$

$$+ V_1 e^{-nt} \left(\cos \omega t - \frac{n}{\omega} \sin \omega t \right) + \frac{2\left(V_1 n_2 + V_2 n_1\right)}{\omega} e^{-nt} \sin \omega t$$

$$+ \frac{V_1 \omega_2^2 + V_2 \omega_1^2}{\omega^2 + n^2} \left[1 - e^{-nt} (\cos \omega t + \frac{n}{\omega} \sin \omega t) \right] \qquad (21.1.15)$$

$$\frac{dx_2}{dt} = \frac{S_2}{\omega} e^{-nt} \left[(n^2 - \omega^2) \sin \omega t - 2n\omega \cos \omega t) \right]$$

$$+ 2nS_2 e^{-nt} \left(\cos \omega t - \frac{n}{\omega} \sin \omega t \right)$$

$$+ \frac{S_2 \omega_1^2 + S_1 \omega_2^2}{\omega^2 + n^2} \left(\frac{1}{\omega} e^{-nt} \sin \omega t \right) + \frac{2(S_2 - S_1)(n_2 \omega_1^2 - n_1 \omega_2^2)}{\omega^2 + n^2}$$

$$\times \left[1 - e^{-nt} \left(\cos \omega t + \frac{n}{\omega} \sin \omega t \right) \right]$$

$$+ V_2 e^{-nt} \left(\cos \omega t - \frac{n}{\omega} \sin \omega t \right) + \frac{2\left(V_2 n_1 + V_1 n_2\right)}{\omega} e^{-nt} \sin \omega t$$

$$+ \frac{V_2 \omega_1^2 + V_1 \omega_2^2}{\omega^2 + n^2} \left[1 - e^{-nt} \left(\cos \omega t + \frac{n}{\omega} \sin \omega t \right) \right] \qquad (21.1.16)$$

Taking in equations (21.1.15) and (21.1.16) that $t = 0$, we obtain $\frac{dx_1}{dt} = V_1$ and $\frac{dx_2}{dt} = V_2$, as expected according to the initial conditions of motion.

21.1.2. *Numerical solution*

Following is a Python program to plot the graph (Figure 21.1.2) of equation (21.1.13), which expresses the law of motion of the first mass of the unrestricted system.

```python
from matplotlib.pyplot import plot, show
from numpy import linspace, sin, cos, exp, sqrt


S1 = 0.05
S2 = 0.06
V1 = 2
V2 = 1
n1 = 5
n2 = 3
n = n1 + n2
omega1 = 20
omega2 = 25
omega0 = sqrt(omega1**2 + omega2**2)
omega = sqrt(omega0**2 - n**2)

t = linspace(0, 1, 1000)

x1 = S1*exp(-n*t)*(cos(omega*t) - (r/omega)*sin(omega*t)) \
  + 2*n*S1*exp(-n*t)*sin(omega*t) \
  + ((S1*omega2**2 + S2*omega1**2)/(omega**2 + n**2)) \
    *(1 - exp(-n*t)*cos(omega*t) + (n/omega)*sin(omega*t)) \
  + ((2*(S1 - S2)*(n2*omega1**2 - n1*omega2**2))/(omega*(omega**2 +
n**2)**2)) \
    *((omega**2 + n**2)*omega*t - 2*n*omega \
    - exp(-n*t)*((omega**2 - n**2)*sin(omega*t) - 2*n*omega*cos(omega*t))) \
  + (V1/omega)*exp(-n*t)*sin(omega*t) \
  + ((2*(V1*n2 + V2*n1))/(omega**2 + n**2)) \
    *(1 - exp(-n*t)*(cos(omega*t) + (n/omega)*sin(omega*t))) \
  + ((V1*omega2**2 + V2*omega1**2)/(omega*(omega**2 + n**2)**2)) \
    *((omega**2 + n**2)*omega*t - 2*n*omega \
    - exp(-n*t)*((omega**2 - n**2)*sin(omega*t) - 2*n*omega*cos(omega*t)))

plot(t, x1)
show()
```

21.1.3. *Critically damped motion*

As mentioned above, critical damping occurs when $\omega^2 = 0$. Therefore, we apply this condition to equations (21.1.11) and (21.1.12),

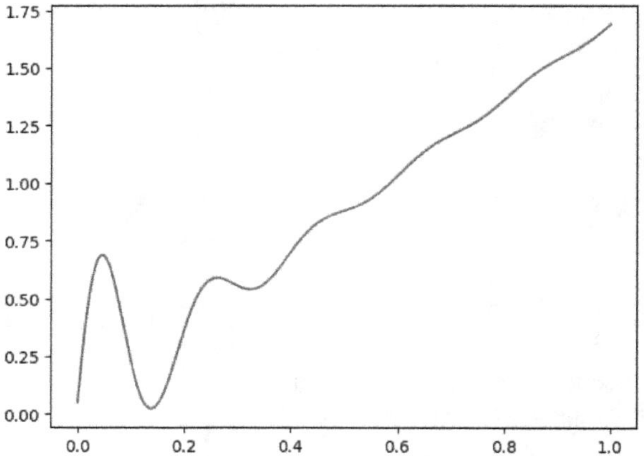

Fig. 21.1.2. Graph of equation (21.1.13).

and we obtain

$$
\begin{aligned}
x_1(s) = {} & \frac{S_1 s^2}{(s+n)^2} + \frac{2nS_1 s}{(s+n)^2} + \frac{S_1\omega_2^2 + S_2\omega_1^2}{(s+n)^2} \\[2mm]
& + \frac{2(S_1 - S_2)(n_1\omega_2^2 - n_2\omega_1^2)}{s(s+n)^2} + \frac{V_1 s}{(s+n)^2} \\[2mm]
& + \frac{2(V_1 n_2 + V_2 n_1)}{(s+n)^2} + \frac{V_1\omega_2^2 + V_2\omega_1^2}{s(s+n)^2}
\end{aligned}
\tag{21.1.17}
$$

$$
\begin{aligned}
x_2(s) = {} & \frac{S_2 s^2}{(s+n)^2} + \frac{2nS_2 s}{(s+n)^2} + \frac{S_2\omega_1^2 + S_1\omega_2^2}{(s+n)^2} \\[2mm]
& + \frac{2(S_2 - S_1)(n_2\omega_1^2 - n_1\omega_2^2)}{s(s+n)^2} + \frac{V_2 s}{(s+n)} \\[2mm]
& + \frac{2(V_2 n_1 + V_1 n_2)}{(s+n)^2} + \frac{V_2\omega_1^2 + V_1\omega_2^2}{s(s+n)^2}
\end{aligned}
\tag{21.1.18}
$$

Applying to equations (21.1.17) and (21.1.18) Laplace Transform
pairs 31, 25, 19, 37, 25, 19, and 37, we invert these equations from
the Laplace domain into the time domain and obtain the solutions
of differential equations (21.1.1) and (21.1.2) with initial conditions

of motion (19.1.3) for the case of critical damping:

$$x_1 = S_1(1 - nt)e^{-nt} + 2nS_1te^{-nt} + \frac{S_1\omega_2^2 + S_2\omega_1^2}{n^2}$$

$$\times \left[1 - e^{-nt}(1 + nt)\right] + \frac{2}{n^2}(S_1 - S_2)\left(n_1\omega_2^2 - n_2\omega_1^2\right)$$

$$\times \left[t - \frac{2}{n} + e^{-nt}\left(\frac{2}{n} + t\right)\right] + V_1te^{-nt}$$

$$+ \frac{2(V_1n_2 + V_2n_1)}{n^2}\left[1 - e^{-nt}(1 + nt)\right]$$

$$+ \frac{V_1\omega_2^2 + V_2\omega_1^2}{n^2}\left[t - \frac{2}{n} + e^{-nt}\left(\frac{2}{n} + t\right)\right] \tag{21.1.19}$$

$$x_2 = S_2(1 - nt)e^{-nt} + 2nS_2te^{-nt} + \frac{S_2\omega_1^2 + S_1\omega_2^2}{n^2}\left[1 - e^{-nt}(1 + nt)\right]$$

$$+ \frac{2}{n^2}(S_1 - S_1)\left(n_2\omega_1^2 - n_1\omega_2^2\right)\left[t - \frac{2}{n} + e^{-nt}\left(\frac{2}{n} + t\right)\right]$$

$$+ V_2te^{-nt} + \frac{2(V_2n_1 + V_1n_2)}{n^2}\left[1 - e^{-nt}(1 + nt)\right]$$

$$+ \frac{V_2\omega_1^2 + V_1\omega_2^2}{n^2}\left[t - \frac{2}{n} + e^{-nt}\left(\frac{2}{n} + t\right)\right] \tag{21.1.20}$$

21.1.4. *Overdamped motion*

In the case of overdamped motion, we have $\omega^2 < 0$. According to this condition, we replace in equations (21.1.11) and (21.1.12) the term ω^2 with its negative value, $-\omega^2$, and we may write

$$x_1(s) = \frac{S_1s^2}{(s + n)^2 - \omega^2} + \frac{2nS_1s}{(s + n)^2 - \omega^2} + \frac{S_1\omega_2^2 + S_2\omega_1^2}{(s + n)^2 - \omega^2}$$

$$+ \frac{2(S_1 - S_2)(n_1\omega_2^2 - n_2\omega_1^2)}{s[(s + n)^2 - \omega^2]} + \frac{V_1s}{(s + n)^2 - \omega^2}$$

$$+ \frac{2(V_1n_2 + V_2n_1)}{(s + n)^2 - \omega^2} + \frac{V_1\omega_2^2 + V_2\omega_1^2}{s[(s + n)^2 - \omega^2]} \tag{21.1.21}$$

$$x_2(s) = \frac{S_2 s^2}{(s+n)^2 - \omega^2} + \frac{2nS_2 s}{(s+n)^2 - \omega^2} + \frac{S_2\omega_1^2 + S_1\omega_2^2}{(s+n)^2 - \omega^2}$$

$$+ \frac{2(S_2 - S_1)(n_2\omega_1^2 - n_1\omega_2^2)}{s[(s+n)^2 - \omega^2]} + \frac{V_2 s}{(s+n)^2 - \omega^2}$$

$$+ \frac{2(V_2 n_1 + V_1 n_2)}{(s+n)^2 - \omega^2} + \frac{V_2\omega_1^2 + V_1\omega_2^2}{s[(s+n)^2 - \omega^2]} \tag{21.1.22}$$

Applying to equations (21.1.21) and (21.1.22) Laplace Transform pairs 36, 28, 22, 46, 28, 22, and 46, we invert these equations from the Laplace domain into the time domain and obtain the solutions of differential equations (21.1.1) and (21.1.2) with initial conditions of motion (19.1.3) for the case of overdamped motion:

$$x_1 = S_1 e^{-nt}\left(\cosh\omega t - \frac{n}{\omega}\sinh\omega t\right) + 2nS_1 e^{-nt}\sinh\omega t$$

$$+ \frac{S_1\omega_2^2 + S_2\omega_1^2}{n^2 - \omega^2}\left[1 - e^{-nt}\left(\cosh\omega t + \frac{n}{\omega}\sinh\omega t\right)\right]$$

$$+ \frac{2(S_1 - S_2)(n_1\omega_2^2 - n_2\omega_1^2)}{\omega(n^2 - \omega^2)^2}\left\{(n^2 - \omega^2)\omega t - 2n\omega + e^{-nt}\right.$$

$$\times \left[(\omega^2 + n^2)\sinh\omega t + 2n\omega\cosh\omega t\right]\right\} + \frac{V_1}{\omega}e^{-nt}\sinh\omega t$$

$$+ \frac{2(V_1 n_2 + V_2 n_1)}{n^2 - \omega^2}\left[1 - e^{-nt}\left(\cosh\omega t + \frac{n}{\omega}\sinh\omega t\right)\right]$$

$$+ \frac{V_1\omega_2^2 + V_2\omega_1^2}{\omega(n^2 - \omega^2)^2}\left\{(n^2 - \omega^2)\omega t - 2n\omega + e^{-nt}\right.$$

$$\times \left[(\omega^2 + n^2)\sinh\omega t + 2n\omega\cosh\omega t\right]\right\} \tag{21.1.23}$$

$$x_2 = S_2 e^{-nt}\left(\cosh\omega t - \frac{n}{\omega}\sinh\omega t\right) + 2nS_2 e^{-nt}\sinh\omega t$$

$$+ \frac{S_2\omega_1^2 + S_1\omega_2^2}{n^2 - \omega^2}\left[1 - e^{-nt}\left(\cosh\omega t + \frac{n}{\omega}\sinh\omega t\right)\right]$$

$$+ \frac{2(S_2 - S_1)(n_2\omega_1^2 - n_1\omega_2^2)}{\omega(n^2 - \omega^2)^2}\left\{(n^2 - \omega^2)\omega t - 2n\omega + e^{-nt}\right.$$

$$\times \left[(\omega^2 + n^2)\sinh\omega t + 2n\omega\cosh\omega t\right]\right\} + \frac{V_2}{\omega}e^{-nt}\sinh\omega t$$

$$+ \frac{2(V_2 n_1 + V_1 n_{21})}{n^2 - \omega^2} \left[1 - e^{-nt} \left(\cosh \omega t + \frac{n}{\omega} \sinh \omega t \right) \right]$$

$$+ \frac{V_2 \omega_1^2 + V_1 \omega_2^2}{\omega (n^2 - \omega^2)^2} \{ (n^2 - \omega^2)\omega t - 2n\omega + e^{-nt}$$

$$\times [(\omega^2 + n^2) \sinh \omega t + 2n\omega \cosh \omega t] \} \tag{21.1.24}$$

Equations (21.1.23) and (21.1.24) describe overdamped (decelerated) motion.

21.2. Motion of an Unrestricted System, the Masses of Which Are Connected by Flexible and Fluid Links in Parallel While Being Subjected to Constant Resisting, Dry Friction, and Constant Active Forces

The masses of the system are in motion since the applied active forces in this case exceed the resisting forces. The motion of the system occurs on a horizontal surface.

The schematic diagram of the unrestricted system is shown in Figure 21.2.1. The notations in this figure are self-explanatory.

Based on the schematic diagram shown in Figure 21.2.1 and the considerations mentioned above, we compose the following pair of

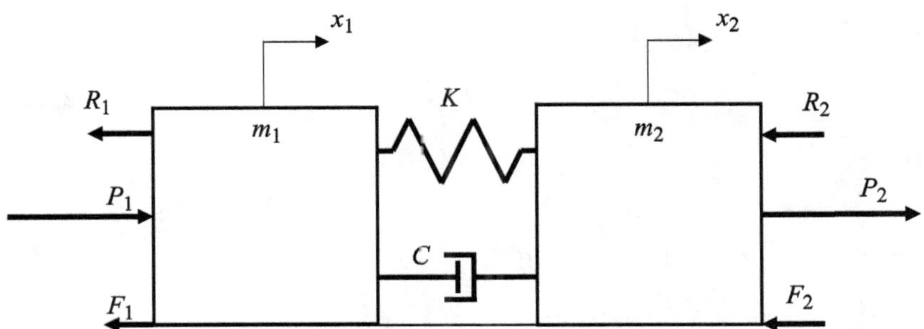

Fig. 21.2.1. Schematic diagram of an unrestricted system, the masses of which are connected by a flexible link and a fluid link and subjected to constant resisting, dry friction, and constant active forces.

simultaneous differential equations of motion:

$$m_1 \frac{d^2 x_1}{dt^2} + C \left(\frac{dx_1}{dt} - \frac{dx_2}{dt} \right) + K(x_1 - x_2) + R_1 + F_1 = P_1$$

$$(21.2.1)$$

$$m_2 \frac{d^2 x_2}{dt^2} + C \left(\frac{dx_2}{dt} - \frac{dx_1}{dt} \right) + K(x_2 - x_1) + R_2 + F_2 = P_2$$

$$(21.2.2)$$

The initial conditions of motion are

$$x_1 = 0, \quad \frac{dx_1}{dt} = 0$$

$$\text{For } t = 0 \qquad\qquad (21.2.3)$$

$$x_2 = 0, \quad \frac{dx_2}{dt} = 0$$

Dividing equations (21.2.1) and (21.2.2) by m_1 and m_2, respectively, and applying to them Laplace Transform pairs 2, 5, 2, 4, 2, 1, 2, and 2, we convert these equations with their initial conditions of motion from the time domain into the Laplace domain and obtain a system of two simultaneous algebraic equations with two unknowns $x_1(s)$ and $x_2(s)$ in each equation:

$$s^2 x_1(s) + 2n_1 s x_1(s) - 2n_1 s x_2(s) + \omega_1^2 x_1(s)$$

$$- \omega_1^2 x_2(s) + r_1 + f_1 = p_1 \qquad (21.2.4)$$

$$s^2 x_2(s) + 2n_2 s x_2(s) - 2n_2 s x_1(s) + \omega_2^2 x_2(s)$$

$$- \omega_2^2 x_1(s) + r_2 + f_2 = p_2 \qquad (21.2.5)$$

where

$$r_1 = \frac{R_1}{m_1}; \quad r_2 = \frac{R_2}{m_2}; \quad f_1 = \frac{F_1}{m_1}; \quad f_2 = \frac{F_2}{m_2};$$

$$p_1 = \frac{P_1}{m_1}; \quad p_2 = \frac{P_2}{m_2} \qquad (21.2.6)$$

Rearranging the terms in equations (21.2.4) and (21.2.5) and applying to them the method of substitutions, we obtain two simultaneous algebraic equations that describe in the Laplace domain the displacements of the first and second masses, respectively:

$$x_1(s) = \frac{p_1 - r_1 - f_1}{s^2 + 2ns + \omega_0^2} + \frac{2[n_2(p_1 - r_1 - f_1) + n_1(p_2 - r_2 - f_2)]}{s(s^2 + 2ns + \omega_0^2)}$$

$$+ \frac{\omega_2^2(p_1 - r_1 - f_1) + \omega_1^2(p_2 - r_2 - f_2)}{s^2(l^2 + 2ns + \omega_0^2)} \tag{21.2.7}$$

$$x_2(s) = \frac{p_2 - r_2 - f_2}{s^2 + 2ns + \omega_0^2} + \frac{2[n_1(p_2 - r_2 - f_2) + n_2(p_1 - r_1 - f_1)]}{s(s^2 + 2ns + \omega_0^2)}$$

$$+ \frac{\omega_1^2(p_2 - r_2 - f_2) + \omega_2^2(p_1 - r_1 - f_1)}{s^2(s^2 + 2ns + \omega_0^2)} \tag{21.2.8}$$

In continuation of the analysis of the current system, we consider the cases of underdamped, critically damped, and overdamped motions.

21.2.1. *Underdamped decaying vibratory motion*

Combining equations (21.1.9) and (21.1.10) with equations (21.2.7) and (21.2.8), we obtain

$$x_1(s) = \frac{p_1 - r_1 - f_1}{(s + n)^2 + \omega^2} + \frac{2[n_2(p_1 - r_1 - f_1) + n_1(p_2 - r_2 - f_2)]}{s[(s + n)^2 + \omega^2]}$$

$$+ \frac{\omega_2^2(p_1 - r_1 - f_1) + \omega_1^2(p_2 - r_2 - f_2)}{s^2[(s + n)^2 + \omega^2]} \tag{21.2.9}$$

$$x_2(s) = \frac{p_2 - r_2 - f_2}{(s + n)^2 + \omega^2} + \frac{2[n_1(p_2 - r_2 - f_2) + n_2(p_1 - r_1 - f_1)]}{s[(s + n)^2 + \omega^2]}$$

$$+ \frac{\omega_1^2(p_2 - r_2 - f_2) + \omega_2^2(p_1 - r_1 - f_1)}{s^2[(s + n)^2 + \omega^2]} \tag{21.2.10}$$

Applying to equations (21.2.9) and (21.2.10) Laplace Transform pairs 21, 45, and 56, we invert these equations from the Laplace domain into the time domain and obtain the solutions of differential equations (21.2.1) and (21.2.2) with their initial conditions of motion. These equations describe the underdamped decaying vibrational motion of the two masses of the system:

$$x_1 = \frac{p_1 - r_1 - f_1}{\omega^2 + n^2} \left[1 - e^{-nt} \left(\cos \omega t + \frac{n}{\omega} \sin \omega t \right) \right]$$

$$+ \frac{2[n_2(p_1 - r_1 - f_1) + n_1(p_2 - r_2 - f_2)}{\omega(\omega^2 + n^2)^2}$$

$$\times \left\{ (\omega^2 + n^2)\omega t - 2n\omega - e^{-nt} \left[(\omega^2 - n^2) \sin \omega t - 2n\omega \cos \omega t \right] \right\}$$

$$+ \frac{\omega_2^2(p_1 - r_1 - f_1) + \omega_1^2(p_2 - r_2 - f_2)}{\omega(\omega^2 + n^2)^2}$$

$$\times \left\{ \frac{\omega t^2}{2}(\omega^2 + n^2) - 2n\omega t - \frac{\omega(\omega^2 - n^2)}{\omega^2 + n^2} \right.$$

$$\times \left[1 - e^{-nt} \left(\cos \omega t + \frac{n}{\omega} \sin \omega t \right) \right]$$

$$\left. - \frac{2n^2\omega}{\omega^2 + n^2} \left[1 + e^{-nt} \left(\frac{\omega}{n} \sin \omega t - \cos \omega t \right) \right] \right\} \tag{21.2.11}$$

$$x_2 = \frac{p_2 - r_2 - f_2}{\omega^2 + n^2} \left[1 - e^{-nt} \left(\cos \omega t + \frac{n}{\omega} \sin \omega t \right) \right]$$

$$+ \frac{2[n_1(p_2 - r_2 - f_2) + n_2(p_1 - r_1 - f_1)}{\omega(\omega^2 + n^2)^2}$$

$$\times \left\{ (\omega^2 + n^2)\omega t - 2n\omega - e^{-nt} \left[(\omega^2 - n^2) \sin \omega t - 2n\omega \cos \omega t \right] \right\}$$

$$+ \frac{\omega_1^2(p_2 - r_2 - f_2) + \omega_2^2(p_1 - r_1 - f_1)}{\omega(\omega^2 + n^2)^2}$$

$$\times \left\{ \frac{\omega t^2}{2}(\omega^2 + n^2) - 2n\omega t - \frac{\omega(\omega^2 - n^2)}{\omega^2 + n^2} \right.$$

$$\times \left[1 - e^{-nt} \left(\cos \omega t + \frac{n}{\omega} \sin \omega t \right) \right]$$

$$\left. - \frac{2n^2\omega}{\omega^2 + n^2} \left[1 + e^{-nt} \left(\frac{\omega}{n} \sin \omega t - \cos \omega t \right) \right] \right\} \tag{21.2.12}$$

Supposing that in equations (21.2.11) and (21.2.12), $t = 0$, then we have that $x_1 = 0$ and $x_2 = 0$, as it should be according to the initial conditions of motion. Taking from these equations the first derivatives, we obtain the velocities of the masses of the system:

$$\frac{dx_1}{dt} = \frac{p_1 - r_1 - f_1}{\omega} e^{-nt} \sin \omega t$$

$$+ \frac{2[n_2(p_1 - r_1 - f_1) + n_1(p_2 - r_2 - f_2)}{\omega^2 + n^2}$$

$$\times \left[1 - e^{-nt}\left(\cos \omega t + \frac{n}{\omega} \sin \omega t\right)\right]$$

$$+ \frac{\omega_2^2(p_1 - r_1 - f_1) + \omega_1^2(p_2 - r_2 - f_2)}{\omega(\omega^2 + n^2)^2}$$

$$\times \left\{(\omega^2 + n^2)\omega t - 2n\omega - e^{-nt}\right.$$

$$\times \left.\left[(\omega^2 - n^2)\sin \omega t - 2n\omega \cos \omega t\right]\right\} \qquad (21.2.13)$$

$$\frac{dx_2}{dt} = \frac{p_2 - r_2 - f_2}{\omega} e^{-nt} \sin \omega t$$

$$+ \frac{2[n_1(p_2 - r_2 - f_2) + n_2(p_1 - r_1 - f_1)}{\omega^2 + n^2}$$

$$\times \left[1 - e^{-nt}\left(\cos \omega t + \frac{n}{\omega} \sin \omega t\right)\right]$$

$$+ \frac{\omega_1^2(p_2 - r_2 - f_2) + \omega_2^2(p_1 - r_1 - f_1)}{\omega(\omega^2 + n^2)^2}$$

$$\times \left\{(\omega^2 + n^2)\omega t - 2n\omega - e^{-nt}\right.$$

$$\times \left.\left[(\omega^2 - n^2)\sin \omega t - 2n\omega \cos \omega t\right]\right\} \qquad (21.2.14)$$

Taking in equations (21.2.13) and (21.2.14) that $t = 0$, we obtain that $\frac{dx_1}{dt} = 0$ and $\frac{dx_2}{dt} = 0$, as expected according to the initial conditions of motion.

21.2.2. *Critically damped motion*

As noted above, critical damping occurs when $\omega_0^2 = 0$. Therefore, accounting for this condition in equations (21.2.7) and (21.2.8), we may write

$$x_1(s) = \frac{p_1 - r_1 - f_1}{(s+n)^2} + \frac{2[n_2(p_1 - r_1 - f_1) + n_1(p_2 - r_2 - f_2)]}{s(s+n)^2}$$

$$+ \frac{\omega_2^2(p_1 - r_1 - f_1) + \omega_1^2(p_2 - r_2 - f_2)}{s^2(s+n)^2} \tag{21.2.15}$$

$$x_2(s) = \frac{p_2 - r_2 - f_2}{(s+n)^2} + \frac{2[n_1(p_2 - r_2 - f_2) + n_2(p_1 - r_1 - f_1)]}{s(s+n)^2}$$

$$+ \frac{\omega_1^2(p_2 - r_2 - f_2) + \omega_2^2(p_1 - r_1 - f_1)}{s^2(s+n)^2} \tag{21.2.16}$$

Applying to equations (21.2.15) and (21.2.1) Laplace Transform pairs 19, 37, and 54, we invert these equations from the Laplace domain into the time domain and obtain the solutions of differential equations (21.2.1) and (21.2.2) with their initial conditions of motion for the case of critical damping:

$$x_1 = \frac{p_1 - r_1 - f_1}{n^2}\left[1 - e^{-nt}(1 + nt)\right]$$

$$+ \frac{2\left[n_2(p_1 - r_1 - f_1) + n_1(p_2 - r_2 - f_2)\right]}{n^2}$$

$$\times \left[t - \frac{2}{n} + e^{-nt}\left(\frac{2}{n} + t\right)\right]$$

$$+ \frac{\omega_2^2(p_1 - r_1 - f_1) + \omega_1^2(p_2 - r_2 - f_2)}{n^2}$$

$$\times \left[\frac{t^2}{2} - \frac{2t}{n} - \frac{3}{n^2} + \frac{e^{-nt}}{n^2}(3 + nt)\right] \tag{21.2.17}$$

$$x_2 = \frac{p_2 - r_2 - f_2}{n^2}\left[1 - e^{-nt}(1 + nt)\right]$$

$$+ \frac{2[n_1(p_2 - r_2 - f_2) + n_2(p_1 - r_1 - f_1)]}{n^2}$$

$$\times \left[t - \frac{2}{n} + e^{-nt}\left(\frac{2}{n} + t\right)\right]$$

$$+ \frac{\omega_1^2 (p_2 - r_2 - f_2) + \omega_2^2 (p_1 - r_1 - f_1)}{n^2}$$

$$\times \left[\frac{t^2}{2} - \frac{2t}{n} - \frac{3}{n^2} + \frac{e^{-nt}}{n^2}(3 + nt) \right] \qquad (21.2.18)$$

21.2.3. *Overdamped motion*

As mentioned above, the overdamped motion occurs when $\omega^2 < 0$. Therefore, we have to replace in equations (21.2.9) and (21.2.10) the signs at the terms ω^2 with the opposite signs:

$$x_1(s) = \frac{p_1 - r_1 - f_1}{(s + n)^2 - \omega^2} + \frac{2[n_2(p_1 - r_1 - f_1) + n_1(p_2 - r_2 - f_2)]}{s[(s + n)^2 - \omega^2]}$$

$$+ \frac{\omega_2^2(p_1 - r_1 - f_1) + \omega_1^2(p_2 - r_2 - f_2)}{s^2[(s + n)^2 - \omega^2]} \qquad (21.2.19)$$

$$x_2(s) = \frac{p_2 - r_2 - f_2}{(s + n)^2 - \omega^2} + \frac{2[n_1(p_2 - r_2 - f_2) + n_2(p_1 - r_1 - f_1)]}{s[(s + n)^2 - \omega^2]}$$

$$+ \frac{\omega_1^2(p_2 - r_2 - f_2) + \omega_2^2(p_1 - r_1 - f_1)}{s^2[(s + n)^2 - \omega^2]} \qquad (21.2.20)$$

Applying to equations (21.2.19) and (21.2.20) Laplace Transform pairs 22, 46, and 57, we invert these equations from the Laplace domain into the time domain and obtain the solutions of differential equations (21.2.1) and (21.2.2) with their initial conditions for the case of overdamped motion:

$$x_1 = \frac{p_1 - r_1 - f_1}{n^2 - \omega^2} \left[1 - e^{-nt} \left(\cosh \omega t + \frac{n}{\omega_v} \sinh \omega t \right) \right]$$

$$+ \frac{2[n_2(p_1 - r_1 - f_1) + n_1(p_2 - r_2 - f_2)]}{\omega(n^2 - \omega^2)^2}$$

$$\times \left\{ (n^2 - \omega^2)\omega t - 2n\omega + e^{-nt} \left[(\omega^2 + n^2)\sinh \omega t \right. \right.$$

$$\left. \left. + 2n\omega \cosh \omega t \right] \right\} + \frac{\omega_2^2 (p_1 - r_1 - f_1) + \omega_1^2 (p_2 - r_2 - f_2)}{\omega(n^2 - \omega^2)^2}$$

$$\times \left\{ \frac{\omega^2 t^2}{2}\left(n^2 - \omega^2\right) - 2n\omega t + \frac{\omega\left(\omega^2 + n^2\right)}{n^2 - \omega^2} \right.$$

$$\times \left[1 - e^{-nt}\left(\cosh \omega t + \frac{n}{\omega}\sinh \omega t\right)\right]$$

$$\left. - \frac{2n^2\omega}{n^2 - \omega^2}\left[1 + e^{-nt}\left(\frac{\omega}{n}\sinh \omega t - \cosh \omega t\right)\right] \right\} \qquad (21.2.21)$$

$$x_2 = \frac{p_2 - r_2 - f_2}{n^2 - \omega^2}\left[1 - e^{-nt}\left(\cosh \omega t + \frac{n}{\omega}\sinh \omega t\right)\right]$$

$$+ \frac{2[n_1(p_2 - r_2 - f_2) + n_2(p_1 - r_1 - f_1)]}{\omega(n^2 - \omega^2)^2}\{(n^2 - \omega^2)\omega t$$

$$- 2n\omega + e^{-nt}[(\omega^2 + n^2)\sinh \omega t + 2n\omega \cosh \omega t]\}$$

$$+ \frac{\omega_1^2(p_2 - r_2 - f_2) + \omega_2^2(p_1 - r_1 - f_1)}{\omega(n^2 - \omega^2)^2}\left\{\frac{\omega t^2}{2}(n^2 - \omega^2)\right.$$

$$- 2n\omega t + \frac{\omega(\omega^2 + n^2)}{n^2 - \omega^2}\left[1 - e^{-nt}\left(\cosh \omega t + \frac{n}{\omega}\sinh \omega t\right)\right]$$

$$\left. - \frac{2n^2\omega}{n^2 - \omega^2}\left[1 + e^{-nt}\left(\frac{\omega}{n}\sinh \omega t - \cosh \omega t\right)\right] \right\} \qquad (21.2.22)$$

21.3. Motion of an Unrestricted System, the Masses of Which Are Connected by a Flexible Link and a Fluid Link in Parallel While Being Subjected to Harmonic Forces

An unrestricted system, the two masses of which are attached to each other by a spring and a dashpot in parallel, is moving on a horizontal frictionless surface due to harmonic forces.

The schematic diagram representing the unrestricted system is shown in Figure 21.3.1, the notations in which are self-explanatory.

The pair of simultaneous differential equations describing the motion of the masses reads:

$$m_1 \frac{d^2 x_1}{dt^2} + C\left(\frac{dx_1}{dt} - \frac{dx_2}{dt}\right) + K\left(x_1 - x_2\right) = A_1 \cos \varphi_1 t$$

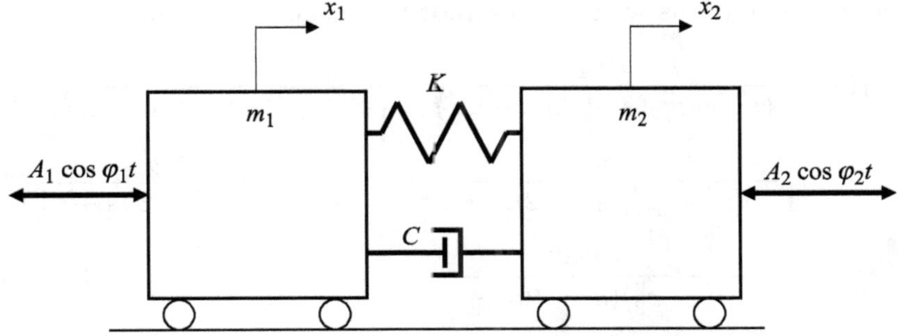

Fig. 21.3.1. Schematic diagram of an unrestricted system with masses connected to each other by a spring and a dashpot while being subjected to harmonic forces.

$$(21.3.1)$$

$$m_2 \frac{d^2 x_2}{dt^2} + C\left(\frac{dx_2}{dt} - \frac{dx_1}{dt}\right) + K\left(x_2 - x_1\right) = A_2 \cos \varphi_2 t$$

$$(21.3.2)$$

The initial conditions of motion are taken according to expression (21.2.3).

Dividing equations (21.3.1) and (21.3.2) by m_1 and m_2, respectively, and applying to them Laplace Transform pairs 2, 5, 2, 4, 2, 1, 2, and 29, we convert these equations from the time domain into the Laplace domain and obtain a system of two simultaneous algebraic equations with two unknowns in each equation that represent the displacements of the masses, respectively:

$$s^2 x_1(s) + 2n_1 s x_1(s) - 2n_1 s x_2(s) + \omega_1^2 x_1(s) - \omega_1^2 x_2(s) = \frac{a_1 s^2}{s^2 + \varphi_1^2}$$

$$(21.3.3)$$

$$s^2 x_2(s) + 2n_2 s x_2(s) - 2n_2 s x_1(s) + \omega_2^2 x_2(s) - \omega_2^2 x_1(s) = \frac{a_2 s^2}{s^2 + \varphi_2^2}$$

$$(21.3.4)$$

Rearranging equations (21.3.3) and (21.3.4) and applying to them the method of substitution, we obtain a system of two simultaneous algebraic equations with one unknown in each equation:

$$x_1(s) = \frac{s^2 a_1}{(s^2 + \varphi_1^2)(s^2 + 2ns + \omega_0^2)} + \frac{2n_2 s a_1}{(s^2 + \varphi_1^2)(s^2 + 2ns + \omega_0^2)}$$

$$+ \frac{a_1 \omega_2^2}{(s^2 + \varphi_1^2)(s^2 + 2ns + \omega_0^2)} + \frac{2n_1 l a_2}{(s^2 + \varphi_2^2)(s^2 + 2ns + \omega_0^2)}$$

$$+ \frac{a_2 \omega_1^2}{(s^2 + \varphi_2^2)(s^2 + 2ns + \omega_0^2)} \tag{21.3.5}$$

$$x_2(s) = \frac{s^2 a_2}{(s^2 + \varphi_2^2)(s^2 + 2ns + \omega_0^2)} + \frac{2n_1 s a_2}{(s^2 + \varphi_2^2)(s^2 + 2ns + \omega_0^2)}$$

$$+ \frac{a_2 \omega_1^2}{(s^2 + \varphi_2^2)(s^2 + 2ns + \omega_0^2)} + \frac{2n_2 s a_1}{(s^2 + \varphi_1^2)(s^2 + 2ns + \omega_0^2)}$$

$$+ \frac{a_1 \omega_2^2}{(s^2 + \varphi_1^2)(s^2 + 2ns + \omega_0^2)} \tag{21.3.6}$$

Combining equations (21.3.5) and (21.3.6) with equations (21.1.9) and (21.1.10), we may write

$$x_1(s) = \frac{s^2 a_1}{(s^2 + \varphi_1^2)[(s + n)^2 + \omega^2]} + \frac{2n_2 s a_1}{(s^2 + \varphi_1^2)[(s + n)^2 + \omega^2]}$$

$$+ \frac{a_1 \omega_2^2}{(s^2 + \varphi_1^2)[(s + n)^2 + \omega^2]} + \frac{2n_1 s a_2}{(s^2 + \varphi_2^2)[(s + n)^2 + \omega^2]}$$

$$+ \frac{a_2 \omega_1^2}{(s^2 + \varphi_2^2)[(l + n)^2 + \omega^2]} \tag{21.3.7}$$

$$x_2(s) = \frac{s^2 a_2}{(s^2 + \varphi_2^2)[(s + n)^2 + \omega^2]} + \frac{2n_1 s a_2}{(s^2 + \varphi_2^2)[(s + n)^2 + \omega^2]}$$

$$+ \frac{a_2 \omega_1^2}{(s^2 + \varphi_2^2)[(s + n)^2 + \omega^2]} + \frac{2n_2 s a_1}{(s^2 + \varphi_1^2)[(s + n)^2 + \omega^2]}$$

$$+ \frac{a_1 \omega_2^2}{(s^2 + \varphi_1^2)[(s + n)^2 + \omega^2]} \tag{21.3.8}$$

As mentioned above, the continuation of the solutions of equations (21.3.7) and (21.3.8) is associated with considering three options that depend on the value of the parameter ω^2.

21.3.1. *Underdamped vibrational motion*

Assuming that $\omega^2 > 0$ and applying to equations (21.3.7) and (21.3.8) Laplace Transform pairs 68, 65, 58, 65, and 58, we invert these equations from the Laplace domain into the time domain and obtain the solutions of differential equations (21.3.1) and (21.3.2) with their initial conditions of motion:

$$
\begin{aligned}
x_1 = {} & \frac{a_1}{4n^2\varphi_1^2 + (\omega^2 + n^2 - \varphi_1^2)^2} \\[2mm]
& \times \left[(\omega^2 + n^2 - \varphi_1^2)^2 (\cos\varphi_1 t - e^{-nt}\cos\omega t) \right. \\[2mm]
& \left. + 2n\varphi_1\sin\varphi_1 t - \frac{n}{\omega}(\omega^2 + n^2 + \varphi_1^2)e^{-nt}\sin\omega t \right] \\[2mm]
& + \frac{2n_2 a_1}{4n^2\varphi_1^2 + (\omega^2 + n^2 - \varphi_1^2)^2}\left[2n(e^{-nt}\cos\omega t - \cos\varphi_1 t) \right. \\[2mm]
& \left. + \frac{1}{\varphi_1}(\omega_v^2 + n^2 - \varphi_1^2)\sin\varphi_1 t - \frac{1}{\omega}(\omega^2 - n^2 - \varphi_1^2)e^{-nt}\sin\omega t \right] \\[2mm]
& + \frac{a_1\omega_2^2}{4n^2\varphi_1^2 + (\omega^2 + n^2 - \varphi_1^2)^2}\left\{ \frac{1}{\varphi_1^2}(\omega^2 + n^2 - \varphi_1^2)(1 - \cos\varphi_1 t) \right. \\[2mm]
& - 2n\left(\frac{1}{\varphi_1}\sin\varphi_1 t - \frac{1}{\omega}e^{-nt}\sin\omega t\right) + \frac{3n^2 + \varphi_1^2 - \omega^2}{\omega^2 + n^2} \\[2mm]
& \left. \times \left[1 - e^{-nt}\left(\cos\omega t + \frac{n}{\omega}\sin\omega t\right) \right] \right\} + \frac{2n_1 a_2}{4n^2\varphi_2^2 + (\omega^2 + n^2 - \varphi_2^2)^2} \\[2mm]
& \times \left[2n(e^{-nt}\cos\omega t - \cos\varphi_2 t) + \frac{1}{\varphi_2}(\omega^2 + n^2 - \varphi_2^2)\sin\varphi_2 t \right. \\[2mm]
& \left. - \frac{1}{\omega}(\omega^2 - n^2 - \varphi_2^2)e^{-nt}\sin\omega t \right] + \frac{a_2\omega_1^2}{4n^2\varphi_2^2 + (\omega^2 + n^2 - \varphi_2^2)^2} \\[2mm]
& \times \left\{ \frac{1}{\varphi_2^2}(\omega^2 + n^2 - \varphi_2^2)(1 - \cos\varphi_2 t) \right. \\[2mm]
& - 2n\left(\frac{1}{\varphi_2}\sin\varphi_2 t - \frac{1}{\omega}e^{-nt}\sin\omega t\right) + \frac{3n^2 + \varphi_2^2 - \omega^2}{\omega^2 + n^2} \\[2mm]
& \left. \times \left[1 - e^{-nt}\left(\cos\omega t + \frac{n}{\omega}\sin\omega t\right) \right] \right\}
\end{aligned}
\tag{21.3.9}
$$

$$x_2 = \frac{a_2}{4n^2\varphi_2^2 + (\omega^2 + n^2 - \varphi_2^2)^2}$$

$$\times \left[(\omega^2 + n^2 - \varphi_2^2)^2 \left(\cos \varphi_2 t - e^{-nt} \cos \omega t \right) \right.$$

$$\left. + 2n\varphi_2 \sin \varphi_2 t - \frac{n}{\omega} \left(\omega^2 + n^2 + \varphi_2^2 \right) e^{-nt} \sin \omega t \right]$$

$$+ \frac{2n_1 a_2}{4n^2\varphi_2^2 + (\omega^2 + n^2 - \varphi_2^2)^2} \left[2n \left(e^{-nt} \cos \omega t - \cos \varphi_2 t \right) \right.$$

$$\left. + \frac{1}{\varphi_2} \left(\omega^2 + n^2 - \varphi_2^2 \right) \sin \varphi_2 t - \frac{1}{\omega} \left(\omega^2 - n^2 - \varphi_2^2 \right) e^{-nt} \sin \omega t \right]$$

$$+ \frac{a_2 \omega_1^2}{4n^2\varphi_2^2 + (\omega^2 + n^2 - \varphi_2^2)^2} \left\{ \frac{1}{\varphi_2^2} (\omega^2 + n^2 - \varphi_2^2)(1 - \cos \varphi_2 t) \right.$$

$$-2n \left(\frac{1}{\varphi_2} \sin \varphi_2 t - \frac{1}{\omega} e^{-nt} \sin \omega t \right)$$

$$\left. + \frac{3n^2 + \varphi_2^2 - \omega^2}{\omega^2 + n^2} \left[1 - e^{-nt} \left(\cos \omega t + \frac{n}{\omega} \sin \omega t \right) \right] \right\}$$

$$+ \frac{2n_2 a_1}{4n^2\varphi_1^2 + (\omega^2 + n^2 - \varphi_1^2)^2} \left[2n(e^{-nt} \cos \omega t - \cos \varphi_1 t) \right.$$

$$\left. + \frac{1}{\varphi_1}(\omega^2 + n^2 - \varphi_1^2) \sin \varphi_1 t - \frac{1}{\omega} \left(\omega^2 - n^2 - \varphi_1^2 \right) e^{-nt} \sin \omega t \right]$$

$$+ \frac{a_1 \omega_1^2}{4n^2\varphi_1^2 + (\omega^2 + n^2 - \varphi_1^2)^2} \left\{ \frac{1}{\varphi_1^2} \left(\omega^2 + n^2 - \varphi_1^2 \right) (1 - \cos \varphi_1 t) \right.$$

$$-2n \left(\frac{1}{\varphi_1} \sin \varphi_1 t - \frac{1}{\omega} e^{-nt} \sin \omega t \right) + \frac{3n^2 + \varphi_1^2 - \omega^2}{\omega^2 + n^2}$$

$$\left. \times \left[1 - e^{-nt}(\cos \omega t + \frac{n}{\omega} \sin \omega t) \right] \right\} \tag{21.3.10}$$

21.3.2. *Critically damped motion*

Substituting into equations (21.3.7) and (21.3.8) that $\omega^2 = 0$, we obtain

$$x_1(s) = \frac{s^2 a_1}{(s^2 + \varphi_1^2)(s+n)^2} + \frac{2n_2 s a_1}{(s^2 + \varphi_1^2)(s+n)^2} + \frac{a_1 \omega_2^2}{(s^2 + \varphi_1^2)(s+n)^2}$$

$$+ \frac{2n_1 a_2}{(s^2 + \varphi_2^2)(s+n)^2} + \frac{a_2 \omega_1^2}{(s^2 + \varphi_2^2)(s+n)^2} \qquad (21.3.11)$$

$$x_2(s) = \frac{s^2 a_2}{(s^2 + \varphi_2^2)(s+n)^2} + \frac{2n_1 s a_2}{(s^2 + \varphi_2^2)(s+n)^2} + \frac{a_2 \omega_1^2}{(s^2 + \varphi_2^2)(s+n)^2}$$

$$+ \frac{2n_2 s a_1}{(s^2 + \varphi_1^2)(s+n)^2} + \frac{a_1 \omega_2^2}{(s^2 + \varphi_1^2)(s+n)^2} \qquad (21.3.12)$$

Applying to equations (21.3.11) and (21.3.12) Laplace Transform pairs 71, 63, 60, 63, and 60, we invert these equations with their initial conditions from the Laplace domain into the time domain and obtain the solutions of differential equations (21.3.1) and (21.3.2) with their initial conditions of motion:

$$x_1 = \frac{a_1}{(\varphi_1^2 + n^2)^2} \left\{ 2n\varphi_1 \sin \varphi_1 t + e^{-nt} \left[\varphi_1^2 - n^2 - nt \left(\varphi_1^2 + n^2 \right) \right] \right.$$

$$\left. - \left(\varphi_1^2 - n^2 \right) \cos \varphi_1 t \right\} + \frac{2n_2 a_1}{(\varphi_1^2 + n^2)^2}$$

$$\times \left\{ 2n\varphi_1 \left[e^{-nt}(1 + nt) - \cos \varphi_1 t \right] \right.$$

$$\left. - (\varphi_1^2 - n^2)(\sin \varphi_1 t - \varphi_1 t e^{-nt}) \right\} + \frac{a_1 \omega_2^2}{(\varphi_1^2 + n^2)^2}$$

$$\times \left\{ \frac{n^2 - \varphi_1^2}{\varphi_1^2} (1 - \cos \varphi_1 t) - \frac{2n}{\varphi_1} \sin \varphi_1 t + \frac{3n^2 + \varphi_1^2}{\varphi_1^2} \right.$$

$$\left. \times \left[1 - e^{-nt}(1 + nt) \right] + 2nt e^{-nt} \right\}$$

$$
+ \frac{2n_1 a_2}{(\varphi_2^2 + n^2)^2} \Big\{ 2n\varphi_2 \left[e^{-nt}(1 + nt) - \cos \varphi_2 t \right]
$$

$$
- (\varphi_2^2 - n^2) \left(\sin \varphi_2 t - \varphi_2 t e^{-nt} \right) \Big\} + \frac{a_2 \omega_1^2}{(\varphi_2^2 + n^2)^2}
$$

$$
\times \left\{ \frac{n^2 - \varphi_2^2}{\varphi_2^2} (1 - \cos \varphi_2 t) - \frac{2n}{\varphi_2} \sin \varphi_2 t \right.
$$

$$
\left. + \frac{3n^2 + \varphi_2^2}{\varphi_2^2} \left[1 - e^{-nt}(1 + nt) \right] + 2nte^{-nt} \right\} \tag{21.3.13}
$$

$$
x_2 = \frac{a_2}{(\varphi_2^2 + n^2)^2} \Big\{ 2n\varphi_2 \sin \varphi_2 t + e^{-nt} \left[\varphi_2^2 - n^2 - nt(\varphi_2^2 + n^2) \right]
$$

$$
- (\varphi_2^2 - n^2) \cos \varphi_2 t \Big\} + \frac{2n_1 a_2}{\left(\varphi_2^2 + n^2 \right)^2}
$$

$$
\times \left\{ 2n\varphi_2 \left[e^{-nt}(1 + nt) - \cos \varphi_2 t \right] \right.
$$

$$
\left. - (\varphi_2^2 - n^2) \left(\sin \varphi_2 t - \varphi_2 t e^{-nt} \right) \right\} + \frac{a_2 \omega_1^2}{(\varphi_1^2 + n^2)^2}
$$

$$
\times \left\{ \frac{n^2 - \varphi_2^2}{\varphi_2^2} (1 - \cos \varphi_2 t) - \frac{2n}{\varphi_2} \sin \varphi_2 t + \frac{3n^2 + \varphi_2^2}{\varphi_2^2} \right.
$$

$$
\left. \times \left[1 - e^{-nt}(1 + nt) \right] + 2nte^{-nt} \right\} + \frac{2n_2 a_1}{\left(\varphi_1^2 + n^2 \right)^2}
$$

$$
\times \left\{ 2n\varphi_1 \left[e^{-nt}(1 + nt) - \cos \varphi_1 t \right] \right.
$$

$$
\left. - (\varphi_1^2 - n^2)(\sin \varphi_1 t - \varphi_1 t e^{-nt}) \right\} + \frac{a_1 \omega_2^2}{(\varphi_1^2 + n^2)^2}
$$

$$
\times \left\{ \frac{n^2 - \varphi_1^2}{\varphi_1^2} (1 - \cos \varphi_1 t) - \frac{2n}{\varphi_1} \sin \varphi_1 t + \frac{3n^2 + \varphi_1^2}{\varphi_1^2} \right.
$$

$$
\left. \times \left[1 - e^{-nt}(1 + nt) \right] + 2nte^{-nt} \right\} \tag{21.3.14}
$$

21.3.3. *Overdamped motion*

Since overdamped motion occurs in the case when $\omega^2 < 0$, we have to replace in equations (21.3.7) and (21.3.8) the parameter ω^2 with its negative value, and then, we obtain the corresponding equations that upon inversion give us the equations describing overdamped motion:

$$x_1(s) = \frac{s^2 a_1}{(s^2 + \varphi_1^2)[(s+n)^2 - \omega^2]} + \frac{2n_2 s a_1}{(s^2 + \varphi_1^2)[(s+n)^2 - \omega^2]}$$

$$+ \frac{a_1 \omega_2^2}{(s^2 + \varphi_1^2)[(s+n)^2 - \omega^2]} + \frac{2n_1 s a_2}{(s^2 + \varphi_2^2)[(s+n)^2 - \omega^2]}$$

$$+ \frac{a_2 \omega_1^2}{(s^2 + \varphi_2^2)[(s+n)^2 - \omega^2]} \tag{21.3.15}$$

$$x_2(s) = \frac{s^2 a_2}{(s^2 + \varphi_2^2)[(s+n)^2 - \omega^2]} + \frac{2n_1 s a_2}{(s^2 + \varphi_2^2)[(s+n)^2 - \omega^2]}$$

$$+ \frac{a_2 \omega_1^2}{(s^2 + \varphi_2^2)[(s+n)^2 - \omega^2]} + \frac{2n_2 s a_1}{(s^2 + \varphi_1^2)[(s+n)^2 - \omega^2]}$$

$$+ \frac{a_1 \omega_2^2}{(s^2 + \varphi_1^2)[(s+n)^2 - \omega^2]} \tag{21.3.16}$$

Applying to equations (21.3.15) and (21.3.16) Laplace Transforms pairs 69, 66, 59, 66, and 59, we invert these equations with their initial conditions from the Laplace domain into the time domain and obtain the solutions of differential equations (21.3.1) and (21.3.2) with their initial conditions of motion:

$$x_1 = \frac{a_1}{4n^2 \varphi_1^2 + (n^2 - \omega^2 - \varphi_1^2)^2}$$

$$\times \left[(n^2 - \omega^2 - \varphi_1^2)^2 (\cos \varphi_1 t - e^{-nt} \cosh \omega t) \right.$$

$$\left. + 2n\varphi_1 \sin \varphi_1 t - \frac{n}{\omega} \left(n^2 - \omega + \varphi_1^2 \right) e^{-nt} \sinh \omega t \right]$$

$$+ \frac{2n_2 a_1}{4n^2 \varphi_1^2 + (n^2 - \omega - \varphi_1^2)^2} \left[2n \left(e^{-nt} \cosh \omega t - \cos \varphi_1 t \right) \right.$$

$$\left. + \frac{1}{\varphi_1} \left(n^2 - \omega^2 - \varphi_1^2 \right) \sin \varphi_1 t + \frac{1}{\omega} \left(\omega^2 + n^2 + \varphi_1^2 \right) e^{-nt} \sinh \omega t \right]$$

$$+ \frac{a_1 \omega_2^2}{4n^2 \varphi_1^2 + (n^2 - \omega^2 - \varphi_1^2)^2} \left\{ \frac{1}{\varphi_1^2} \left(n^2 - \omega^2 - \varphi_1^2 \right) (1 - \cos \varphi_1 t) \right.$$

$$-2n\left(\frac{1}{\varphi_1}\sin\varphi_1 t-\frac{1}{\omega}e^{-nt}\sinh\omega t\right)+\frac{3n^2+\varphi_1^2+\omega^2}{n^2-\omega^2}$$

$$\times\left[1-e^{-nt}\left(\cosh\omega t+\frac{n}{\omega}\sinh\omega t\right)\right]\Bigg\}$$

$$+\frac{2n_1 a_2}{4n^2\varphi_2^2+(n^2-\omega_v^2-\varphi_2^2)^2}\left[2n(e^{-nt}\cosh\omega t-\cos\varphi_2 t)\right.$$

$$+\frac{1}{\varphi_2}(n^2-\omega^2-\varphi_2^2)\sin\varphi_2 t+\frac{1}{\omega}\left(\omega^2+n^2+\varphi_2^2\right)e^{-nt}\sinh\omega t\bigg]$$

$$+\frac{a_2\omega_1^2}{4n^2\varphi_2^2+(n^2-\omega_v^2-\varphi_2^2)^2}\left\{\frac{1}{\varphi_2^2}\left(n^2-\omega_v^2-\varphi_2^2\right)\right.$$

$$\times(1-\cos\varphi_2 t)-2n\left(\frac{1}{\varphi_2}\sin\varphi_2 t-\frac{1}{\omega}e^{-nt}\sinh\omega t\right)$$

$$+\frac{3n^2+\varphi_2^2+\omega^2}{n^2-\omega^2}\left[1-e^{-nt}\left(\cosh\omega t+\frac{n}{\omega}\sinh\omega t\right)\right]\Bigg\}\qquad(21.3.17)$$

$$x_2=\frac{a_2}{4n^2\varphi_2^2+(n^2-\omega^2-\varphi_2^2)^2}$$

$$\times\left[(n^2-\omega^2-\varphi_2^2)^2\left(\cos\varphi_2 t-e^{-nt}\cosh\omega t\right)\right.$$

$$+2n\varphi_2\sin\varphi_2 t-\frac{n}{\omega}(n^2-\omega^2+\varphi_2^2)e^{-nt}\sinh\omega t\bigg]$$

$$+\frac{2n_1 a_2}{4n^2\varphi_2^2+(n^2-\omega^2-\varphi_2^2)^2}\left[2n(e^{-nt}\cosh\omega t-\cos\varphi_2 t)\right.$$

$$+\frac{1}{\varphi_2}\left(n^2-\omega^2-\varphi_2^2\right)\sin\varphi_2 t+\frac{1}{\omega}\left(\omega^2+n^2+\varphi_2^2\right)e^{-nt}\sinh\omega t\bigg]$$

$$+\frac{a_2\omega_1^2}{4n^2\varphi_2^2+(n^2-\omega^2-\varphi_2^2)^2}\left\{\frac{1}{\varphi_2^2}(n^2-\omega^2-\varphi_2^2)(1-\cos\varphi_2 t)\right.$$

$$-2n\left(\frac{1}{\varphi_2}\sin\varphi_2 t-\frac{1}{\omega}e^{-nt}\sinh\omega t\right)+\frac{3n^2+\varphi_2^2+\omega^2}{n^2-\omega^2}$$

$$\times\left[1-e^{-nt}\left(\cosh\omega t+\frac{n}{\omega}\sinh\omega t\right)\right]\Bigg\}$$

$$+ \frac{2n_2 a_1}{4n^2 \varphi_1^2 + (n^2 - \omega^2 - \varphi_1^2)^2} \left[2n \left(e^{-nt} \cosh \omega t - \cos \varphi_1 t \right) \right.$$

$$\left. + \frac{1}{\varphi_1} \left(n^2 - \omega^2 - \varphi_1^2 \right) \sin \varphi_1 t + \frac{1}{\omega} \left(\omega^2 + n^2 + \varphi_1^2 \right) e^{-nt} \sinh \omega t \right]$$

$$+ \frac{a_1 \omega_1^2}{4n^2 \varphi_1^2 + (n^2 - \omega^2 - \varphi_1^2)^2} \left\{ \frac{1}{\varphi_1^2} (n^2 - \omega^2 - \varphi_1^2)(1 - \cos \varphi_1 t) \right.$$

$$- 2n \left(\frac{1}{\varphi_1} \sin \varphi_1 t - \frac{1}{\omega} e^{-nt} \sinh \omega t \right) + \frac{3n^2 + \varphi_1^2 + \omega^2}{n^2 - \omega^2}$$

$$\left. \times \left[1 - e^{-nt} \left(\cosh \omega t + \frac{n}{\omega} \sinh \omega t \right) \right] \right\} \qquad (21.3.18)$$

CHAPTER 22

RESTRICTED TWO-DEGREE-OF-FREEDOM SYSTEMS WITH TWO SEQUENCES OF FLEXIBLE LINKS

22.1. Motion of the Masses of a Restricted System with Two Sequences of Flexible Links, Which Occurs Due to the Initial Displacements and Velocities of the Masses

We consider the operational process of a restricted two-degree-of-freedom system, the masses of which are connected to each other by a flexible link (spring), while one of the masses is attached by a flexible link (spring) to a non-movable support. The masses of the system are moving on a horizontal frictionless surface. The air resistance to the motion of the masses is negligible. Figure 22.1.1 shows a schematic diagram of the system, where K_1 and K_2 are the stiffness coefficients of the respective springs, while the rest of the notations in this figure are self-explanatory.

Based on the schematic diagram shown in Figure 22.1.1 and the considerations mentioned above, we compose a pair of simultaneous differential equations describing the motion of the system:

$$m_1 \frac{d^2 x_1}{dt^2} + (K_1 + K_2)(x_1 - x_2) = 0 \qquad (22.1.1)$$

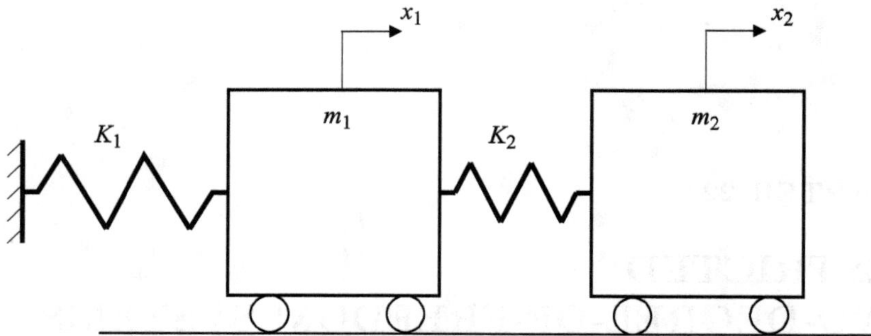

Fig. 22.1.1. Schematic diagram of a restricted two-degree-of-freedom system, the masses of which are connected by a flexible link (spring), while one of the masses is attached by a flexible link (spring) to a non-movable support.

$$m_2 \frac{d^2 x_2}{dt^2} + (K_1 + K_2)(x_2 - x_1) = 0 \qquad (22.1.2)$$

The initial conditions of motion are

$$x_1 = S_1; \quad \frac{dx_1}{dt} = V_1;$$

$$\text{for } t = 0 \qquad\qquad\qquad (22.1.3)$$

$$x_2 = S_2; \quad \frac{dx_2}{dt} = V_2$$

Dividing equations (22.1.1) and (22.1.2) by m_1 and m_2, respectively, while applying to them Laplace Transform pairs 2, 5, 2, and 1, we convert these equations with initial conditions of motion (22.1.3) from the time domain into a system of two simultaneous algebraic equations with two unknowns in each equation in the Laplace domain:

$$s^2 x_1(s) - sV_1 - s^2 S_1 + \omega_{11}^2 x_1(s)$$
$$-\omega_{11}^2 x_2(s) + \omega_{21}^2 x_1(s) - \omega_{21}^2 x_2(s) = 0 \qquad (22.1.4)$$
$$s^2 x_2(s) - sV_2 - s^2 S_2 + \omega_{12}^2 x_2(s)$$
$$-\omega_{12}^2 x_1(s) + \omega_{22}^2 x_2(s) - \omega_{22}^2 x_1(s) = 0 \qquad (22.1.5)$$

Where, just for this chapter, we denote

$$\omega_{11}^2 = \frac{K_1}{m_1}; \quad \omega_{21}^2 = \frac{K_2}{m_1}; \qquad \omega_{12}^2 = \frac{K_1}{m_2}; \quad \omega_{22}^2 = \frac{K_2}{m_2} \qquad (22.1.6)$$

Rearranging equations (22.1.4) and (22.1.5), we obtain

$$x_1(s)(s^2 + \omega_{11}^2 + \omega_{21}^2) = sV_1 + s^2 S_1 + x_2(s)(\omega_{11}^2 + \omega_{21}^2) \tag{22.1.7}$$

$$x_2(s)(s^2 + \omega_{12}^2 + \omega_{22}^2) = sV_2 + s^2 S_2 + x_1(s)(\omega_{12}^2 + \omega_{22}^2) \tag{22.1.8}$$

The system of two simultaneous algebraic equations (22.1.7) and (22.1.8) contains two unknowns $x_1(s)$ and $x_2(s)$ in each equation. Applying to these equations the method of substitutions, we get two equations with one unknown in each equation. These unknowns, respectively, represent the displacements of the masses in Laplace domain:

$$x_1(s) = \frac{sV_1}{s^2 + \omega^2} + \frac{V_1(\omega_{12}^2 + \omega_{22}^2) + V_2(\omega_{11}^2 + \omega_{21}^2)}{s(s^2 + \omega^2)} + \frac{s^2 S_1}{s^2 + \omega^2}$$

$$+ \frac{S_1(\omega_{12}^2 + \omega_{22}^2) + S_2(\omega_{11}^2 + \omega_{21}^2)}{s^2 + \omega^2} \tag{22.1.9}$$

$$x_2(s) = \frac{sV_2}{s^2 + \omega^2} + \frac{V_2(\omega_{11}^2 + \omega_{21}^2) + V_1(\omega_{12}^2 + \omega_{22}^2)}{s(s^2 + \omega^2)}$$

$$+ \frac{s^2 S_2}{s^2 + \omega^2} + \frac{S_2(\omega_{11}^2 + \omega_{21}^2) + S_1(\omega_{12}^2 + \omega_{22}^2)}{s^2 + \omega^2} \tag{22.1.10}$$

where ω is the natural compound frequency of the system, while

$$\omega^2 = \omega_{11}^2 + \omega_{21}^2 + \omega_{12}^2 + \omega_{22}^2 \tag{22.1.11}$$

Applying to equations (22.1.9) and (22.1.10) Laplace Transform pairs 23, 39, 29, and 17, we invert these equations from the Laplace domain into the time domain and obtain the solutions of differential equations (22.1.1) and (22.1.2) for initial conditions of motion (22.1.3):

$$x_1 = \frac{V_1}{\omega} \sin \omega t + \frac{1}{\omega^2} \left[V_1 \left(\omega_{12}^2 + \omega_{22}^2 \right) + V_2 \left(\omega_{11}^2 + \omega_{21}^2 \right) \right] \left(t - \frac{1}{\omega} \sin \omega t \right)$$

$$+ S_1 \cos \omega t + \frac{1}{\omega^2} \left[S_1 \left(\omega_{12}^2 + \omega_{22}^2 \right) + S_2 \left(\omega_{11}^2 + \omega_{21}^2 \right) \right] \left(1 - \cos \omega t \right)$$

$$(22.1.12)$$

$$x_2 = \frac{V_2}{\omega} \sin \omega t + \frac{1}{\omega^2} \left[V_2 \left(\omega_{11}^2 + \omega_{21}^2 \right) + V_1 \left(\omega_{12}^2 + \omega_{22}^2 \right) \right] \left(t - \frac{1}{\omega} \sin \omega t \right)$$

$$+ S_2 \cos \omega t + \frac{1}{\omega^2} \left[S_2 \left(\omega_{11}^2 + \omega_{21}^2 \right) + S_1 \left(\omega_{12}^2 + \omega_{22}^2 \right) \right] \left(1 - \cos \omega t \right)$$

$$(22.1.13)$$

Simplifying equations (22.1.12) and (22.1.13), we obtain

$$x_1 = \frac{1}{\omega^2} (S_1 - S_2)(\omega_{11}^2 + \omega_{21}^2) \cos \omega t + \frac{1}{\omega^3} (V_1 - V_2)$$

$$\times (\omega_{11}^2 + \omega_{21}^2) \sin \omega t + \frac{1}{\omega^2} [S_1(\omega_{12}^2 + \omega_{22}^2) + S_2(\omega_{11}^2 + \omega_{21}^2)]$$

$$+ \frac{1}{\omega^2} [V_1(\omega_{12}^2 + \omega_{22}^2) + V_2(\omega_{11}^2 + \omega_{21}^2)]t \qquad (22.1.14)$$

$$x_2 = \frac{1}{\omega^2} (S_2 - S_1)(\omega_{12}^2 + \omega_{22}^2) \cos \omega t + \frac{1}{\omega^3} (V_2 - V_1)(\omega_{12}^2 + \omega_{22}^2) \sin \omega t$$

$$+ \frac{1}{\omega^2} [S_1(\omega_{12}^2 + \omega_{22}^2) + S_2(\omega_{11}^2 + \omega_{21}^2)]$$

$$+ \frac{1}{\omega^2} [V_1(\omega_{12}^2 + \omega_{22}^2) + V_2(\omega_{11}^2 + \omega_{21}^2)]t \qquad (22.1.15)$$

Equations (22.1.14) and (22.1.15) show that if $S_1 \neq S_2$ and $V_1 \neq V_2$, then the first two terms in both equations will have, respectively, opposite algebraic values; therefore, the two masses are in anti-phase vibratory motion. The next two terms in these equations are, respectively, equal and do not influence the mode of vibration. In the cases when $S_1 = S_2$ and $V_1 \neq V_2$ or $S_1 \neq S_2$ and $V_1 = V_2$, one of these two terms is canceled out, and the remaining terms have opposite algebraic values. Therefore, in these cases, the two masses are in anti-phase vibrational motion. In addition, in the case when $S_1 = S_2$ and

$V_1 = V_2$, then the masses perform uniform motion with the same velocity. First of all, this is physically impossible since one of the masses is connected by a spring to a non-movable support. Second, if both masses move with the same velocity, then their motions are not relative; they move as one body, and the system does not have two degrees of freedom. Therefore, the analyzed expressions are not applicable to the case when $S_1 = S_2$ and $V_1 = V_2$.

Taking in equations (22.1.14) and (22.1.15) that $S_1 = S_2 = 0$, we obtain

$$x_1 = \frac{1}{\omega^3}(V_1 - V_2)(\omega_{11}^2 + \omega_{21}^2)\sin \omega t$$

$$+ \frac{1}{\omega^2}[V_1(\omega_{12}^2 - \omega_{22}^2) + V_2(\omega_{11}^2 + \omega_{21}^2)]t \qquad (22.1.16)$$

$$x_2 = \frac{1}{\omega^3}(V_2 - V_1)(\omega_{12}^2 + \omega_{22}^2)\sin \omega t$$

$$+ \frac{1}{\omega^2}[V_1(\omega_{12}^2 + \omega_{22}^2) + V_2(\omega_{11}^2 + \omega_{21}^2)]t \qquad (22.1.17)$$

Equations (22.1.16) and (22.1.17) show that if $V_1 \neq V_2$, then the masses perform anti-phased vibrational motion. Taking in equations (22.1.14) and (22.1.15) that $V_1 = V_2 = 0$ and $S_1 \neq S_2$, we obtain

$$x_1 = \frac{1}{\omega^2}(S_1 - S_2)(\omega_{11}^2 + \omega_{21}^2)\cos \omega t$$

$$+ \frac{1}{\omega^2}[S_1(\omega_{12}^2 + \omega_{22}^2) + S_2(\omega_{11}^2 + \omega_{21}^2)] \qquad (22.1.18)$$

$$x_2 = \frac{1}{\omega^2}(S_2 - S_1)(\omega_{12}^2 + \omega_{22}^2)\cos \omega t$$

$$+ \frac{1}{\omega^2}[S_1(\omega_{12}^2 + \omega_{22}^2) + S_2(\omega_{11}^2 + \omega_{21}^2)] \qquad (22.1.19)$$

Equations (22.1.18) and (22.1.19) show that the masses are in anti-phase vibrational motion.

In conclusion, it is justifiable to emphasize that due to the initial conditions of motion of the masses of a restricted two-degree-of-freedom system, the masses of which are connected by a spring, while one of the masses is attached to a non-movable support, the two masses perform anti-phase vibrational motion with a single value of its frequency.

22.1.1. *Velocities and accelerations of the masses*

Taking the first derivatives from equations (22.1.14) and (22.1.15), we obtain the equations describing the velocities of the masses:

$$\frac{dx_1}{dt} = \frac{1}{\omega^2}(V_1 - V_2)(\omega_{11}^2 + \omega_{21}^2)\cos\omega t$$

$$- \frac{1}{\omega}(S_1 - S_2)(\omega_{11}^2 + \omega_{21}^2)\sin\omega t$$

$$+ \frac{1}{\omega^2}[V_1(\omega_{12}^2 + \omega_{22}^2) + V_2(\omega_{11}^2 + \omega_{21}^2)] \qquad (22.1.20)$$

$$\frac{dx_2}{dt} = \frac{1}{\omega^2}(V_2 - V_1)(\omega_{12}^2 + \omega_{22}^2)\cos\omega t$$

$$- \frac{1}{\omega}(S_2 - S_1)(\omega_{12}^2 + \omega_{22}^2)\sin\omega t$$

$$+ \frac{1}{\omega^2}[V_1(\omega_{12}^2 + \omega_{22}^2) + V_2(\omega_{11}^2 + \omega_{21}^2)] \qquad (22.1.21)$$

Supposing for equations (22.1.14), (22.1.15), (22.1.20), and (22.1.21) that $t = 0$, then we have $x_1 = S_1$, $x_2 = S_2$, $\frac{dx_1}{dt} = V_1$, and $\frac{dx_2}{dt} = V_2$, as it should be according to their initial conditions of motion.

Taking from equations (22.1.20) and (22.1.21) the first derivatives, we obtain the equations for the accelerations of the masses, respectively:

$$\frac{d^2x_1}{dt^2} = -\frac{1}{\omega}(V_1 - V_2)(\omega_{11}^2 + \omega_{21}^2)\sin\omega t$$

$$- (S_1 - S_2)(\omega_{11}^2 + \omega_{21}^2)\cos\omega t \qquad (22.1.22)$$

$$\frac{d^2x_2}{dt^2} = -\frac{1}{\omega}(V_2 - V_1)(\omega_{12}^2 + \omega_{22}^2)\sin\omega t$$

$$- (S_2 - S_1)(\omega_{12}^2 + \omega_{22}^2)\cos\omega t \qquad (22.1.23)$$

Taking in equations (22.1.22) and (22.3.23) that $t = 0$, we determine the initial accelerations of the masses at the beginning of the

motion:

$$\frac{d^2 x_1}{dt^2} = -(S_1 - S_2)(\omega_{11}^2 + \omega_{21}^2) \qquad (22.1.24)$$

$$\frac{d^2 x_2}{dt^2} = -(S_2 - S_1)(\omega_{12}^2 + \omega_{22}^2) \qquad (22.1.25)$$

22.2. A Restricted System with Two Sequences of Flexible Links While the Masses of the System Are Subjected to Constant Resisting and Constant Active Forces

Consider the operational process of a system associated with the motion of the masses on a horizontal frictionless surface, while the masses of the system are subjected to constant resisting and constant active forces. The air resistance to the motion of the system is negligible. Figure 22.2.1 shows a schematic diagram of the described system, the masses of which are, respectively, subjected to the constant resisting forces R_1 and R_2 and the constant active forces P_1 and P_2.

Based on the schematic diagram shown in Figure 22.2.1 and the related considerations, we compose the following pair of simultaneous

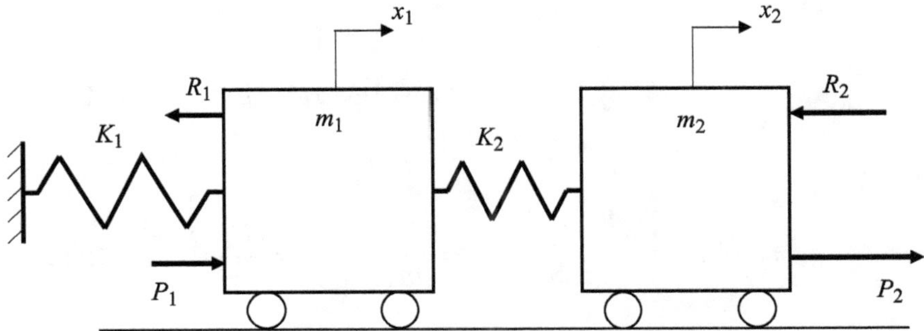

Fig. 22.2.1. Schematic diagram of a restricted two-degree-of-freedom system, the masses of which are connected by springs, while one of the masses is attached to a non-movable support by another spring, while these masses are subjected to constant resisting and constant active forces.

differential equations of motion of the system:

$$m_1\frac{d^2x_1}{dt^2} + (K_1 + K_2)(x_1 - x_2) + R_1 = P_1 \qquad (22.2.1)$$

$$m_2\frac{d^2x_2}{dt^2} + (K_1 + K_2)(x_2 - x_1) + R_2 = P_2 \qquad (22.2.2)$$

The initial conditions of motion are

$$x_1 = 0, \qquad \frac{dx_1}{dt} = 0$$

$$\text{For } t = 0 \qquad\qquad\qquad\qquad (22.2.3)$$

$$x_2 = 0, \qquad \frac{dx_2}{dt} = 0$$

Dividing differential equations (22.2.1) and (22.2.2) by m_1 and m_2, respectively, and applying to them Laplace Transform pairs 2, 5, 2, 1, 2, and 2, we convert these equations with the initial conditions of motion according to expression (22.2.3) from the time domain into a system of two simultaneous algebraic equations with two unknowns in the Laplace domain:

$$s^2x_1(s) + \omega_{11}^2x_1(s) - \omega_{11}^2x_2(s)$$

$$+ \omega_{21}^2x_1(s) - \omega_{21}^2x_2(s) + r_1 = p_1 \qquad (22.2.4)$$

$$s^2x_2(s) + \omega_{12}^2x_2(s) - \omega_{12}^2x_1(s)$$

$$+ \omega_{22}^2x_2(s) - \omega_{22}^2x_1(s) + r_2 = p_2 \qquad (22.2.5)$$

where

$$r_1 = \frac{R_1}{m_1}; \quad r_2 = \frac{R_2}{m_2}; \quad p_1 = \frac{P_1}{m_1}; \quad p_2 = \frac{P_2}{m_2} \qquad (22.2.6)$$

while we have to keep in mind that $P_1 > R_1$ and $P_2 > R_2$. Rearranging the pair of simultaneous algebraic equations (22.2.4) and (22.2.5) and applying to them the method of substitutions, we obtain the equations describing the displacements $x_1(s)$ and $x_2(s)$

in the Laplace domain:

$$x_1(s) = \frac{p_1 - r_1}{s^2 + \omega^2} + \frac{(\omega_{12}^2 + \omega_{22}^2)(p_1 - r_1) + (\omega_{11}^2 + \omega_{21}^2)(p_2 - r_2)}{s^2(s^2 + \omega^2)}$$

$$(22.2.7)$$

$$x_2(s) = \frac{p_2 - r_2}{l^2 + \omega^2} + \frac{(\omega_{11}^2 + \omega_{21}^2)(p_2 - r_2) + (\omega_{12}^2 + \omega_{22}^2)(p_1 - r_1)}{s^2(s^2 + \omega^2)}$$

$$(22.2.8)$$

Applying to equations (22.2.7) and (22.2.8) Laplace Transform pairs 17 and 51, we invert these equations from the Laplace domain into the time domain and obtain the solutions of the differential equations of motion (22.2.1) and (22.2.2) with their initial conditions of motion:

$$x_1 = \frac{p_1 - r_1}{\omega^2}(1 - \cos \omega t) + \left[(\omega_{12}^2 + \omega_{22}^2)(p_1 - r_1)\right.$$
$$\left. + (\omega_{11}^2 + \omega_{21}^2)(p_2 - r_2)\right]\left[\frac{1}{\omega^4}(\cos \omega t - 1) + \frac{1}{2\omega^2}t^2\right]$$

$$(22.2.9)$$

$$x_2 = \frac{p_2 - r_2}{\omega^2}(1 - \cos \omega t) + \left[(\omega_{11}^2 + \omega_{21}^2)(p_2 - r_2)\right.$$
$$\left. + (\omega_{12}^2 + \omega_{22}^2)(p_1 - r_1)\right]\left[\frac{1}{\omega^4}(\cos \omega t - 1) + \frac{1}{2\omega^2}t^2\right]$$

$$(22.2.10)$$

Assuming for equations (22.2.9) and (22.2.10) that $t = 0$, we determine that $x_1 = 0$ and $x_2 = 0$, which corresponds to the initial conditions of motion.

Taking the first derivatives from equations (22.2.9) and (22.2.10), we determine the velocities of the two masses:

$$\frac{dx_1}{dt} = \frac{p_1 - r_1}{\omega}\sin \omega t + \left[(\omega_{12}^2 + \omega_{22}^2)(p_1 - r_1)\right.$$
$$\left. + (\omega_{11}^2 + \omega_{21}^2)(p_2 - r_2)\right]\left[\frac{1}{\omega^2}\left(t - \frac{1}{\omega}\sin \omega t\right)\right]$$

$$(22.2.11)$$

$$\frac{dx_2}{dt} = \frac{p_2 - r_2}{\omega} \sin \omega t + \left[\left(\omega_{11}^2 + \omega_{21}^2 \right) (p_2 - r_2) \right.$$

$$\left. + \left(\omega_{12}^2 + \omega_{22}^2 \right) (p_1 - r_1) \right] \left[\frac{1}{\omega^2} \left(t - \frac{1}{\omega} \sin \omega t \right) \right]$$

$$(22.2.12)$$

Supposing in equations (22.2.11) and (22.2.12) that $t = 0$, we calculate that $\frac{dx_1}{dt} = 0$ and $\frac{dx_2}{dt} = 0$, as it should be according to the initial conditions of motion.

Taking the first derivatives from equations (22.2.11) and (22.2.12), we determine the accelerations of the masses, respectively:

$$\frac{d^2 x_1}{dt^2} = (p_1 - r_1) \cos \omega t + [(\omega_{12}^2 + \omega_{22}^2)(p_1 - r_1)$$

$$+ (\omega_{11}^2 + \omega_{21}^2)(p_2 - r_2)] \frac{1}{\omega^2} (1 - \cos \omega t) \quad (22.2.13)$$

$$\frac{d^2 x_2}{dt^2} = (p_2 - r_2) \cos \omega t + [(\omega_{11}^2 + \omega_{21}^2)(p_2 - r_2)$$

$$+ (\omega_{12}^2 + \omega_{22}^2)(p_1 - r_1)] \frac{1}{\omega^2} (1 - \cos \omega t) \quad (22.2.14)$$

Assuming in equations (22.2.13) and (22.2.14) that $t = 0$, we determine the initial accelerations of the masses, respectively, at the beginning of the motion:

$$\frac{d^2 x_1}{dt^2} = p_1 - r_1 \qquad (22.2.15)$$

$$\frac{d^2 x_2}{dt^2} = p_2 - r_2 \qquad (22.2.16)$$

22.2.1. *Numerical solution*

Following is a Python program to plot the graph (Figure 22.2.2) of equation (22.2.9).

```python
from matplotlib.pyplot import plot, show
from numpy import linspace, cos, sqrt

p1 = 100
p2 = 75
r1 = 25
r2 = 15
omega11 = 30
omega12 = 25
omega21 = 20
omega22 = 15
omega = sqrt(omega11**2 + omega12**2 + omega21**2 + omega22**2)

t = linspace(0, 1, 1000)
x1  = ((p1 - r1)/omega**2)*(1 - cos(omega*t)) \
    + ( (omega12**2 + omega22**2)*(p1 - r1) \
    + (omega11**2 + omega21**2)*(p2 - r2) \
    ) \
    *((1/omega**4)*cos(omega*t - 1) \
    + (1/(2*omega**2))*t**2)

plot(t, x1)
show()
```

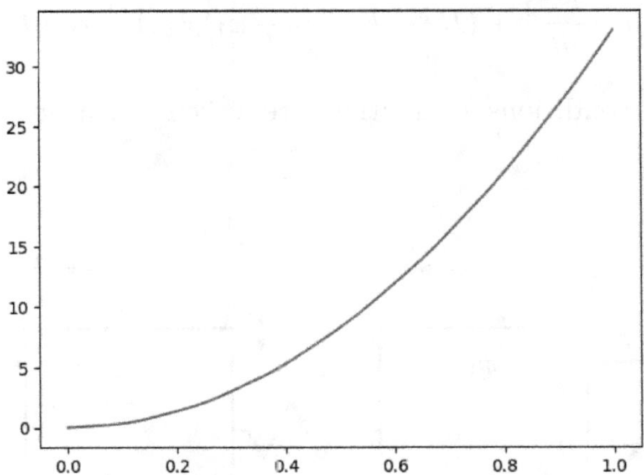

Fig. 22.2.2. Graph of equation (22.2.9).

22.3. A Restricted System, the Masses of Which Are Connected to Each Other by a Spring and Subjected to Harmonic Forces, While One of the Masses Is Attached to a Non-movable Support by an Additional Spring

The schematic diagram that represents the restricted system subjected to the actions of harmonic forces is shown in Figure 22.3.1, where A_1 and A_2 are the amplitudes of the harmonic forces, while φ_1 and φ_2 are, respectively, the frequencies of the harmonic functions. The rest of the notations in this figure are self-explanatory. The system is moving on a horizontal frictionless surface. The air resistance to the motion of the system is negligible.

Based on the schematic diagram and the considerations mentioned above, we compose the pair of simultaneous differential equations of motion of the system:

$$m_1 \frac{d^2 x_1}{dt^2} + (K_1 + K_2)(x_1 - x_2) = A_1 \cos \varphi_1 t \qquad (22.3.1)$$

$$m_2 \frac{d^2 x_2}{dt^2} + (K_1 + K_2)(x_2 - x_1) = A_2 \cos \varphi_2 t \qquad (22.3.2)$$

The initial conditions of motion are taken according to expression (22.1.3).

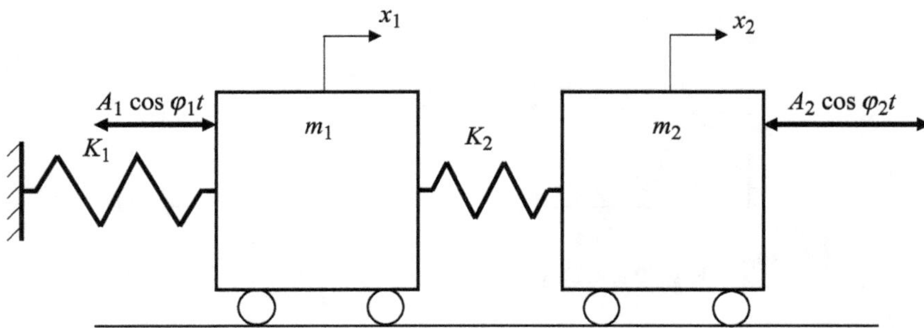

Fig. 22.3.1. Schematic diagram of a restricted two-degree-of-freedom system, the masses of which are connected by a spring and subjected to harmonic forces, while one of the masses is attached to a non-movable support by another spring.

Dividing equations (22.3.1) and (22.3.2) by m_1 and m_2, while applying to them Laplace Transform pairs 2, 5, 2, 1, 2, and 29, we convert these equations with their initial conditions of motion from the time domain into a system of two simultaneous algebraic equations with two unknowns in each of the equations in the Laplace domain:

$$s^2 x_1(s) - sV_1 - s^2 S_1 + (\omega_{11}^2 + \omega_{21}^2)x_1(s)$$

$$- (\omega_{11}^2 + \omega_{21}^2)x_2(s) = a_1 \frac{s^2}{s^2 + \varphi_1^2} \tag{22.3.3}$$

$$s^2 x_2(s) - sV_2 - s^2 S_2 + (\omega_{12}^2 + \omega_{22}^2)x_2(s)$$

$$- (\omega_{12}^2 + \omega_{22}^2)x_1(s) = a_2 \frac{s^2}{s^2 + \varphi_2^2} \tag{22.3.4}$$

Rearranging equations (22.3.3) and (22.3.4) and applying to them the method of substitutions, we eliminate one unknown in each of the equations and obtain expressions that allow us to determine the displacements $x_1(s)$ and $x_2(s)$ in the Laplace domain:

$$x_1(s) = \frac{sV_1}{s^2 + \omega^2} + \frac{V_1(\omega_{12}^2 + \omega_{22}^2) + V_2(\omega_{11}^2 + \omega_{21}^2)}{s(s^2 + \omega^2)} + \frac{s^2 S_1}{s^2 + \omega^2}$$

$$+ \frac{S_1(\omega_{12}^2 + \omega_{22}^2) + S_2(\omega_{11}^2 + \omega_{21}^2)}{s^2 + \omega^2} + \frac{s^2 a_1}{(s^2 + \varphi_1^2)(s^2 + \omega^2)}$$

$$+ \frac{a_1(\omega_{12}^2 + \omega_{22}^2)}{(s^2 + \varphi_1^2)(s^2 + \omega^2)} + \frac{a_2(\omega_{11}^2 + \omega_{21}^2)}{(s^2 + \varphi_2^2)(s^2 + \omega^2)} \tag{22.3.5}$$

$$x_2(s) = \frac{sV_2}{s^2 + \omega^2} + \frac{V_2(\omega_{11}^2 + \omega_{21}^2) + V_1(\omega_{12}^2 + \omega_{22}^2)}{s(s^2 + \omega^2)} + \frac{s^2 S_2}{s^2 + \omega^2}$$

$$+ \frac{S_2(\omega_{11}^2 + \omega_{21}^2) + S_1(\omega_{12}^2 + \omega_{22}^2)}{s^2 + \omega^2} + \frac{s^2 a_2}{(s^2 + \varphi_2^2)(s^2 + \omega^2)}$$

$$+ \frac{a_2(\omega_{11}^2 + \omega_{21}^2)}{(s^2 + \varphi_2^2)(s^2 + \omega^2)} + \frac{a_1(\omega_{12}^2 + \omega_{22}^2)}{(s^2 + \varphi_1^2)(s^2 + \omega^2)} \tag{22.3.6}$$

where

$$a_1 = \frac{A_1}{m_1}; \quad a_2 = \frac{A_2}{m_2} \tag{22.3.7}$$

Applying to equations (22.3.5) and (22.3.6) Laplace Transform pairs 23, 39, 29, 17, 70, 55, and 55, we invert these equations from the Laplace domain into the time domain and obtain the solutions of differential equations (22.3.1) and (22.3.2) with their initial conditions of motion:

$$
x_1 = \frac{V_1}{\omega}\sin\omega t + \frac{V_1(\omega_{12}^2 + \omega_{22}^2) + V_2(\omega_{11}^2 + \omega_{21}^2)}{\omega^2}\left(t - \frac{1}{\omega}\sin\omega t\right)
$$

$$
+ S_1\cos\omega t + \frac{S_1(\omega_{12}^2 + \omega_{22}^2) + S_2(\omega_{11}^2 + \omega_{21}^2)}{\omega^2}(1 - \cos\omega t)
$$

$$
+ a_1\frac{\cos\omega t - \cos\varphi_1 t}{\varphi_1^2 - \omega^2} + \frac{a_1(\omega_{12}^2 + \omega_{22}^2)}{\varphi_1^2 - \omega^2}
$$

$$
\times\left[\frac{1}{\omega^2}(1 - \cos\omega t) - \frac{1}{\varphi_1^2}(1 - \cos\varphi_1 t)\right] + \frac{a_2(\omega_{11}^2 + \omega_{21}^2)}{\varphi_2^2 - \omega^2}
$$

$$
\times\left[\frac{1}{\omega^2}(1 - \cos\omega t) - \frac{1}{\varphi_2^2}(1 - \cos\varphi_2 t)\right] \tag{22.3.8}
$$

$$
x_2 = \frac{V_2}{\omega}\sin\omega t + \frac{V_2(\omega_{11}^2 + \omega_{21}^2) + V_1(\omega_{12}^2 + \omega_{22}^2)}{\omega^2}\left(t - \frac{1}{\omega}\sin\omega t\right)
$$

$$
+ S_2\cos\omega t + \frac{S_2(\omega_{11}^2 + \omega_{21}^2) + S_1(\omega_{12}^2 + \omega_{22}^2)}{\omega^2}(1 - \cos\omega t)
$$

$$
+ a_2\frac{\cos\omega t - \cos\varphi_2 t}{\varphi_2^2 - \omega^2} + \frac{a_2(\omega_{11}^2 + \omega_{21}^2)}{\varphi_2^2 - \omega^2}
$$

$$
\times\left[\frac{1}{\omega^2}(1 - \cos\omega t) - \frac{1}{\varphi_2^2}(1 - \cos\varphi_2 t)\right] + \frac{a_1(\omega_{12}^2 + \omega_{22}^2)}{\varphi_1^2 - \omega^2}
$$

$$
\times\left[\frac{1}{\omega^2}(1 - \cos\omega t) - \frac{1}{\varphi_1^2}(1 - \cos\varphi_1 t)\right] \tag{22.3.9}
$$

Assuming for equations (22.3.8) and (22.3.9) that $t = 0$, we determine that $x_1 = S_1$ and $x_2 = S_2$, as it should be according to the initial conditions of motion.

Taking from equations (22.3.8) and (22.3.9) the first derivatives, we determine the velocities of the masses:

$$\frac{dx_1}{dt} = V_1 \cos \omega t + \frac{V_1(\omega_{12}^2 + \omega_{22}^2) + V_2(\omega_{11}^2 + \omega_{21}^2)}{\omega^2}(1 - \cos \omega t)$$

$$- S_1 \omega \sin \omega t + \frac{1}{\omega}[S_1(\omega_{12}^2 + \omega_{22}^2) + S_2(\omega_{11}^2 + \omega_{21}^2)] \sin \omega t$$

$$+ \frac{a_1}{\varphi_1^2 - \omega^2}(\varphi_1 \sin \varphi_1 t - \omega \sin \omega t) + \frac{a_1(\omega_{12}^2 + \omega_{22}^2)}{\varphi_1^2 - \omega^2}$$

$$\times \left(\frac{1}{\omega} \sin \omega t - \frac{1}{\varphi_1} \sin \varphi_1 t\right) + \frac{a_2(\omega_{11}^2 + \omega_{21}^2)}{\varphi_2^2 - \omega^2}$$

$$\times \left(\frac{1}{\omega} \sin \omega t - \frac{1}{\varphi_2} \sin \varphi_2 t\right) \qquad (22.3.10)$$

$$\frac{dx_2}{dt} = V_2 \cos \omega t + \frac{V_2(\omega_{11}^2 + \omega_{21}^2) + V_1(\omega_{12}^2 + \omega_{22}^2)}{\omega^2}(1 - \cos \omega t)$$

$$- S_2 \omega \sin \omega t + \frac{1}{\omega}[S_2(\omega_{11}^2 + \omega_{21}^2) + S_1(\omega_{12}^2 + \omega_{22}^2)] \sin \omega t$$

$$+ \frac{a_2}{\varphi_2^2 - \omega^2}(\varphi_2 \sin \varphi_2 t - \omega \sin \omega t) + \frac{a_2(\omega_{11}^2 + \omega_{21}^2)}{\varphi_2^2 - \omega^2}$$

$$\times \left(\frac{1}{\omega} \sin \omega t - \frac{1}{\varphi_2} \sin \varphi_2 t\right) + \frac{a_1(\omega_{12}^2 + \omega_{22}^2)}{\varphi_1^2 - \omega^2}$$

$$\times \left(\frac{1}{\omega} \sin \omega t - \frac{1}{\varphi_1} \sin \varphi_1 t\right) \qquad (22.3.11)$$

By accepting in equations (22.3.10) and (22.3.11) that $t = 0$, we determine that $\frac{dx_1}{dt} = V_1$ and $\frac{dx_2}{dt} = V_2$, as expected according to the initial conditions of motion.

22.3.1. *Resonance of a restricted system subjected to harmonic forces*

The phenomenon of resonance is possible if the compound natural frequency is equal to the frequency of the exciting force. Therefore, we consider that in equations (22.3.8) and (22.3.9) the compound natural frequency ω, introduced in equation (22.1.11), equals to the frequency of the exiting force applied to one of the masses, for instance to the first mass; therefore, we may write

$$\omega = \varphi_1 \tag{22.3.12}$$

A similar investigation on resonance is carried out for an unrestricted two-degree-of-freedom system that is presented in Chapter 19.

According to equation (22.3.12), several denominators of the fractional members of equations (22.3.8) and (22.3.9) become equal to zero. These members have the potential to tend to infinity and cause resonance of the system. The rest of the members of these equations cannot cause a resonance and are eliminated from the current analysis. Therefore, for the purpose of investigation on resonance, we restructure equations (22.3.8) and (22.3.9) by keeping in them only the terms, the values of which may potentially tend to infinity:

$$x'_1 = \frac{a_1(\cos \omega t - \cos \varphi_1 t)}{\varphi_1^2 - \omega^2} + \frac{a_1(\omega_{12}^2 + \omega_{22}^2)}{\varphi_1^2 - \omega^2}$$
$$\times \left[\frac{1}{\omega^2}(1 - \cos \omega t) - \frac{1}{\varphi_1^2}(1 - \cos \varphi_1 t) \right] \tag{22.3.13}$$

$$x'_2 = \frac{a_1(\omega_{12}^2 + \omega_{22}^2)}{\varphi_1^2 - \omega^2} \left[\frac{1}{\omega^2}(1 - \cos \omega t) - \frac{1}{\varphi_1^2}(1 - \cos \varphi_1 t) \right] \tag{22.3.14}$$

where x'_1 and x'_2 represent the displacements that tend to infinity. According to equation (23.3.12), both terms of equations (22.3.13) and (22.3.14) represent indeterminates of the type $\frac{0}{0}$.

Rearranging these equations, we have

$$x'_1 = \frac{a_1(\cos \omega t - \cos \varphi_1 t)}{\varphi_1^2 - \omega^2}$$
$$+ \frac{a_1(\omega_{12}^2 + \omega_{22}^2)(\varphi_1^2 - \omega^2 - \varphi_1^2 \cos \omega t + \omega^2 \cos \varphi_1 t)}{\omega^2 \varphi_1^2(\varphi_1^2 - \omega^2)}$$
$$\tag{22.3.15}$$

$$x'_2 = \frac{a_1(\omega_{12}^2 + \omega_{22}^2)(\varphi_1^2 - \omega^2 - \varphi_1^2 \cos \omega t + \omega^2 \cos \varphi_1 t)}{\omega^2 \varphi_1^2(\varphi_1^2 - \omega^2)} \tag{22.3.16}$$

where x'_1 and x'_2 represent the displacements of the respective masses.

Applying to equations (22.3.15) and (22.3.16) L'Hopital's rule, we evaluate the values of the mentioned indeterminates:

$$x'_1 = \lim_{\varphi_1^2 \to \omega^2} \frac{a_1(\cos \omega t - \cos \varphi_1 t)}{\varphi_1^2 - \omega^2}$$

$$+ \lim_{\varphi_1^2 \to \omega^2} \frac{a_1(\omega_{12}^2 + \omega_{22}^2)(\varphi_1^2 - \omega^2 - \varphi_1^2 \cos \omega t + \omega^2 \cos \varphi_1 t)}{\omega^2 \varphi_1^2 (\varphi_1^2 - \omega^2)}$$

$$(22.3.17)$$

$$x'_2 = \lim_{\varphi_1^2 \to \omega^2} \frac{a_1(\omega_{12}^2 + \omega_{22}^2)(\varphi_1^2 - \omega^2 - \varphi_1^2 \cos \omega t + \omega^2 \cos \varphi_1 t)}{\omega^2 \varphi_1^2 (\varphi_1^2 - \omega^2)}$$

$$(22.3.18)$$

Differentiating the numerators and denominators of equations (22.3.17) and (22.3.18) with respect to ω, we have

$$x'_1 = \lim_{\varphi_1^2 \to \omega^2} \frac{a_1(-t \sin \omega t)}{-2\omega}$$

$$+ \lim_{\varphi_1^2 \to \omega^2} \frac{a_1(\omega_{12}^2 + \omega_{22}^2)(-2\omega + \varphi_1^2 t \sin \omega t + 2\omega \cos \varphi_1 t)}{2\omega \varphi_1^4 - 4\omega^3 \varphi_1^2}$$

$$(22.3.19)$$

$$x'_2 = \lim_{\varphi_1^2 \to \omega^2} \frac{a_1(\omega_{12}^2 + \omega_{22}^2)(-2\omega + \varphi_1^2 t \sin \omega t + 2\omega \cos \varphi_1 t)}{2\omega \varphi_1^4 - 4\omega^3 \varphi_1^2}$$

$$(22.3.20)$$

Continuing with the evaluation of the indeterminate, we obtain

$$x'_1 = \frac{a_1 t \sin \omega t}{2\omega} + \frac{a_1(\omega_{12}^2 + \omega_{22}^2)(-2 + \omega t \sin \omega t + 2 \cos \varphi_1 t)}{-2\omega^4}$$

$$(22.3.21)$$

$$x'_2 = \frac{a_1(\omega_{12}^2 + \omega_{22}^2)(-2 + \omega t \sin \omega t + 2 \cos \varphi_1 t)}{-2\omega^4}$$

$$(22.3.22)$$

Eliminating from equations (22.3.21) and (22.3.22) the terms that do not influence the resonance, we have

$$x'_1 = \frac{a_1 t \sin \omega t}{2\omega} - \frac{a_1(\omega_{12}^2 + \omega_{22}^2)t \sin \omega t}{2\omega^3} \tag{22.3.23}$$

$$x'_2 = -\frac{a_1(\omega_{12}^2 + \omega_{22}^2)t \sin \omega t}{2\omega^3} \tag{22.3.24}$$

Further analysis of equation (22.3.23) yields

$$\begin{aligned}
x'_1 &= \frac{a_1 \omega^2 t \sin \omega t}{2\omega^3} - \frac{a_1(\omega_{12}^2 + \omega_{22}^2)t \sin \omega t}{2\omega^3} \\
&= \frac{a_1(\omega_{11}^2 + \omega_{21}^2 + \omega_{12}^2 + \omega_{22}^2 - \omega_{12}^2 - \omega_{22}^2)t \sin \omega t}{2\omega^3} \\
&= \frac{a_1(\omega_{11}^2 + \omega_{21}^2)t \sin \omega t}{2\omega^3}
\end{aligned} \tag{22.3.25}$$

Equations (22.3.24) and (23.3.25) allow us to evaluate the above-mentioned indeterminates and indicate that the resonance can occur in the considered system:

$$X_1 = \frac{a_1(\omega_{11}^2 + \omega_{21}^2)t \sin \omega t}{2\omega^3} \tag{22.3.26}$$

$$X_2 = \frac{a_1(\omega_{12}^2 + \omega_{22}^2)t \sin \omega t}{2\omega^3} \tag{22.3.27}$$

where X_1 and X_2 represent the respective evaluations of the indeterminates.

It should be emphasized that the multiplier t in equations (22.3.26) and (22.3.27) causes the amplitudes of both masses to tend to infinity, bringing the system to a state of resonance.

CHAPTER 23

RESTRICTED TWO-DEGREE-OF-FREEDOM SYSTEMS WITH TWO SEQUENCES OF FLUID LINKS

These systems are widely used in pneumatically operated impact machinery, hydraulic impact tools, and other industrial applications.

23.1. Motion of the Masses of a Restricted System with Two Sequences of Fluid Links, Which Occurs Due to the Initial Velocities of the Masses

Consider the operational process of a restricted system associated with the motion of its masses on a horizontal frictionless surface due to their initial velocities. The schematic diagram representing the restricted system is shown in Figure 23.1.1, where C_1 and C_2 are the damping coefficients of the dashpots, while the masses are not subjected to any external forces. The rest of the notations in this figure are self-explanatory.

The notations of the parameters are only for this chapter.

Based on the schematic diagram shown in Figure 23.1.1 and the considerations mentioned above, we compose a pair of simultaneous

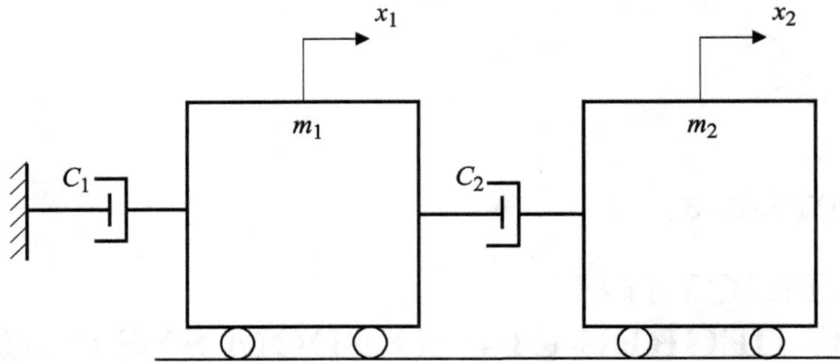

Fig. 23.1.1. Schematic diagram of a restricted system with two sequences of fluid links, moving on a horizontal frictionless surface.

differential equations describing the motion of the system:

$$m_1\frac{d^2x_1}{dt^2} + (C_1 + C_2)\left(\frac{dx_1}{dt} - \frac{dx_2}{dt}\right) = 0 \qquad (23.1.1)$$

$$m_2\frac{d^2x_2}{dt^2} + (C_1 + C_2)\left(\frac{dx_2}{dt} - \frac{dx_1}{dt}\right) = 0 \qquad (23.1.2)$$

The initial conditions of motion are taken according to expression (22.1.3).

Dividing, respectively, equations (23.1.1) and (23.1.2) by m_1 and m_2, while applying to them Laplace Transform pairs 2, 5, 2, and 4, we convert these equations from the time domain into a system of two simultaneous algebraic equations with two unknowns in each equation, describing the displacements of the masses in the Laplace domain:

$$s^2x_1(s) - sV_1 - s^2S_1 + 2n_{11}sx_1(s) - 2n_{11}sS_1 + 2n_{21}sx_1(s)$$

$$-2n_{21}sS_1 - 2n_{11}sx_2(s) + 2n_{11}sS_2 - 2n_{21}sx_2(s) + 2n_{21}sS_2 = 0$$

$$(23.1.3)$$

$$s^2x_2(s) - sV_2 - s^2S_2 + 2n_{12}sx_2(s) - 2n_{12}sS_2 + 2n_{22}sx_2(s)$$

$$-2n_{22}sS_2 - 2n_{12}sx_1(s) + 2n_{12}sS_1 - 2n_{22}sx_1(s) + 2n_{22}sS_1 = 0$$

$$(23.1.4)$$

where

$$2n_{11} = \frac{C_1}{m_1}; \quad 2n_{21} = \frac{C_2}{m_1}; \quad 2n_{12} = \frac{C_1}{m_2}; \quad 2n_{22} = \frac{C_2}{m_2} \qquad (23.1.5)$$

Rearranging the terms in equations (23.1.3) and (23.1.4) and applying to them the method of substitutions, we obtain a pair of two simultaneous algebraic equations with one unknown in each equation, respectively, describing the displacements of the masses in the Laplace domain:

$$x_1(s) = \frac{V_1}{s + 2n} + \frac{2[V_1(n_{12} + n_{22}) + V_2(n_{11} + n_{21})]}{s(s + 2n)}$$

$$+ \frac{sS_1}{s + 2n} + \frac{2nS_1}{s + 2n} \qquad (23.1.6)$$

$$x_2(s) = \frac{V_2}{s + 2n} + \frac{2[V_2(n_{11} + n_{21}) + V_1(n_{12} + n_{22})]}{s(s + 2n)}$$

$$+ \frac{sS_2}{s + 2n} + \frac{2nS_2}{s + 2n} \qquad (23.1.7)$$

where

$$n = n_{11} + n_{21} + n_{12} + n_{22} \qquad (23.1.8)$$

Applying to equations (23.1.6) and (23.1.7) Laplace Transform pairs 8, 16, 14, and 8, we invert these equations from the Laplace domain into the time domain and obtain the solutions of the mentioned differential equations (23.1.1) and (23.1.2) with their initial conditions of motion according to expression (22.1.3):

$$x_1 = S_1 + \frac{V_1}{2n}(1 - e^{-2nt}) + \frac{V_1(n_{12} + n_{22}) + V_2(n_{11} + n_{21})}{n}$$

$$\times \left[t + \frac{1}{2n}(e^{-2nt} - 1) \right] \qquad (23.1.9)$$

$$x_2 = S_2 + \frac{v_2}{2n}(1 - e^{-2nt}) + \frac{V_2(n_{11} + n_{21}) + V_1(n_{12} + n_{22})}{n}$$

$$\times \left[t + \frac{1}{2n}(e^{-2nt} - 1) \right] \qquad (23.1.10)$$

Substituting into equations (23.1.9) and (23.1.10) that $t = 0$, we obtain $x_1 = S_1$ and $x_2 = S_2$, as it should be according to initial conditions of motion (22.1.3). Taking the first derivatives from equations (23.1.9) and (23.1.10), we determine the velocities of the masses:

$$\frac{dx_1}{dt} = V_1 e^{-2nt} + \frac{V_1(n_{12} + n_{22}) + V_2(n_{11} + n_{21})}{n}(1 - e^{-2nt})$$

$$(23.1.11)$$

$$\frac{dx_2}{dt} = V_2 e^{-2nt} + \frac{V_2(n_{11} + n_{21}) + V_1(n_{12} + n_{22})}{n}(1 - e^{-2nt})$$

$$(23.1.12)$$

Assuming in equations (23.1.11) and (23.1.12) that $t = 0$, we determine $\frac{dx_1}{dt} = V_1$ and $\frac{dx_2}{dt} = V_2$, as expected according to initial conditions of motion (22.1.3). Taking the first derivatives from equations (23.1.11) and (23.1.12), we determine the accelerations of the masses:

$$\frac{d^2 x_1}{dt^2} = -2(V_1 - V_2)(n_{11} + n_{21})e^{-2nt} \qquad (23.1.13)$$

$$\frac{d^2 x_2}{dt^2} = -2(V_2 - V_1)(n_{12} + n_{22})e^{-2nt} \qquad (23.1.14)$$

Hence, equations (23.1.9)–(23.1.14) describe the basic parameters of motion of the masses.

23.2. Motion of a Restricted System with Two Sequences of Fluid Links While Subjected to Constant Resisting, Dry Friction, and Constant Active Forces

The masses m_1 and m_2 of the restricted system are moving on a horizontal surface and are, respectively, subjected to the actions of the constant resisting forces R_1 and R_2, the dry friction forces F_1 and F_2, and the constant active forces P_1 and P_2.

The schematic diagram of the system is shown in Figure 23.2.1, the notations of which are self-explanatory.

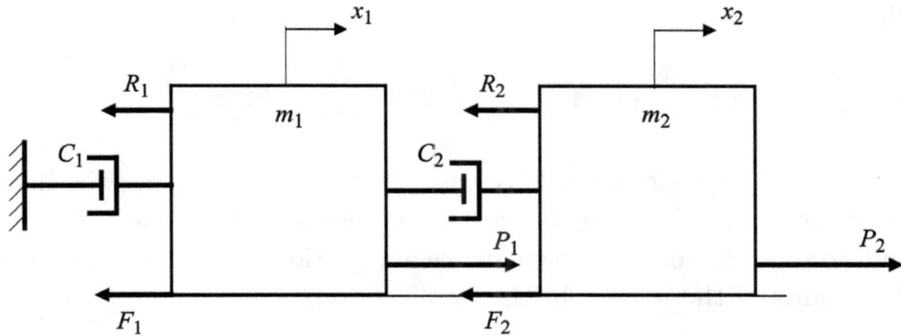

Fig. 23.2.1. Schematic diagram of a restricted system with two sequences of fluid links moving on a horizontal surface while being subjected to constant resisting, dry friction, and constant active forces.

Based on the schematic diagram shown in Figure 23.2.1 and the above-mentioned considerations, we compose the pair of simultaneous differential equations describing the motion of the system:

$$m_1 \frac{d^2 x_1}{dt^2} + (C_1 + C_2)\left(\frac{dx_1}{dt} - \frac{dx_2}{dt}\right) + R_1 + F_1 = P_1 \quad (23.2.1)$$

$$m_2 \frac{d^2 x_2}{dt^2} + (C_1 + C_2)\left(\frac{dx_2}{dt} - \frac{dx_1}{dt}\right) + R_2 + F_2 = P_2 \quad (23.2.2)$$

The initial conditions of motion are taken according to expression (22.1.3).

Dividing equations (23.2.1) and (23.2.2) by m_1 and m_2, respectively, while applying to them Laplace Transform pairs 2, 5, 2, 4, 2, 2, and 2, we convert these equations with their initial conditions of motion from the time domain into a system of two simultaneous algebraic equations with two unknowns in each equation in the Laplace domain:

$$s^2 x_1(s) - sV_1 - s^2 S_1 + 2n_{11}sx_1(s) - 2n_{11}sS_1 - 2n_{11}sx_2(s)$$

$$+ 2n_{11}sS_2 + 2n_{21}sx_1(s) - 2n_{21}sS_1$$

$$- 2n_{21}sx_2(s) + 2n_{21}sS_2 + r_1 + f_1 = p_1 \quad (23.2.3)$$

$$s^2 x_2(s) - sV_2 - s^2 S_2 + 2n_{12}sx_2(s) - 2n_{12}sS_2 - 2n_{12}sx_1(s)$$

$$+ 2n_{12}sS_1 + 2n_{22}sx_2(s) - 2n_{22}sS_2$$

$$- 2n_{22}sx_1(s) + 2n_{22}sS_1 + r_2 + f_2 = p_2 \quad (23.2.4)$$

where

$$r_1 = \frac{R_1}{m_1}; \quad r_2 = \frac{R_2}{m_2}; \quad f_1 = \frac{F_1}{m_1}; \quad f_2 = \frac{F_2}{m_2}; \quad p_1 = \frac{P_1}{m_1}; \quad p_2 = \frac{P_2}{m_2}$$

$$(23.2.5)$$

Rearranging equations (23.2.3) and (23.2.4) and applying to them the method of substitutions, we obtain two simultaneous algebraic equations with one unknown in each equation describing the displacements of the masses in the Laplace domain:

$$x_1(s) = \frac{V_1 + 2nS_1}{s + 2n}$$

$$+ \frac{2[V_1(n_{12} + n_{22}) + V_2(n_{11} + n_{21})] + p_1 - r_1 - f_1}{s(s + 2n)} + \frac{sS_1}{s + 2n}$$

$$+ \frac{2[(n_{12} + n_{22})(p_1 - r_1 - f_1) + (n_{11} + n_{21})(p_2 - r_2 - f_2)]}{s^2(s + 2n)}$$

$$(23.2.6)$$

$$x_2(s) = \frac{V_2 + 2nS_2}{s + 2n}$$

$$+ \frac{2[V_2(n_{11} + n_{21}) + V_1(n_{12} + n_{22})] + p_2 - r_2 - f_2}{s(s + 2n)} + \frac{sS_2}{s + 2n}$$

$$+ \frac{2[(n_{11} + n_{21})(p_2 - r_2 - f_2) + (n_{12} + n_{22})(p_1 - r_1 - f_1)]}{s^2(s + 2n)}$$

$$(23.2.7)$$

Applying to equations (23.2.6) and (23.2.7) Laplace Transform pairs 8, 16, 14, and 48, we invert them from the Laplace domain into the time domain and obtain the solutions of differential equations (23.2.1) and (23.2.2) with initial conditions of motion (22.1.3):

$$x_1 = \frac{V_1 + 2nS_1}{2n}(1 - e^{-2nt})$$

$$+ \frac{2[V_1(n_{12} + n_{22}) + V_2(n_{11} + n_{21})] + p_1 - r_1 - f_1}{2n}$$

$$\times \left[t + \frac{1}{2n}(e^{-2nt} - 1)\right] + S_1 e^{-2nt}$$

$$+ \frac{(n_{12} + n_{22})(p_1 - r_1 - f_1) + (n_{11} + n_{21})(p_2 - r_2 - f_2)}{2n}$$

$$\times \left[t^2 - \frac{t}{n} - \frac{1}{2n^2}(e^{-2nt} - 1) \right] \qquad (23.2.8)$$

$$x_2 = \frac{V_2 + 2nS_2}{2n}(1 - e^{-2nt})$$

$$+ \frac{2[V_2(n_{11} + n_{21}) + V_1(n_{12} + n_{22})] + p_2 - r_2 - f_2}{2n}$$

$$\times \left[t + \frac{1}{2n}(e^{-2nt} - 1) \right] + S_2 e^{-2nt}$$

$$+ \frac{(n_{11} + n_{21})(p_2 - r_2 - f_2) + (n_{12} + n_{22})(p_1 - r_1 - f_1)}{2n}$$

$$\times \left[t^2 - \frac{t}{n} - \frac{1}{2n^2}(e^{-2nt} - 1) \right] \qquad (23.2.9)$$

Assuming in equations (23.2.8) and (23.2.9) that $t = 0$, we determine that $x_1 = S_1$ and $x_2 = S_2$, as expected according to initial conditions of motion (22.1.3). Taking from equations (23.2.8) and (23.2.9) the first derivatives, we, respectively, obtain the equations of the velocities of the masses:

$$\frac{dx_1}{dt} = (V_1 + 2nS_1)e^{-2nt}$$

$$+ \frac{2[V_1(n_{12} + n_{22}) + V_2(n_{11} + n_{21})] + p_1 - r_1 - f_1}{2n}$$

$$\times (1 - e^{-2nt}) - 2nS_1 e^{-2nt}$$

$$+ \frac{(n_{12} + n_{22})(p_1 - r_1 - f_1) + (n_{11} + n_{21})(p_2 - r_2 - f_2)}{2n}$$

$$\times \left(2t - \frac{1}{n} + \frac{1}{n}e^{-2nt} \right) \qquad (23.2.10)$$

$$\frac{dx_2}{dt} = (V_2 + 2nS_2)e^{-2nt}$$

$$+ \frac{2[V_2(n_{11} + n_{21}) + V_1(n_{12} + n_{22})] + p_2 - r_2 - f_2}{2n}$$

$$\times (1 - e^{-2nt}) - 2nS_2 e^{-2nt}$$

$$+ \frac{(n_{11} + n_{21})(p_2 - r_2 - f_2) + (n_{12} + n_{22})(p_1 - r_1 - f_1)}{2n}$$

$$\times \left(2t - \frac{1}{n} + \frac{1}{n} e^{-2nt} \right) \tag{23.2.11}$$

Supposing for the equations (23.2.10) and (23.2.11) that $t = 0$, we determine that $\frac{dx_1}{dt} = V_1$ and $\frac{dx_2}{dt} = V_2$, as it should be according to the initial conditions of motion.

Taking the first derivatives from equations (23.2.10) and (23.2.11), we determine the respective accelerations of the masses:

$$\frac{d^2 x_1}{dt^2} = -2n(V_1 + 2nS_1)e^{-2nt} + \{2[V_1(n_{12} + n_{22}) + V_2(n_{11} + n_{21})]$$

$$+ p_1 - r_1 - f_1\}e^{-2nt} + 4n^2 S_1 e^{-2nt} + \frac{1}{n}[(n_{12} + n_{22})$$

$$\times (p_1 - r_1 - f_1) + (n_{11} + n_{21})(p_2 - r_2 - f_2)](1 - e^{-2nt})$$

$$\tag{23.2.12}$$

$$\frac{d^2 x_2}{dt^2} = -2n(V_2 + 2nS_2)e^{-2nt} + \{2[V_1(n_{12} + n_{22}) + V_2(n_{11} + n_{21})]$$

$$+ p_2 - r_2 - f_2\}e^{-2nt} + 4n^2 S_2 e^{-2nt} + \frac{1}{n}[(n_{11} + n_{21})$$

$$\times (p_2 - r_2 - f_2) + (n_{12} + n_{22})(p_1 - r_1 - f_1)](1 - e^{-2nt})$$

$$\tag{23.2.13}$$

Hence, equations (23.2.8)–(23.2.13) describe the basic parameters of motion of the system.

Taking in equations (23.2.12) and (23.2.13) that $t = 0$, we obtain the respective values of the accelerations of the masses at the beginning of the process of motion:

$$\frac{d^2 x_1}{dt^2} = -2(V_1 - V_2)(n_{11} + n_{21}) + p_1 - r_1 - f_1 \tag{23.2.14}$$

$$\frac{d^2 x_2}{dt^2} = -2(V_2 - V_1)(n_{12} + n_{22}) + p_2 - r_2 - f_2 \tag{23.2.15}$$

23.3. Motion of a Restricted System with Two Sequences of Fluid Links While Subjected to Harmonic Forces

The system is moving on a horizontal surface and subjected to harmonic forces.

Figure 23.3.1 shows a schematic diagram of the restricted system, the masses of which are connected to each other by a fluid link (dashpot), while one of the masses is attached to a non-movable support by another dashpot. The system is subjected to the actions of harmonic forces. The notations in this figure are self-explanatory.

Based on the schematic diagram shown in Figure 23.3.1 and the related considerations, we compose the pair of simultaneous differential equations describing the motion of the masses of this system:

$$m_1 \frac{d^2 x_1}{dt^2} + (C_1 + C_2)\left(\frac{dx_1}{dt} - \frac{dx_2}{dt}\right) = A_1 \cos \varphi_1 t \quad (23.3.1)$$

$$m_2 \frac{d^2 x_2}{dt^2} + (C_1 + C_2)\left(\frac{dx_2}{dt} - \frac{dx_1}{dt}\right) = A_2 \cos \varphi_2 t \quad (23.3.2)$$

The initial conditions of motion are taken according to expression (22.1.3).

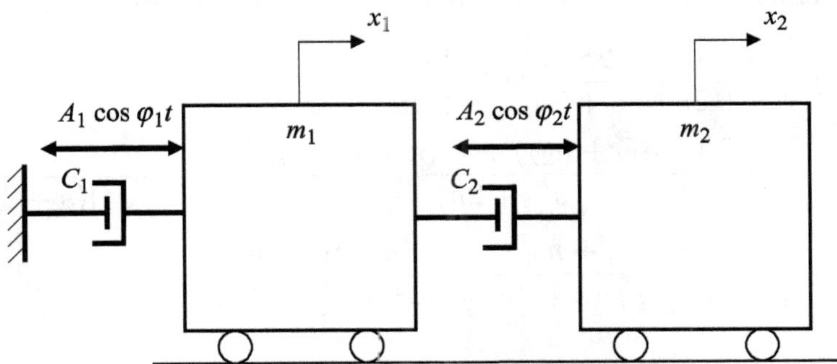

Fig. 23.3.1. Schematic diagram of a restricted system with two sequences of fluid links moving on a horizontal surface while being subjected to harmonic forces.

Dividing equations (23.3.1) and (23.3.2) by m_1 and m_2, respectively, while applying to them Laplace Transform pairs 2, 5, 2, 4, 2, and 29, we convert these equations with their initial conditions of motion from the time domain into the Laplace domain and obtain a corresponding system of two simultaneous algebraic equations with two unknowns in each equation in the Laplace domain:

$$s^2 x_1(s) - sV_1 - s^2 S_1 + 2n_{11}sx_1(s) - 2n_{11}sS_1$$

$$- 2n_{11}x_2(s) + 2n_{11}sS_2 + 2n_{21}sx_1(s) - 2n_{21}sS_1 - 2n_{21}x_2(s)$$

$$+ 2n_{21}sS_2 = \frac{a_1 s^2}{s^2 + \varphi_1^2} \tag{23.3.3}$$

$$s^2 x_2(s) - sV_2 - s^2 S_2 + 2n_{12}sx_2(s) - 2n_{12}sS_2$$

$$- 2n_{12}x_1(s) + 2n_{12}sS_1 + 2n_{22}sx_2(s) - 2n_{22}sS_2 - 2n_{22}x_1(s)$$

$$+ 2n_{22}sS_1 = \frac{a_2 s^2}{s^2 + \varphi_2^2} \tag{23.3.4}$$

where

$$a_1 = \frac{A_1}{m_1} \quad \text{and} \quad a_2 = \frac{A_2}{m_2} \tag{23.3.5}$$

Rearranging equations (23.3.3) and (23.3.4) and applying to them the method of substitutions, we obtain two simultaneous algebraic equations with one unknown in each equation describing, respectively, the displacements of the two masses in the Laplace domain:

$$x_1(s) = \frac{sS_1}{s + 2n} + \frac{2S_1 n}{s + 2n} + \frac{V_1}{s + 2n}$$

$$+ \frac{2[V_1(n_{12} + n_{22}) + V_2(n_{11} + n_{21})]}{s(s + 2n)} + \frac{a_1 s}{(s^2 + \varphi_1^2)(s + 2n)}$$

$$+ \frac{2a_1(n_{12} + n_{22})}{(s^2 + \varphi_1^2)(s + 2n)} + \frac{2a_2(n_{11} + n_{21})}{(s^2 + \varphi_2^2)(s + 2n)} \tag{23.3.6}$$

$$x_2(s) = \frac{sS_2}{s+2n} + \frac{2S_2 n}{s+2n} + \frac{V_2}{s+2n}$$

$$+ \frac{2[V_2(n_{11}+n_{21}) + V_1(n_{12}+n_{22})]}{s(s+2n)} + \frac{a_2}{(s^2+\varphi_1^2)(s+2n)}$$

$$+ \frac{2_1 a_2(n_{11}+n_{21})}{(s^2+\varphi_2^2)(s+2n)} + \frac{2a_1(n_{12}+n_{22})}{(s^2+\varphi_1^2)(s+2n)} \tag{23.3.7}$$

Applying to equations (23.3.6) and (23.3.7) Laplace Transform pairs 14, 8, 8, 16, 44, 42, and 42, we invert these equations from the Laplace domain into the time domain and obtain the solutions of differential equations (23.3.1) and (23.3.2) with the initial conditions of motion according to expression (22.1.3):

$$x_1 = S_1 + \frac{V_1}{2n}(1 - e^{-2nt}) + \frac{V_1(n_{12}+n_{22}) + V_2(n_{11}+n_{21})}{n}$$

$$\times \left[t + \frac{1}{2n}(1 - e^{-2nt}) \right] + \frac{a_1}{\varphi_1^2 + 4n^2}$$

$$\times \left(e^{-2nt} + \frac{2n}{\varphi_1} \sin \varphi_1 t - \cos \varphi_1 t \right) + \frac{2a_1(n_{12}+n_{22})}{\varphi_1^2 + 4n^2}$$

$$\times \left[\frac{2n}{\varphi_1^2}(1 - \cos \varphi_1 t) - \frac{1}{\varphi_1} \sin \varphi_1 t + \frac{1}{2n}(1 - e^{-2nt}) \right]$$

$$+ \frac{2a_2(n_{11}+n_{21})}{\varphi_2^2 + 4n^2} \left[\frac{2n}{\varphi_2^2}(1 - \cos \varphi_2 t) \right.$$

$$\left. - \frac{1}{\varphi_2} \sin \varphi_2 t + \frac{1}{2n}(1 - e^{-2nt}) \right] \tag{23.3.8}$$

$$x_2 = S_2 + \frac{V_1}{2n}(1 - e^{-2nt}) + \frac{V_2(n_{12}+n_{22}) + V_1(n_{11}+n_{21})}{n}$$

$$\times \left[t + \frac{1}{2n}(1 - e^{-2nt}) \right] + \frac{a_2}{\varphi_2^2 + 4n^2}$$

$$\times \left(e^{-2nt} + \frac{2n}{\varphi_2} \sin \varphi_2 t - \cos \varphi_2 t \right) + \frac{2a_2(n_{11}+n_{21})}{\varphi_2^2 + 4n^2}$$

$$\times \left[\frac{2n}{\varphi_2^2}(1 - \cos \varphi_2 t) - \frac{1}{\varphi_2} \sin \varphi_2 t + \frac{1}{2n}(1 - e^{-2nt}) \right]$$

$$+ \frac{2a_1(n_{12} + n_{22})}{\varphi_1^2 + 4n^2} \left[\frac{2n}{\varphi_1^2}(1 - \cos \varphi_1 t) \right.$$

$$\left. - \frac{1}{\varphi_1} \sin \varphi_1 t + \frac{1}{2n}(1 - e^{-2nt}) \right] \tag{23.3.9}$$

In equations (23.3.8) and (23.3.9), equating the running time to zero, we obtain that $x_1 = S_1$ and $x_2 = S_2$, which conforms with the initial conditions of motion according to (22.1.3). Taking from these equations the first derivatives, we determine the velocities of the first and second masses, respectively:

$$\frac{dx_1}{dt} = V_1 e^{-2nt} + \frac{1}{n}[V_1(n_{12} + n_{22}) + V_2(n_{11} + n_{21})](1 - e^{-2nt})$$

$$+ \frac{a_1}{\varphi_1^2 + 4n^2}(\varphi_1 \sin \varphi_1 t + 2n \cos \varphi_1 t - 2ne^{-2nt})$$

$$+ \frac{2a_1(n_{12} + n_{22})}{\varphi_1^2 + 4n^2} \left(\frac{2n}{\varphi_1} \sin \varphi_1 t - \cos \varphi_1 t + e^{-2nt} \right)$$

$$+ \frac{2a_2(n_{11} + n_{21})}{\varphi_2^2 + 4n^2} \left(\frac{2n}{\varphi_2} \sin \varphi_2 t - \cos \varphi_2 t + e^{-2nt} \right)$$

$$\tag{23.3.10}$$

$$\frac{dx_2}{dt} = V_2 e^{-2nt} + \frac{1}{n}[V_2(n_{11} + n_{21}) + V_1(n_{12} + n_{22})](1 - e^{-2nt})$$

$$+ \frac{a_2}{\varphi_2^2 + 4n^2}(\varphi_2 \sin \varphi_2 t + 2n \cos \varphi_2 t - 2ne^{-2nt})$$

$$+ \frac{2a_2(n_{11} + n_{21})}{\varphi_2^2 + 4n^2} \left(\frac{2n}{\varphi_2} \sin \varphi_2 t - \cos \varphi_2 t + e^{-2nt} \right)$$

$$+ \frac{2a_1(n_{12} + n_{22})}{\varphi_1^2 + 4n^2} \left(\frac{2n}{\varphi_1} \sin \varphi_1 t - \cos \varphi_1 t + e^{-2nt} \right)$$

$$\tag{23.3.11}$$

Taking in equations (23.3.10) and (23.3.11) that $t = 0$, we obtain that $\frac{dx_1}{dt} = V_1$ and $\frac{dx_2}{dt} = V_2$, as it is supposed to be according to the initial conditions of motion.

The first derivatives from equations (23.3.10) and (23.3.11) yield the accelerations of the masses:

$$\frac{d^2x_1}{dt^2} = -2nV_1 e^{-2nt} + 2[V_1(n_{12} + n_{22}) + V_2(n_{11} + n_{21})]e^{-2nt}$$

$$+ \frac{a_1}{\varphi_1^2 + 4n^2}(\varphi_1^2 \cos \varphi_1 t - 2n\varphi_1 \sin \varphi_1 t + 4n^2 e^{-2nt})$$

$$+ \frac{2a_1(n_{12} + n_{22})}{\varphi_1^2 + 4n^2}(2n \cos \varphi_1 t + \varphi_1 \sin \varphi_1 t - 2n e^{-2nt})$$

$$+ \frac{2a_2(n_{11} + n_{21})}{\varphi_2^2 + 4n^2}(2n \cos \varphi_2 t + \varphi_2 \sin \varphi_2 t - 2n e^{-2nt})$$

$$(23.3.12)$$

$$\frac{d^2x_2}{dt^2} = -2nV_2 e^{-2nt} + 2[V_2(n_{11} + n_{21}) + V_1(n_{12} + n_{22})]e^{-2nt}$$

$$+ \frac{a_2}{\varphi_2^2 + 4n^2}(\varphi_2^2 \cos \varphi_2 t - 2n\varphi_2 \sin \varphi_2 t + 4n^2 e^{-2nt})$$

$$+ \frac{2a_2(n_{11} + n_{21})}{\varphi_2^2 + 4n^2}(2n \cos \varphi_2 t + \varphi_2 \sin \varphi_2 t - 2n e^{-2nt})$$

$$+ \frac{2a_1(n_{12} + n_{22})}{\varphi_1^2 + 4n^2}(2n \cos \varphi_1 t + \varphi_1 \sin \varphi_1 t - 2n e^{-2nt})$$

$$(23.3.13)$$

Hence, equations (23.3.8)–(23.3.13) represent the basic parameters of motion of the considered system.

23.3.1. *Numerical solution*

The following is a Python program to plot the graph (Figure 23.3.2) of equation (23.3.8).

```python
from matplotlib.pyplot import plot, show
from numpy import linspace, sin, cos, exp

S1 = 0.1
V1 = 0.15
V2 = 0.25
n12 = 3
n22 = 4
n11 = 2
n21 = 5
n = n12 + n22 + n11 + n21
phi1 = 20
phi2 = 15
a1 = 15
a2 = 25

t = linspace(0, 4, 1000)
x1  = S1 + (V1/(2*n))*(1 - exp(-2*n*t)) \
    + ((V1*(n12 + n22) + V2*(n11 + n21))/n) \
   * (t + (1/(2*n))*(1 - exp(-2*n*t))) \
    + (a1/(phi1**2 + 4*n**2))*(exp(-2*n*t) \
                    + (2*n/phi1)*(sin(phi1*t) \
                    - cos(phi1*t))) \
    + ((2*a1*(n12 + n22))/(phi1**2 + 4*n**2)) \
     *((2*n/phi1**2)*(1 - cos(phi1*t)) \
      - (1/phi1)*sin(phi1*t) \
       + (1/(2*n))*(1 - exp(-2*n*t))) \
    + ((2*a2*(n11 + n21))/(phi2**2 + 4*n**2)) \
     *((2*n/phi2**2)*(1 - cos(phi2*t)) \
      - (1/phi2**2)*sin(phi2*t) \
       + (1/(2*n))*(1 - exp(-2*n*t)))

plot(t, x1)
show()
```

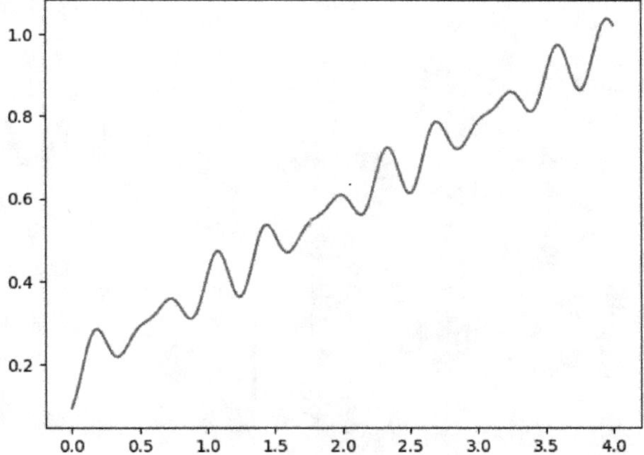

Fig. 23.3.2. Graph of equation (23.3.8).

RESTRICTED SYSTEMS WITH MASSES CONNECTED TO EACH OTHER AND TO A NON-MOVABLE SUPPORT BY TWO SEQUENCES OF FLEXIBLE AND FLUID LINKS IN PARALLEL

In this chapter, we present the analyses of the operational process of a restricted system, the masses of which are connected to each other and to a non-movable support by combined, parallel flexible and fluid links, while the motion of the masses is caused by their initial velocities.

The notations of the parameters of motion are applicable only to this chapter.

24.1. Motion of a Restricted System Due to Initial Velocities of Its Masses That Are Attached to Each Other and to a Non-movable Support by Springs and Dashpots in Parallel

Consider the operational process of a system, the two masses of which are moving on a horizontal frictionless surface due to their initial velocities.

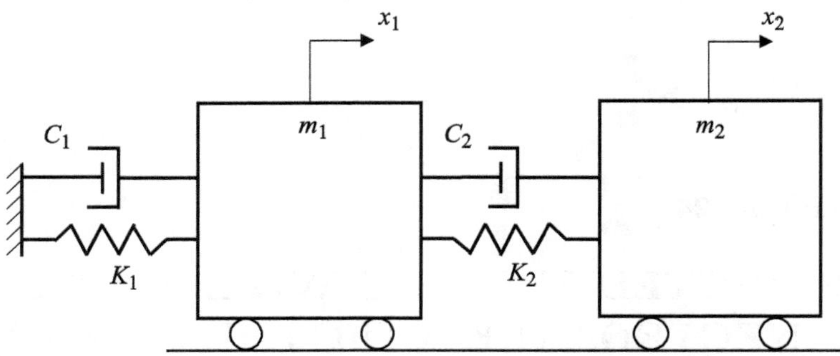

Fig. 24.1.1. Schematic diagram of a restricted system, the two masses of which are connected to each other by a dashpot and a spring in parallel, while one of the masses is connected to a non-movable support by an identical pair of links.

Figure 24.1.1 shows a schematic diagram of the system where C_1 and C_2 are the damping coefficients of the dashpots, while K_1 and K_2 are the stiffness coefficients of the springs.

Based on the schematic diagram in Figure 24.1.1 and the related considerations, we compose a pair of simultaneous differential equations of motion describing the operational process of the system:

$$m_1 \frac{d^2 x_1}{dt^2} + (C_1 + C_2)\left(\frac{dx_1}{dt} - \frac{dx_2}{dt}\right) + (K_1 + K_2)(x_1 - x_2) = 0$$

$$(24.1.1)$$

$$m_2 \frac{d^2 x_2}{dt^2} + (C_1 + C_2)\left(\frac{dx_2}{dt} - \frac{dx_1}{dt}\right) + (K_1 + K_2)(x_2 - x_1) = 0$$

$$(24.1.2)$$

The initial conditions of motion are

$$x_1 = 0; \quad \frac{dx_1}{dt} = V_1$$

For $t = 0$ $(24.1.3)$

$$x_2 = 0; \quad \frac{dx_2}{dt} = V_2$$

Dividing, respectively, equations (24.1.1) and (24.1.2) by m_1 and m_2, while applying to them Laplace Transform pairs 5, 2, 4, 2, and 1, we convert these equations with their initial conditions of motion from the time domain into a system of two simultaneous algebraic equations with two unknowns in each equation in the Laplace domain:

$$s^2 x_1(s) - sV_1 + 2n_{11}sx_1(s) - 2n_{11}sx_2(s) + 2n_{21}sx_1(s) - 2n_{21}sx_2(s)$$

$$+ \omega_{11}^2 x_1(s) - \omega_{11}^2 x_2(s) + \omega_{21}^2 x_1(s) - \omega_{21}^2 x_2(s) = 0 \tag{24.1.4}$$

$$s^2 x_2(s) - sV_2 + 2n_{12}sx_2(s) - 2n_{12}sx_1(s) + 2n_{22}sx_2(s) - 2n_{22}sx_1(s)$$

$$+ \omega_{12}^2 x_2(s) - \omega_{12}^2 x_1(s) + \omega_{22}^2 x_2(s) - \omega_{22}^2 x_1(s) = 0 \tag{24.1.5}$$

where

$$2n_{11} = \frac{C_1}{m_1}; \quad 2n_{21} = \frac{C_2}{m_1}; \quad 2n_{12} = \frac{C_1}{m_2}; \quad 2n_{22} = \frac{C_2}{m_2} \tag{24.1.6}$$

and

$$\omega_{11}^2 = \frac{K_1}{m_1}; \quad \omega_{21}^2 = \frac{K_2}{m_1}; \quad \omega_{12}^2 = \frac{K_1}{m_2}; \quad \omega_{22}^2 = \frac{K_2}{m_2} \tag{24.1.7}$$

Rearranging the terms in equations (24.1.4) and (24.1.5) and applying to them the method of substitutions, we obtain two simultaneous algebraic equations with two unknowns in each equation that describe in the Laplace domain the displacements of the first and second masses, respectively:

$$x_1(s) = \frac{V_1 s}{s^2 + 2ns + \omega_0^2} + \frac{2[V_1(n_{12} + n_{22}) + V_2(n_{11} + n_{21})]}{s^2 + 2ns + \omega_0^2}$$

$$+ \frac{V_1(\omega_{12}^2 + \omega_{22}^2) + V_2(\omega_{11}^2 + \omega_{21}^2)}{s(s^2 + 2ns + \omega_0^2)} \tag{24.1.8}$$

$$x_2(s) = \frac{V_2 s}{s^2 + 2ns + \omega_0^2} + \frac{2[V_2(n_{11} + n_{21}) + V_1(n_{12} + n_{22})]}{s^2 + 2ns + \omega_0^2}$$

$$+ \frac{V_2(\omega_{11}^2 + \omega_{21}^2) + V_1(\omega_{12}^2 + \omega_{22}^2)}{s(s^2 + 2ns + \omega_0^2)} \tag{24.1.9}$$

where

$$n = n_{11} + n_{21} + n_{12} + n_{22} \tag{24.1.10}$$

and

$$\omega_0^2 = \omega_{11}^2 - \omega_{21}^2 + \omega_{12}^2 + \omega_{22}^2 \tag{24.1.11}$$

In order to determine the characteristics of the damped motion of the masses according to equations (24.1.8) and (24.1.9), we should rearrange the denominators of the fractions of these equations in the following way:

$$s^2 + 2ns + \omega_0^2 = s^2 + 2ns + n^2 - n^2 + \omega_0^2 = s^2 + 2ns + n^2 + \omega^2$$

$$= (s + n)^2 + \omega^2 \tag{24.1.12}$$

where

$$\omega^2 = \omega_0^2 - n^2 \tag{24.1.13}$$

In order to continue solving these equations, we need to determine the value of the parameter ω^2 for each case. Hence, in the case when $\omega^2 > 0$, the masses perform underdamped vibrational motions; however, when $\omega^2 = 0$, then we have the case of critical damping, and the masses are in a state of non-vibrational motion; and finally, in the case when $\omega^2 < 0$, we have over-damped non-vibrational motion. These three cases are considered as follows.

24.1.1. *Underdamped vibrational motion*

Assuming that $\omega^2 > 0$, we continue the solutions by combining equations (24.1.12) and (24.1.13) with equations (24.1.8) and (24.1.9), and as a result, we obtain

$$x_1(s) = \frac{V_1 s}{(s + n)^2 + \omega^2} + \frac{2[V_1(n_{12} + n_{22}) + V_2(n_{11} + n_{21})]}{(s + n)^2 + \omega^2}$$

$$+ \frac{V_1(\omega_{12}^2 + \omega_{22}^2) + V_2(\omega_{11}^2 + \omega_{21}^2)}{s[(s + n)^2 + \omega^2]} \tag{24.1.14}$$

$$x_2(s) = \frac{V_2 s}{(s + n)^2 + \omega^2} + \frac{2[V_2(n_{11} + n_{21}) + V_1(n_{12} + n_{22})]}{(s + n)^2 + \omega^2}$$

$$+ \frac{V_2(\omega_{11}^2 + \omega_{21}^2) + V_1(\omega_{12}^2 + \omega_{22}^2)}{s[((s + n)^2 + \omega^2]} \tag{24.1.15}$$

Applying to equations (24.1.14) and (24.1.15) Laplace Transform pairs 27, 21, and 45, we invert these equations from the Laplace domain into the time domain and obtain the solutions of differential equations (24.1.1) and (24.1.2) with initial conditions of motion (24.1.3):

$$x_1 = \frac{V_1}{\omega} e^{-nt} \sin \omega t + \frac{2[V_1(n_{12} + n_{22}) + V_2(n_{11} + n_{21})]}{\omega^2 + n^2}$$

$$\times \left[1 - e^{-nt} \left(\cos \omega t + \frac{n}{\omega} \sin \omega t \right) \right.$$

$$+ \frac{V_1(\omega_{12}^2 + \omega_{22}^2) + V_2(\omega_{11}^2 + \omega_{21}^2)}{\omega(\omega^2 + n^2)^2}$$

$$\times \{(\omega^2 + n^2)\omega t - 2n\omega - e^{-nt}[(\omega^2 - n^2)\sin \omega t - 2n\omega \cos \omega t]\}$$

$$(24.1.16)$$

$$x_2 = \frac{V_2}{\omega} e^{-nt} \sin \omega t + \frac{2[V_2(n_{11} + n_{21}) + V_1(n_{12} + n_{22})]}{\omega^2 + n^2}$$

$$\times \left[1 - e^{-nt} \left(\cos \omega t + \frac{n}{\omega} \sin \omega t \right) \right]$$

$$+ \frac{V_2(\omega_{11}^2 + \omega_{21}^2) + V_1(\omega_{12}^2 + \omega_{22}^2)}{\omega(\omega^2 + n^2)^2}$$

$$\times \{(\omega^2 + n^2)\omega t - 2n\omega - e^{-nt}[(\omega^2 - n^2)\sin \omega t - 2n\omega \cos \omega t]\}$$

$$(24.1.17)$$

Assuming that in equations (24.1.16) and (24.1.17), $t = 0$, we will have $x_1 = 0$ and $x_2 = 0$, as it should be according to the initial conditions of motion. Taking from these equations the first derivatives, we determine the velocities of the two masses, respectively:

$$\frac{dx_1}{dt} = V_1 e^{-nt} \left(\cos \omega t - \frac{n}{\omega} \sin \omega t \right)$$

$$+ \frac{2[V_1(n_{12} + n_{22}) + V_2(n_{11} + n_{21})]}{\omega} e^{-nt} \sin \omega t$$

$$+ \frac{V_1(\omega_{12}^2 + \omega_{22}^2) + V_2(\omega_{11}^2 + \omega_{21}^2)}{\omega^2 + n^2}$$

$$\times \left[1 - e^{-nt} \left(\cos \omega t + \frac{n}{\omega} \sin \omega t \right) \right] \qquad (24.1.18)$$

$$\frac{dx_2}{dt} = V_2 e^{-nt}\left(\cos\omega t - \frac{n}{\omega}\sin\omega t\right)$$
$$+ \frac{2[V_2(n_{11} + n_{21}) + V_1(n_{12} + n_{22})]}{\phi_v} e^{-nt}\sin\omega t$$
$$+ \frac{V_2(\omega_{11}^2 + \omega_{21}^2) + V_1(\omega_{12}^2 + \omega_{22}^2)}{\omega^2 + n^2}$$
$$\times \left[1 - e^{-nt}\left(\cos\omega t + \frac{n}{\omega}\sin\omega t\right)\right] \tag{24.1.19}$$

Taking in equations (24.1.18) and (24.1.19) that $t = 0$, we obtain $\frac{dx_1}{dt} = V_1$ and $\frac{dx_2}{dt} = V_2$, as expected according to the initial conditions of motion.

24.1.2. *Critically damped motion*

As mentioned above, critical damping occurs when $\omega^2 = 0$. Therefore, we apply this condition to equations (24.1.14) and (24.1.15), and we obtain

$$x_1(s) = \frac{V_1 s}{(s+n)^2} + \frac{2[V_1(n_{12} + n_{22}) + V_2(n_{11} + n_{21})]}{(s+n)^2}$$
$$+ \frac{V_1(\omega_{12}^2 + \omega_{22}^2) + V_2(\omega_{11}^2 + \omega_{21}^2)}{s(s+n)^2} \tag{24.1.20}$$

$$x_2(s) = \frac{V_2 s}{(s+n)^2} + \frac{2[V_2(n_{11} + n_{21}) + V_1(n_{12} + n_{22})]}{(s+n)^2}$$
$$+ \frac{V_2(\omega_{11}^2 + \omega_{21}^2) + V_1(V_{12}^2 + V_{22}^2)}{s(s+n)^2} \tag{24.1.21}$$

Applying to equations (24.1.20) and (24.1.21) Laplace Transform pairs 25, 19, and 37, we invert these equations from the Laplace domain into the time domain and obtain the solutions of differential equations (24.1.1) and (24.1.2) with the initial conditions of motion (24.1.3) for the case of critical damping:

$$x_1 = V_1 t e^{-nt} + \frac{2[V_1(n_{12} + n_{22}) + V_2(n_{11} + n_{21})]}{n^2}[1 - e^{-nt}(1 + nt)]$$
$$+ \frac{V_1(\omega_{12}^2 + \omega_{22}^2) + V_2(\omega_{11}^2 + \omega_{21}^2)}{n^2}\left[t - \frac{2}{n} + e^{-nt}\left(\frac{2}{n} + t\right)\right] \tag{24.1.22}$$

$$x_2 = v_1 t e^{-nt} + \frac{2[V_2(n_{11} + n_{21}) + V_1(n_{12} + n_{22})]}{n^2}[1 - e^{-nt}(1 + nt)]$$

$$+ \frac{V_2(\omega_{11}^2 + \omega_{21}^2) + V_1(\omega_{12}^2 + \omega_{22}^2)}{n^2}\left[t - \frac{2}{n} + e^{-nt}\left(\frac{2}{n} + t\right)\right]$$

$$(24.1.23)$$

24.1.3. *Overdamped motion*

In the case of overdamped motion, we have $\omega^2 < 0$. According to this condition, we replace in equations (24.1.14) and (24.1.15) the term ω^2 with its negative value $-\omega^2$, and we may write

$$x_1(s) = \frac{V_1 s}{(s+n)^2 - \omega^2} + \frac{2[V_1(n_{12} + n_{22}) + V_2(n_{11} + n_{21})}{(s+n)^2 - \omega^2}$$

$$+ \frac{V_1(\omega_{12}^2 + \omega_{22}^2) + V_2(\omega_{11}^2 + \omega_{21}^2)}{s[(s+n)^2 - \omega^2]} \qquad (24.1.24)$$

$$x_2(s) = \frac{V_2 s}{(s+n)^2 - \omega^2} + \frac{2[V_2(n_{11} + n_{21}) + V_1(n_{12} + n_{22})]}{(s+n)^2 - \omega^2}$$

$$+ \frac{V_2(\omega_{11}^2 + \omega_{21}^2) + V_1(\omega_{12}^2 + \omega_{22}^2)}{s[(s+n)^2 - \omega^2]} \qquad (24.1.25)$$

Applying to equations (24.1.24) and (24.1.25) Laplace Transform pairs 28, 22, and 46, we invert these equations from the Laplace domain into the time domain and obtain the solutions of differential equations (24.1.1) and (24.1.2) with initial conditions of motion (24.1.3) for the case of overdamped motion:

$$x_1 = \frac{V_1}{\omega} e^{-nt} \sinh \omega t + \frac{2[V_1(n_{12} + n_{22}) + V_2(n_{11} + n_{21})]}{n^2 - \omega^2}$$

$$\times \left[1 - e^{-nt}\left(\cosh \omega t + \frac{n}{\omega} \sinh \omega t\right)\right]$$

$$+ \frac{V_1(\omega_{12}^2 + \omega_{22}^2) + V_2(\omega_{11}^2 + \omega_{21}^2)}{\omega(n^2 - \omega^2)^2}\{(n^2 - \omega^2)\omega t - 2n\omega + e^{-nt}$$

$$\times [(\omega^2 + n^2)\sinh \omega t + 2n\omega \cosh \omega t]\} \qquad (24.1.26)$$

$$x_2 = \frac{V_2}{\omega}e^{-nt}\sinh\omega t + \frac{2[V_2(n_{11}+n_{21})+V_1(n_{12}+n_{22})]}{n^2-\omega^2}$$

$$\times\left[1-e^{-nt}\left(\cosh\omega t + \frac{n}{\omega}\sinh\omega t\right)\right]$$

$$+\frac{V_2(\omega_{11}^2+\omega_{21}^2)+V_1(\omega_{12}^2+\omega_{22}^2)}{\omega(n^2-\omega^2)^2}\{(n^2-\omega^2)\omega t - 2n\omega + e^{-nt}$$

$$\times[(\omega^2+n^2)\sinh\omega t + 2n\omega\cosh\omega t]\}\qquad(24.1.27)$$

24.2. Motion of a Restricted System Due to Constant Active Forces Applied to the Masses Which Are Attached to Each Other and to a Non-movable Support by Two Sequences of Springs and Dashpots

The system is moving on a horizontal frictionless surface and subjected to constant active forces. The schematic diagram of the system is shown in Figure 24.2.1, the notations in which are self-explanatory.

Based on the schematic diagram shown in Figure 24.2.1 and the related considerations, we compose a pair of simultaneous differential

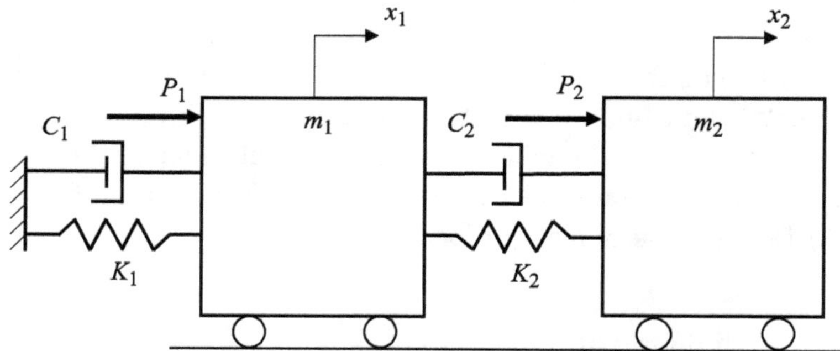

Fig. 24.2.1. Schematic diagram of a restricted system subjected to constant active forces, the two masses of which are connected to each other by a dashpot and a spring in parallel and one of the masses is connected to a non-movable support by an identical pair of links.

equations of motion:

$$m_1 \frac{d^2 x_1}{dt^2} + (C_1 + C_2)\left(\frac{dx_1}{dt} - \frac{dx_2}{dt}\right) + (K_1 + K_2)(x_1 - x_2) = P_1$$

$$(24.2.1)$$

$$m_2 \frac{d^2 x_2}{dt^2} + (C_1 + C_2)\left(\frac{dx_2}{dt} - \frac{dx_1}{dt}\right) + (K_1 + K_2)(x_2 - x_1) = P_2$$

$$(24.2.2)$$

where P_1 and P_2 are the respective constant active forces applied to the masses.

The initial conditions of motion are

$$x_1 = 0; \quad \frac{dx_1}{dt} = 0$$

For $\quad t = 0$ $$(24.2.3)$$

$$x_2 = 0: \quad \frac{dx_2}{dt} = 0$$

Dividing, respectively, equations (24.2.1) and (24.2.2) by m_1 and m_2, while applying to them Laplace Transform pairs 2, 5, 2, 4, 2, 1, and 2, we convert these equations with their initial conditions of motion from the time domain into the Laplace domain and obtain a system of two simultaneous algebraic equations with two unknowns $x_1(s)$ and $x_2(s)$ in each equation:

$$s^2 x_1(s) + 2n_{11}sx_1(s) - 2n_{11}sx_2(s) + 2n_{21}sx_1(s) - 2n_{21}sx_2(s)$$

$$+ \omega_{11}^2 x_1(s) - \omega_{11}^2 x_2(s) + \omega_{21}^2 x_1(s) - \omega_{21}^2 x_2(s) = p_1 \qquad (24.2.4)$$

$$s^2 x_2(s) + 2n_{12}sx_2(s) - 2n_{11}sx_1(s)_1 + 2n_{22}lx_2(s) - 2n_{22}sx_1(s)$$

$$+ \omega_{12}^2 x_2(s) - \omega_{12}^2 x_1(s) + \omega_{22}^2 x_2(s) - \omega_{22}^2 x_1(s) = p_2 \qquad (24.2.5)$$

Rearranging the terms in equations (24.2.34) and (24.2.5) and applying to them the method of substitutions, we obtain two simultaneous algebraic equations with one unknown in each equation that describe in the Laplace domain the displacements of the first and second masses:

$$x_1(s) = \frac{p_1}{s^2 + 2ns + \omega^2} + \frac{2[(n_{12} + n_{22})p_1 + (n_{11} + n_{21})p_2]}{s(s^2 + 2ns + \omega^2)}$$

$$+ \frac{(\omega_{12}^2 + \omega_{22}^2)p_1 + (\omega_{11}^2 + \omega_{21}^2)p_2}{s^2(s^2 + 2ns + \omega^2)} \tag{24.2.6}$$

$$x_2(s) = \frac{p_2}{s^2 + 2ns + \omega^2} + \frac{2[(n_{11} + n_{21})p_2 + (n_{12} + n_{22})p_1]}{s(s^2 + 2ns + \omega^2)}$$

$$+ \frac{(\omega_{11}^2 + \omega_{21}^2)p_2 + (\omega_{12}^2 + \omega_{22}^2)p_1}{s^2(s^2 + 2ns + \omega^2)} \tag{24.2.7}$$

In continuation of the analysis of the current system, we consider the cases of underdamped, critically damped, and overdamped motions.

24.2.1. *Underdamped vibrational motion*

Combining equations (24.2.6) and (24.2.7) with equations (24.1.12) and (24.1.13), we obtain

$$x_1(s) = \frac{p_1}{(s + n)^2 + \omega^2} + \frac{2[(n_{12} + n_{22})p_1 + (n_{11} + n_{21})p_2]}{s[(s + n)^2 + \omega^2]}$$

$$+ \frac{(\omega_{12}^2 + \omega_{22}^2)p_1 + (\omega_{11}^2 + \omega_{01}^2)p_2}{s^2[(s + n)^2 + \omega^2]} \tag{24.2.8}$$

$$x_2(s) = \frac{p_2}{(s + n)^2 + \omega^2} + \frac{2[(n_{11} + n_{21})p_2 + (n_{12} + n_{22})p_1]}{s[(s + n)^2 + \omega^2]}$$

$$+ \frac{(\omega_{11}^{22} + \omega_{21}^2)p_2 + (\omega_{12}^2 + \omega_{22}^2)p_1}{s^2[(s + n)^2 + \omega^2]} \tag{24.2.9}$$

Applying to equations (24.2.8) and (24.2.9) Laplace Transform pairs 21, 45, and 56, we invert these equations from the Laplace domain into the time domain and obtain the solutions of differential equations (24.2.1) and (24.2.2) with their initial conditions of motion. These equations describe the underdamped vibrational motion of the two masses of the system:

$$x_1 = \frac{p_1}{\omega^2 + n^2}\left[1 - e^{-nt}\left(\cos\omega t + \frac{n}{\omega}\sin\omega t\right)\right]$$

$$+ \frac{2[(n_{12} + n_{22})p_1 + (n_{11} + n_{21})p_2]}{\omega(\omega^2 + n^2)^2}$$

$$\times \left\{(\omega^2 + n^2)\omega t - 2n\omega - e^{-nt}[(\omega^2 - n^2)\sin\omega t - 2n\omega\cos\omega t]\right\}$$

$$+ \frac{(\omega_{12}^2 + \omega_{22}^2)p_1 + (\omega_{11}^2 + \omega_{21}^2)p_2}{\omega(\omega^2 + n^2)^2}\left\{\frac{\omega t^2}{2}(\omega^2 + n^2) - 2n\omega t\right.$$

$$- \frac{\omega(\omega^2 - n^2)}{\omega^2 + n^2}\left[1 - e^{-nt}\left(\cos\omega t + \frac{n}{\omega}\sin\omega t\right)\right]$$

$$\left. - \frac{2n^2\omega}{\omega^2 + n^2}\left[1 + e^{-nt}\left(\frac{\omega}{n}\sin\omega t - \cos\omega t\right)\right]\right\} \tag{24.2.10}$$

$$x_2 = \frac{p_2}{\omega^2 + n^2}\left[1 - e^{-nt}\left(\cos\omega t + \frac{n}{\omega}\sin\omega t\right)\right]$$

$$+ \frac{2[(n_{11} + n_{21})p_2 + (n_{12} + n_{22})p_1]}{\omega(\omega^2 + n^2)^2}$$

$$\times \left\{(\omega^2 + n^2)\omega t - 2n\omega - e^{-nt}[(\omega^2 - n^2)\sin\omega t - 2n\omega\cos\omega t]\right\}$$

$$+ \frac{(\omega_{11}^2 + \omega_{21}^2)p_2 + (\omega_{12}^2 + \omega_{22}^2)p_1}{\omega(\omega^2 + n^2)^2}\left\{\frac{\omega t^2}{2}(\omega^2 + n^2) - 2n\omega t\right.$$

$$- \frac{\omega(\omega^2 - n^2)}{\omega^2 + n^2}\left[1 - e^{-nt}\left(\cos\omega t + \frac{n}{\omega}\sin\omega t\right)\right]$$

$$\left. - \frac{2n^2\omega}{\omega^2 + n^2}\left[1 + e^{-nt}\left(\frac{\omega}{n}\sin\omega t - \cos\omega t\right)\right]\right\} \tag{24.2.11}$$

Supposing that in equations (24.2.10) and (24.2.11), $t = 0$, we have that $x_1 = 0$ and $x_2 = 0$, as it should be according to the initial conditions of motion. Taking from these equations the first derivatives, we obtain the velocities of the masses of the system:

$$\frac{dx_1}{dt} = \frac{p_1}{\omega} e^{-nt} \sin \omega t + \frac{2[(n_{12} + n_{22})p_1 + (n_{11} + n_{21})p_2]}{\omega^2 + n^2}$$

$$\times \left[1 - e^{-nt} \left(\cos \omega t + \frac{n}{\omega} \sin \omega t \right) \right]$$

$$+ \frac{(\omega_{12}^2 + \omega_{22}^2)p_1 + (\omega_{11}^2 + \omega_{21}^2)p_2}{\omega(\omega^2 + n^2)^2} \{(\omega^2 + n^2)\omega t - 2n\omega - e^{-nt}$$

$$\times [(\omega^2 - n^2) \sin \omega t - 2n\omega \cos \omega t]\} \tag{24.2.12}$$

$$\frac{dx_2}{dt} = \frac{p_2}{\omega} e^{-nt} \sin \omega t + \frac{2[(n_{11} + n_{21})p_2 + (n_{12} + n_{22})p_1]}{\omega^2 + n^2}$$

$$\times \left[1 - e^{-nt} \left(\cos \omega t + \frac{n}{\omega} \sin \omega t \right) \right]$$

$$+ \frac{(\omega_{11}^2 + \omega_{21}^2)p_2 + (\omega_{12}^2 + \omega_{22}^2)p_1}{\omega(\omega^2 + n^2)^2} \{(\omega^2 + n^2)\omega t - 2n\omega - e^{-nt}$$

$$\times [(\omega^2 - n^2) \sin \omega t - 2n\omega \cos \omega t]\} \tag{24.2.13}$$

Supposing for equations (24.2.12) and (24.2.13) that $t = 0$, we obtain that $\frac{dx_1}{dt} = 0$ and $\frac{dx_2}{dt} = 0$, as expected according to the initial conditions of motion.

Taking from equations (24.2.12) and (24.2.13) the first derivatives, we determine the accelerations of the masses:

$$\frac{d^2 x_1}{dt^2} = p_1 e^{-nt} \left(\cos \omega t - \frac{n}{\omega} \sin \omega t \right)$$

$$+ \frac{2}{\omega}[(n_{12} + n_{22})p_1 + (n_{11} + n_{21})p_2]e^{-nt} \sin \omega t$$

$$+ \frac{1}{\omega^2 + n^2}[(\omega_{12}^2 + \omega_{22}^2)p_1 + (\omega_{11}^2 + \omega_{21}^2)p_2]$$

$$\times \left[1 - \left(\cos \omega t + \frac{n}{\omega} \sin \omega t \right) \right] \sin \omega t \tag{24.2.14}$$

$$\frac{d^2 x_2}{dt^2} = p_2 e^{-nt} \left(\cos \omega t - \frac{n}{\omega} \sin \omega t \right)$$

$$+ \frac{2}{\omega} [(n_{11} + n_{21})p_2 + (n_{12} + n_{22})p_1] e^{-nt} \sin \omega t$$

$$+ \frac{1}{\omega^2 + n^2} [(\omega_{11}^2 + \omega_{21}^2)p_2 + (\omega_{12}^2 + \omega_{22}^2)p_1]$$

$$\times \left[1 - \left(\cos \omega t + \frac{n}{\omega} \sin \omega t \right) \right] \sin \omega t \qquad (24.2.15)$$

Equations (24.2.9)–(24.2.15) represent the basic parameters of motion of the system and allow for performing the analysis of the considered operational process.

24.2.2. *Critically damped motion*

As noted above, critical damping occurs when $\omega^2 = 0$. Therefore, accounting for this condition in equations (24.2.6) and (24.2.8), we may write

$$x_1(s) = \frac{p_1}{(s+n)^2} + \frac{2[(n_{12} + n_{22})p_1 + (n_{11} + n_{21})p_2]}{s(s+n)^2}$$

$$+ \frac{(\omega_{12}^2 + \omega_{22}^2)p_1 + (\omega_{11}^2 + \omega_{21}^2)p_2}{s^2(s+n)^2} \qquad (24.2.16)$$

$$x_2(s) = \frac{p_2}{(s+n)^2} + \frac{2[(n_{11} + n_{21})p_2 + (n_{12} + n_{22})p_1]}{s(s+n)^2}$$

$$+ \frac{(\omega_{11}^2 + \omega_{21}^2)p_2 + (\omega_{12}^2 + \omega_{22}^2)p_1}{s^2(s+n)^2} \qquad (24.2.17)$$

Applying to equations (24.2.16) and (24.2.17) Laplace Transform pairs 19, 37, and 54, we invert these equations from the Laplace domain into the time domain and obtain the solutions of differential equations (24.2.1) and (24.2.2) with their initial conditions of motion for the case of critical damping:

$$x_1 = \frac{p_1}{n^2}[1 - e^{-nt}(1 + nt)] + \frac{2[(n_{12} + n_{22})p_1 + (n_{11} + n_{21})p_2]}{n^2}$$

$$\times \left[t - \frac{2}{n} + e^{-nt}\left(\frac{2}{n} + t\right)\right] + \frac{(\omega_{12}^2 + \omega_{22}^2)p_1 + (\omega_{11}^2 + \omega_{21}^2)p_2}{n^2}$$

$$\times \left[\frac{t^2}{2} - \frac{2t}{n} - \frac{3}{n^2} + \frac{1}{n^2}e^{-nt}(3 + nt)\right] \tag{24.2.18}$$

$$x_2 = \frac{p_1}{n^2}[1 - e^{-nt}(1 + nt)] + \frac{2[(n_{11} + n_{21})p_2 + (n_{12} + n_{22})p_1]}{n^2}$$

$$\times \left[t - \frac{2}{n} + e^{-nt}\left(\frac{2}{n} + t\right)\right] + \frac{(\omega_{11}^2 + \omega_{21}^2)p_2 + (\omega_{12}^2 + \omega_{22}^2)p_1}{n^2}$$

$$\times \left[\frac{t^2}{2} - \frac{2t}{n} - \frac{3}{n^2} + \frac{1}{n^2}e^{-nt}(3 + nt)\right] \tag{24.2.19}$$

Assuming that in equations (24.2.18) and (24.2.19), $t = 0$, we have $x_1 = 0$ and $x_2 = 0$, as it should be according to the initial conditions of motion. Taking from these equations the first derivatives, we obtain the velocities of the masses of the system:

$$\frac{dx_1}{dt} = p_1 t e^{-nt} + \frac{2[(n_{12} + n_{22})p_1 + (n_{11} + n_{21})p_2]}{n^2}[1 - e^{-nt}(1 + nt)]$$

$$+ \frac{(\omega_{12}^2 + \omega_{22}^2)p_1 + (\omega_{11}^2 + \omega_{21}^2)p_2}{n^2}\left[t - \frac{2}{n} + e^{-nt}\left(\frac{2}{n} + t\right)\right] \tag{24.2.20}$$

$$\frac{dx_2}{dt} = p_2 t e^{-nt} + \frac{2[(n_{11} + n_{21})p_2(n_{12} + n_{22})p_1]}{n^2}[1 - e^{-nt}(1 + nt)]$$

$$+ \frac{(\omega_{11}^2 + \omega_{21}^2)p_2 + (\omega_{12}^2 + \omega_{22}^2)p_1}{n^2}\left[t - \frac{2}{n} + e^{-nt}\left(\frac{2}{n} + t\right)\right] \tag{24.2.21}$$

Supposing for equations (24.2.20) and (24.2.21) that $t = 0$, we obtain that $\frac{dx_1}{dt} = 0$ and $\frac{dx_2}{dt} = 0$, as expected according to the initial conditions of motion.

24.2.3. *Numerical solution*

The following is a Python program to plot the graph (Figure 24.2.2) of equation (24.2.18), which describes critically damped motion.

```python
from matplotlib.pyplot import plot, show
from numpy import linspace, exp


p1 = 100
p2 = 75
n11 = 3
n12 = 6
n21 = 4
n22 = 2
n = n11 + n12 + n21 + n22
omega11 = 30
omega12 = 25
omega21 = 20
omega22 = 15

t = linspace(0, 2, 1000)
x1  = (p1/n**2)*(1 - exp(-n*t)*(1 + n*t)) \
    + (2*((n12 + n22)*p1 - (n11 + n21)*p2)/n**2) \
      *(t - 2/n + exp(-n*t)*2/n + t)) \
    + (((omega11**2 + omega21**2)*p2 \
      + (omega12**2 + omega22**2)*p1)/n**2) \
      *(t**2/2 - 2*t/n - 3/n**2 + (1/n**2)*exp(-n*t)*(3+n*t))

plot(t, x1)
show()
```

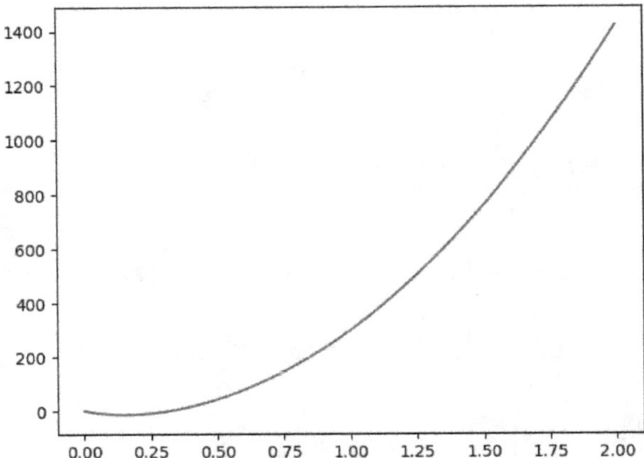

Fig. 24.2.2. Graph of equation (24.2.18).

24.2.4. *Overdamped motion*

As mentioned above, overdamped motion occurs when $\omega^2 < 0$. Therefore, we have to replace in equations (24.2.8) and (24.2.9) the signs at the terms ω^2 with the opposite signs, and we obtain

$$x_1(s) = \frac{p_1}{(s+n)^2 - \omega^2} + \frac{2[(n_{12} + n_{22})p_1 + (n_{11} + n_{21})p_2]}{s[(s+n)^2 - \omega^2]}$$

$$+ \frac{(\omega_{12}^2 + \omega_{22}^2)p_1 + (\omega_{11}^2 + \omega_{21}^2)p_2}{s^2[(s+n)^2 - \omega^2]} \tag{24.2.22}$$

$$x_2(s) = \frac{p_2}{(s+n)^2 - \omega^2} + \frac{2[(n_{11} + n_{21})p_2 + (n_{12} + n_{22})p_1]}{s[(s+n)^2 - \omega^2]}$$

$$+ \frac{(\omega_{11}^{22} + \omega_{21}^2)p_2 + (\omega_{12}^2 + \omega_{22}^2)p_1}{s^2[(s+n)^2 - \omega^2]} \tag{24.2.23}$$

Applying to equations (24.2.22) and (24.2.23) Laplace Transform pairs, 22, 46, and 57, we invert these equations from the Laplace domain into the time domain and obtain the solutions of differential equations (24.2.1) and (24.2.2) with their initial conditions of motion for the case of overdamped motion:

$$x_1 = \frac{p_1}{n^2 - \omega^2}\left[1 - e^{-nt}\left(\cosh \omega t + \frac{n}{\omega}\sinh \omega t\right)\right]$$

$$+ \frac{2[(n_{12} + n_{22})p_1 + (n_{11} + n_{21})p_2]}{\omega(n^2 - \omega^2)^2}$$

$$\times \{\omega t(n^2 - \omega^2) - 2n\omega + e^{-nt}[\omega t(n^2 + \omega^2)\sinh \omega t + 2n\omega \cosh \omega t]\}$$

$$+ \frac{(\omega_{12}^2 + \omega_{22}^2)p_1 + (\omega_{11}^2 + \omega_{21}^2)p_2}{\omega(n^2 - \omega^2)^2}\left\{\frac{\omega t^2}{2}(n^2 - \omega^2) - 2n\omega t\right.$$

$$+ \frac{\omega(n^2 + \omega^2)}{n^2 - \omega^2}\left[1 - e^{-nt}\left(\cosh \omega t + \frac{n}{\omega}\sinh \omega t\right)\right]$$

$$\left.- \frac{2n^2\omega}{n^2 - \omega^2}\left[1 + e^{-nt}\left(\frac{n}{\omega}\sinh \omega t - \cosh \omega t\right)\right]\right\} \tag{24.2.24}$$

$$x_2 = \frac{p_2}{n^2 - \omega^2} \left[1 - e^{-nt} \left(\cosh \omega t + \frac{n}{\omega} \sinh \omega t \right) \right]$$

$$+ \frac{2[(n_{11} + n_{21})p_2 + (n_{12} + n_{22})p_1]}{\omega(n^2 - \omega^2)^2}$$

$$\times \{\omega t(n^2 - \omega^2) - 2n\omega + e^{-nt}[\omega t(n^2 + \omega^2)\sinh \omega t + 2n\omega \cosh \omega t]\}$$

$$+ \frac{(\omega_{11}^2 + \omega_{21}^2)p_2 + (\omega_{12}^2 + \omega_{22}^2)p_1}{\omega(n^2 - \omega^2)^2} \left\{ \frac{\omega t^2}{2}(n^2 - \omega^2) - 2n\omega t \right.$$

$$+ \frac{\omega(n^2 + \omega^2)}{n^2 - \omega^2} \left[1 - e^{-nt} \left(\cosh \omega t + \frac{n}{\omega} \sinh \omega t \right) \right]$$

$$\left. - \frac{2n^2\omega}{n^2 - \omega^2} \left[1 + e^{-nt} \left(\frac{n}{\omega} \sinh \omega t - \cosh \omega t \right) \right] \right\} \qquad (24.2.25)$$

Assuming that in equations (24.2.24) and (24.2.25), $t = 0$, then we obtain that $x_1 = 0$ and $x_2 = 0$, as expected according to the initial conditions of motion of the masses, which are performing a decelerated translation.

24.3. Motion of a Restricted System with Two Sequences of Flexible and Fluid Links, Where One of the Masses Is Attached by a Sequence of These Links to a Non-movable Support and Just One Mass Is Subjected to a Harmonic Force

Consider the operational process related to a restricted system, the two masses of which are connected to each other by a spring and a dashpot in parallel, while one of the masses is attached to a non-movable support and subjected to a harmonic force. The system is moving on a horizontal frictionless surface.

The schematic diagram representing the restricted system is shown in Figure 24.3.1, the notations in which are self-explanatory.

Based on the schematic diagram shown in Figure 24.3.1 and the above considerations, we compose a pair of simultaneous differential

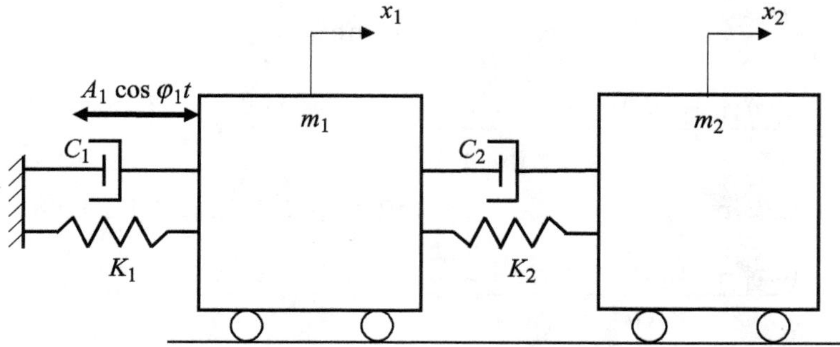

Fig. 24.3.1. Schematic diagram of a restricted system, the two masses of which are connected to each other by a dashpot and a spring in parallel, while one of the masses is connected to a non-movable support by a similar pair of links and subjected to a harmonic force.

equations describing the motion of the masses:

$$m_1 \frac{d^2 x_1}{dt^2} + (C_1 + C_2)\left(\frac{dx_1}{dt} - \frac{dx_2}{dt}\right) + (K_1 + K_2)(x_1 - x_2)$$

$$= A_1 \cos \varphi_1 t \tag{24.3.1}$$

$$m_2 \frac{d^2 x_2}{dt^2} + (C_1 + C_2)\left(\frac{dx_2}{dt} - \frac{dx_1}{dt}\right) + (K_1 + K_2)(x_2 - x_1) = 0$$

$$\tag{24.3.2}$$

The initial conditions of motion are taken according to expression (24.2.3).

Dividing, respectively, equations (24.3.1) and (24.3.2) by m_1 and m_2, we may write

$$\frac{d^2 x_1}{dt^2} + 2(n_{11} + n_{21})\left(\frac{dx_1}{dt} - \frac{dx_2}{dt}\right) + (\omega_{11}^2 + \omega_{21}^2)(x_1 - x_2)$$

$$= a_1 \cos \varphi_1 t \tag{24.3.3}$$

$$\frac{d^2 x_2}{dt^2} + 2(n_{12} + n_{22})\left(\frac{dx_2}{dt} - \frac{dx_1}{dt}\right) + (\omega_{12}^2 + \omega_{22}^2)(x_2 - x_1) = 0$$

$$\tag{24.3.4}$$

Applying, respectively, to equations (24.3.3) and (24.3.4) Laplace Transform pairs 5, 2, 4, 2, 1, 2, and 29 and 5, 2, 4, 2, and 1, we convert these equations from the time domain into the Laplace domain and obtain a system of two simultaneous algebraic equations with two unknowns in each equation that represent the displacements of the respective masses:

$$s^2 x_1(s) + 2n_{11}sx_1(s) - 2n_{11}sx_2(s) + 2n_{21}sx_1(s) - 2n_{21}sx_2(s)$$

$$+ \omega_{11}^2 x_1(s) - \omega_{11}^2 x_2(s) + \omega_{21}^2 x_1(s) - \omega_{21}^2 x_2(s) = \frac{a_1 s^2}{s^2 + \varphi_1^2}$$

$$\tag{24.3.5}$$

$$s^2 x_2(s) + 2n_{12}sx_2(s) - 2n_{12}sx_1(s) + 2n_{22}sx_2(s) - 2n_{22}sx_1(s)$$

$$+ \omega_{12}^2 sx_2(s) - \omega_{12}^2 sx_1(s) + \omega_{22}^2 sx_2(s) - \omega_{22}^2 sx_1(s) = 0 \tag{24.3.6}$$

Rearranging equations (24.3.5) and (24.3.6), we write

$$x_1(s)(s^2 + 2n_{11}s + 2n_{21}s + \omega_{11}^2 + \omega_{21}^2)$$

$$= \frac{a_1 s^2}{s^2 + \varphi_1^2} + x_2(s)(2n_{11}s + 2n_{21}s + \omega_{11}^2 + \omega_{21}^2) \tag{24.3.7}$$

$$x_2(s)(s^2 + 2n_{12}s + 2n_{22}s + \omega_{12}^2 + \omega_{22}^2)$$

$$= x_1(s)(2n_{12}s + 2n_{22}s + \omega_{12}^2 + \omega_{22}^2) \tag{24.3.8}$$

Solving equation (24.3.8) for $x_2(s)$, we obtain

$$x_2(s) = \frac{x_1(s)(2n_{12}s + 2n_{22}s + \omega_{12}^2 + \omega_{22}^2)}{s^2 + 2n_{12}s + 2n_{22}s + \omega_{12}^2 + \omega_{22}^2} \tag{24.3.9}$$

Using the method of substitutions, we combine equations (24.3.7) and (24.3.9) and eliminate one unknown from equation (24.3.7):

$$x_1(s)(s^2 + 2n_{11}s + 2n_{21}s + \omega_{11}^2 + \omega_{21}^2) = \frac{a_1 s^2}{s^2 + \varphi_1^2}$$

$$+ \frac{x_1(s)(2n_{12}s + 2n_{22}s + \omega_{12}^2 + \omega_{22}^2)(2n_{11}s + 2n_{21}s + \omega_{11}^2 + \omega_{21}^2)}{s^2 + 2n_{12}s + 2n_{22}s + \omega_{12}^2 + \omega_{22}^2}$$

$$\tag{24.3.10}$$

Hence, the algebraic equation (24.3.10) contains one unknown $x_1(s)$, representing the displacement of the first mass in the Laplace domain. Performing the conventional algebraic procedures, we have

$$x_1(s)[s^2 + 2s(n_{11} + n_{21} + n_{12} + n_{22}) + \omega_{11}^2 + \omega_{21}^2 + \omega_{12}^2 + \omega_{22}^2]$$
$$= \frac{a_1(s^2 + 2n_{12}s + 2n_{22}s + \omega_{12}^2 + \omega_{22}^2)}{s^2 + \varphi_1^2} \tag{24.3.11}$$

Solving equation (24.3.11) for $x_1(s)$, we obtain

$$x_1(s) = \frac{s^2 a_1}{(s^2 + \varphi_1^2)(s^2 + 2ns + \omega_0^2)} + \frac{2sa_1(n_{12} + n_{22})}{(s^2 + \varphi_1^2)(s^2 + 2ns + \omega_0^2)}$$
$$+ \frac{a_1(\omega_{12}^2 + \omega_{22}^2)}{(s^2 + \varphi_1^2)(s^2 + 2ns + \omega_0^2)} \tag{24.3.12}$$

Applying similar algebraic procedures, we determine displacement of the second mass in the Laplace domain:

$$x_2(s) = \frac{2n_2 sa_1}{(s^2 + \varphi_1^2)(s^2 + 2ns + \omega_0^2)} + \frac{a_1(\omega_{12}^2 + \omega_{22}^2)}{(s^2 + \varphi_1^2)(s^2 + 2ns + \omega_0^2)} \tag{24.3.13}$$

Combining equations (24.3.12) and (24.3.13) with equations (24.1.12) and (24.1.13), we may write

$$x_1(s) = \frac{s^2 a_1}{(s^2 + \varphi_1^2)[(s + n)^2 + \omega^2]} + \frac{2sa_1(n_{12} + n_{22})}{(s^2 + \varphi_1^2)[(s + n)^2 + \omega^2]}$$
$$+ \frac{a_1(\omega_{12}^2 + \omega_{22}^2)}{(s^2 + \varphi_1^2)[(s + n)^2 + \omega^2]} \tag{24.3.14}$$

$$x_2(s) = \frac{2sa_1(n_{12} + n_{22})}{(s^2 + \varphi_1^2)[(s + n)^2 + \omega^2]} + \frac{a_1(\omega_{12}^2 + \omega_{22}^2)}{(s^2 + \varphi_1^2)[(s + n)^2 + \omega^2]} \tag{24.3.15}$$

24.3.1. *Underdamped vibrational motion*

Assuming that $\omega^2 > 0$ and applying to equations (24.3.14) and (24.3.15), respectively, Laplace Transform pairs 68, 65, and 58, and 65 and 58, we invert these equations from the Laplace domain into the time domain and obtain the solutions of differential equations

(24.3.1) and (24.3.2) with their initial conditions of motion:

$$x_1 = \frac{a_1}{4n^2\varphi_1^2 + (\omega^2 + n^2 - \varphi_1^2)^2}\Bigg[(\omega^2 + n^2 - \varphi_1^2)^2$$
$$\times (\cos\varphi_1 t - e^{-nt}\cos\omega t) + 2n\varphi_1\sin\varphi_1 t$$
$$- \frac{n}{\omega}(\omega^2 + n^2 + \varphi_1^2)e^{-nt}\sin\omega t\Bigg]$$
$$+ \frac{2a_1(n_{12} + n_{22})}{4n^2\varphi_1^2 + (\omega^2 + n^2 - \varphi_1^2)^2}\Bigg[2n(e^{-nt}\cos\omega t - \cos\varphi_1 t)$$
$$+ \frac{1}{\varphi_1}(\omega_v^2 + n^2 - \varphi_1^2)\sin\varphi_1 t - \frac{1}{\omega}(\omega^2 - n^2 - \varphi_1^2)e^{-nt}\sin\omega t\Bigg]$$
$$+ \frac{a_1(\omega_{12}^2 + \omega_{22}^2)}{4n^2\varphi_1^2 + (\omega^2 + n^2 - \varphi_1^2)^2}\Bigg\{\frac{1}{\varphi_1^2}(\omega^2 + n^2 - \varphi_1^2)(1 - \cos\varphi_1 t)$$
$$- 2n\left(\frac{1}{\varphi_1}\sin\varphi_1 t - \frac{1}{\omega}e^{-nt}\sin\omega t\right) + \frac{3n^2 + \varphi_1^2 - \omega^2}{\omega^2 + n^2}$$
$$\times \left[1 - e^{-nt}\left(\cos\omega t + \frac{n}{\omega}\sin\omega t\right)\right. \tag{24.3.16}$$

$$x_2 = \frac{2a_1(n_{12} + n_{22})}{4n^2\varphi_1^2 + (\omega^2 + n^2 - \varphi_1^2)^2}\Bigg[2n(e^{-nt}\cos\omega t - \cos\varphi_1 t)$$
$$+ \frac{1}{\varphi_1}(\omega^2 + n^2 - \varphi_1^2)\sin\varphi_1 t - \frac{1}{\omega}(\omega^2 - n^2 - \varphi_1^2)e^{-nt}\sin\omega t\Bigg]$$
$$+ \frac{a_1(\omega_{12}^2 + \omega_{22}^2)}{4n^2\varphi_1^2 + (\omega^2 + n^2 - \varphi_1^2)^2}\Bigg\{\frac{1}{\varphi_1^2}(\omega^2 + n^2 - \varphi_1^2)(1 - \cos\varphi_1 t)$$
$$- 2n\left(\frac{1}{\varphi_1}\sin\varphi_1 t - \frac{1}{\omega}e^{-nt}\sin\omega t\right) + \frac{3n^2 + \varphi_1^2 - \omega^2}{\omega^2 + n^2}$$
$$\times \left[1 - e^{-nt}\left(\cos\omega t + \frac{n}{\omega}\sin\omega t\right)\right]\Bigg\} \tag{24.3.17}$$

24.3.2. *Critically damped motion*

Substituting into equations (24.3.7) and (24.3.8) $\omega^2 = 0$, we obtain

$$x_1(s) = \frac{s^2 a_1}{(s^2 + \varphi_1^2)(s + n)^2} + \frac{2sa_1(n_{12} + n_{22})}{(s^2 + \varphi_1^2)(s + n)^2} + \frac{a_1(\omega_{12}^2 + \omega_{22}^2)}{(s^2 + \varphi_1^2)(s + n)^2}$$
$$\tag{24.3.18}$$

$$x_2(s) = \frac{2sa_1(n_{12} + n_{22})}{(s^2 + \varphi_1^2)(s + n)^2} + \frac{a_1(\omega_{12}^2 + \omega_{22}^2)}{(s^2 + \varphi_1^2)(s + n)^2} \qquad (24.3.19)$$

Applying, respectively, to equations (24.3.18) and (24.3.19) Laplace Transform pairs 71, 63, 60, 63, and 60, we invert these equations with their initial conditions from the Laplace domain into the time domain and obtain the solutions of differential equations (24.3.1) and (24.3.2) with their initial conditions of motion:

$$x_1 = \frac{a_1}{(\varphi_1^2 + n^2)^2}\{2n\varphi_1 \sin \varphi_1 t + e^{-nt}[\varphi_1^2 - n^2 - nt(\varphi_1^2 + n^2)]$$

$$- (\varphi_1^2 - n^2)\cos \varphi_1 t\} + \frac{2a_1(n_{12} + n_{22})}{(\varphi_1^2 + n^2)^2}$$

$$\times \{2n\varphi_1[e^{-nt}(1 + nt) - \cos \varphi_1 t] - (\varphi_1^2 - n^2)(\sin \varphi_1 t - \varphi_1 te^{-nt})\}$$

$$+ \frac{a_1(\omega_{12}^2 + \omega_{22}^2)}{(\varphi_1^2 + n^2)^2}\left\{\frac{n^2 - \varphi_1^2}{\varphi_1^2}(1 - \cos \varphi_1 t) - \frac{2n}{\varphi_1}\sin \varphi_1 t\right.$$

$$\left. + \frac{3n^2 + \varphi_1^2}{\varphi_1^2}[1 - e^{-nt}(1 + nt)] + 2nte^{-nt}\right\} \qquad (24.3.20)$$

$$x_2 = \frac{2a_1(n_{12} + n_{22})}{(\varphi_1^2 + n^2)^2}\{2n\varphi_1[e^{-nt}(1 + nt) - \cos \varphi_1 t]$$

$$- (\varphi_1^2 - n^2)(\sin \varphi_1 t - \varphi_1 te^{-nt})\} + \frac{a_1(\omega_{12}^2 + \omega_{22}^2)}{(\varphi_1^2 + n^2)^2}$$

$$\times \left\{\frac{n^2 - \varphi_1^2}{\varphi_1^2}(1 - \cos \varphi_1 t) - \frac{2n}{\varphi_1}\sin \varphi_1 t\right.$$

$$\left. + \frac{3n^2 + \varphi_1^2}{\varphi_1^2}[1 - e^{-nt}(1 + nt)] + 2nte^{-nt}\right\} \qquad (24.3.21)$$

24.3.3. *Overdamped motion*

Since overdamped motion occurs in the case when $\omega^2 < 0$, we have to replace in equations (24.3.7) and (24.3.8) the parameter ω^2 with its negative value, and then, we obtain the corresponding equations that upon inversion give us the equations describing overdamped

motion:

$$x_1(s) = \frac{s^2 a_1}{(s^2 + \varphi_1^2)[(s+n)^2 - \omega^2]} + \frac{2sa_1(n_{12} + n_{22})}{(s^2 + \varphi_1^2)[(s+n)^2 - \omega^2]}$$

$$+ \frac{a_1(\omega_{12}^2 + \omega_{22}^2)}{(s^2 + \varphi_1^2)[(s+n)^2 - \omega^2]} \tag{24.3.22}$$

$$x_2(s) = \frac{2sa_1(n_{12} + n_{22})}{(s^2 + \varphi_1^2)[(s+n)^2 - \omega^2]} + \frac{a_1(\omega_{12}^2 + \omega_{22}^2)}{(s^2 + \varphi_1^2)[(s+n)^2 - \omega^2]}$$

$$\tag{24.3.23}$$

Applying, respectively, to equations (24.3.22) and (24.3.23) Laplace Transform pairs 69, 66, and 59, 66, and 59, we invert these equations with their initial conditions from the Laplace domain into the time domain and obtain the solutions of differential equations (24.3.1) and (24.3.2) with their initial conditions of motion:

$$x_1 = \frac{a_1}{4n^2\varphi_1^2 + (n^2 - \omega^2 - \varphi_1^2)^2} \left[(n^2 - \omega^2 - \varphi_1^2)^2 \right.$$

$$\times (\cos \varphi_1 t - e^{-nt} \cosh \omega t) + 2n\varphi_1 \sin \varphi_1 t$$

$$\left. - \frac{n}{\omega}(n^2 - \omega + \varphi_1^2)e^{-nt} \sinh \omega t \right] + \frac{2a_1(n_{12} + n_{22})}{4n^2\varphi_1^2 + (n^2 - \omega - \varphi_1^2)^2}$$

$$\times \left[2n(e^{-nt}\cosh \omega t - \cos \varphi_1 t) + \frac{1}{\varphi_1}(n^2 - \omega^2 - \varphi_1^2)\sin \varphi_1 t \right.$$

$$\left. + \frac{1}{\omega}(\omega^2 + n^2 + \varphi_1^2)e^{-nt}\sinh \omega t \right] + \frac{a_1(\omega_{12}^2 + \omega_{22}^2)}{4n^2\varphi_1^2 + (n^2 - \omega^2 - \varphi_1^2)^2}$$

$$\times \left\{ \frac{1}{\varphi_1^2}(n^2 - \omega^2 - \varphi_1^2)(1 - \cos \varphi_1 t) \right.$$

$$- 2n\left(\frac{1}{\varphi_1}\sin \varphi_1 t - \frac{1}{\omega}e^{-nt}\sinh \omega t \right)$$

$$\left. + \frac{3n^2 + \varphi_1^2 + \omega^2}{n^2 - \omega^2}\left[1 - e^{-nt}\left(\cosh \omega t + \frac{n}{\omega}\sinh \omega t \right) \right] \right\}$$

$$\tag{24.3.24}$$

$$x_2 = \frac{2a_1(n_{12} + n_{22})}{4n^2\varphi_1^2 + (n^2 - \omega^2 - \varphi_1^2)^2}\left[2n(e^{-nt}\cosh\omega t - \cos\varphi_1 t)\right.$$

$$\left. + \frac{1}{\varphi_1}(n^2 - \omega^2 - \varphi_1^2)\sin\varphi_1 t + \frac{1}{\omega}(\omega^2 + n^2 + \varphi_1^2)e^{-nt}\sinh\omega t\right]$$

$$+ \frac{a_1(\omega_{12}^2 + \omega_{22}^2)}{4n^2\varphi_1^2 + (n^2 - \omega^2 - \varphi_1^2)^2}\left\{\frac{1}{\varphi_1^2}(n^2 - \omega^2 - \varphi_1^2)(1 - \cos\varphi_1 t)\right.$$

$$- 2n\left(\frac{1}{\varphi_1}\sin\varphi_1 t - \frac{1}{\omega}e^{-nt}\sinh\omega t\right) + \frac{3n^2 + \varphi_1^2 + \omega^2}{n^2 - \omega^2}$$

$$\left.\times\left[1 - e^{-nt}\left(\cosh\omega t + \frac{n}{\omega}\sinh\omega t\right)\right]\right\} \tag{24.3.25}$$

CHAPTER 25

TWO-DEGREE-OF-FREEDOM RESTRICTED SYSTEMS WITH THREE SEQUENCES OF FLEXIBLE LINKS

25.1. Motion of the Masses of a Restricted System with Three Sequences of Flexible Links Due to the Initial Displacements and Initial Velocities of the Masses

We consider the operational process of a restricted two-degree-of-freedom system, the masses of which are connected to each other by a spring, while each mass of the system is attached by a spring to a non-movable support. The system is moving on a horizontal frictionless surface due to the initial conditions of motion. The air resistance to the motion of the system is negligible.

It should be noted that the notations of all involved parameters are applicable only to the current chapter.

Figure 25.1.1 shows a schematic diagram of the system, where K_1, K_2, and K_3 are the stiffness coefficients of the respective springs, while the rest of the notations in this figure are self-explanatory.

Based on the schematic diagram shown in Figure 25.1.1 and the considerations mentioned above, we compose a pair of simultaneous differential equations describing the motion of the system:

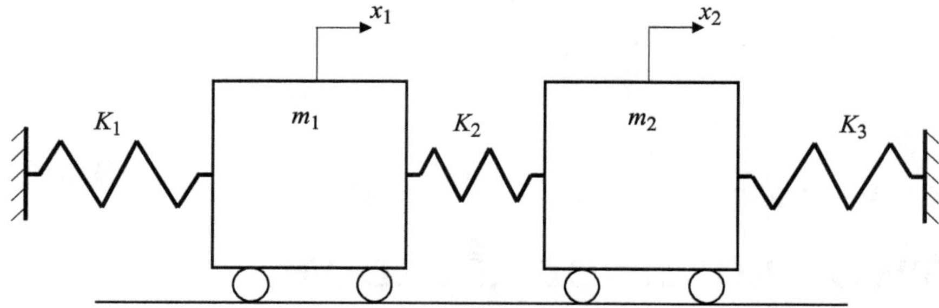

Fig. 25.1.1. Schematic diagram of a restricted two-degree-of-freedom system with three sequences of flexible links.

$$m_1\frac{d^2x_1}{dt^2} + (K_1 + K_2 + K_3)(x_1 - x_2) = 0 \qquad (25.1.1)$$

$$m_2\frac{d^2x_2}{dt^2} + (K_1 + K_2 + K_3)(x_2 - x_1) = 0 \qquad (25.1.2)$$

The initial conditions of motion are

for $t = 0$

$$x_1 = S_1; \quad \frac{dx_1}{dt} = V_1;$$

$$x_2 = S_2; \quad \frac{dx_2}{dt} = V_2 \qquad (25.1.3)$$

Dividing equations (25.1.1) and (25.1.2) by m_1 and m_2 while applying to these equations Laplace Transform pairs 2, 5, 2, and 1, we convert these equations with initial conditions of motion (25.1.3) from the time domain into a system of two simultaneous algebraic equations with two unknowns in each equation in the Laplace domain:

$$s^2x_1(s) - sV_1 - s^2S_1 + \omega_{11}^2x_1(s) - \omega_{11}^2x_2(s) + \omega_{21}^2x_1(s)$$
$$- \omega_{21}^2x_2(s) + \omega_{31}^2x_1(s) - \omega_{31}^2x_2(s) = 0 \qquad (25.1.4)$$
$$s^2x_2(s) - sV_2 - s^2S_2 + \omega_{12}^2x_2(s) - \omega_{12}^2x_1(s) + \omega_{22}^2x_2(s)$$
$$- \omega_{22}^2x_1(s) + \omega_{32}^2x_2(s) - \omega_{32}^2x_1(s) = 0 \qquad (25.1.5)$$

where, just for this chapter, we denote

$$\omega_{11}^2 = \frac{K_1}{m_1}; \quad \omega_{21}^2 = \frac{K_2}{m_1}; \quad \omega_{31}^2 = \frac{K_3}{m_1} \quad \omega_{12}^2 = \frac{K_1}{m_2};$$

$$\omega_{22}^2 = \frac{K_2}{m_2}; \quad \omega_{32}^2 = \frac{K_3}{m_2} \qquad (25.1.6)$$

Rearranging equations (25.1.4) and (25.1.5), we may write

$$x_1(s)(s^2 + \omega_{11}^2 + \omega_{21}^2 + \omega_{31}^2) = sV_1 + s^2 S_1 + x_2(s)(\omega_{11}^2 + \omega_{21}^2 + \omega_{31}^2)$$
$$(25.1.7)$$

$$x_2(s)(s^2 + \omega_{12}^2 + \omega_{22}^2 + \omega_{32}^2) = sV_2 + s^2 S_2 + x_1(s)(\omega_{12}^2 + \omega_{22}^2 + \omega_{32}^2)$$
$$(25.1.8)$$

The system of two simultaneous algebraic equations (22.1.7) and (22.1.8) contains two unknowns $x_1(s)$ and $x_2(s)$ in each equation. Applying to these equations the method of substitutions, we get two equations with one unknown in each equation. These unknowns, respectively, represent the displacements of the masses in the Laplace domain:

$$x_1(s) = \frac{sV_1}{s^2 + \omega^2} + \frac{V_1(\omega_{12}^2 + \omega_{22}^2 + \omega_{32}^2) + V_2(\omega_{11}^2 + \omega_{21}^2 + \omega_{31}^2)}{s(s^2 + \omega^2)}$$
$$+ \frac{s^2 S_1}{s^2 + \omega^2} + \frac{S_1(\omega_{12}^2 + \omega_{22}^2 + \omega_{32}^2) + S_2(\omega_{11}^2 + \omega_{21}^2 + \omega_{31}^2)}{s^2 + \omega^2}$$
$$(25.1.9)$$

$$x_2(s) = \frac{sV_2}{s^2 + \omega^2} + \frac{V_2(\omega_{11}^2 + \omega_{21}^2 + \omega_{31}^2) + V_1(\omega_{12}^2 + \omega_{22}^2 + \omega_{32}^2)}{s(s^2 + \omega^2)}$$
$$+ \frac{s^2 S_2}{s^2 + \omega^2} + \frac{S_2(\omega_{11}^2 - \omega_{21}^2 + \omega_{31}^2) + S_1(\omega_{12}^2 + \omega_{22}^2 + \omega_{32}^2)}{s^2 + \omega^2}$$
$$(25.1.10)$$

where ω is the natural compound frequency of the system, while

$$\omega^2 = \omega_{11}^2 + \omega_{21}^2 + \omega_{31}^2 + \omega_{12}^2 + \omega_{22}^2 + \omega_{32}^2 \qquad (25.1.11)$$

Applying to equations (25.1.9) and (25.1.10) Laplace Transform pairs 23, 39, 29, and 17, we invert these equations from the Laplace domain into the time domain and obtain the solutions of differential equations (25.1.1) and (25.1.2) for the initial conditions of motion (25.1.3):

$$x_1 = \frac{V_1}{\omega}\sin\omega t + \frac{1}{\omega^2}[V_1(\omega_{12}^2 + \omega_{22}^2 + \omega_{32}^2) + V_2(\omega_{11}^2 + \omega_{21}^2 + \omega_{31}^2)]$$

$$\times\left(t - \frac{1}{\omega}\sin\omega t\right) + S_1\cos\omega t + \frac{1}{\omega^2}[S_1(\omega_{12}^2 + \omega_{22}^2 + \omega_{32}^2)$$

$$+ S_2(\omega_{11}^2 + \omega_{21}^2 + \omega_{31}^2)](1 - \cos\omega t) \tag{25.1.12}$$

$$x_2 = \frac{V_2}{\omega}\sin\omega t + \frac{1}{\omega^2}[V_2(\omega_{11}^2 + \omega_{21}^2 + \omega_{31}^2) + V_1(\omega_{12}^2 + \omega_{22}^2 + \omega_{32}^2)]$$

$$\times\left(t - \frac{1}{\omega}\sin\omega t\right) + S_2\cos\omega t + \frac{1}{\omega^2}[S_2(\omega_{11}^2 + \omega_{21}^2 + \omega_{31}^2)$$

$$+ S_1(\omega_{12}^2 + \omega_{22}^2 + \omega_{32}^2)](1 - \cos\omega t) \tag{25.1.13}$$

Simplifying equations (22.1.12) and (22.1.13), we obtain

$$x_1 = \frac{1}{\omega^2}(S_1 - S_2)(\omega_{11}^2 + \omega_{21}^2 + \omega_{31}^2)\cos\omega t$$

$$+ \frac{1}{\omega^3}(V_1 - V_2)(\omega_{11}^2 + \omega_{21}^2 + \omega_{31}^2)\sin\omega t + S_1(\omega_{12}^2 + \omega_{22}^2 + \omega_{32}^2)$$

$$+ S_2(\omega_{11}^2 + \omega_{21}^2 + \omega_{31}^2)] + \frac{1}{\omega^2}[V_1(\omega_{12}^2 + \omega_{22}^2 + \omega_{32}^2)$$

$$+ V_2(\omega_{11}^2 + \omega_{21}^2 + \omega_{31}^2)]t \tag{25.1.14}$$

$$x_2 = \frac{1}{\omega^2}(S_2 - S_1)(\omega_{12}^2 + \omega_{22}^2 + \omega_{32}^2)\cos\omega t$$

$$+ \frac{1}{\omega^3}(V_2 - V_1)(\omega_{12}^2 + \omega_{22}^2 + \omega_{32}^2)\sin\omega t + S_1(\omega_{12}^2 + \omega_{22}^2 + \omega_{32}^2)$$

$$+ S_2(\omega_{11}^2 + \omega_{21}^2 + \omega_{31}^2)] + \frac{1}{\omega^2}[V_1(\omega_{12}^2 + \omega_{22}^2 + \omega_{32}^2)$$

$$+ V_2(\omega_{11}^2 + \omega_{21}^2 + \omega_{31}^2)]t \tag{25.1.15}$$

Equations (25.1.14) and (25.1.15) show that if $S_1 \neq S_2$ and $V_1 \neq V_2$, then the first two terms in both equations will have opposite algebraic values, respectively; therefore, the two masses are in anti-phase vibratory motion. The next two respective terms in these equations are equal and do not influence the mode of vibration. In the cases when $S_1 = S_2$ and $V_1 \neq V_2$ or $S_1 \neq S_2$ and $V_1 = V_2$, one of these two terms is canceled out and the remaining terms have opposite algebraic values. Therefore, in these cases, the two masses are in anti-phase vibrational motion. In addition, in the case when $S_1 = S_2$ and $V_1 = V_2$, the masses should perform uniform motion with the same velocity. First of all, this is physically impossible since the masses are connected by springs to non-movable supports. Second, if both masses move with the same velocity, then their motion is not relative, and they move as one body, and the system does not have two degrees of freedom. Therefore, these analyzed expressions are not applicable to the case when $S_1 = S_2$ and $V_1 = V_2$.

Taking in equations (25.1.14) and (25.1.15) that $S_1 = S_2 = 0$, we obtain

$$x_1 = \frac{1}{\omega^3}(V_1 - V_2)(\omega_{11}^2 + \omega_{21}^2 + \omega_{31}^2)\sin\omega t$$
$$+ \frac{1}{\omega^2}[V_1(\omega_{12}^2 + \omega_{22}^2 + \omega_{32}^2) + V_2(\omega_{11}^2 + \omega_{21}^2 + \omega_{31}^2)]t \quad (25.1.16)$$

$$x_2 = \frac{1}{\omega^3}(V_2 - V_1)(\omega_{12}^2 + \omega_{22}^2 + \omega_{32}^2)\sin\omega t$$
$$+ \frac{1}{\omega^2}[V_1(\omega_{12}^2 + \omega_{22}^2 + \omega_{32}^2) + V_2(\omega_{11}^2 + \omega_{21}^2 + \omega_{31}^2)]t \quad (25.1.17)$$

Equations (25.1.16) and (25.1.17) show that if $V_1 \neq V_2$, then the masses perform anti-phase vibrational motion. Taking in equations (25.1.14) and (25.1.15) that $V_1 = V_2 = 0$ and $S_1 \neq S_2$, we obtain

$$x_1 = \frac{1}{\omega^2}(S_1 - S_2)(\omega_{11}^2 + \omega_{21}^2 + \omega_{31}^2)\cos\omega t$$
$$+ \frac{1}{\omega^2}[S_1(\omega_{12}^2 + \omega_{22}^2 + \omega_{32}^2) + S_2(\omega_{11}^2 + \omega_{21}^2 + \omega_{31}^2)] \quad (25.1.18)$$

$$x_2 = \frac{1}{\omega^2}(S_2 - S_1)(\omega_{12}^2 + \omega_{22}^2 + \omega_{32}^2)\cos\omega t$$
$$+ \frac{1}{\omega^2}[S_1(\omega_{12}^2 + \omega_{22}^2 + \omega_{32}^2) + S_2(\omega_{11}^2 + \omega_{21}^2 + \omega_{31}^2)] \quad (25.1.19)$$

Equations (25.1.18) and (25.1.19) show that the masses are in anti-phase vibrational motion.

In conclusion, it is justifiable to emphasize that due to the initial conditions of motion of the masses of a restricted two-degree-of-freedom system, where the masses are connected by a spring and are attached by springs to non-movable supports, then the two masses perform anti-phase vibrational motion.

25.1.1. *Velocities and accelerations*

Taking the first derivatives from equations (25.1.14) and (25.1.15), we obtain the equations describing the velocities of the masses:

$$\frac{dx_1}{dt} = \frac{1}{\omega^2}(V_1 - V_2)(\omega_{11}^2 + \omega_{21}^2 + \omega_{31}^2)\cos\omega t$$

$$-\frac{1}{\omega}(S_1 - S_2)(\omega_{11}^2 + \omega_{21}^2 + \omega_{31}^2)\sin\omega t$$

$$+\frac{1}{\omega^2}[V_1(\omega_{12}^2 + \omega_{22}^2 + \omega_{32}^2) + V_2((\omega_{11}^2 + \omega_{21}^2 + \omega_{31}^2)]$$

$$(25.1.20)$$

$$\frac{dx_2}{dt} = \frac{1}{\omega^2}(V_2 - V_1)(\omega_{12}^2 + \omega_{22}^2 + \omega_{32}^2)\cos\omega t$$

$$-\frac{1}{\omega}(S_2 - S_1)(\omega_{12}^2 + \omega_{22}^2 + \omega_{32}^2)\sin\omega t$$

$$+\frac{1}{\omega^2}[V_1(\omega_{12}^2 + \omega_{22}^2 + \omega_{32}^2) + V_2(\omega_{11}^2 + \omega_{21}^2 + \omega_{31}^2)] \quad (25.1.21)$$

Supposing for equations (25.1.14), (25.1.15), (25.1.20), and (25.1.21) that $t = 0$, we determine the initial displacements and velocities of the masses, which, respectively, are $x_1 = S_1$, $x_2 = S_2$, $\frac{dx_1}{dt} = V_1$, and $\frac{dx_2}{dt} = V_2$, as it should be according to the initial conditions of motion.

Taking from equations (25.1.20) and (25.1.21) the first derivatives, we obtain the equations for the accelerations of the masses, respectively:

$$\frac{d^2x_1}{dt^2} = -\frac{1}{\omega}(V_1 - V_2)(\omega_{11}^2 + \omega_{21}^2 + \omega_{31}^2)\sin\omega t$$

$$-(S_1 - S_2)(\omega_{11}^2 + \omega_{21}^2 + \omega_{31}^2)\cos\omega t \quad (25.1.22)$$

$$\frac{d^2 x_2}{dt^2} = -\frac{1}{\omega}(V_2 - V_1)(\omega_{12}^2 + \omega_{22}^2 + \omega_{32}^2)\sin \omega t$$

$$- (S_2 - S_1)(\omega_{12}^2 + \omega_{22}^2 + \omega_{32}^2)\cos \omega t \qquad (25.1.23)$$

Assuming in equations (25.1.22) and (25.3.23) that $t = 0$, we determine the initial accelerations of the masses at the beginning of the motion:

$$\frac{d^2 x_1}{dt^2} = -(S_1 - S_2)(\omega_{11}^2 + \omega_{21}^2 + \omega_{31}^2) \qquad (25.1.24)$$

$$\frac{d^2 x_2}{dt^2} = -(S_2 - S_1)(\omega_{12}^2 + \omega_{22}^2 + \omega_{32}^2) \qquad (25.1.25)$$

25.2. Restricted Systems with Three Sequences of Flexible Links, the Masses of Which Are Subjected to Constant Resisting and Constant Active Forces

Consider the operational process of those systems associated with the motion of the masses on a horizontal frictionless surface, while the masses of the system are subjected to constant resisting and constant active forces. The air resistance to the motion of the systems is negligible. Figure 25.2.1 shows a schematic diagram of the described system, the masses of which are subjected to the constant resisting forces R_1 and R_2 and the constant active forces P_1 and P_2.

The rest of the notations in this figure are self-explanatory.

Based on the schematic diagram shown in Figure 25.2.1 and the related considerations, we compose the following pair of simultaneous differential equations of motion of the system:

$$m_1 \frac{d^2 x_1}{dt^2} + (K_1 + K_2 + K_3)(x_1 - x_2) + R_1 = P_1 \qquad (25.2.1)$$

$$m_2 \frac{d^2 x_2}{dt^2} + (K_1 + K_2 + K_3)(x_2 - x_1) + R_2 = P_2 \qquad (25.2.2)$$

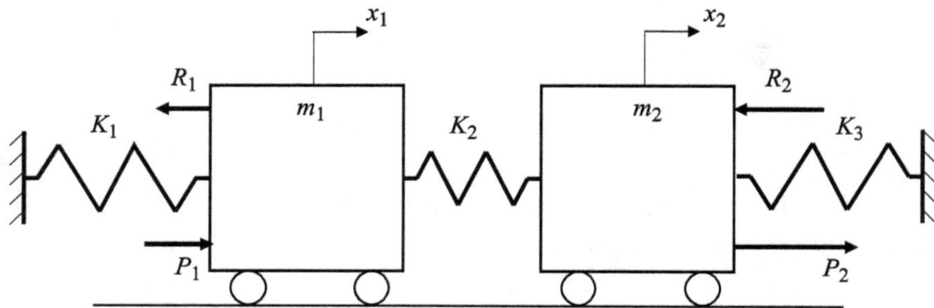

Fig. 25.2.1. Schematic diagram of a restricted two-degree-of-freedom system, the masses of which are connected to each other by a spring, while both masses are attached to non-movable supports by additional springs and the masses are subjected to constant resisting and constant active forces.

The initial conditions of motion are

$$x_1 = 0, \quad \frac{dx_1}{dt} = 0$$

For $t = 0$ (25.2.3)

$$x_2 = 0, \quad \frac{dx_2}{dt} = 0$$

Dividing differential equations (25.2.1) and (25.2.2) by m_1 and m_2, respectively, while applying to them Laplace Transform pairs 2, 5, 2, 1, 2, and 2, we convert these equations with the initial conditions of motion according to expression (25.2.3) from the time domain into a system of two simultaneous algebraic equations with two unknowns in the Laplace domain:

$$s^2 x_1(s) + \omega_{11}^2 x_1(s) - \omega_{11}^2 x_2(s) + \omega_{21}^2 x_1(s)$$

$$- \omega_{21}^2 x_2(s) + \omega_{31}^2 x_1(s) - \omega_{31}^2 x_2(s) + r_1 = p_1 \tag{25.2.4}$$

$$s^2 x_2(s) + \omega_{12}^2 x_2(s) - \omega_{12}^2 x_1(s) + \omega_{22}^2 x_2(s)$$

$$- \omega_{22}^2 x_1(s) + \omega_{32}^2 x_2(s) - \omega_{32}^2 x_1(s) + r_2 = p_2 \tag{25.2.5}$$

where

$$r_1 = \frac{R_1}{m_1}; \quad r_2 = \frac{R_2}{m_2}; \quad p_1 = \frac{P_1}{m_1}; \quad p_2 = \frac{P_2}{m_2} \tag{25.2.6}$$

while we must keep in mind that $P_1 > R_1$ and $P_2 > R_2$. Rearranging the pair of simultaneous algebraic equations (22.2.4) and (22.2.5) and applying to them the method of substitutions, we obtain the equations describing the displacements $x_1(s)$ and $x_2(s)$ in the Laplace domain:

$$x_1(s) = \frac{p_1 - r_1}{s^2 + \omega^2}$$
$$+ \frac{(\omega_{12}^2 + \omega_{22}^2 + \omega_{32}^2)(p_1 - r_1) + (\omega_{11}^2 + \omega_{21}^2 + \omega_{31}^2)(p_2 - r_2)}{s^2(s^2 + \omega^2)}$$

$$(25.2.7)$$

$$x_2(s) = \frac{p_2 - r_2}{l^2 + \omega^2}$$
$$+ \frac{(\omega_{11}^2 + \omega_{21}^2 + \omega_{31}^2)(p_2 - r_2) + (\omega_{12}^2 + \omega_{22}^2 + \omega_{32}^2)(p_1 - r_1)}{s^2(s^2 + \omega^2)}$$

$$(25.2.8)$$

Applying to equations (25.2.7) and (25.2.8) Laplace Transform pairs 1, 17, and 51, we invert these equations from the Laplace domain into the time domain and obtain the solutions of differential equations of motion (25.2.1) and (25.2.2) with their initial conditions of motion:

$$x_1 = \frac{p_1 - r_1}{\omega^2}(1 - \cos \omega t) + [(\omega_{12}^2 + \omega_{22}^2 + \omega_{32}^2)(p_1 - r_1)$$
$$+ (\omega_{11}^2 + \omega_{21}^2 + \omega_{31}^2)(p_2 - r_2)]\left[\frac{1}{\omega^4}(\cos \omega t - 1) + \frac{1}{2\omega^2}t^2\right]$$

$$(25.2.9)$$

$$x_2 = \frac{p_2 - r_2}{\omega^2}(1 - \cos \omega t) + [(\omega_{11}^2 + \omega_{21}^2 + \omega_{31}^2)(p_2 - r_2)$$
$$+ (\omega_{12}^2 + \omega_{22}^2 + \omega_{32}^2)(p_1 - r_1)]\left[\frac{1}{\omega^4}(\cos \omega t - 1) + \frac{1}{2\omega^2}t^2\right]$$

$$(25.2.10)$$

Assuming for equations (25.2.9) and (25.2.10) that $t = 0$, we determine that $x_1 = 0$ and $x_2 = 0$, which correspond to initial conditions of motion (25.2.3).

Taking the first derivatives from equations (25.2.9) and (25.2.10), we determine the velocities of the two masses:

$$\frac{dx_1}{dt} = \frac{p_1 - r_1}{\omega} \sin \omega t + [(\omega_{12}^2 + \omega_{22}^2 + \omega_{32}^2)(p_1 - r_1)$$

$$+ (\omega_{11}^2 + \omega_{21}^2 + \omega_{31}^2)(p_2 - r_2)] \left[\frac{1}{\omega^2} \left(t - \frac{1}{\omega} \sin \omega t \right) \right]$$

$$(25.2.11)$$

$$\frac{dx_2}{dt} = \frac{p_2 - r_2}{\omega} \sin \omega t + [(\omega_{11}^2 + \omega_{21}^2 + \omega_{31}^2)(p_2 - r_2)$$

$$+ (\omega_{12}^2 + \omega_{22}^2 + \omega_{32}^2)(p_1 - r_1)] \left[\frac{1}{\omega^2} \left(t - \frac{1}{\omega} \sin \omega t \right) \right]$$

$$(25.2.12)$$

Supposing in equations (25.2.11) and (25.2.12) that $t = 0$, we calculate that $\frac{dx_1}{dt} = 0$ and $\frac{dx_2}{dt} = 0$, as it should be according to initial conditions of motion (25.2.3).

Taking the first derivatives from equations (25.2.11) and (25.2.12), we determine the accelerations of the masses, respectively:

$$\frac{d^2 x_1}{dt^2} = (p_1 - r_1) \cos \omega t + [(\omega_{12}^2 + \omega_{22}^2 + \omega_{32}^2)(p_1 - r_1)$$

$$+ (\omega_{11}^2 + \omega_{21}^2 + \omega_{31}^2)(p_2 - r_2)] \frac{1}{\omega^2}(1 - \cos \omega t) \qquad (25.2.13)$$

$$\frac{d^2 x_2}{dt^2} = (p_2 - r_2) \cos \omega t + [(\omega_{11}^2 + \omega_{21}^2 + \omega_{31}^2)(p_2 - r_2)$$

$$+ (\omega_{12}^2 + \omega_{22}^2 + \omega_{32}^2)(p_1 - r_1)] \frac{1}{\omega^2}(1 - \cos \omega t) \qquad (25.2.14)$$

Taking in equations (25.2.13) and (25.2.14) that $t = 0$, we determine the initial accelerations of the masses, respectively, at the beginning of the motion:

$$\frac{d^2 x_1}{dt^2} = p_1 - r_1 \qquad (25.2.15)$$

$$\frac{d^2 x_2}{dt^2} = p_2 - r_2 \qquad (25.2.16)$$

25.3. Restricted Systems with Three Sequences of Flexible Links and Subjected to Harmonic Forces

A schematic diagram representing a restricted system subjected to the actions of harmonic forces is shown in Figure 25.3.1, where A_1 and A_2 are the amplitudes of the harmonic forces, while φ_1 and φ_2 are, respectively, the frequencies of the harmonic functions. The rest of the notations in this figure are self-explanatory. The system is moving on a horizontal frictionless surface. The air resistance to the motion of the system is negligible.

Based on the schematic diagram shown in Figure 25.3.1 and the considerations mentioned above, we compose a pair of simultaneous differential equations of motion of the system:

$$m_1 \frac{d^2 x_1}{dt^2} + (K_1 + K_2 + K_3)(x_1 - x_2) = A_1 \cos \varphi_1 t \qquad (25.3.1)$$

$$m_2 \frac{d^2 x_2}{dt^2} + (K_1 + K_2 + K_3)(x_2 - x_1) = A_2 \cos \varphi_2 t \qquad (25.3.2)$$

Dividing equations (25.3.1) and (25.3,2) by m_1 and m_2, respectively, while applying Laplace Transform pairs 2, 5, 2, 1, 2, and 29 to these equations with initial conditions of motion (25.3.3), we convert

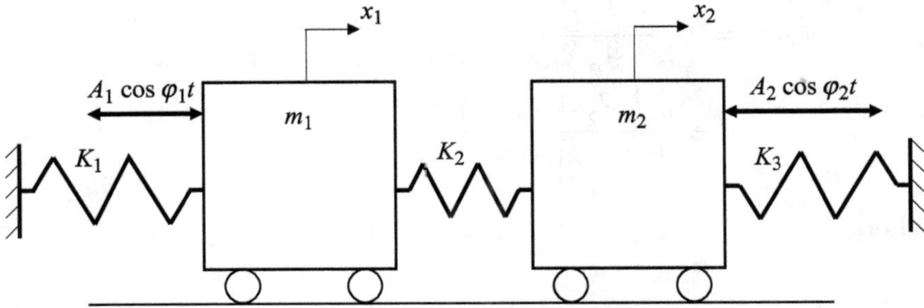

Fig. 25.3.1. Schematic diagram of a restricted two-degree-of-freedom system, the masses of which are connected to each other by a spring, while both masses are attached to non-movable supports by additional springs and the masses are subjected to harmonic forces.

these equations with their initial conditions of motion from the time domain into a system of two simultaneous algebraic equations with two unknowns in each of the equations in the Laplace domain:

$$s^2 x_1(s) + (\omega_{11}^2 + \omega_{21}^2 + \omega_{31}^2) x_1(s)$$

$$- (\omega_{11}^2 + \omega_{21}^2 + \omega_{31}^2) x_2(s) = a_1 \frac{s^2}{s^2 + \varphi_1^2} \tag{25.3.3}$$

$$s^2 x_2(s) + (\omega_{12}^2 + \omega_{22}^2 + \omega_{32}^2) x_2(s)$$

$$- (\omega_{12}^2 + \omega_{22}^2 + \omega_{32}^2) x_1(s) = a_2 \frac{s^2}{s^2 + \varphi_2^2} \tag{25.3.4}$$

Rearranging equations (25.3.3) and (25.3.4) and applying to them the method of substitutions, we eliminate one unknown in each of the equations and obtain expressions that allow us to determine the displacements $x_1(s)$ and $x_2(s)$ in the Laplace domain:

$$x_1(s) = \frac{s^2 a_1}{(s^2 + \varphi_1^2)(s^2 + \omega^2)} + \frac{a_1(\omega_{12}^2 + \omega_{22}^2 + \omega_{32}^2)}{(s^2 + \varphi_1^2)(s^2 + \omega^2)}$$

$$+ \frac{a_2(\omega_{11}^2 + \omega_{21}^2 + \omega_{31}^2)}{(s^2 + \varphi_2^2)(s^2 + \omega^2)} \tag{25.3.5}$$

$$x_2(s) = + \frac{s^2 a_2}{(s^2 + \varphi_2^2)(s^2 + \omega^2)} + \frac{a_2(\omega_{11}^2 + \omega_{21}^2 + \omega_{31}^2)}{(s^2 + \varphi_2^2)(s^2 + \omega^2)}$$

$$+ \frac{a_1(\omega_{12}^2 + \omega_{22}^2 + \omega_{32}^2)}{(s^2 + \varphi_1^2)(s^2 + \omega^2)} \tag{25.3.6}$$

where

$$a_1 = \frac{A_1}{m_1}; \quad a_2 = \frac{A_2}{m_2} \tag{25.3.7}$$

Applying to equations (25.3.5) and (25.3.6) Laplace Transform pairs 70, 55, and 55, we invert these equations from the Laplace domain into the time domain and obtain the solutions of differential equations (25.3.1) and (25.3.2) with their initial conditions of

motion:

$$x_1 = a_1 \frac{\cos \omega t - \cos \varphi_1 t}{\varphi_1^2 - \omega^2} + \frac{a_1(\omega_{12}^2 + \omega_{22}^2 + \omega_{32}^2)}{\varphi_1^2 - \omega^2}$$

$$\times \left[\frac{1}{\omega^2}(1 - \cos \omega t) - \frac{1}{\varphi_1^2}(1 - \cos \varphi_1 t) \right]$$

$$+ \frac{a_2(\omega_{11}^2 + \omega_{21}^2 + \omega_{31}^2)}{\varphi_2^2 - \omega^2} \left[\frac{1}{\omega^2}(1 - \cos \omega t) - \frac{1}{\varphi_2^2}(1 - \cos \varphi_2 t) \right]$$

$$\tag{25.3.8}$$

$$x_2 = a_2 \frac{\cos \omega t - \cos \varphi_2 t}{\varphi_2^2 - \omega^2} + \frac{a_2(\omega_{11}^2 + \omega_{21}^2 + \omega_{31}^2)}{\varphi_2^2 - \omega^2}$$

$$\times \left[\frac{1}{\omega^2}(1 - \cos \omega t) - \frac{1}{\varphi_2^2}(1 - \cos \varphi_2 t) \right]$$

$$+ \frac{a_1(\omega_{12}^2 + \omega_{22}^2 + \omega_{32}^2)}{\varphi_1^2 - \omega^2} \left[\frac{1}{\omega^2}(1 - \cos \omega t) - \frac{1}{\varphi_1^2}(1 - \cos \varphi_1 t) \right]$$

$$\tag{25.3.9}$$

Assuming for equations (25.3.8) and (29.3.9) that $t = 0$, we determine that $x_1 = 0$ and $x_2 = 0$, as it should be according to the initial conditions of motion.

Taking from equations (25.3.8) and (25.3.9) the first derivatives, we determine the velocities of the masses:

$$\frac{dx_1}{dt} = \frac{a_1}{\varphi_1^2 - \omega^2}(\varphi_1 \sin \varphi_1 t - \omega \sin \omega t) + \frac{a_1(\omega_{12}^2 + \omega_{22}^2 + \omega_{32}^2)}{\varphi_1^2 - \omega^2}$$

$$\times \left(\frac{1}{\omega} \sin \omega t - \frac{1}{\varphi_1} \sin \varphi_1 t \right) + \frac{a_2(\omega_{11}^2 + \omega_{21}^2 + \omega_{31}^2)}{\varphi_2^2 - \omega^2}$$

$$\times \left(\frac{1}{\omega} \sin \omega t - \frac{1}{\varphi_2} \sin \varphi_2 t \right) \tag{25.3.10}$$

$$\frac{dx_2}{dt} = \frac{a_2}{\varphi_2^2 - \omega^2}(\varphi_2 \sin \varphi_2 t - \omega \sin \omega t) + \frac{a_2(\omega_{11}^2 + \omega_{21}^2 + \omega_{31}^2)}{\varphi_2^2 - \omega^2}$$

$$\times \left(\frac{1}{\omega} \sin \omega t - \frac{1}{\varphi_2} \sin \varphi_2 t \right) + \frac{a_1(\omega_{12}^2 + \omega_{22}^2 + \omega_{32}^2)}{\varphi_1^2 - \omega^2}$$

$$\times \left(\frac{1}{\omega} \sin \omega t - \frac{1}{\varphi_1} \sin \varphi_1 t \right) \tag{25.3.11}$$

By accepting in equations (22.3.10) and (22.3.11) that $t = 0$, we determine that $\frac{dx_1}{dt} = 0$ and $\frac{dx_2}{dt} = 0$, as expected according to the initial conditions of motion.

25.3.1. *Resonance of a restricted system subjected to harmonic forces*

The phenomenon of resonance is possible if the compound natural frequency is equal to the frequency of the exciting force. Therefore, we consider a case when in equations (25.3.8) and (25.3.9) the compound natural frequency ω, introduced in equation (25.1.11), is equal to the frequency of the exiting force applied to one of the masses, for instance, to the first mass; therefore, we may write

$$\omega = \varphi_1 \tag{25.3.12}$$

Similar investigations on resonance carried out for unrestricted two-degree-of-freedom systems were presented in Chapters 19 and 22.

According to equation (25.3.12), several denominators of the fractional members of equations (25.3.8) and (25.3.9) become equal to zero. These members have the potential to tend to infinity and cause resonance of the system. The rest of the members of these equations cannot cause resonance and are eliminated from the current analysis. Therefore, for the purpose of investigation on resonance, we restructure equations (25.3.8) and (25.3.9) by keeping in them only the terms whose values may potentially tend to infinity:

$$x_1' = \frac{a_1(\cos\omega t - \cos\varphi_1 t)}{\varphi_1^2 - \omega^2} + \frac{a_1(\omega_{12}^2 + \omega_{22}^2 + \omega_{32}^2)}{\varphi_1^2 - \omega^2}$$

$$\times \left[\frac{1}{\omega^2}(1 - \cos\omega t) - \frac{1}{\varphi_1^2}(1 - \cos\varphi_1 t) \right] \tag{25.3.13}$$

$$x_2' = \frac{a_1(\omega_{12}^2 + \omega_{22}^2 + \omega_{32}^2)}{\varphi_1^2 - \omega^2} \left[\frac{1}{\omega^2}(1 - \cos\omega t) - \frac{1}{\varphi_1^2}(1 - \cos\varphi_1 t) \right]$$

$$\tag{25.3.14}$$

where x_1' and x_2' represent the displacements that may tend to infinity. According to equation (25.3.12), both terms in equations (25.3.13) and (25.3.14) represent indeterminates of the type $\frac{0}{0}$.

Rearranging these equations, we have

$$x_1' = \frac{a_1(\cos\omega t - \cos\varphi_1 t)}{\varphi_1^2 - \omega^2}$$
$$+ \frac{a_1(\omega_{12}^2 + \omega_{22}^2 + \omega_{32}^2)(\varphi_1^2 - \omega^2 - \varphi_1^2\cos\omega t + \omega^2\cos\varphi_1 t)}{\omega^2\varphi_1^2(\varphi_1^2 - \omega^2)}$$

$$(25.3.15)$$

$$x_2' = \frac{a_1(\omega_{12}^2 + \omega_{22}^2 + \omega_{32}^2)(\varphi_1^2 - \omega^2 - \varphi_1^2\cos\omega t + \omega^2\cos\varphi_1 t)}{\omega^2\varphi_1^2(\varphi_1^2 - \omega^2)}$$

$$(25.3.16)$$

where x_1' and x_2' represent the displacements of the respective masses.

Applying to equations (25.3.15) and (25.3.16) L'Hopital's rule, we evaluate the values of the mentioned indeterminates:

$$x_1' = \lim_{\varphi_1^2 \to \omega^2} \frac{a_1(\cos\omega t - \cos\varphi_1 t)}{\varphi_1^2 - \omega^2}$$
$$+ \lim_{\varphi_1^2 \to \omega^2} \frac{a_1(\omega_{12}^2 + \omega_{22}^2 + \omega_{32}^2)(\varphi_1^2 - \omega^2 - \varphi_1^2\cos\omega t + \omega^2\cos\varphi_1 t)}{\omega^2\varphi_1^2(\varphi_1^2 - \omega^2)}$$

$$(25.3.17)$$

$$x_2' = \lim_{\varphi_1^2 \to \omega^2} \frac{a_1(\omega_{12}^2 + \omega_{22}^2 + \omega_{32}^2)(\varphi_1^2 - \omega^2 - \varphi_1^2\cos\omega t + \omega^2\cos\varphi_1 t)}{\omega^2\varphi_1^2(\varphi_1^2 - \omega^2)}$$

$$(25.3.18)$$

Differentiating the numerators and denominators of equations (25.3.17) and (25.3.18) with respect to ω, we have

$$x_1' = \lim_{\varphi_1^2 \to \omega^2} \frac{a_1(-t\sin\omega t)}{-2\omega}$$
$$+ \lim_{\varphi_1^2 \to \omega^2} \frac{a_1(\omega_{12}^2 + \omega_{22}^2 + \omega_{32}^2)(-2\omega + \varphi_1^2 t\sin\omega t + 2\omega\cos\varphi_1 t)}{2\omega\varphi_1^4 - 4\omega^3\varphi_1^2}$$

$$(25.3.19)$$

$$x_2' = \lim_{\varphi_1^2 \to \omega^2} \frac{a_1(\omega_{12}^2 + \omega_{22}^2 + \omega_{32}^2)(-2\omega + \varphi_1^2 t\sin\omega t + 2\omega\cos\varphi_1 t)}{2\omega\varphi_1^4 - 4\omega^3\varphi_1^2}$$

$$(25.3.20)$$

Continuing with the evaluation of the indeterminates, we obtain

$$x_1' = \frac{a_1 t \sin \omega t}{2\omega} + \frac{a_1(\omega_{12}^2 + \omega_{22}^2 + \omega_{32}^2)(-2 + \omega t \sin \omega t + 2\cos \varphi_1 t)}{-2\omega^4}$$

$$(25.3.21)$$

$$x_2' = \frac{a_1(\omega_{12}^2 + \omega_{22}^2 + \omega_{32}^2)(-2 + \omega t \sin \omega t + 2\cos \varphi_1 t)}{-2\omega^4} \tag{25.3.22}$$

Eliminating from equations (25.3.21) and (25.3.22) the terms that do not influence the resonance, we have

$$x_1' = \frac{a_1 t \sin \omega t}{2\omega} - \frac{a_1(\omega_{12}^2 + \omega_{22}^2 + \omega_{32}^2)t \sin \omega t}{2\omega^3} \tag{25.3.23}$$

$$x_2' = -\frac{a_1(\omega_{12}^2 + \omega_{22}^2 + \omega_{32}^2)t \sin \omega t}{2\omega^3} \tag{25.3.24}$$

Further analysis of equation (25.3.23) yields

$$
\begin{aligned}
x_1' &= \frac{a_1 \omega^2 t \sin \omega t}{2\omega^3} - \frac{a_1(\omega_{12}^2 + \omega_{22}^2 + \omega_{32}^2)t \sin \omega t}{2\omega^3} \\
&= \frac{a_1(\omega_{11}^2 + \omega_{21}^2 + \omega_{31}^2 + \omega_{12}^2 + \omega_{22}^2 + \omega_{32}^2 - \omega_{12}^2 - \omega_{22}^2 - \omega_{32}^2)t \sin \omega t}{2\omega^3} \\
&= \frac{a_1(\omega_{11}^2 + \omega_{21}^2 + \omega_{32}^2)t \sin \omega t}{2\omega^3} \tag{25.3.25}
\end{aligned}
$$

Equations (25.3.24) and (25.3.25) allow us to evaluate the above-mentioned indeterminates and indicate that the resonance will occur in the considered system:

$$X_1 = \frac{a_1(\omega_{11}^2 + \omega_{21}^2 + \omega_{31}^2)t \sin \omega t}{2\omega^3} \tag{25.3.26}$$

$$X_2 = \frac{a_1(\omega_{12}^2 + \omega_{22}^2 + \omega_{32}^2)t \sin \omega t}{2\omega^3} \tag{25.3.27}$$

where X_1 and X_2 represent the respective evaluations of the indeterminates.

It should be emphasized that the multiplier t in equations (25.3.26) and (25.3.27) causes the amplitudes of both masses to tend to infinity, bringing the system to a state of resonance.

25.3.2. *Numerical solution*

The following is a Python program to plot equation (25.3.8).
The resulting graphs are plotted in Figures 25.3.2 and 25.3.3.

```python
from matplotlib.pyplot import plot, show
from numpy import linspace, exp, cos, sqrt


omega12 = 30
omega22 = 25
omega32 = 20
omega11 = 30
omega21 = 25
omega31 = 20
omega = sqrt(omega12**2 + omega22**2 + omega32**2 + \
        omega11**2 + omega21**2 + omega31**2)
phi1 = 60
phi2 = 60

x1 = lambda t, a1, a2 : \
  a1*((cos(omega*t) - cos(phi1*t))/(phi1**2 - omega**2)) \
  + ((a1*(omega12**2 + omega22**2 + omega32**2))/(phi1**2 - omega**2)) \
   *((1/omega**2)*(1 - cos(omega*t) - (1/phi1**2)*(1 - cos(phi1*t)))) \
  + ((a2*(omega11**2 + omega21**2 + omega31**2))/(phi2**2 - omega**2)) \
   *((1/omega**2)*(1 - cos(omega*t)) - (1/phi2**2)*(1 - cos(phi2*t)))

t = linspace(0, 4, 1000)
plot(t, x1(t, a1=10, a2=1))
show()

plot(t, x1(t, a1=10, a2=10))
show()
```

Fig. 25.3.2. Graph of equation (25.3.8). Effect of beating: $a1 = 10$ and $a2 = 1$.

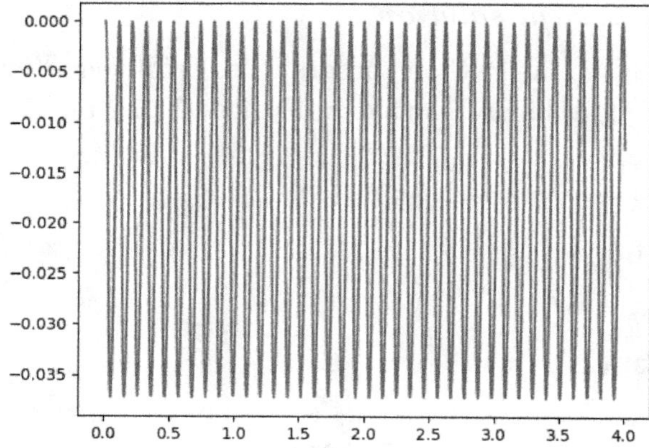

Fig. 25.3.3. Graph of equation (25.3.8). Harmonic vibration: $a1 = a2 = 10$.

CHAPTER 26

RESTRICTED TWO-DEGREE-OF-FREEDOM SYSTEMS WITH THREE SEQUENCES OF FLUID LINKS

26.1. Masses of a Restricted System with Three Sequences of Fluid Links Moving on a Horizontal Frictionless Surface Due to Their Initial Velocities

The operational process of the restricted system is associated with the motion of the masses on a horizontal frictionless surface due to their initial velocities. The schematic diagram representing the restricted system is shown in Figure 26.1.1, where C_1, C_2, and C_3 are the damping coefficients of the dashpots, while the masses are not subjected to any external forces. The rest of the notations in this figure are self-explanatory.

The notations of the parameters are only for this chapter.

Based on the schematic diagram shown in Figure 26.1.1 and the considerations mentioned above, we compose a pair of simultaneous differential equations describing the motion of the system:

$$m_1 \frac{d^2 x_1}{dt^2} + (C_1 + C_2 + C_3)\left(\frac{dx_1}{dt} - \frac{dx_2}{dt}\right) = 0 \qquad (26.1.1)$$

$$m_2 \frac{d^2 x_2}{dt^2} + (C_1 + C_2 + C_3)\left(\frac{dx_2}{dt} - \frac{dx_1}{dt}\right) = 0 \qquad (26.1.2)$$

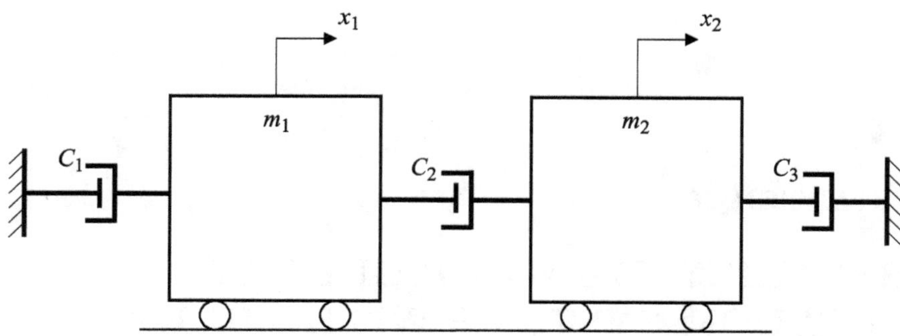

Fig. 26.1.1. Schematic diagram of a two-degree-of-freedom restricted system with three sequences of fluid links is moving on a horizontal frictionless surface due to the initial velocities of the masses.

The initial conditions of motion are

$$x_1 = S_1; \quad \frac{dx_1}{dt} = V_1;$$

for $\quad t = 0$ $\hspace{6cm}$ (26.1.3)

$$x_2 = S_2; \quad \frac{dx_2}{dt} = V_2$$

Dividing equations (26.1.1) and (26.1.2) by m_1 and m_2, respectively, and applying to them the Laplace Transform pairs 2, 5, 2, and 4, we convert these equations from the time domain into a system of two simultaneous algebraic equations with two unknowns in each equation, describing the displacements of the masses in the Laplace domain:

$$s^2 x_1(s) - sV_1 - s^2 S_1 + 2n_{11}sx_1(s) - 2n_{11}sS_1$$

$$+ 2n_{21}sx_1(s) - 2n_{21}sS_1 + 2n_{31}sx_1(s) - 2n_{31}sS_1$$

$$- 2n_{11}sx_2(s) + 2n_{11}sS_2 - 2n_{21}sx_2(s) + 2n_{21}sS_2$$

$$- 2n_{31}sx_2(s) + 2n_{31}sS_2 = 0 \hspace{3cm} (26.1.4)$$

$$s^2 x_2(s) - sV_2 - s^2 S_2 + 2n_{12}sx_2(s) - 2n_{12}sS_2$$

$$+ 2n_{22}sx_2(s) - 2n_{22}sS_2 + 2n_{32}sx_2(s) - 2n_{32}sS_2$$

$$- 2n_{12}sx_1(s) + 2n_{12}sS_1 - 2n_{22}sx_1(s) + 2n_{22}sS_1$$

$$- 2n_{32}sx_1(s) + 2n_{32}sS_1 = 0 \hspace{3cm} (26.1.5)$$

where

$$2n_{11} = \frac{C_1}{m_1}; \quad 2n_{21} = \frac{C_2}{m_1}; \quad 2n_{31} = \frac{C_3}{m_1};$$

$$2n_{12} = \frac{C_1}{m_2}; \quad 2n_{22} = \frac{C_2}{m_2}; \quad 2n_{32} = \frac{C_3}{m_2} \qquad (26.1.6)$$

Rearranging the terms in equations (26.1.4) and (26.1.5) and applying to them the method of substitutions, we obtain a pair of two simultaneous algebraic equations with one unknown in each equation, respectively, describing the displacements of the masses in the Laplace domain:

$$x_1(s) = \frac{V_1}{s + 2n} + \frac{2[V_1(n_{12} + n_{22} + n_{32}) + V_2(n_{11} + n_{21} + n_{31})]}{s(s + 2n)}$$

$$+ \frac{sS_1}{s + 2n} + \frac{2nS_1}{s + 2n} \qquad (26.1.7)$$

$$x_2(s) = \frac{V_2}{s + 2n} + \frac{2[V_2(n_{11} + n_{21} + n_{31}) + V_1(n_{12} + n_{22} + n_{32})]}{s(s + 2n)}$$

$$+ \frac{sS_2}{s + 2n} + \frac{2nS_2}{s + 2n} \qquad (26.1.8)$$

where

$$n = n_{11} + n_{21} + n_{31} + n_{12} + n_{22} + n_{32} \qquad (26.1.9)$$

Applying to equations (26.1.7) and (26.1.8) Laplace Transform pairs 8, 16, 14, and 8, we invert these equations from the Laplace domain into the time domain and obtain the solutions of the mentioned differential equations (26.1.1) and (26.1.2) with their initial conditions of motion according to expression (26.1.3):

$$x_1 = S_1 + \frac{V_1}{2n}(1 - e^{-2nt})$$

$$+ \frac{V_1(n_{12} + n_{22} + n_{32}) + V_2(n_{11} + n_{21} + n_{31})}{n}$$

$$\times \left[t + \frac{1}{2n}(e^{-2nt} - 1) \right] \qquad (26.1.10)$$

$$x_2 = S_2 + \frac{v_2}{2n}(1 - e^{-2nt})$$

$$+ \frac{V_2(n_{11} + n_{21} + n_{31}) + V_1(n_{12} + n_{22} + n_{32})}{n}$$

$$\times \left[t + \frac{1}{2n}(e^{-2nt} - 1) \right] \qquad (26.1.11)$$

Substituting into equations (26.1.10) and (26.1.11) that $t = 0$, we obtain $x_1 = S_1$ and $x_2 = S_2$, as it should be according to initial conditions of motion (26.1.3). Taking the first derivatives from equations (26.1.10) and (26.1.11), we determine the velocities of the masses:

$$\frac{dx_1}{dt} = V_1 e^{-2nt} + \frac{V_1(n_{12} + n_{22} + n_{32}) + V_2(n_{11} + n_{21} + n_{31})}{n}$$

$$\times (1 - e^{-2nt}) \qquad (26.1.12)$$

$$\frac{dx_2}{dt} = V_2 e^{-2nt} + \frac{V_2(n_{11} + n_{21} + n_{31}) + V_1(n_{12} + n_{22} + n_{32})}{n}$$

$$\times (1 - e^{-2nt}) \qquad (26.1.13)$$

Assuming in equations (26.1.12) and (26.1.13) that $t = 0$, we determine $\frac{dx_1}{dt} = V_1$ and $\frac{dx_2}{dt} = V_2$, as expected according to initial conditions of motion (26.1.3). Taking the first derivatives from equations (26.1.12) and (26.1.13), we determine the accelerations of the masses:

$$\frac{d^2 x_1}{dt^2} = -2(V_1 - V_2)(n_{11} + n_{21} + n_{31})e^{-2nt} \qquad (26.1.14)$$

$$\frac{d^2 x_2}{dt^2} = -2(V_2 - V_1)(n_{12} + n_{22} + n_{32})e^{-2nt} \qquad (26.1.15)$$

Hence, equations (26.1.10)–(26.1.15) describe the basic parameters of motion of the system.

26.2. Restricted System with Three Sequences of Fluid Links Is Moving on a Horizontal Surface and Subjected to Constant Resisting, Dry Friction, and Constant Active Forces

Consider an operational process of a restricted system with three sequences of fluid links, moving on a horizontal surface and being subjected to the action of constant resisting forces R_1 and R_2, dry friction forces F_1 and F_2, and constant active forces P_1 and P_2. The schematic diagram of the system is shown in Figure 26.2.1, the notations in which are self-explanatory.

Based on the schematic diagram shown in Figure 26.2.1 and the above-mentioned considerations, we compose the pair of simultaneous differential equations describing the motion of the system:

$$m_1 \frac{d^2 x_1}{dt^2} + (C_1 + C_2 + C_3) \left(\frac{dx_1}{dt} - \frac{dx_2}{dt} \right) + R_1 + F_1 = P_1$$

$$(26.2.1)$$

$$m_2 \frac{d^2 x_2}{dt^2} + (C_1 + C_2 + C_3) \left(\frac{dx_2}{dt} - \frac{dx_1}{dt} \right) + R_2 + F_2 = P_2$$

$$(26.2.2)$$

The initial conditions of motion are taken according to expression (26.1.3).

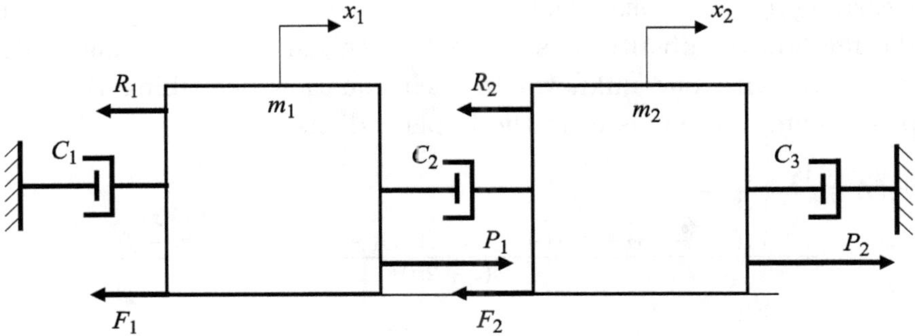

Fig. 26.2.1. Schematic diagram of a restricted system with three sequences of fluid links moving on a horizontal surface and being subjected to constant resisting, dry friction, and constant active forces.

Dividing equations (26.2.1) and (26.2.2) by m_1 and m_2, respectively, and applying Laplace Transform pairs 2, 5, 2, 4, 2, 2, and 2, we convert these equations with their initial conditions of motion from the time domain into a system of two simultaneous algebraic equations with two unknowns in each equation in the Laplace domain:

$$s^2 x_1(s) - sV_1 - s^2 S_1 + 2n_{11}sx_1(s) - 2n_{11}sS_1 - 2n_{11}sx_2(s)$$

$$+ 2n_{11}sS_2 + 2n_{21}sx_1(s) - 2n_{21}sS_1 - 2n_{21}sx_2(s)$$

$$+ 2n_{21}sS_2 + 2n_{31}sx_1(s) - 2n_{31}sS_1 - 2n_{31}sx_2(s)$$

$$+ 2n_{31}sS_2 + r_1 + f_1 = p_1 \tag{26.2.3}$$

$$s^2 x_2(s) - sV_2 - s^2 S_2 + 2n_{12}sx_2(s) - 2n_{12}sS_2 - 2n_{12}sx_1(s)$$

$$+ 2n_{12}sS_1 + 2n_{22}sx_2(s) - 2n_{22}sS_2 - 2n_{22}sx_1(s)$$

$$+ 2n_{22}sS_1 + 2n_{32}sx_2(s) - 2n_{32}sS_2 - 2n_{32}sx_1(s)$$

$$+ 2n_{32}sS_1 + r_2 + f_2 = p_2 \tag{26.2.4}$$

where

$$r_1 = \frac{R_1}{m_1}; \quad r_2 = \frac{R_2}{m_2}; \quad f_1 = \frac{F_1}{m_1};$$
$$f_2 = \frac{F_2}{m_2}; \quad p_1 = \frac{P_1}{m_1}; \quad p_2 = \frac{P_2}{m_2} \tag{26.2.5}$$

Rearranging equations (26.2.3) and (26.2.4) and applying to them the method of substitutions, we obtain two simultaneous algebraic equations with one unknown in each equation describing the displacements of the masses in the Laplace domain:

$$x_1(s) = \frac{V_1 + 2nS_1}{s + 2n}$$
$$+ \frac{2[V_1(n_{12} + n_{22} + n_{32}) + V_2(n_{11} + n_{21} + n_{31})] + p_1 - r_1 - f_1}{s(s + 2n)}$$
$$+ \frac{sS_1}{s + 2n}$$
$$+ \frac{2[(n_{12} + n_{22} + n_{32})(p_1 - r_1 - f_1) + (n_{11} + n_{21} + n_{31})(p_2 - r_2 - f_2)]}{s^2(s + 2n)}$$

$$\tag{26.2.6}$$

$$x_2(s) = \frac{V_2 + 2nS_2}{s + 2n}$$
$$+ \frac{2[V_2(n_{11} + n_{21} + n_{31}) + V_1(n_{12} + n_{22} + n_{32})] + p_2 - r_2 - f_2}{s(s + 2n)}$$
$$+ \frac{sS_2}{s + 2n}$$
$$+ \frac{2[(n_{11} + n_{21} + n_{31})(p_2 - r_2 - f_2) + (n_{12} + n_{22} + n_{32})(p_1 - r_1 - f_1)]}{s^2(l + 2n)}$$

$$(26.2.7)$$

Applying to equations (26.2.6) and (26.2.7) Laplace Transform pairs 8, 16, 14, and 48, we invert them from the Laplace domain into the time domain and obtain the solutions of differential equations (26.2.1) and (26.2.2) with initial conditions of motion (26.1.3):

$$x_1 = \frac{V_1 + 2nS_1}{2n}(1 - e^{-2nt})$$
$$+ \frac{2[V_1(n_{12} + n_{22} + n_{32}) + V_2(n_{11} + n_{21} + n_{31})] + p_1 - r_1 - f_1}{2n}$$
$$\times \left[t + \frac{1}{2n}\left(e^{-2nt} - 1\right) \right] + S_1 e^{-2nt}$$
$$+ \frac{(n_{12} + n_{22} + n_{32})(p_1 - r_1 - f_1) + (n_{11} + n_{21} + n_{31})(p_2 - r_2 - f_2)}{2n}$$
$$+ \left[t^2 - \frac{t}{n} - \frac{1}{2n^2}\left(e^{-2nt} - 1\right) \right]$$

$$(26.2.8)$$

$$x_2 = \frac{V_2 + 2nS_2}{2n}(1 - e^{-2nt})$$
$$+ \frac{2[V_2(n_{11} + n_{21} + n_{31}) + V_1(n_{12} + n_{22} + n_{32})] + p_2 - r_2 - f_2}{2n}$$
$$\times \left[t + \frac{1}{2n}\left(e^{-2nt} - 1\right) \right] + S_2 e^{-2nt}$$
$$+ \frac{(n_{11} + n_{21} + n_{31})(p_2 - r_2 - f_2) + (n_{12} + n_{22} + n_{32})(p_1 - r_1 - f_1)}{2n}$$
$$\times \left[t^2 - \frac{t}{n} - \frac{1}{2n^2}\left(e^{-2nt} - 1\right) \right]$$

$$(26.2.9)$$

Assuming in equations (26.2.8) and (26.2.9) that $t = 0$, we determine that $x_1 = S_1$ and $x_2 = S_2$, as expected according to initial conditions of motion (26.1.3). Taking from equations (26.2.8) and (26.2.9) the first derivatives, we obtain the equations of the velocities of the masses:

$$\frac{dx_1}{dt} = (V_1 + 2nS_1)\,e^{-2nt}$$
$$+ \frac{2[V_1(n_{12} + n_{22} + n_{32}) + V_2(n_{11} + n_{21} + n_{31})] + p_1 - r_1 - f_1}{2n}$$
$$\times \left(1 - e^{-2nt}\right) - 2nS_1 e^{-2nt}$$
$$+ \frac{(n_{12} + n_{22} + n_{32})(p_1 - r_1 - f_1) + (n_{11} + n_{21} + n_{31})\,(p_2 - r_2 - f_2)}{2n}$$
$$\times \left(2t - \frac{1}{n} + \frac{1}{n}e^{-2nt}\right) \tag{26.2.10}$$

$$\frac{dx_2}{dt} = (V_2 + 2nS_2)e^{-2nt}$$
$$+ \frac{2[V_2(n_{11} + n_{21} + n_{31}) + V_1(n_{12} + n_{22} + n_{32})] + p_2 - r_2 - f_2}{2n}$$
$$\times \left(1 - e^{-2nt}\right) - 2nS_2 e^{-2nt}$$
$$+ \frac{(n_{11} + n_{21} + n_{31})\,(p_2 - r_2 - f_2) + (n_{12} + n_{22} + n_{32})\,(p_1 - r_1 - f_1)}{2n}$$
$$\times \left(2t - \frac{1}{n} + \frac{1}{n}e^{-2nt}\right) \tag{26.2.11}$$

Supposing for the equations (26.2.10) and (26.2.11) that $t = 0$, we determine that $\frac{dx_1}{dt} = V_1$ and $\frac{dx_2}{dt} = V_2$, as it should be according to the initial conditions of motion. Taking the first derivatives from equations (26.2.10) and (26.2.11), we determine the accelerations of the masses:

$$\frac{d^2 x_1}{dt^2} = -2n(V_1 + 2nS_1)e^{-2nt} + \{2[V_1(n_{12} + n_{22} + n_{32})$$
$$+ V_2(n_{11} + n_{21} + n_{31})] + p_1 - r_1 - f_1\}e^{-2nt} + 4n^2 S_1 e^{-2nt}$$
$$+ \frac{1}{n}[(n_{12} + n_{22} + n_{32})(p_1 - r_1 - f_1) + (n_{11} + n_{21} + n_{31})$$
$$\times (p_2 - r_2 - f_2)](1 - e^{-2nt}) \tag{26.2.12}$$

$$\frac{d^2 x_2}{dt^2} = -2n(V_2 + 2nS_2)e^{-2nt} + \{2[V_1(n_{12} + n_{22})$$

$$+ V_2(n_{11} + n_{21})] + p_2 - r_2 - f_2\}e^{-2nt} + 4n^2 S_2 e^{-2nt}$$

$$+ \frac{1}{n}[(n_{11} + n_{21})(p_2 - r_2 - f_2) + (n_{12} + n_{22})(p_1 - r_1 - f_1)]$$

$$\times \left(1 - e^{-2nt}\right) \tag{26.2.13}$$

Hence, equations (26.2.8)–(26.2.13) describe the basic parameters of motion of the system.

Taking in equations (26.2.12) and (26.2.13) that $t = 0$, we determine the accelerations of the masses at the beginning of the process of motion:

$$\frac{d^2 x_1}{dt^2} = -2(V_1 - V_2)(n_{11} + n_{21} + n_{31}) + p_1 - r_1 - f_1 \tag{26.2.14}$$

$$\frac{d^2 x_2}{dt^2} = -2(V_2 - V_1)(n_{12} + n_{22} + n_{32}) + p_2 - r_2 - f_2 \tag{26.2.15}$$

26.3. Restricted System with Three Sequences of Fluid Links Being Subjected to Harmonic Forces

The operational process is associated with the motion of a restricted system with three sequences of fluid links that are moving on a horizontal frictionless surface and are subjected to harmonic forces.

Figure 26.3.1 shows a schematic diagram of the restricted system, the masses of which are connected to each other by a fluid link (dashpot), while both masses are appropriately attached by dashpots to non-movable supports. The system is subjected to the actions of harmonic forces. The notations on this figure are self-explanatory.

Based on the schematic diagram shown in Figure 26.3.1 and related considerations, we compose a pair of simultaneous differential

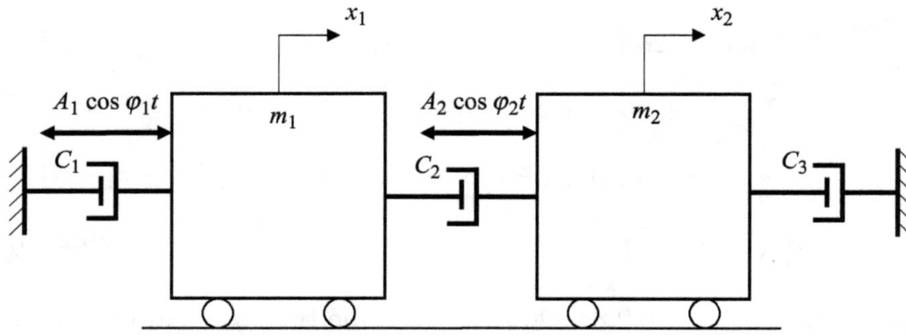

Fig. 26.3.1. Schematic diagram of a restricted system with three sequences of fluid links moving on a horizontal frictionless surface while being subjected to harmonic forces.

equations describing the motion of the masses of this system:

$$m_1 \frac{d^2 x_1}{dt^2} + (C_1 + C_2 + C_3)\left(\frac{dx_1}{dt} - \frac{dx_2}{dt}\right) = A_1 \cos \varphi_1 t \qquad (26.3.1)$$

$$m_2 \frac{d^2 x_2}{dt^2} + (C_1 + C_2 + C_3)\left(\frac{dx_2}{dt} - \frac{dx_1}{dt}\right) = A_2 \cos \varphi_2 t \qquad (26.3.2)$$

The initial conditions of motion are taken according to expression (26.1.3).

Dividing equations (26.3.1) and (26.3.2) by m_1 and m_2, respectively, and applying to them Laplace Transform pairs 2, 5, 2, 4, 2, and 29, we convert these equations with their initial conditions of motion from the time domain into the Laplace domain and obtain a corresponding system of two simultaneous algebraic equations with two unknowns in each equation in the Laplace domain:

$$s^2 x_1(s) - sV_1 - s^2 S_1 + 2n_{11} s x_1(s) - 2n_{11} s S_1 - 2n_{11} s x_2(s)$$

$$+ 2n_{11} s S_2 + 2n_{21} s x_1(s) - 2n_{21} s S_1 - 2n_{21} s x_2(s)$$

$$+ 2n_{21} s S_2 + 2n_{31} s x_1(s) - 2n_{31} s S_1 - 2n_{31} s x_2(s)$$

$$+ 2n_{31} s S_2 = \frac{a_1 s^2}{s^2 + \varphi_1^2} \qquad (26.3.3)$$

$$s^2 x_2(s) - sV_2 - s^2 S_2 + 2n_{12}sx_2(s) - 2n_{12}sS_2 - 2n_{12}sx_1(s)$$

$$+ 2n_{12}sS_1 + 2n_{22}sx_2(s) - 2n_{22}sS_2 - 2n_{22}sx_1(s)$$

$$+ 2n_{22}sS_1 + 2n_{32}sx_2(s) - 2n_{32}sS_2 - 2n_{32}sx_1(s)$$

$$+ 2n_{32}sS_1 = \frac{a_2 s^2}{s^2 + \varphi_2^2} \tag{26.3.4}$$

where

$$a_1 = \frac{A_1}{m_1} \quad \text{and} \quad a_2 = \frac{A_2}{m_2} \tag{26.3.5}$$

Rearranging equations (26.3.3) and (26.3.4) and applying to them the method of substitutions, we obtain two simultaneous algebraic equations with one unknown in each equation describing, respectively, the displacements of the two masses in the Laplace domain:

$$x_1(s) = \frac{sS_1}{s + 2n} + \frac{2S_1 n}{s + 2n} + \frac{V_1}{s + 2n}$$

$$+ \frac{2[V_1(n_{12} + n_{22} + n_{32}) + V_2(n_{11} + n_{21} + n_{31})]}{s(s + 2n)}$$

$$+ \frac{a_1 s}{(s^2 + \varphi_1^2)(s + 2n)} + \frac{2a_1(n_{12} + n_{22} + n_{32})}{(s^2 + \varphi_1^2)(s + 2n)}$$

$$+ \frac{2a_2(n_{11} + n_{21} + n_{31})}{(s^2 + \varphi_2^2)(s + 2n)} \tag{26.3.6}$$

$$x_2(s) = \frac{sS_2}{s + 2n} + \frac{2S_2 n}{s + 2n} + \frac{V_2}{s + 2n}$$

$$+ \frac{2[V_2(n_{11} + n_{21} + n_{31}) + V_1(n_{12} + n_{22} + n_{32})]}{s(s + 2n)}$$

$$+ \frac{a_2}{(s^2 + \varphi_1^2)(s + 2n)} + \frac{2_1 a_2(n_{11} + n_{21} + n_{31})}{(s^2 + \varphi_2^2)(s + 2n)}$$

$$+ \frac{2a_1(n_{12} + n_{22} + n_{32})}{(s^2 + \varphi_1^2)(s - 2n)} \tag{26.3.7}$$

Applying to equations (26.3.6) and (26.3.7) Laplace Transform pairs 14, 8, 8, 16, 44, 42, and 42, we invert these equations from the Laplace domain into the time domain and obtain the solutions of differential equations (26.3.1) and (26.3.2) with the initial conditions of motion according to expression (26.1.3):

$$
\begin{aligned}
x_1 = {}& S_1 + \frac{V_1}{2n}\left(1 - e^{-2nt}\right) \\[2mm]
& + \frac{V_1(n_{12} + n_{22} + n_{32}) + V_2(n_{11} + n_{21} + n_{31})}{n} \\[2mm]
& \times \left[t + \frac{1}{2n}\left(1 - e^{-2nt}\right)\right] + \frac{a_1}{\varphi_1^2 + 4n^2} \\[2mm]
& \times \left(e^{-2nt} + \frac{2n}{\varphi_1}\sin\varphi_1 t - \cos\varphi_1 t\right) + \frac{2a_1(n_{12} + n_{22} + n_{32})}{\varphi_1^2 + 4n^2} \\[2mm]
& \times \left[\frac{2n}{\varphi_1^2}\left(1 - \cos\varphi_1 t\right) - \frac{1}{\varphi_1}\sin\varphi_1 t + \frac{1}{2n}\left(1 - e^{-2nt}\right)\right] \\[2mm]
& + \frac{2a_2(n_{11} + n_{21} + n_{31})}{\varphi_2^2 + 4n^2}\left[\frac{2n}{\varphi_2^2}\left(1 - \cos\varphi_2 t\right)\right. \\[2mm]
& \left. - \frac{1}{\varphi_2}\sin\varphi_2 t + \frac{1}{2n}\left(1 - e^{-2nt}\right)\right]
\end{aligned} \tag{26.3.8}
$$

$$
\begin{aligned}
x_2 = {}& S_2 + \frac{V_2}{2n}\left(1 - e^{-2nt}\right) \\[2mm]
& + \frac{V_2(n_{11} + n_{21} + n_{31}) + V_1(n_{12} + n_{22} + n_{32})}{n_{11}} \\[2mm]
& \times \left[t + \frac{1}{2n}\left(1 - e^{-2nt}\right)\right] + \frac{a_2}{\varphi_2^2 + 4n^2} \\[2mm]
& \times \left(e^{-2nt} + \frac{2n}{\varphi_2}\sin\varphi_2 t - \cos\varphi_2 t\right) + \frac{2a_2(n_{11} + n_{21} + n_{31})}{\varphi_2^2 + 4n^2} \\[2mm]
& \times \left[\frac{2n}{\varphi_2^2}\left(1 - \cos\varphi_2 t\right) - \frac{1}{\varphi_2}\sin\varphi_2 t + \frac{1}{2n}\left(1 - e^{-2nt}\right)\right]
\end{aligned}
$$

$$+ \frac{2a_1\left(n_{12} + n_{22} + n_{32}\right)}{\varphi_1^2 + 4n^2} \left[\frac{2n}{\varphi_1^2}\left(1 - \cos\varphi_1 t\right)\right.$$

$$\left. - \frac{1}{\varphi_1}\sin\varphi_1 t + \frac{1}{2n}\left(1 - e^{-2nt}\right)\right] \tag{26.3.9}$$

In equations (26.3.8) and (26.3.9), equating the running time to zero, we obtain that $x_1 = S_1$ and $x_2 = S_2$, which conforms with the initial conditions of motion according to (26.1.3). Taking from these equations the first derivatives, we determine the velocities of the first and second masses, respectively:

$$\frac{dx_1}{dt} = V_1 e^{-2nt} + \frac{1}{n}\left[V_1\left(n_{12} + n_{22} + n_{32}\right) + V_2\left(n_{11} + n_{21} + n_{31}\right)\right]$$

$$\times\left(1 - e^{-2nt}\right) + \frac{a_1}{\varphi_1^2 + 4n^2}\left(\varphi_1\sin\varphi_1 t + 2n\cos\varphi_1 t - 2ne^{-2nt}\right)$$

$$+ \frac{2a_1\left(n_{12} + n_{22} + n_{32}\right)}{\varphi_1^2 + 4n^2}\left(\frac{2n}{\varphi_1}\sin\varphi_1 t - \cos\varphi_1 t + e^{-2nt}\right)$$

$$+ \frac{2a_2\left(n_{11} + n_{21} + n_{31}\right)}{\varphi_2^2 + 4n^2}\left(\frac{2n}{\varphi_2}\sin\varphi_2 t - \cos\varphi_2 t + e^{-2nt}\right)$$

$$\tag{26.3.10}$$

$$\frac{dx_2}{dt} = V_2 e^{-2nt} + \frac{1}{n}\left[V_2\left(n_{11} + n_{21} + n_{31}\right) + V_1\left(n_{12} + n_{22} + n_{32}\right)\right]$$

$$\times\left(1 - e^{-2nt}\right) + \frac{a_2}{\varphi_2^2 + 4n^2}\left(\varphi_2\sin\varphi_2 t + 2n\cos\varphi_2 t - 2ne^{-2nt}\right)$$

$$+ \frac{2a_2\left(n_{11} + n_{21} + n_{31}\right)}{\varphi_2^2 + 4n^2}\left(\frac{2n}{\varphi_2}\sin\varphi_2 t - \cos\varphi_2 t + e^{-2nt}\right)$$

$$+ \frac{2a_1\left(n_{12} + n_{22} + n_{32}\right)}{\varphi_1^2 + 4n^2}\left(\frac{2n}{\varphi_1}\sin\varphi_1 t - \cos\varphi_1 t + e^{-2nt}\right)$$

$$\tag{26.3.11}$$

Taking in equations (26.3.10) and (26.3.11) that $t = 0$, we obtain that $\frac{dx_1}{dt} = V_1$ and $\frac{dx_2}{dt} = V_2$, as supposed to be according to the initial conditions of motion.

The first derivatives from equations (26.3.10) and (26.3.11) yield the accelerations of the masses:

$$\frac{d^2 x_1}{dt^2} = -2nV_1 e^{-2nt} + 2[V_1(n_{12} + n_{22} + n_{32})$$

$$+ V_2(n_{11} + n_{21} + n_{31})]e^{-2nt} + \frac{a_1}{\varphi_1^2 + 4n^2}$$

$$\times (\varphi_1^2 \cos \varphi_1 t - 2n\varphi_1 \sin \varphi_1 t + 4n^2 e^{-2nt})$$

$$+ \frac{2a_1(n_{12} + n_{22} + n_{32})}{\varphi_1^2 + 4n^2}(2n \cos \varphi_1 t + \varphi_1 \sin \varphi_1 t - 2ne^{-2nt})$$

$$+ \frac{2a_2(n_{11} + n_{21} + n_{31})}{\varphi_2^2 + 4n^2}(2n \cos \varphi_2 t + \varphi_2 \sin \varphi_2 t - 2ne^{-2nt})$$

$$(26.3.12)$$

$$\frac{d^2 x_2}{dt^2} = -2nV_2 e^{-2nt} + 2[V_2(n_{11} + n_{21} + n_{31})$$

$$+ V_1(n_{12} + n_{22} + n_{32})]e^{-2nt} + \frac{a_2}{\varphi_2^2 + 4n^2}$$

$$\times (\varphi_2^2 \cos \varphi_2 t - 2n\varphi_2 \sin \varphi_2 t + 4n^2 e^{-2nt})$$

$$+ \frac{2a_2(n_{11} + n_{21} + n_{31})}{\varphi_2^2 + 4n^2}(2n \cos \varphi_2 t + \varphi_2 \sin \varphi_2 t - 2ne^{-2nt})$$

$$+ \frac{2a_1(n_{12} + n_{22} + n_{32})}{\varphi_1^2 + 4n^2}(2n \cos \varphi_1 t + \varphi_1 \sin \varphi_1 t - 2ne^{-2nt})$$

$$(26.3.13)$$

Hence, equations (26.3.8)–(26.3.13) represent the basic parameters of motion of the considered system.

26.3.1. *Numerical solution*

A Python program to plot the graph of equation (26.3.8) is presented as follows. The graph of equation (26.3.8) is shown in Figure 26.3.2.

```python
from matplotlib.pyplot import plot, show
from numpy import linspace, exp, sin, cos


S1 = 0.1
V1 = 0.15
V2 = 0.25
n11 = 3
n12 = 6
n21 = 4
n22 = 2
n31 = 5
n32 = 7
n = n11 + n12 + n21 + n22 + n31 + n32
phi1 = 20
phi2 = 15
a1 = 15
a2 = 25

t = linspace(0, 4, 1000)
x1  = S1 + (V1/(2*n))*(1 - exp(-2*n*t)) \
    + ((V1*(n12 + n22 + n32) + V2*(n11 + n21 + n31))/n) \
      *(t + (1/(2*n))*(1 - exp(-2*n*t))) \
    + (a1/(phi1**2 + 4*n**2))*(exp(-2*n*t) + ((2*n)/phi1)*sin(phi1*t) \
                    - cos(phi1*t)) \
    + ((2*a1*(n12 + n22 + n32))/(phi1**2 + 4*n**2)) \
      *(((2*n)/phi1**2)*(1 - cos(phi1*t)) - (1/phi1)*sin(phi1*t) \
        + (1/(2*n))*(1 - exp(-2*n*t))) \
    + ((2*a2*(n11 + n21 + n31))/(phi2**2 + 4*n**2)) \
      *((2*n/phi2**2)*(1 - cos(phi2*t)) - (1/phi2)*sin(phi2*t) \
        + (1/(2*n))*(1 - exp(-2*n*t)))

plot(t, x1)
show()
```

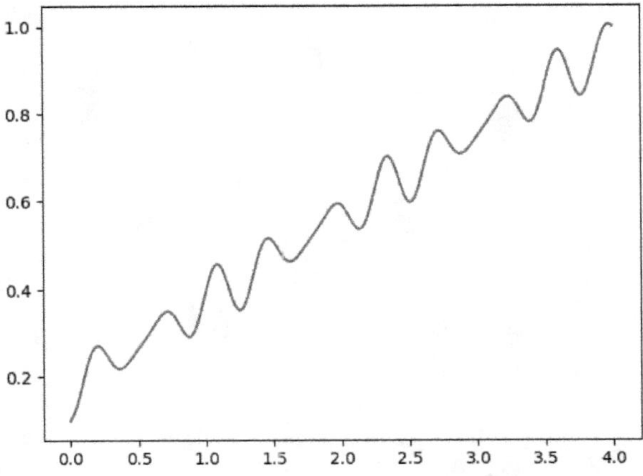

Fig. 26.3.2. Graph of equation (26.3.8).

RESTRICTED TWO-DEGREE-OF-FREEDOM SYSTEM WITH THREE SEQUENCES OF FLEXIBLE AND FLUID LINKS

This chapter is devoted to the motion of restricted systems with three sequences of flexible and fluid links, while the masses of the systems are moving on a horizontal surface due to common combinations of initial conditions of motion and forces. Following is an analysis of an operational process of a restricted system, the masses of which are connected to each other and to two non-movable supports by a combination of parallel flexible and fluid links, while the motion of the masses is caused by their initial velocities.

The notations of the parameters of motion are applicable only to this chapter.

27.1. Restricted System with Three Sequences of Flexible and Fluid Links, While the Motion of the Masses Occurs Due to Their Initial Velocities

Consider the operational process of a system, the two masses of which are moving on a horizontal frictionless surface due to their initial velocities.

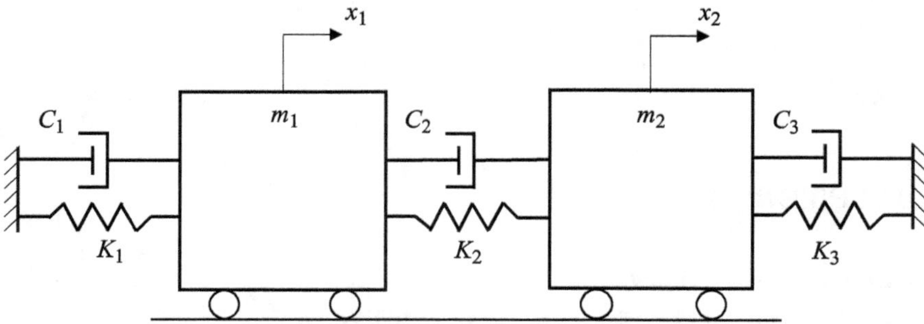

Fig. 27.1.1. Schematic diagram of a restricted system, the two masses of which are connected to each other by a dashpot and a spring in parallel, while both masses are attached to non-movable supports by two additional pairs of these links.

Figure 27.1.1 shows a schematic diagram of the system where C_1, C_2, and C_3 are the damping coefficients of the dashpots, while K_1, K_2, and K_3 are the stiffness coefficients of the springs.

Based on the schematic diagram shown in Figure 27.1.1 and the related considerations, we compose a pair of simultaneous differential equations of motion describing the operational process of the system:

$$m_1\frac{d^2x_1}{dt^2} + (C_1 + C_2 + C_3)\left(\frac{dx_1}{dt} - \frac{dx_2}{dt}\right)$$

$$+ (K_1 + K_2 + K_3)(x_1 - x_2) = 0 \qquad (27.1.1)$$

$$m_2\frac{d^2x_2}{dt^2} + (C_1 + C_2 + C_3)\left(\frac{dx_2}{dt} - \frac{dx_1}{dt}\right)$$

$$+ (K_1 + K_2 + K_3)(x_2 - x_1) = 0 \qquad (27.1.2)$$

The initial conditions of motion are

$$x_1 = 0; \quad \frac{dx_1}{dt} = V_1$$

For $\quad t = 0$ $\qquad\qquad\qquad\qquad\qquad\qquad\qquad (27.1.3)$

$$x_2 = 0; \quad \frac{dx_2}{dt} = V_2$$

Dividing, respectively, equations (27.1.1) and (27.1.2) by m_1 and m_2, while applying to them Laplace Transform pairs 2, 5, 2, 4, 2, and 1, we convert these equations with their initial conditions of motion from the time domain into a system of two simultaneous algebraic equations with two unknowns in each equation in the Laplace domain:

$$s^2 x_1(s) - sV_1 + 2n_{11}sx_1(s) - 2n_{11}sx_2(s) + 2n_{21}sx_1(s)$$

$$- 2n_{21}sx_2(s) + 2n_{31}sx_1(s) - 2n_{31}sx_2(s) + \omega_{11}^2 x_1(s)$$

$$- \omega_{11}^2 x_2(s) + \omega_{21}^2 x_1(s) - \omega_{21}^2 x_2(s) + \omega_{31}^2 x_1(s)$$

$$- \omega_{31}^2 x_2(s) = 0 \tag{27.1.4}$$

$$s^2 x_2(s) - sV_2 + 2n_{12}sx_2(s) - 2n_{12}sx_1(s)$$

$$+ 2n_{22}sx_2(s) - 2n_{22}sx_1(s) + 2n_{32}sx_2(s) - 2n_{32}sx_1(s) + \omega_{12}^2 x_2(s)$$

$$- \omega_{12}^2 x_1(s) + \omega_{22}^2 x_2(s) - \omega_{22}^2 x_1(s) + \omega_{32}^2 x_2(s)$$

$$- \omega_{32}^2 x_1(s) = 0 \tag{27.1.5}$$

where

$$2n_{11} = \frac{C_1}{m_1}; \quad 2n_{21} = \frac{C_2}{m_1}; \quad 2n_{31} = \frac{C_3}{m_1}; \quad 2n_{12} = \frac{C_1}{m_2};$$

$$2n_{22} = \frac{C_2}{m_2}; \quad 2n_{32} = \frac{C_3}{m_2} \tag{27.1.6}$$

and

$$\omega_{11}^2 = \frac{K_1}{m_1}; \quad \omega_{21}^2 = \frac{K_2}{m_1}; \quad \omega_{31}^2 = \frac{K_3}{m_1}; \quad \omega_{12}^2 = \frac{K_1}{m_2};$$

$$\omega_{22}^2 = \frac{K_2}{m_2}; \quad \omega_{32}^2 = \frac{K_3}{m_2} \tag{27.1.7}$$

Rearranging the terms in equations (27.1.4) and (27.1.5) and applying to them the method of substitutions, we obtain two simultaneous algebraic equations with two unknowns in each equation that describe in the Laplace domain the displacements of the first and second masses, respectively:

$$x_1(s) = \frac{V_1 s}{s^2 + 2ns + \omega_0^2}$$

$$+ \frac{2[V_1(n_{12} + n_{22} + n_{32}) + V_2(n_{11} + n_{21} + n_{31})]}{s^2 + 2ns + \omega_0^2}$$

$$+ \frac{V_1(\omega_{12}^2 + \omega_{22}^2 + \omega_{32}^2) + V_2(\omega_{11}^2 + \omega_{21}^2 + \omega_{31}^2)}{s(l^2 + 2ns + \omega_0^2)} \qquad (27.1.8)$$

$$x_2(s) = \frac{V_2 s}{s^2 + 2ns + \omega_0^2}$$

$$+ \frac{2[V_2(n_{11} + n_{21} + n_{31}) + V_1(n_{12} + n_{22} + n_{32})]}{s^2 + 2ns + \omega_0^2}$$

$$+ \frac{V_2(\omega_{11}^2 + \omega_{21}^2 + \omega_{31}^2) + V_1(\omega_{12}^2 + \omega_{22}^2 + \omega_{32}^2)}{s(s^2 + 2ns + \omega_0^2)} \qquad (27.1.9)$$

where

$$n = n_{11} + n_{21} + n_{31} + n_{12} + n_{22} + n_{32} \qquad (27.1.10)$$

and

$$\omega_0^2 = \omega_{11}^2 + \omega_{21}^2 + \omega_{31}^2 + \omega_{12}^2 + \omega_{22}^2 + \omega_{32}^2 \qquad (27.1.11)$$

In order to determine the characteristics of the damped motion of the masses according to equations (27.1.8) and (27.1.9), we should rearrange the denominators of the fractions of these equations in the following way:

$$s^2 + 2ns + \omega_0^2 = s^2 + 2ns + n^2 - n^2 + \omega_0^2 = s^2 + 2ns + n^2 + \omega^2$$

$$= (s + n)^2 + \omega^2 \qquad (27.1.12)$$

where

$$\omega^2 = \omega_0^2 - n^2 \qquad (27.1.13)$$

In order to continue the solutions of these equations, we need to determine the value of the parameter ω^2 for each case. Hence, in the case when $\omega^2 > 0$, the masses perform underdamped vibrational motion; however, when $\omega^2 = 0$, we have critical damping, and the

27.1.2. *Critically damped motion*

As mentioned above, critical damping occurs when $\omega^2 = 0$. Therefore, we apply this condition to equations (27.1.14) and (27.1.15), and we obtain

$$
\begin{aligned}
x_1(s) = {} & \frac{V_1 s}{(s+n)^2} + \frac{2[V_1(n_{12} + n_{22} + n_{32}) + V_2(n_{11} + n_{21} + n_{31})]}{(s+n)^2} \\
& + \frac{V_1(\omega_{12}^2 + \omega_{22}^2 + \omega_{32}^2) + V_2(\omega_{11}^2 + \omega_{21}^2 + \omega_{31}^2)}{s(s+n)^2}
\end{aligned}
\tag{27.1.20}
$$

$$
\begin{aligned}
x_2(s) = {} & \frac{V_2 s}{(s+n)} + \frac{2[V_2(n_{11} + n_{21} + n_{31}) + V_1(n_{12} + n_{22} + n_{32})}{(s+n)^2} \\
& + \frac{V_2(\omega_{11}^2 + \omega_{21}^2 + \omega_{31}^2) + V_1(\omega_{12}^2 + \omega_{22}^2 + \omega_{32}^2)}{s(s+n)^2}
\end{aligned}
\tag{27.1.21}
$$

Applying to equations (27.1.20) and (27.1.21) Laplace Transform pairs 25, 19, and 37, we invert these equations from the Laplace domain into the time domain and obtain the solutions of differential equations (27.1.1) and (27.1.2) with initial conditions of motion (27.1.3) for the case of critical damping:

$$
\begin{aligned}
x_1 = {} & V_1 t e^{-nt} + \frac{2[V_1(n_{12} + n_{22} + n_{32}) + V_2(n_{11} + n_{21} + n_{31})]}{n^2} \\
& \times [1 - e^{-nt}(1 + nt)] + \frac{V_1(\omega_{12}^2 + \omega_{22}^2 + \omega_{32}^2) + V_2(\omega_{11}^2 + \omega_{21}^2 + \omega_{31}^2)}{n^2} \\
& \times \left[t - \frac{2}{n} + e^{-nt}\left(\frac{2}{n} + t\right) \right]
\end{aligned}
\tag{27.1.22}
$$

$$
\begin{aligned}
x_2 = {} & V_1 t e^{-nt} + \frac{2[V_2(n_{11} + n_{21} + n_{31}) + V_1(n_{12} + n_{22} + n_{32})]}{n^2} \\
& \times [1 - e^{-nt}(1 + nt)] + \frac{V_2(\omega_{11}^2 + \omega_{21}^2 + \omega_{31}^2) + V_1(\omega_{12}^2 + \omega_{22}^2 + \omega_{32}^2)}{n^2} \\
& \times \left[t - \frac{2}{n} + e^{-nt}\left(\frac{2}{n} + t\right) \right]
\end{aligned}
\tag{27.1.23}
$$

27.1.3. *Overdamped motion*

In the case of overdamped motion, we have $\omega^2 < 0$. According to this condition, we replace in equations (27.1.14) and (27.1.15) the term ω^2 with its negative value $-\omega^2$, and we may write

$$x_1(s) = \frac{V_1 s}{(s+n)^2 - \omega^2} + \frac{2[V_1(n_{12} + n_{22} + n_{32}) + V_2(n_{11} + n_{21} + n_{31})}{(s+n)^2 - \omega^2}$$

$$+ \frac{V_1(\omega_{12}^2 + \omega_{22}^2 + \omega_{32}^2) + V_2(\omega_{11}^2 + \omega_{21}^2 + \omega_{31}^2)}{s[(s+n)^2 - \omega^2]} \tag{27.1.24}$$

$$x_2(s) = \frac{V_2 s}{(s+n)^2 - \omega^2} + \frac{2[V_2(n_{11} + n_{21} + n_{31}) + V_1(n_{12} + n_{22} + n_{32})]}{(s+n)^2 - \omega^2}$$

$$+ \frac{V_2(\omega_{11}^2 + \omega_{21}^2 + \omega_{31}^2) + V_1(\omega_{12}^2 + \omega_{22}^2 + \omega_{32}^2)}{s[(s+n)^2 - \omega^2]} \tag{27.1.25}$$

Applying to equations (27.1.24) and (27.1.25) Laplace Transform pairs 28, 22, and 46, we invert these equations from the Laplace domain into the time domain and obtain the solutions of differential equations (27.1.1) and (27.1.2) with initial conditions of motion (27.1.3) for the case of overdamped motion:

$$x_1 = \frac{V_1}{\omega} e^{-nt} \sinh \omega t + \frac{2[V_1(n_{12} + n_{22} + n_{32}) + V_2(n_{11} + n_{21} + n_{31})]}{n^2 - \omega^2}$$

$$\times \left[1 - e^{-nt}\left(\cosh \omega t + \frac{n}{\omega} \sinh \omega t\right)\right]$$

$$+ \frac{V_1(\omega_{12}^2 + \omega_{22}^2 + \omega_{32}^2) + V_2(\omega_{11}^2 + \omega_{21}^2 + \omega_{31}^2)}{\omega(n^2 - \omega^2)^2}$$

$$\times \{(n^2 - \omega^2)\omega t - 2n\omega + e^{-nt}[(\omega^2 + n^2)\sinh \omega t + 2n\omega \cosh \omega t]\}$$

$$\tag{27.1.26}$$

$$x_2 = \frac{V_2}{\omega} e^{-nt} \sinh \omega t + \frac{2[V_2(n_{11} + n_{21} + n_{31}) + V_1(n_{12} + n_{22} + n_{32})]}{n^2 - \omega^2}$$

$$\times \left[1 - e^{-nt}\left(\cosh \omega t + \frac{n}{\omega} \sinh \omega t\right)\right]$$

$$+ \frac{V_2(\omega_{11}^2 + \omega_{21}^2 + \omega_{31}^2) + V_1(\omega_{12}^2 + \omega_{22}^2 + \omega_{32}^2)}{\omega(n^2 - \omega^2)}$$

$$\times \{((n^2 - \omega^2)\omega t - 2n\omega + e^{-nt}[(\omega^2 + n^2)\sinh\omega t + 2n\omega\cosh\omega t]\}$$

$$(27.1.27)$$

27.2. Motion of a Restricted System Due to Constant Active Forces Applied to the Masses, Which Are Attached to Each Other and to Two Non-movable Supports by Three Sequences of Springs and Dashpots

The system is moving on a horizontal frictionless surface and being subjected to constant active forces. The schematic diagram of the system is shown in Figure 27.2.1, the notations in which are self-explanatory.

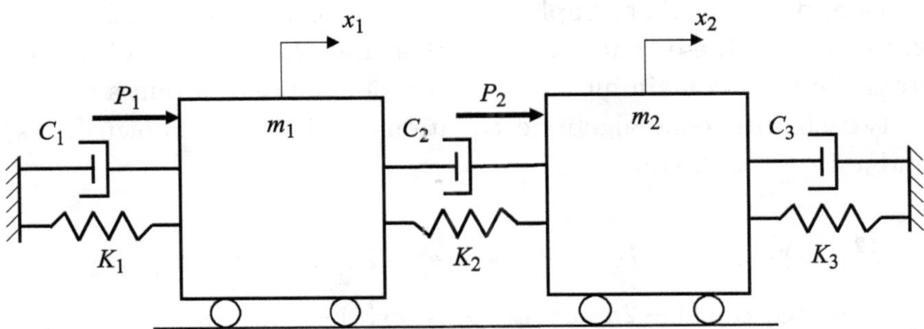

Fig. 27.2.1. Schematic diagram of a restricted system, the two masses of which are connected to each other by a pair of a fluid and an elastic links in parallel, while both masses are attached to non-movable supports by similar pairs of links.

Based on the schematic diagram shown in Figure 27.2.1 and the related considerations, we compose the corresponding pair of simultaneous differential equations of motion:

$$m_1\frac{d^2x_1}{dt^2} + (C_1 + C_2 + C_3)\left(\frac{dx_1}{dt} - \frac{dx_2}{dt}\right)$$

$$+ (K_1 + K_2 + K_3)(x_1 - x_2) = P_1 \qquad (27.2.1)$$

$$m_2 \frac{d^2 x_2}{dt^2} + (C_1 + C_2 + C_3)\left(\frac{dx_2}{dt} - \frac{dx_1}{dt}\right)$$

$$+ (K_1 + K_2 + K_3)(x_2 - x_1) = P_2 \qquad (27.2.2)$$

where P_1 and P_2 are the respective constant active forces applied to the masses.

The initial conditions of motion are

$$x_1 = 0; \quad \frac{dx_1}{dt} = 0$$

For $\quad t = 0$ $\qquad\qquad\qquad\qquad\qquad\qquad (27.2.3)$

$$x_2 = 0; \quad \frac{dx_2}{dt} = 0$$

Dividing, respectively, equations (24.2.1) and (24.2.2) by m_1 and m_2, while applying to them Laplace Transform pairs 2, 5, 2, 4, 2, 1, and 2, we convert these equations with their initial conditions of motion from the time domain into the Laplace domain and obtain a system of two simultaneous algebraic equations with two unknowns $x_1(s)$ and $x_2(s)$ in each equation:

$$s^2 x_1(s) + 2n_{11}sx_1(s) - 2n_{11}sx_2(s) + 2n_{21}sx_1(s) - 2n_{21}sx_2(s)$$

$$+ 2n_{31}sx_1(s) - 2n_{31}sx_2(s) + \omega_{11}^2 x_1(s) - \omega_{11}^2 x_2(s)$$

$$+ \omega_{21}^2 x_1(s) - \omega_{21}^2 x_2(s) + \omega_{31}^2 x_1(s) - \omega_{31}^2 x_2(s) = p_1 \qquad (27.2.4)$$

$$s^2 x_2(s) + 2n_{12}sx_2(s) - 2n_{11}sx_1(s)_1 + 2n_{22}sx_2(s) - 2n_{22}sx_1(s)$$

$$+ 2n_{32}sx_2(s) - 2n_{32}sx_1(s) + \omega_{12}^2 x_2(s) - \omega_{12}^2 x_1(s)$$

$$+ \omega_{22}^2 x_2(s) - \omega_{22}^2 x_1(s) + \omega_{32}^2 x_2(s) - \omega_{32}^2 x_1(s) = p_2 \qquad (27.2.5)$$

Rearranging the terms in equations (27.2.4) and (27.2.5) and applying to them the method of substitutions, we obtain two simultaneous algebraic equations with one unknown in each equation that describe in the Laplace domain the displacements of the first and second masses:

$$x_1(s) = \frac{p_1}{s^2 + 2ns + \omega^2}$$

$$+ \frac{2[(n_{12} + n_{22} + n_{32})p_1 + (n_{11} + n_{21} + n_{31})p_2]}{s(s^2 + 2ns + \omega^2)}$$

$$+ \frac{(\omega_{12}^2 + \omega_{22}^2 + \omega_{32}^2)p_1 + (\omega_{11}^2 + \omega_{21}^2 + \omega_{31}^2)p_2}{s^2(s^2 + 2ns + \omega^2)} \qquad (27.2.6)$$

$$x_2(s) = \frac{p_2}{s^2 + 2ns + \omega^2}$$

$$+ \frac{2[(n_{11} + n_{21} + n_{31})p_2 + (n_{12} + n_{22} + n_{32})p_1]}{s(s^2 + 2ns + \omega^2)}$$

$$+ \frac{(\omega_{11}^2 + \omega_{21}^2 + \omega_{31}^2)p_2 + (\omega_{12}^2 + \omega_{22}^2 + \omega_{32}^2)p_1}{s^2(s^2 + 2ns + \omega^2)} \qquad (27.2.7)$$

In continuation of the analysis of the current system, we consider the cases of underdamped, critically damped, and overdamped motions.

27.2.1. *Underdamped vibrational motion*

Combining equation (27.2.6) and (27.2.7) with equations (27.1.12) and (27.1.13), we obtain

$$x_1(s) = \frac{p_1}{(s + n)^2 + \omega^2}$$

$$+ \frac{2[(n_{12} + n_{22} + n_{32})p_1 + (n_{11} + n_{21} + n_{31})p_2]}{s[(s + n)^2 + \omega^2]}$$

$$+ \frac{(\omega_{12}^2 + \omega_{22}^2 + \omega_{32}^2)p_1 + (\omega_{11}^2 + \omega_{21}^2 + \omega_{31}^2)p_2}{s^2[(s + n)^2 + \omega^2]} \qquad (27.2.8)$$

$$x_2(s) = \frac{p_2}{(s + n)^2 + \omega^2}$$

$$+ \frac{2[(n_{11} + n_{21} + n_{31})p_2 + (n_{12} + n_{22} + n_{32})p_1]}{s[(s + n)^2 + \omega^2]}$$

$$+ \frac{(\omega_{11}^2 + \omega_{21}^2 + \omega_{31}^2)p_2 + (\omega_{12}^2 + \omega_{22}^2 + \omega_{32}^2)p_1}{s^2[(s + n)^2 + \omega^2]} \qquad (27.2.9)$$

Applying to equations (27.2.8) and (27.2.9) Laplace Transform pairs 21, 45, and 56, we invert these equations from the Laplace domain into the time domain and obtain the solutions of differential equations (27.2.1) and (27.2.2) with their initial conditions of motion. These equations describe the underdamped vibrational motion of the two masses of the system:

$$
\begin{aligned}
x_1 =\ & \frac{p_1}{\omega^2 + n^2}\left[1 - e^{-nt}\left(\cos\omega t + \frac{n}{\omega}\sin\omega t\right)\right] \\
& + \frac{2[(n_{12} + n_{22} + n_{32})p_1 + (n_{11} + n_{21} + n_{31})p_2]}{\omega(\omega^2 + n^2)^2} \\
& \times \left\{(\omega^2 + n^2)\omega t - 2n\omega - e^{-nt}[(\omega^2 - n^2)\sin\omega t - 2n\omega\cos\omega t]\right\} \\
& + \frac{(\omega_{12}^2 + \omega_{22}^2 + \omega_{32}^2)p_1 + (\omega_{11}^2 + \omega_{21}^2 + \omega_{31}^2)p_2}{\omega(\omega^2 + n^2)^2} \\
& \times \left\{\frac{\omega t^2}{2}(\omega^2 + n^2) - 2n\omega t - \frac{\omega(\omega^2 - n^2)}{\omega^2 + n^2}\right. \\
& \times \left[1 - e^{-nt}\left(\cos\omega t + \frac{n}{\omega}\sin\omega t\right)\right] - \frac{2n^2\omega}{\omega^2 + n^2} \\
& \times \left.\left[1 + e^{-nt}\left(\frac{\omega}{n}\sin\omega t - \cos\omega t\right)\right]\right\}
\end{aligned}
\tag{27.2.10}
$$

$$
\begin{aligned}
x_2 =\ & \frac{p_2}{\omega^2 + n^2}\left[1 - e^{-nt}\left(\cos\omega t + \frac{n}{\omega}\sin\omega t\right)\right] \\
& + \frac{2[(n_{11} + n_{21} + n_{31})p_2 + (n_{12} + n_{22} + n_{32})p_1]}{\omega(\omega^2 + n^2)^2} \\
& \times \left\{(\omega^2 + n^2)\omega t - 2n\omega - e^{-nt}[(\omega^2 - n^2)\sin\omega t - 2n\omega\cos\omega t]\right\} \\
& + \frac{(\omega_{11}^2 + \omega_{21}^2 + \omega_{31}^2)p_2 + (\omega_{12}^2 + \omega_{22}^2 + \omega_{32}^2)p_1}{\omega(\omega^2 + n^2)^2} \\
& \times \left\{\frac{\omega t^2}{2}(\omega^2 + n^2) - 2n\omega t - \frac{\omega(\omega^2 - n^2)}{\omega^2 + n^2}\right. \\
& \times \left[1 - e^{-nt}\left(\cos\omega t + \frac{n}{\omega}\sin\omega t\right)\right] - \frac{2n^2\omega}{\omega^2 + n^2} \\
& \times \left.\left[1 + e^{-nt}\left(\frac{\omega}{n}\sin\omega t - \cos\omega t\right)\right]\right\}
\end{aligned}
\tag{27.2.11}
$$

Supposing that in equations (27.2.10) and (27.2.11), $t = 0$, we have that $x_1 = 0$ and $x_2 = 0$, as it should be according to the initial conditions of motion. Taking from these equations the first derivatives, we obtain the velocities of the masses of the system:

$$
\begin{aligned}
\frac{dx_1}{dt} =\ & \frac{p_1}{\omega} e^{-nt} \sin \omega t \\
& + \frac{2[(n_{12} + n_{22} + n_{32})p_1 + (n_{11} + n_{21} + n_{31})p_2]}{\omega^2 + n^2} \\
& \times \left[1 - e^{-nt} \left(\cos \omega t + \frac{n}{\omega} \sin \omega t \right) \right] \\
& + \frac{\left(\omega_{12}^2 + \omega_{22}^2 + \omega_{32}^2\right) p_1 + \left(\omega_{11}^2 + \omega_{21}^2 + \omega_{31}^2\right) p_2}{\omega \left(\omega^2 + n^2\right)^2} \\
& \times \left\{ (\omega^2 + n^2)\omega t - 2n\omega \right. \\
& \left. - e^{-nt} \left[\left(\omega^2 - n^2\right) \sin \omega t - 2n\omega \cos \omega t \right] \right\}
\end{aligned}
\tag{27.2.12}
$$

$$
\begin{aligned}
\frac{dx_2}{dt} =\ & \frac{p_2}{\omega} e^{-nt} \sin \omega t \\
& + \frac{2\left[(n_{11} + n_{21} + n_{31}) p_2 + (n_{12} + n_{22} + n_{32}) p_1\right]}{\omega^2 + n^2} \\
& \times \left[1 - e^{-nt} \left(\cos \omega t + \frac{n}{\omega} \sin \omega t \right) \right] \\
& + \frac{\left(\omega_{11}^2 + \omega_{21}^2 + \omega_{31}^2\right) p_2 + \left(\omega_{12}^2 + \omega_{22}^2 + \omega_{32}^2\right) p_1}{\omega \left(\omega^2 + n^2\right)^2} \\
& \times \left\{ \left(\omega^2 + n^2\right) \omega t - 2n\omega \right. \\
& \left. - e^{-nt} \left[\left(\omega^2 - n^2\right) \sin \omega t - 2n\omega \cos \omega t \right] \right\}
\end{aligned}
\tag{27.2.13}
$$

Supposing for equations (27.2.12) and (27.2.13) that $t = 0$, we obtain $\frac{dx_1}{dt} = 0$ and $\frac{dx_2}{dt} = 0$, as expected according to the initial conditions of motion.

Taking from equations (27.2.12) and (27.2.13) the first derivatives, we determine the accelerations of the masses:

$$\frac{d^2 x_1}{dt^2} = p_1 e^{-nt} \left(\cos \omega t - \frac{n}{\omega} \sin \omega t \right)$$

$$+ \frac{2}{\omega} \left[(n_{12} + n_{22} + n_{32})\, p_1 + (n_{11} + n_{21} + n_{31})\, p_2 \right] e^{-nt} \sin \omega t$$

$$+ \frac{1}{\omega^2 + n^2} [(\omega_{12}^2 + \omega_{22}^2 + \omega_{32}^2)p_1 + (\omega_{11}^2 + \omega_{21}^2 + \omega_{31}^2)p_2]$$

$$\times \left[1 - \left(\cos \omega t + \frac{n}{\omega} \sin \omega t \right) \right] \sin \omega t \qquad (27.2.14)$$

$$\frac{d^2 x_2}{dt^2} = p_2 e^{-nt} \left(\cos \omega t - \frac{n}{\omega} \sin \omega t \right)$$

$$+ \frac{2}{\omega} [(n_{11} + n_{21} + n_{31})p_2 + (n_{12} + n_{22} + n_{52})p_1] e^{-nt} \sin \omega t$$

$$+ \frac{1}{\omega^2 + n^2} \left[(\omega_{11}^2 + \omega_{21}^2 + \omega_{31}^2)p_2 + (\omega_{12}^2 + \omega_{22}^2 + \omega_{32}^2)\, p_1 \right]$$

$$\times \left[1 - \left(\cos \omega t + \frac{n}{\omega} \sin \omega t \right) \right] \sin \omega t \qquad (27.2.15)$$

Equations (27.2.10)–(27.2.15) represent the basic parameters of motion of the system and allow us to perform the analysis of the considered operational process.

27.2.2. *Critically damped motion*

As noted above, critical damping occurs when $\omega^2 = 0$. Therefore, accounting for this condition in equations (27.2.8) and (27.2.9), we may write

$$x_1(s) = \frac{p_1}{(s+n)^2} + \frac{2[(n_{12} + n_{22} + n_{32})p_1 + (n_{11} + n_{21} + n_{31})p_2]}{s(s+n)^2}$$

$$+ \frac{(\omega_{12}^2 + \omega_{22}^2 + \omega_{32}^2)p_1 + (\omega_{11}^2 + \omega_{21}^2 + \omega_{31}^2)p_2}{s^2(s+n)^2} \qquad (27.2.16)$$

$$x_2(s) = \frac{p_2}{(s+n)^2} + \frac{2[(n_{11} + n_{21} + n_{31})p_2 + (n_{12} + n_{22} + n_{32})p_1]}{s(s+n)^2}$$

$$+ \frac{(\omega_{11}^2 + \omega_{21}^2 + \omega_{31}^2)p_2 + (\omega_{12}^2 + \omega_{22}^2 + \omega_{32}^2)p_1}{s^2(s+n)^2} \qquad (27.2.17)$$

Applying to equations (27.2.16) and (27.2.17) Laplace Transform pairs 19, 37, and 54, we invert these equations from the Laplace domain into the time domain and obtain the solutions of differential equations (27.2.1) and (27.2.2) with their initial conditions of motion for the case of critical damping:

$$x_1 = \frac{p_1}{n^2}\left[1 - e^{-nt}(1 + nt)\right]$$

$$+ \frac{2[(n_{12} + n_{22} + n_{32})p_1 + (n_{11} + n_{21} + n_{31})p_2]}{n^2}$$

$$\times \left[t - \frac{2}{n} + e^{-nt}\left(\frac{2}{n} + t\right)\right]$$

$$+ \frac{(\omega_{12}^2 + \omega_{22}^2 + \omega_{32}^2)p_1 + (\omega_{11}^2 + \omega_{21}^2 + \omega_{31}^2)p_2}{n^2}$$

$$\times \left[\frac{t^2}{2} - \frac{2t}{n} - \frac{3}{n^2} + \frac{1}{n^2}e^{-nt}(3 + nt)\right] \qquad (27.2.18)$$

$$x_2 = \frac{p_2}{n^2}\left[1 - e^{-nt}(1 + nt)\right]$$

$$+ \frac{2[(n_{11} + n_{21} + n_{31})p_2 + (n_{12} + n_{22} + n_{32})p_1]}{n^2}$$

$$\times \left[t - \frac{2}{n} + e^{-nt}\left(\frac{2}{n} + t\right)\right]$$

$$+ \frac{(\omega_{11}^2 + \omega_{21}^2 + \omega_{31}^2)p_2 + (\omega_{12}^2 + \omega_{22}^2 + \omega_{32}^2)p_1}{n^2}$$

$$\times \left[\frac{t^2}{2} - \frac{2t}{n} - \frac{3}{n^2} + \frac{1}{n^2}e^{-nt}(3 + nt)\right] \qquad (27.2.19)$$

Assuming that in equations (27.2.18) and (27.2.19), $t = 0$, we have that $x_1 = 0$ and $x_2 = 0$, as it should be according to the initial conditions of motion. Taking from these equations the first derivatives, we obtain the velocities of the masses of the system:

$$\frac{dx_1}{dt} = p_1 t e^{-nt} + \frac{2[(n_{12} + n_{22} + n_{32})p_1 + (n_{11} + n_{21} + n_{31})p_2]}{n^2}$$

$$\times \left[1 - e^{-nt}(1 + nt)\right]$$

$$+ \frac{(\omega_{12}^2 + \omega_{22}^2 + \omega_{32}^2)p_1 + (\omega_{11}^2 + \omega_{21}^2 + \omega_{31}^2)p_2}{n^2}$$

$$\times \left[t - \frac{2}{n} + e^{-nt}\left(\frac{2}{n} + t\right)\right] \tag{27.2.20}$$

$$\frac{dx_2}{dt} = p_2 t e^{-nt} + \frac{2[(n_{11} + n_{21} + n_{31})p_2 + (n_{12} + n_{22} + n_{32})p_1]}{n^2}$$

$$\times \left[1 - e^{-nt}(1 + nt)\right]$$

$$+ \frac{(\omega_{11}^2 + \omega_{21}^2 + \omega_{31}^2)p_2 + (\omega_{12}^2 + \omega_{22}^2 + \omega_{32})p_1}{n^2}$$

$$\times \left[t - \frac{2}{n} + e^{-nt}\left(\frac{2}{n} + t\right)\right] \tag{27.2.21}$$

Supposing for equations (27.2.20) and (27.2.21) that $t = 0$, we obtain that $\frac{dx_1}{dt} = 0$ and $\frac{dx_2}{dt} = 0$, as expected according to the initial conditions of motion.

27.2.3. *Numerical solution*

The following is a Python program to plot the graph (Figure 27.2.2) of equation (27.2.18).

```python
from matplotlib.pyplot import plot, show
from numpy import linspace, exp, sin, cos, sqrt

p1 = 100
p2 = 75
n11 = 8
n12 = 6
n21 = 18
n22 = 4
n31 = 28
n32 = 4.1909
n = n11 + n12 + n21 + n22 + n31 + n32
omega11 = 30
omega12 = 25
omega21 = 20
omega22 = 15
omega31 = 40
omega32 = 30

t = linspace(0, 0.5, 1000)

x1 = (p1/n**2)*(1 - exp(-n*t)*(1 + n*t)) \
    + (2*((n12 + n22 + n32)*p1 + (n11 + n21 + n31)*p2)/n**2) \
    *(t - 2/n + exp(-n*t)*(2/n + t)) \
    + (((omega12**2 + omega22**2 + omega32**2)*p1 \
     + (omega11**2 + omega21**2 + omega31**2)*p2)/n**2) \
    *(t**2/2 - 2*t/n - 3/n**2 + (1/n**2)*exp(-n*t)*(3 + n*t))

plot(t, x1)
show()
```

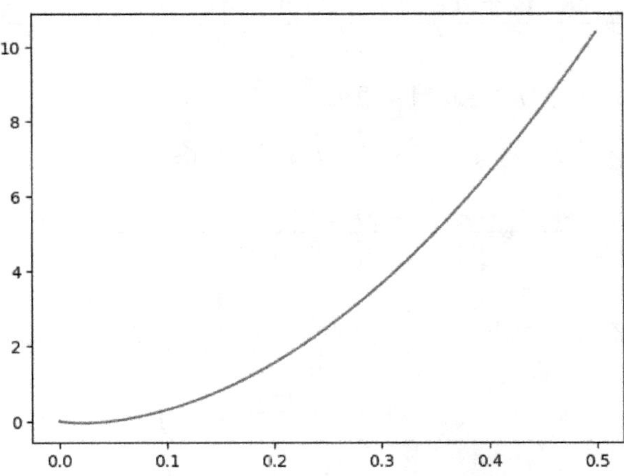

Fig. 27.2.2. Graph of equation (27.2.18).

27.2.4. *Overdamped motion*

As mentioned above, the overdamped motion occurs when $\omega^2 < 0$. Therefore, we have to replace in equations (27.2.8) and (27.2.9) the signs at the terms ω^2 with the opposite signs, and we obtain

$$x_1(s) = \frac{p_1}{(s+n)^2 - \omega^2} + \frac{2[(n_{12} + n_{22} + n_{32})p_1 + (n_{11} + n_{21} + n_{31})p_2]}{s[(s+n)^2 - \omega^2]}$$

$$+ \frac{(\omega_{12}^2 + \omega_{22}^2 + \omega_{32}^2)p_1 + (\omega_{11}^2 + \omega_{21}^2 + \omega_{31}^2)p_2}{s^2[(s+n)^2 - \omega^2]} \tag{27.2.22}$$

$$x_2(s) = \frac{p_2}{(s+n)^2 - \omega^2} + \frac{2[(n_{11} + n_{21} + n_{31})p_2 + (n_{12} + n_{22} + n_{32})p_1]}{s[(s+n)^2 - \omega^2]}$$

$$+ \frac{(\omega_{11}^2 + \omega_{21}^2\omega_{31}^2)p_2 + (\omega_{12}^2 + \omega_{22}^2 + \omega_{32}^2)p_1}{s^2[(s+n)^2 - \omega^2]} \tag{27.2.23}$$

Applying to equations (27.2.22) and (27.2.23) Laplace Transform pairs 22, 46, and 57, we invert these equations from the Laplace domain into the time domain and obtain the solutions of differential equations (27.2.1) and (27.2.2) with their initial conditions for the case of overdamped motion:

$$x_1 = \frac{p_1}{n^2 - \omega^2}\left[1 - e^{-nt}\left(\cosh\omega t + \frac{n}{\omega}\sinh\omega t\right)\right]$$

$$+ \frac{2[(n_{12} + n_{22} + n_{32})p_1 + (n_{11} + n_{21} + n_{31})p_2]}{\omega(n^2 - \omega^2)^2}$$

$$\times \left\{\omega t(n^2 - \omega^2) - 2n\omega + e^{-nt}\right.$$

$$\times \left[\omega t(n^2 + \omega^2)\sinh\omega t + 2n\omega\cosh\omega t\right]\}$$

$$+ \frac{(\omega_{12}^2 + \omega_{22}^2 + \omega_{32}^2)p_1 + (\omega_{11}^2 + \omega_{21}^2 + \omega_{31}^2)p_2}{\omega(n^2 - \omega^2)^2}$$

$$\times \left\{\frac{\omega t^2}{2}(n^2 - \omega^2) - 2n\omega t + \frac{\omega(n^2 + \omega^2)}{n^2 - \omega^2}\right.$$

$$\times \left[1 - e^{-nt}\left(\cosh\omega t + \frac{n}{\omega}\sinh\omega t\right)\right]$$

$$\left.- \frac{2n^2\omega}{n^2 - \omega^2}\left[1 + e^{-nt}\left(\frac{n}{\omega}\sinh\omega t - \cosh\omega t\right)\right]\right\} \tag{27.2.24}$$

$$x_2 = \frac{p_2}{n^2 - \omega^2}\left[1 - e^{-nt}\left(\cosh\omega t + \frac{n}{\omega}\sinh\omega t\right)\right]$$

$$+ \frac{2[(n_{11} + n_{21} + n_{31})p_2 + (n_{12} + n_{22} + n_{32})p_1]}{\omega(n^2 - \omega^2)^2}$$

$$\times\ \{\omega t(n^2 - \omega^2) - 2n\omega + e^{-nt}$$

$$\times\ [\omega t(n^2 + \omega^2)\sin\omega t + 2n\omega\cosh\omega t]\}$$

$$+ \frac{(\omega_{11}^2 + \omega_{21}^2 + \omega_{31}^2)p_2 + (\omega_{12}^2 + \omega_{22}^2 + \omega_{32}^2)p_1}{\omega(n^2 - \omega^2)^2}$$

$$\times\ \left\{\frac{\omega t^2}{2}(n^2 - \omega^2) - 2n\omega t + \frac{\omega(n^2 + \omega^2)}{n^2 - \omega^2}\right.$$

$$\times\ \left[1 - e^{-nt}\left(\cosh\omega t + \frac{n}{\omega}\sinh\omega t\right)\right]$$

$$\left.- \frac{2n^2\omega}{n^2 - \omega^2}\left[1 + e^{-nt}\left(\frac{n}{\omega}\sinh\omega t - \cosh\omega t\right)\right]\right\} \qquad (27.2.25)$$

Assuming that in equations (27.2.24) and (27.2.25), $t = 0$, we obtain that $x_1 = 0$ and $x_2 = 0$, as expected according to the initial conditions of motion of the masses, which are performing a decelerated translation.

27.3. Motion of a Restricted System with Three Sequences of Flexible and Fluid Links, While Each Mass Is Attached by a Sequence of These Links to a Non-movable Support, and Just One of the Masses Is Subjected to a Harmonic Force

The operational process of the restricted system considered in the following represents the motion of two masses connected to each other and to two non-movable supports by three sequences of springs and dashpots, while just one mass is subjected to a harmonic force. The system is moving on a horizontal frictionless surface.

The schematic diagram representing the restricted system is shown in Figure 27.3.1, the notations in which are self-explanatory.

Based on the schematic diagram shown in Figure 27.3.1 and the above considerations, we compose a pair of simultaneous differential

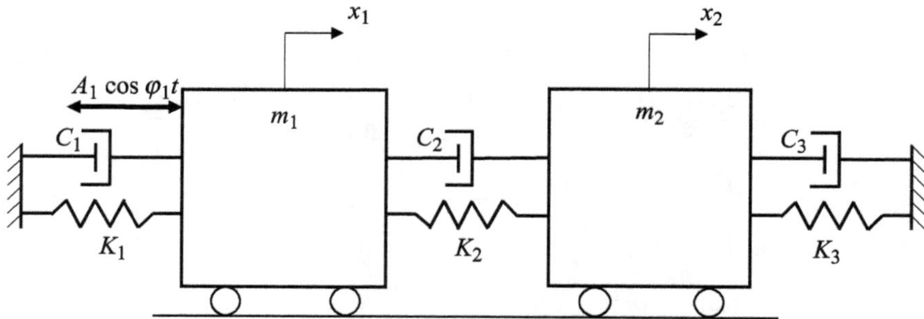

Fig. 27.3.1. Schematic diagram of a restricted system, the two masses of which are connected to each other and to two non-movable supports by three sequences of dashpots and springs in parallel, while just the first mass is subjected to a harmonic force.

equations describing the motion of the masses:

$$m_1\frac{d^2x_1}{dt^2} + (C_1 + C_2 + C_3)\left(\frac{dx_1}{dt} - \frac{dx_2}{dt}\right)$$

$$+ (K_1 + K_2 + K_3)(x_1 - x_2) = A_1 \cos\varphi_1 t \qquad (27.3.1)$$

$$m_2\frac{d^2x_2}{dt^2} + (C_1 + C_2 + C_3)\left(\frac{dx_2}{dt} - \frac{dx_1}{dt}\right)$$

$$+ (K_1 + K_2 + K_3)(x_2 - x_1) = 0 \qquad (27.3.2)$$

where C_1, C_2, and C_3 are the damping coefficients of the dashpots, while K_1, K_2, K_3 are the stiffness coefficients of the springs.

The initial conditions of motion are taken according to expression (27.2.3).

Dividing, respectively, equations (27.3.1) and (27.3.2) by m_1 and m_2, we may write

$$\frac{d^2x_1}{dt^2} + 2(n_{11} + n_{21} + n_{31})\left(\frac{dx_1}{dt} - \frac{dx_2}{dt}\right)$$

$$+ (\omega_{11}^2 + \omega_{21}^2 + \omega_{31}^2)(x_1 - x_2) = a_1 \cos\varphi_1 t \qquad (27.3.3)$$

$$\frac{d^2x_2}{dt^2} + 2(n_{12} + n_{22} + n_{32})\left(\frac{dx_2}{dt} - \frac{dx_1}{dt}\right)$$

$$+ (\omega_{12}^2 + \omega_{22}^2 + \omega_{23}^2)(x_2 - x_1) = 0 \qquad (27.3.4)$$

Applying, respectively, to equations (27.3.3) and (27.3.4) Laplace Transform pairs 5, 2, 4, 2, 1, 2, and 29, and 5, 2, 4, 2, and 1, we convert these equations from the time domain into the Laplace domain and obtain a system of two simultaneous algebraic equations with two unknowns in each equation that represent the displacements of the respective masses:

$$s^2 x_1(s) + 2n_{11}sx_1(s) - 2n_{11}sx_2(s) + 2n_{21}sx_1(s)$$

$$- 2n_{21}sx_2(s) + 2n_{31}sx_1 - 2n_{31}sx_2(s)$$

$$+ \omega_{11}^2 x_1(s) - \omega_{11}^2 x_2(s) + \omega_{21}^2 x_1(s)$$

$$- \omega_{21}^2 x_2(s) + \omega_{31}^2 sx_1(s) - \omega_{31}^2 sx_2(s) = \frac{a_1 s^2}{s^2 + \varphi_1^2} \qquad (27.3.5)$$

$$s^2 x_2(s) + 2n_{12}sx_2(s) - 2n_{12}sx_1(s) + 2n_{22}sx_2(s)$$

$$- 2n_{22}sx_1(s) + 2n_{32}sx_2(s) - 2n_{32}sx_1(s)$$

$$+ \omega_{12}^2 sx_2(s) - \omega_{12}^2 sx_1(s) + \omega_{22}^2 sx_2(s)$$

$$- \omega_{22}^2 sx_1(s) + \omega_{32}^2 sx_2(s) - \omega_{32}^2 sx_1(s) = 0 \qquad (27.3.6)$$

Rearranging equations (27.3.5) and (27.3.6), we write

$$x_1(s)(s^2 + 2n_{11}s + 2n_{21}s + 2n_{31}s + \omega_{11}^2 + \omega_{21}^2 + \omega_{31}^2) = \frac{a_1 s^2}{s^2 + \varphi_1^2}$$

$$+ x_2(s)(2n_{11}s + 2n_{21}s + 2n_{31}s + \omega_{11}^2 + \omega_{21}^2 + \omega_{31}^2) \qquad (27.3.7)$$

$$x_2(s)(s^2 + 2n_{12}s + 2n_{22}s + 2n_{32}s + \omega_{12}^2 + \omega_{22}^2 + \omega_{32}^2) = x_1(s)$$

$$\times (2n_{12}s + 2n_{22}s + 2n_{32}s + \omega_{12}^2 + \omega_{22}^2 + \omega_{32}^2) \qquad (27.3.8)$$

Solving equation (27.3.8) for $x_2(s)$, we obtain

$$x_2(s) = \frac{x_1(s)(2n_{12}s + 2n_{22}s + 2n_{32}s + \omega_{12}^2 + \omega_{22}^2 + \omega_{32}^2)}{s^2 + 2n_{12}s + 2n_{22}s + 2n_{32}s + \omega_{12}^2 + \omega_{22}^2 + \omega_{32}^2} \qquad (27.3.9)$$

Using the method of substitutions, we combine equations (27.3.7) and (27.3.9) and eliminate one unknown from equation (27.3.7):

$$x_1(s)(s^2 + 2n_{11}s + 2n_{21}s + 2n_{31}s + \omega_{11}^2 + \omega_{21}^2 + \omega_{31}^2)$$

$$= \frac{a_1 s^2}{s^2 + \varphi_1^2} + \frac{x_1(s)(2n_{12}s + 2n_{22}s + 2n_{32}s + \omega_{12}^2 + \omega_{22}^2 + \omega_{32}^2)}{s^2 + 2n_{12}s + 2n_{22}s + 2n_{32}s + \omega_{12}^2 + \omega_{22}^2 + \omega_{32}^2}$$

$$\times (2n_{11}s + 2n_{21}s + 2n_{31}s + \omega_{11}^2 + \omega_{21}^2 + \omega_{31}^2) \qquad (27.3.10)$$

Hence, the algebraic equation (27.3.10) contains one unknown $x_1(s)$, representing the displacement of the first mass in the Laplace domain. Performing the conventional algebraic procedures, we have

$$x_1(s)\left[s^2 + 2s(n_{11} + n_{21} + n_{31} + n_{12} + n_{22} + n_{32})\right.$$

$$\left. + \omega_{11}^2 + \omega_{21}^2 + \omega_{31}^2 + \omega_{12}^2 + \omega_{22}^2 + \omega_{32}^2\right]$$

$$= \frac{a_1(s^2 + 2n_{12}s + 2n_{22}s + 2n_{32}s + \omega_{12}^2 + \omega_{22}^2 + \omega_{32}^2)}{s^2 + \varphi_1^2} \qquad (27.3.11)$$

Solving equation (27.3.11) for $x_1(s)$, we obtain

$$x_1(s) = \frac{s^2 a_1}{(s^2 + \varphi_1^2)(s^2 + 2ns + \omega_0^2)} + \frac{2sa_1(n_{12} + n_{22} + n_{32})}{(s^2 + \varphi_1^2)(s^2 + 2ns + \omega_0^2)}$$

$$+ \frac{a_1(\omega_{12}^2 + \omega_{22}^2 + \omega_{32}^2)}{(s^2 + \varphi_1^2)(s^2 + 2ns + \omega_0^2)} \qquad (27.3.12)$$

Applying similar algebraic procedures, we determine the displacement of the second mass in the Laplace domain:

$$x_2(s) = \frac{2a_1 s(n_{12} + n_{22} + n_{32})}{(s^2 + \varphi_1^2)(s^2 + 2ns + \omega_0^2)} + \frac{a_1(\omega_{12}^2 + \omega_{22}^2 + \omega_{32}^2)}{(s^2 + \varphi_1^2)(s^2 + 2ns + \omega_0^2)}$$

$$(27.3.13)$$

Combining equations (27.3.12) and (27.3.13) with equations (27.1.12) and (27.1.13), we may write

$$x_1(s) = \frac{s^2 a_1}{(s^2 + \varphi_1^2)[(s+n)^2 + \omega^2]} + \frac{2sa_1(n_{12} + n_{22} + n_{32})}{(s^2 + \varphi_1^2)[(s+n)^2 + \omega^2]}$$

$$+ \frac{a_1(\omega_{12}^2 + \omega_{22}^2 + \omega_{32}^2)}{(s^2 + \varphi_1^2)[(s+n)^2 + \omega^2]} \tag{27.3.14}$$

$$x_2(s) = \frac{2a_1 s(n_{12} + n_{22} + n_{32})}{(s^2 + \varphi_1^2)[(s+n)^2 + \omega^2]} + \frac{a_1(\omega_{12}^2 + \omega_{22}^2 + \omega_{32}^2)}{(s^2 + \varphi_1^2)[(s+n)^2 + \omega^2]}$$

$$\tag{27.3.15}$$

27.3.1.　*Underdamped vibrational motion*

If $\omega^2 > 0$, applying to equations (27.3.14) and (27.3.15), respectively, Laplace Transform pairs 68, 65, and 58, and 65 and 58, we invert these equations from the Laplace domain into the time domain and obtain the solutions of differential equations (27.3.1) and (27.3.2) with their initial conditions of motion:

$$x_1 = \frac{a_1}{4n^2\varphi_1^2 + (\omega^2 + n^2 - \varphi_1^2)^2}$$

$$\times \left[(\omega^2 + n^2 - \varphi_1^2)^2 (\cos \varphi_1 t - e^{-nt} \cos \omega t) \right.$$

$$\left. + 2n\varphi_1 \sin \varphi_1 t - \frac{n}{\omega}(\omega^2 + n^2 + \varphi_1^2)e^{-nt} \sin \omega t \right]$$

$$+ \frac{2a_1(n_{12} + n_{22} + n_{32})}{4n^2\varphi_1^2 + (\omega^2 + n^2 - \varphi_1^2)^2} \left[2n(e^{-nt} \cos \omega t - \cos \varphi_1 t) \right.$$

$$\left. + \frac{1}{\varphi_1}(\omega_v^2 + n^2 - \varphi_1^2) \sin \varphi_1 t - \frac{1}{\omega}(\omega^2 - n^2 - \varphi_1^2)e^{-nt} \sin \omega t \right]$$

$$+ \frac{a_1(\omega_{12}^2 + \omega_{22}^2 + \omega_{32}^2)}{4n^2\varphi_1^2 + (\omega^2 + n^2 - \varphi_1^2)^2} \left\{ \frac{1}{\varphi_1^2}(\omega^2 + n^2 - \varphi_1^2)(1 - \cos \varphi_1 t) \right.$$

$$- 2n\left(\frac{1}{\varphi_1} \sin \varphi_1 t - \frac{1}{\omega} e^{-nt} \sin \omega t \right) + \frac{3n^2 + \varphi_1^2 - \omega^2}{\omega^2 + n^2}$$

$$\left. \times \left[1 - e^{-nt}\left(\cos \omega t + \frac{n}{\omega} \sin \omega t \right) \right] \right\} \tag{27.3.16}$$

$$x_2 = \frac{2a_1(n_{12} + n_{22} + n_{32})}{4n^2\varphi_1^2 + (\omega^2 + n^2 - \varphi_1^2)^2}\left[2n(e^{-nt}\cos\omega t - \cos\varphi_1 t)\right.$$

$$\left. + \frac{1}{\varphi_1}(\omega^2 + n^2 - \varphi_1^2)\sin\varphi_1 t - \frac{1}{\omega}(\omega^2 - n^2 - \varphi_1^2)e^{-nt}\sin\omega t\right]$$

$$+ \frac{a_1(\omega_{12}^2 + \omega_{22}^2 + \omega_{32}^2)}{4n^2\varphi_1^2 + (\omega^2 + n^2 - \varphi_1^2)^2}\left\{\frac{1}{\varphi_1^2}(\omega^2 + n^2 - \varphi_1^2)(1 - \cos\varphi_1 t)\right.$$

$$- 2n\left(\frac{1}{\varphi_1}\sin\varphi_1 t - \frac{1}{\omega}e^{-nt}\sin\omega t\right) + \frac{3n^2 + \varphi_1^2 - \omega^2}{\omega^2 + n^2}$$

$$\left. \times \left[1 - e^{-nt}\left(\cos\omega t + \frac{n}{\omega}\sin\omega t\right)\right]\right\} \tag{27.3.17}$$

27.3.2. *Critically damped motion*

Substituting into equations (27.3.14) and (27.3.15) that $\omega^2 = 0$, we obtain

$$x_1(s) = \frac{s^2 a_1}{(s^2 + \varphi_1^2)(s + n)^2} + \frac{2sa_1(n_{12} + n_{22} + n_{32})}{(s^2 + \varphi_1^2)(s + n)^2}$$

$$+ \frac{a_1(\omega_{12}^2 + \omega_{22}^2 + \omega_{32}^2)}{(s^2 + \varphi_1^2)(s + n)^2} \tag{27.3.18}$$

$$x_2(s) = \frac{2a_1 s(n_{12} + n_{22} + n_{32})}{(s^2 + \varphi_1^2)(s + n)^2} + \frac{a_1(\omega_{12}^2 + \omega_{22}^2 + \omega_{32}^2)}{(s^2 + \varphi_1^2)(s + n)^2} \tag{27.3.19}$$

Applying, respectively, to equations (27.3.18) and (27.3.19) Laplace Transform pairs 71, 63 and 60, and 63 and 60, we invert these equations with their initial conditions from the Laplace domain into the time domain and obtain the solutions of differential equations (27.3.1) and (27.3.2) with their initial conditions of motion:

$$x_1 = \frac{a_1}{(\varphi_1^2 + n^2)^2}\left\{2n\varphi_1\sin\varphi_1 t + e^{-nt}\left[\varphi_1^2 - n^2 - nt(\varphi_1^2 + n^2)\right]\right.$$

$$\left. - (\varphi_1^2 - n^2)\cos\varphi_1 t\right\} + \frac{2a_1(n_{12} + n_{22} + n_{32})}{(\varphi_1^2 + n^2)^2}$$

$$\times \left\{2n\varphi_1\left[e^{-nt}(1 + nt) - \cos\varphi_1 t\right]\right.$$

$$\left. - (\varphi_1^2 - n^2)\left(\sin\varphi_1 t - \varphi_1 t e^{-nt}\right)\right\}$$

$$+ \frac{a_1(\omega_{12}^2 + \omega_{22}^2 + \omega_{32}^2)}{(\varphi_1^2 + n^2)^2} \left\{ \frac{n^2 - \varphi_1^2}{\varphi_1^2} (1 - \cos\varphi_1 t) \right.$$

$$\left. - \frac{2n}{\varphi_1} \sin\varphi_1 t + \frac{3n^2 + \varphi_1^2}{\varphi_1^2} \left[1 - e^{-nt}(1 + nt) \right] + 2nte^{-nt} \right\}$$

$$(27.3.20)$$

$$x_2 = \frac{2a_1(n_{12} + n_{22} + n_{32})}{(\varphi_1^2 + n^2)^2} \left\{ 2n\varphi_1 \left[e^{-nt}(1 + nt) - \cos\varphi_1 t \right] \right.$$

$$\left. - (\varphi_1^2 - n^2)(\sin\varphi_1 t - \varphi_1 t e^{-nt}) \right\} + \frac{a_1(\omega_{12}^2 + \omega_{22}^2 + \omega_{32}^2)}{(\varphi_1^2 + n^2)^2}$$

$$\times \left\{ \frac{n^2 - \varphi_1^2}{\varphi_1^2} (1 - \cos\varphi_1 t) - \frac{2n}{\varphi_1} \sin\varphi_1 t \right.$$

$$\left. + \frac{3n^2 + \varphi_1^2}{\varphi_1^2} \left[1 - e^{-nt}(1 + nt) \right] + 2nte^{-nt} \right\}$$

$$(27.3.21)$$

27.3.3. *Overdamped motion*

Since overdamped motion occurs in the case when $\omega^2 < 0$, we have to replace in equations (27.3.14) and (27.3.15) the parameter ω^2 with its negative value, and then, we obtain the corresponding equations that upon inversion give us the equations describing overdamped motion:

$$x_1(s) = \frac{s^2 a_1}{(s^2 + \varphi_1^2)[(s + n)^2 - \omega^2]} + \frac{2sa_1(n_{12} + n_{22} + n_{32})}{(s^2 + \varphi_1^2)[(s + n)^2 - \omega^2]}$$

$$+ \frac{a_1(\omega_{12}^2 + \omega_{22}^2 + \omega_{32}^2)}{(s^2 + \varphi_1^2)[(s + n)^2 - \omega^2]} \qquad (27.3.22)$$

$$x_2(s) = \frac{2a_1 s(n_{12} + n_{22} + n_{32})}{(s^2 + \varphi_1^2)[(s + n)^2 - \omega^2]} + \frac{a_1(\omega_{12}^2 + \omega_{22}^2 + \omega_{32}^2)}{(s^2 + \varphi_1^2)[(s + n)^2 - \omega^2]}$$

$$(27.3.23)$$

Applying, respectively, to equations (27.3.22) and (27.3.23) Laplace Transform pairs 69, 66, and 59, and 66 and 59, we invert these equations with their initial conditions from the Laplace domain into the time domain and obtain the solutions of differential equations (27.3.1) and (27.3.2) with their initial conditions of motion:

$$
x_1 = \frac{a_1}{4n^2\varphi_1^2 + (n^2 - \omega^2 - \varphi_1^2)^2}
$$
$$
\times \left[(n^2 - \omega^2 - \varphi_1^2)^2 (\cos\varphi_1 t - e^{-nt}\cosh\omega t) \right.
$$
$$
\left. + 2n\varphi_1 \sin\varphi_1 t - \frac{n}{\omega}(n^2 - \omega + \varphi_1^2)e^{-nt}\sinh\omega t \right]
$$
$$
+ \frac{2a_1(n_{12} + n_{22} + n_{32})}{4n^2\varphi_1^2 + (n^2 - \omega - \varphi_1^2)^2} \left[2n\left(e^{-nt}\cosh\omega t - \cos\varphi_1 t\right) \right.
$$
$$
\left. + \frac{1}{\varphi_1}\left(n^2 - \omega^2 - \varphi_1^2\right)\sin\varphi_1 t + \frac{1}{\omega}\left(\omega^2 + n^2 + \varphi_1^2\right)e^{-nt}\sinh\omega t \right]
$$
$$
+ \frac{a_1(\omega_{12}^2 + \omega_{22}^2 + \omega_{32}^2)}{4n^2\varphi_1^2 + (n^2 - \omega^2 - \varphi_1^2)^2} \left\{ \frac{1}{\varphi_1^2}(n^2 - \omega^2 - \varphi_1^2)(1 - \cos\varphi_1 t) \right.
$$
$$
- 2n\left(\frac{1}{\varphi_1}\sin\varphi_1 t - \frac{1}{\omega}e^{-nt}\sinh\omega t\right) + \frac{3n^2 + \varphi_1^2 + \omega^2}{n^2 - \omega^2}
$$
$$
\left. \times \left[1 - e^{-nt}\left(\cosh\omega t + \frac{n}{\omega}\sinh\omega t\right)\right] \right\} \tag{27.3.24}
$$

$$
x_2 = \frac{2a_1(n_{12} + n_{22} + n_{32})}{4n^2\varphi_1^2 + (n^2 - \omega^2 - \varphi_1^2)^2} \left[2n(e^{-nt}\cosh\omega t - \cos\varphi_1 t) \right.
$$
$$
\left. + \frac{1}{\varphi_1}(n^2 - \omega^2 - \varphi_1^2)\sin\varphi_1 t + \frac{1}{\omega}(\omega^2 + n^2 + \varphi_1^2)e^{-nt}\sinh\omega t \right]
$$
$$
+ \frac{a_1(\omega_{12}^2 + \omega_{22}^2 + \omega_{32}^2)}{4n^2\varphi_1^2 + (n^2 - \omega^2 - \varphi_1^2)^2} \left\{ \frac{1}{\varphi_1^2}\left(n^2 - \omega^2 - \varphi_1^2\right)(1 - \cos\varphi_1 t) \right.
$$
$$
- 2n\left(\frac{1}{\varphi_1}\sin\varphi_1 t - \frac{1}{\omega}e^{-nt}\sinh\omega t\right) + \frac{3n^2 + \varphi_1^2 + \omega^2}{n^2 - \omega^2}
$$
$$
\left. \times \left[1 - e^{-nt}\left(\cosh\omega t + \frac{n}{\omega}\sinh\omega t\right)\right] \right\} \tag{27.3.25}
$$

CHAPTER 2

PROBLEMS

Provide all mathematical procedures that lead to the answer of the problems

1. Determine the basic parameters of motion of a system, shown in Figure 2.1.1, moving on a horizontal frictionless surface in the absence of resisting and active forces, while the air resistance is negligible. The initial conditions of motion are as follows: for $t = 0$, $x = 0$; $\frac{dx}{dt} = V$.

   Answer:

   $$x = Vt; \quad \frac{dx}{dt} = V; \quad \frac{d^2x}{dt^2} = 0.$$

2. Determine the basic parameters of motion of a system, shown in Figure 2.2.1, moving on a horizontal frictionless surface in the absence of resisting forces while being subjected to a constant active force. The air resistance force is negligible. The initial conditions of motion are as follows: for $t = 0$, $x = 0$; $\frac{dx}{dt} = 0$.

   Answer:

   $$x = \frac{1}{2}pt^2; \quad \frac{dx}{dt} = pt; \quad \frac{d^2x}{dt^2} = p.$$

3. Determine the basic parameters of motion of a system, shown in Figure 2.2.1, moving on a horizontal frictionless surface in the absence of resisting forces while being subjected to a constant active force. The air resistance force is negligible. The initial conditions of motion are as follows: for $t = 0$, $x = S$; $\frac{dx}{dt} = 0$.

   Answer:

   $$x = S + \frac{1}{2}pt^2; \quad \frac{dx}{dt} = pt; \quad \frac{d^2x}{dt^2} = p.$$

4. Determine the basic parameters of motion of a system, shown in Figure 2.2.1, moving on a horizontal frictionless surface in the absence of resisting forces while being subjected to a constant active force. The air resistance force is negligible. The initial conditions of motion are as follows: for $t = 0$, $x = 0$; $\frac{dx}{dt} = V$.

   Answer:

   $$x = Vt + \frac{1}{2}pt^2; \quad \frac{dx}{dt} = V + pt; \quad \frac{d^2x}{dt^2} = p.$$

5. Determine the basic parameters of motion of a system, shown in Figure 2.3.1, moving on a horizontal frictionless surface in the absence of resisting forces while being subjected to a harmonic force. The air resistance is negligible. The initial conditions of motion are as follows: for $t = 0$, $x = 0$; $\frac{dx}{dt} = 0$.

   Answer:

   $$x = \frac{a}{\varphi^2}(1 - \cos \varphi t); \quad \frac{dx}{dt} = \frac{a}{\varphi} \sin \varphi t; \quad \frac{d^2x}{dt^2} = a \cos \varphi t.$$

6. Determine the basic parameters of motion of a system, shown in Figure 2.3.1, moving on a horizontal frictionless surface in the absence of resisting forces while being subjected to a harmonic force. The air resistance is negligible. The initial conditions of motion are as follows: for $t = 0$, $x = S$; $\frac{dx}{dt} = 0$.

Answer:

$$x = S + \frac{a}{\varphi^2}(1 - \cos \varphi t); \quad \frac{dx}{dt} = \frac{a}{\varphi} \sin \varphi t; \quad \frac{d^2 x}{dt^2} = a \cos \varphi t.$$

7. Determine the basic parameters of motion of a system, shown in Figure 2.3.1, moving on a horizontal frictionless surface in the absence of resisting forces while being subjected to a harmonic force. The air resistance is negligible. The initial conditions of motion are as follows: for $t = 0$, $x = 0$; $\frac{dx}{dt} = V$.

Answer:

$$x = Vt + \frac{a}{\varphi^2}(1 - \cos \varphi t); \quad \frac{dx}{dt} = V + \frac{a}{\varphi} \sin \varphi t; \quad \frac{d^2 x}{dt^2} = a \cos \varphi t.$$

8. Determine the basic parameters of motion of a system, shown in Figure 2.4.1, moving on a horizontal frictionless surface in the absence of resisting forces while being subjected to constant active and harmonic forces. The air resistance is negligible. The initial conditions of motion are as follows: for $t = 0$, $x = S$; $\frac{dx}{dt} = 0$.

Answer:

$$x = S + \frac{1}{2}pt^2 + \frac{a}{\varphi^2}(1 - \cos \varphi t); \quad \frac{dx}{dt} = pt + \frac{a}{\varphi} \sin \varphi t;$$

$$\frac{d^2 x}{dt^2} = p + a \cos \varphi t.$$

9. Determine the basic parameters of motion of a system, shown in Figure 2.4.1 moving on a horizontal frictionless surface in the absence of resisting forces while being subjected to constant active and harmonic forces. The air resistance is negligible. The initial conditions of motion are as follows: for $t = 0$, $x = 0$; $\frac{dx}{dt} = V$.

Answer:

$$x = Vt + \frac{1}{2}pt^2 + \frac{a}{\varphi^2}(1 - \cos \varphi t); \quad \frac{dx}{dt} = V + pt + \frac{a}{\varphi}\sin \varphi t;$$

$$\frac{d^2x}{dt^2} = p + a\cos \varphi t.$$

10. Determine the basic parameters of motion of a system, shown in Figure 2.4.1, moving on a horizontal frictionless surface in the absence of resisting forces while being subjected to constant active and harmonic forces. The air resistance is negligible. The initial conditions of motion are as follows: for $t = 0$, $x = 0$; $\frac{dx}{dt} = 0$.

Answer:

$$x = \frac{1}{2}pt^2 + \frac{a}{\varphi^2}(1 - \cos \varphi t); \quad \frac{dx}{dt} = pt + \frac{a}{\varphi}\sin \varphi t;$$

$$\frac{d^2x}{dt^2} = p + a\cos \varphi t.$$

CHAPTER 3

PROBLEMS

---

1. Determine the basic parameters of motion of a system, shown in Figure 3.1.1, moving on a horizontal frictionless surface in the absence of active forces while being subjected to a constant resisting force. The air resistance is negligible. The initial conditions of motion are as follows: for $t = 0$, $x = 0$; $\frac{dx}{dt} = V$.

   Answer:

$$x = Vt - \frac{1}{2}rt^2; \quad \frac{dx}{dt} = V - rt; \quad \frac{d^2x}{dt^2} = -r.$$

2. Determine the basic parameters of motion of a system, shown in Figure 3.2.1, moving on a horizontal frictionless surface while being subjected to a constant resisting force and a constant active force. The air resistance is negligible. The initial conditions of motion are as follows: for $t = 0$, $x = 0$; $\frac{dx}{dt} = 0$.

   Answer:

$$x = \frac{1}{2}(p - r)t^2; \quad \frac{dx}{dt} = (p - r)t; \quad \frac{d^2x}{dt^2} = p - r.$$

3. Determine the basic parameters of motion of a system, shown in Figure 3.2.1, moving on a horizontal frictionless surface while being subjected to a constant resisting force and a constant

active force. The air resistance is negligible. The initial conditions of motion are as follows: for $t = 0$, $x = S$; $\frac{dx}{dt} = 0$.

Answer:

$$x = S + \frac{1}{2}(p - r)t^2; \quad \frac{dx}{dt} = (p - r)t; \quad \frac{d^2 x}{dt^2} = p - r.$$

4. Determine the basic parameters of motion of a system, shown in Figure 3.2.1, moving on a horizontal frictionless surface while being subjected to a constant resisting force and a constant active force. The air resistance is negligible. The initial conditions of motion are as follows: for $t = 0$, $x = 0$; $\frac{dx}{dt} = V$.

Answer:

$$x = Vt + \frac{1}{2}(p - r)t^2; \quad \frac{dx}{dt} = V + (p - r)t; \quad \frac{d^2 x}{dt^2} = p - r.$$

5. Determine the basic parameters of motion of a system, shown in Figure 3.3.1, moving on a horizontal frictionless surface while being subjected to a constant resisting force and a harmonic force. The air resistance is negligible. The initial conditions of motion are as follows: for $t = 0$, $x = 0$; $\frac{dx}{dt} = 0$.

Answer:

$$x = \frac{a}{\varphi^2}(1 - \cos \varphi t) - \frac{1}{2}rt^2; \quad \frac{dx}{dt} = a \sin \varphi t - rt;$$

$$\frac{d^2 x}{dt^2} = a \sin \varphi t - r.$$

6. Determine the basic parameters of motion of a system, shown in Figure 3.3.1, moving on a horizontal frictionless surface while being subjected to a constant resisting force and a harmonic force. The air resistance is negligible. The initial conditions of motion are as follows: for $t = 0$, $x = S$; $\frac{dx}{dt} = 0$.

Answer:

$$x = S + \frac{a}{\varphi^2}(1 - \cos\varphi t) - \frac{1}{2}rt^2; \qquad \frac{dx}{dt} = \frac{a}{\varphi}\sin\varphi t - rt;$$

$$\frac{d^2x}{dt^2} = a\cos\varphi t - r.$$

7. Determine the basic parameters of motion of a system, shown in Figure 3.3.1, moving on a horizontal frictionless surface while being subjected to a constant resisting force and a harmonic force. The air resistance is negligible. The initial conditions of motion are as follows: for $t = 0$, $x = 0$; $\frac{dx}{dt} = V$.

Answer:

$$x = Vt + \frac{a}{\varphi^2}(1 - \cos\varphi t) - \frac{1}{2}rt^2; \qquad \frac{dx}{dt} = V + \frac{a}{\varphi}\sin\varphi t - rt;$$

$$\frac{d^2x}{dt^2} = a\cos\varphi t - r.$$

8. Determine the basic parameters of motion of a system, shown in Figure 3.4.1, moving on a horizontal frictionless surface while being subjected to a constant resisting force, a constant active force, and a harmonic force. The air resistance is negligible. The initial conditions of motion are as follows: for $t = 0$, $x = 0$; $\frac{dx}{dt} = 0$.

Answer:

$$x = \frac{a}{\varphi^2}(1 - \cos\varphi t) + \frac{1}{2}(p - r)t^2; \qquad \frac{dx}{dt} = \frac{a}{\varphi}\sin\varphi t + (p - r)t;$$

$$\frac{d^2x}{dt^2} = a\cos\varphi t + p - r.$$

9. Determine the basic parameters of motion of a system, shown in Figure 3.4.1, moving on a horizontal frictionless surface while being subjected to a constant resisting force, a constant active

force, and a harmonic force. The air resistance is negligible. The initial conditions of motion are as follows: for $t = 0$, $x = S$; $\frac{dx}{dt} = 0$.

Answer:

$$x = S + \frac{a}{\varphi^2}(1 - \cos \varphi t) + \frac{1}{2}(p - r)t^2;$$

$$\frac{dx}{dt} = \frac{a}{\varphi} \sin \varphi t + (p - r)t; \qquad \frac{d^2x}{dt^2} = a \cos \varphi t + p - r.$$

10. Determine the basic parameters of motion of a system, shown in Figure 3.4.1, moving on a horizontal frictionless surface while being subjected to a constant resisting force, a constant active force, and a harmonic force. The air resistance is negligible. The initial conditions of motion are as follows: for $t = 0$, $x = 0$; $\frac{dx}{dt} = V$.

Answer:

$$x = Vt + \frac{a}{\varphi^2}(1 - \cos \varphi t) + \frac{1}{2}(p - r)t^2;$$

$$\frac{dx}{dt} = V + \frac{a}{\varphi} \sin \varphi t + (p - r)t; \qquad \frac{d^2x}{dt^2} = a \cos \varphi t + p - r.$$

CHAPTER 4

PROBLEMS

---

1. Determine the basic parameters of motion of a system, shown in Figure 4.1.1, moving on a horizontal surface in the absence of active forces while being subjected to a dry friction force. The air resistance is negligible. The initial conditions of motion are as follows: for $t = 0$, $x = 0$; $\frac{dx}{dt} = V$.

   Answer:

   $$x = Vt - \frac{1}{2}ft^2; \quad \frac{dx}{dt} = V - ft; \quad \frac{d^2x}{dt^2} = -f.$$

2. Determine the basic parameters of motion of a system, shown in Figure 4.2.1, moving on a horizontal surface while being subjected to a dry friction force and a constant active force. The air resistance is negligible. The initial conditions of motion are as follows: for $t = 0$, $x = 0$; $\frac{dx}{dt} = 0$.

   Answer:

   $$x = \frac{1}{2}(p - f)t^2; \quad \frac{dx}{dt} = (p - f)t; \quad \frac{d^2x}{dt^2} = p - f.$$

3. Determine the basic parameters of motion of a system, shown in Figure 4.2.1, moving on a horizontal surface while being subjected to a dry friction force and a constant active force.

The air resistance is negligible. The initial conditions of motion are as follows: for $t = 0$, $x = 0$; $\frac{dx}{dt} = 0$.

Answer:

$$x = \frac{1}{2}(p - f)t^2; \quad \frac{dx}{dt} = (p - f)t; \quad \frac{d^2x}{dt^2} = p - f.$$

4. Determine the basic parameters of motion of a system, shown in Figure 4.2.1, moving on a horizontal surface while being subjected to a dry friction force and a constant active force. The air resistance is negligible. The initial conditions of motion are as follows: for $t = 0$, $x = 0$; $\frac{dx}{dt} = V$.

Answer:

$$x = Vt + \frac{1}{2}(p - r)t^2; \quad \frac{dx}{dt} = V + (p - r)t; \quad \frac{d^2x}{dt^2} = p - r.$$

5. Determine the basic parameters of motion of a system, shown in Figure 4.3.1, moving on a horizontal surface while being subjected to a dry fiction force and a harmonic force. The air resistance is negligible. The initial conditions of motion are as follows: for $t = 0$, $x = 0$; $\frac{dx}{dt} = 0$.

Answer:

$$x = \frac{a}{\varphi^2}(1 - \cos \varphi t) - \frac{1}{2}ft^2; \quad \frac{dx}{dt} = \frac{a}{\varphi}\sin \varphi t - ft;$$

$$\frac{d^2x}{dt^2} = a \cos \varphi t - f.$$

6. Determine the basic parameters of motion of a system, shown in Figure 4.3.1, moving on a horizontal surface while being subjected to a dry friction force and a harmonic force. The air resistance is negligible. The initial conditions of motion are as follows: for $t = 0$, $x = S$; $\frac{dx}{dt} = 0$.

Answer:

$$x = S + \frac{a}{\varphi^2}(1 - \cos \varphi t) - \frac{1}{2}ft^2; \quad \frac{dx}{dt} = \frac{a}{\varphi}\sin \varphi t - ft;$$

$$\frac{d^2 x}{dt^2} = a \cos \varphi t - f.$$

7. Determine the basic parameters of motion of a system, shown in Figure 4.3.1, moving on a horizontal surface while being subjected to a dry friction force and a harmonic force. The air resistance is negligible. The initial conditions of motion are as follows: for $t = 0$, $x = 0$; $\frac{dx}{dt} = V$.

Answer:

$$x = Vt + \frac{a}{\varphi^2}(1 - \cos \varphi t) - \frac{1}{2}ft^2; \quad \frac{dx}{dt} = V + \frac{a}{\varphi}\sin \varphi t - ft;$$

$$\frac{d^2 x}{dt^2} = a \cos \varphi t - f.$$

8. Determine the basic parameters of motion of a system, shown in Figure 4.4.1, moving on a horizontal surface while being subjected to a dry friction force, a constant active force, and a harmonic force. The air resistance is negligible. The initial conditions of motion are as follows: for $t = 0$, $x = 0$; $\frac{dx}{dt} = 0$.

Answer:

$$x = \frac{a}{\varphi^2}(1 - \cos \varphi t) + \frac{1}{2}(p - f)t^2; \quad \frac{dx}{dt} = \frac{a}{\varphi}\sin \varphi t + (p - f)t;$$

$$\frac{d^2 x}{dt^2} = a \cos \varphi t + p - f.$$

9. Determine the basic parameters of motion of a system, shown in Figure 4.4.1, moving on a horizontal surface while being subjected to a dry friction force, a constant active force, and

a harmonic force. The air resistance is negligible. The initial conditions of motion are as follows: for $t = 0$, $x = S$; $\frac{dx}{dt} = 0$.

Answer:

$$x = S + \frac{a}{\varphi^2}(1 - \cos \varphi t) + \frac{1}{2}(p - f)t^2;$$

$$\frac{dx}{dt} = \frac{a}{\varphi}\sin \varphi t + (p - f)t; \quad \frac{d^2x}{dt^2} = a\cos \varphi t + p - f.$$

10. Determine the basic parameters of motion of a system, shown in Figure 4.4.1, moving on a horizontal surface while being subjected to a dry friction force, a constant active force, and a harmonic force. The air resistance is negligible. The initial conditions of motion are as follows: for $t = 0$ $x = 0$; $\frac{dx}{dt} = V$.

Answer:

$$x = Vt + \frac{a}{\varphi^2}(1 - \cos \varphi t) + \frac{1}{2}(p - f)t^2;$$

$$\frac{dx}{dt} = V + \frac{a}{\varphi}\sin \varphi t + (p - f)t; \quad \frac{d^2x}{dt^2} = a\cos \varphi t + p - f.$$

CHAPTER 5

PROBLEMS

---

1. Determine the basic parameters of motion of a system, shown in Figure 5.1.1, moving on a horizontal surface in the absence of active forces while being subjected to a constant resisting force and a dry friction force. The air resistance is negligible. The initial conditions of motion are as follows: for $t = 0$, $x = 0$; $\frac{dx}{dt} = V$.

   Answer:

   $$x = Vt - \frac{1}{2}(r+f)t^2; \quad \frac{dx}{dt} = V - (r+f)t; \quad \frac{d^2x}{dt^2} = -r - f.$$

2. Determine the basic parameters of motion of a system, shown in Figure 5.2.1, moving on a horizontal surface while being subjected to a dry friction force, a constant resisting force, and a constant active force. The air resistance is negligible. The initial conditions of motion are as follows: for $t = 0$, $x = 0$; $\frac{dx}{dt} = 0$.

   Answer:

   $$x = \frac{1}{2}(p-r-f)t^2; \quad \frac{dx}{dt} = (p-r-f)t; \quad \frac{d^2x}{dt^2} = p - r - f.$$

3. Determine the basic parameters of motion of a system, shown in Figure 5.2.1, moving on a horizontal surface while being subjected to a dry friction force, a constant resisting force, and a constant active force. The air resistance is negligible. The initial conditions of motion are as follows: for $t = 0$, $x = S$; $\frac{dx}{dt} = 0$.

Answer:

$$x = S + \frac{1}{2}(p - r - f)t^2; \quad \frac{dx}{dt} = (p - r - f)t; \quad \frac{d^2x}{dt^2} = p - r - f.$$

4. Determine the basic parameters of motion of a system, shown in Figure 5.2.1, moving on a horizontal surface while being subjected to a dry friction force, a constant resisting force, and a constant active force. The air resistance is negligible. The initial conditions of motion are as follows: for $t = 0$, $x = 0$; $\frac{dx}{dt} = V$.

Answer:

$$x = Vt + \frac{1}{2}(p - r - f)t^2; \quad \frac{dx}{dt} = V + (p - r - f)t; \quad \frac{d^2x}{dt^2} = p - r - f.$$

5. Determine the basic parameters of motion of a system, shown in Figure 5.3.1, moving on a horizontal surface while being subjected to a dry friction force, a constant resisting force, and a harmonic force. The air resistance is negligible. The initial conditions of motion are as follows: for $t = 0$, $x = 0$; $\frac{dx}{dt} = 0$.

Answer:

$$x = \frac{a}{\varphi^2}(1 - \cos \varphi t) - \frac{1}{2}(r + f)t^2; \quad \frac{dx}{dt} = \frac{a}{\varphi}\sin \varphi t - (r + f)t;$$

$$\frac{d^2x}{dt^2} = a \cos \varphi t - r - f.$$

6. Determine the basic parameters of motion of a system, shown in Figure 5.3.1, moving on a horizontal surface while being subjected to a constant resisting force, a dry friction force, and a harmonic force. The air resistance is negligible. The initial conditions of motion are as follows: for $t = 0$, $x = S$; $\frac{dx}{dt} = 0$.

Answer:

$$x = S + \frac{a}{\varphi^2}(1 - \cos \varphi t) - \frac{1}{2}(r + f)t^2;$$

$$\frac{dx}{dt} = \frac{a}{\varphi} \sin \varphi t - (r + f)t; \quad \frac{d^2 x}{dt^2} = a \cos \varphi t - r - f.$$

7. Determine the basic parameters of motion of a system, shown in Figure 5.3.1, moving on a horizontal surface while being subjected to a dry friction force, a constant resisting force, and a harmonic force. The air resistance is negligible. The initial conditions of motion are as follows: for $t = 0$, $x = 0$; $\frac{dx}{dt} = V$.

Answer:

$$x = Vt + \frac{a}{\varphi^2}(1 - \cos \varphi t) - \frac{1}{2}(r + f)t^2;$$

$$\frac{dx}{dt} = V + \frac{a}{\varphi} \sin \varphi t - (r + f)t; \quad \frac{d^2 x}{dt^2} = a \cos \varphi t - r - f.$$

8. Determine the basic parameters of motion of a system, shown in Figure 5.4.1, moving on a horizontal surface while being subjected to a dry friction force, a constant resisting force, a constant active force, and a harmonic force. The air resistance is negligible. The initial conditions of motion are as follows: for $t = 0$, $x = 0$; $\frac{dx}{dt} = 0$.

Answer:

$$x = \frac{a}{\varphi^2}(1 - \cos \varphi t) + \frac{1}{2}(p - r - f)t^2;$$

$$\frac{dx}{dt} = \frac{a}{\varphi} \sin \varphi t + (p - r - f)t; \quad \frac{d^2 x}{dt^2} = a \cos \varphi t + p - r - f.$$

9. Determine the basic parameters of motion of a system, shown in Figure 5.4.1, moving on a horizontal surface while being subjected to a dry friction force, a constant resisting force, a constant active force, and a harmonic force. The air resistance

is negligible. The initial conditions of motion are as follows: for $t = 0$, $x = S$; $\frac{dx}{dt} = 0$.

Answer:

$$x = S + \frac{a}{\varphi^2}(1 - \cos \varphi t) + \frac{1}{2}(p - r - f)t^2;$$

$$\frac{dx}{dt} = \frac{a}{\varphi} \sin \varphi t + (p - r - f)t; \quad \frac{d^2 x}{dt^2} = a \cos \varphi t + p - r - f.$$

10. Determine the basic parameters of motion of a system, shown in Figure 5.4.1, moving on a horizontal surface while being subjected to a dry friction force, a constant resisting force, a constant active force, and a harmonic force. The air resistance is negligible. The initial conditions of motion are as follows: for $t = 0$, $x = 0$; $\frac{dx}{dt} = V$.

Answer:

$$x = Vt + \frac{a}{\varphi^2}(1 - \cos \varphi t) + \frac{1}{2}(p - r - f)t^2;$$

$$\frac{dx}{dt} = V + \frac{a}{\varphi} \sin \varphi t + (p - r - f)t; \quad \frac{d^2 x}{dt^2} = a \cos \varphi t + p - r - f.$$

CHAPTER 6

PROBLEMS

---

1. Determine the basic parameters of motion of a system, shown in Figure 6.1.1, moving on a horizontal frictionless surface due to the initial velocity of the system in the absence of active forces. The mass of the system is connected to a non-movable support by a flexible link. The air resistance is negligible. The initial conditions of motion are as follows: for $t = 0$, $x = 0$; $\frac{dx}{dt} = V$.

   Answer:

   $$x = \frac{V}{\omega}\sin\omega t; \quad \frac{dx}{dt} = V\cos\omega t; \quad \frac{d^2x}{dt^2} = -V\omega\sin\omega t.$$

2. Determine the basic parameters of motion of a system, shown in Figure 6.2.1, moving on a horizontal frictionless surface while being subjected to a constant active force. The mass of the system is connected to a non-movable support by a flexible link. The air resistance is negligible. The initial conditions of motion are as follows: for $t = 0$, $x = 0$; $\frac{dx}{dt} = 0$.

   Answer:

   $$x = \frac{p}{\omega^2}(1 - \cos\omega t); \quad \frac{dx}{dt} = \frac{p}{\omega}\sin\omega t; \quad \frac{d^2x}{dt^2} = p\cos\omega t.$$

3. Determine the basic parameters of motion of a system, shown in Figure 6.2.1, moving on a horizontal frictionless surface while being subjected to a constant active force. The mass of the system is connected to a non-movable support by a flexible link. The air resistance is negligible. The initial conditions of motion are as follows: for $t = 0$, $x = S$; $\frac{dx}{dt} = 0$.

Answer:

$$x = S \cos \omega t + \frac{p}{\omega^2}(1 - \cos \omega t); \quad \frac{dx}{dt} = -S\omega \sin \omega t + \frac{p}{\omega} \sin \omega t;$$

$$\frac{d^2 x}{dt^2} = -S\omega^2 \cos \omega t + p \cos \omega t.$$

4. Determine the basic parameters of motion of a system, shown in Figure 6.2.1, moving on a horizontal frictionless surface while being subjected to a constant active force. The mass of the system is connected to a non-movable support by a flexible link. The air resistance is negligible. The initial conditions of motion are as follows: for $t = 0$, $x = 0$; $\frac{dx}{dt} = V$.

Answer:

$$x = \frac{p}{\omega^2}(1 - \cos \omega t) + \frac{V}{\omega} \sin \omega t; \quad \frac{dx}{dt} = V \cos \omega t + \frac{p}{\omega} \sin \omega t;$$

$$\frac{d^2 x}{dt^2} = p \cos \omega t - V\omega \sin \omega t.$$

5. Determine the basic parameters of motion of a system, shown in Figure 6.3.1, moving on a horizontal frictionless surface while being subjected to a harmonic force. The mass of the system is connected to a non-movable support by a flexible link. The air resistance is negligible. The initial conditions of motion are as follows: for $t = 0$. $x = 0$; $\frac{dx}{dt} = 0$.

Answer:

$$x = \frac{a(\cos \omega t - \cos \varphi t)}{\varphi^2 - \omega^2}; \quad \frac{dx}{dt} = \frac{a(\varphi \sin \varphi t - \omega \sin \omega t)}{\varphi^2 - \omega^2};$$

$$\frac{d^2 x}{dt^2} = \frac{a(\varphi^2 \cos \varphi t - \omega^2 \cos \omega t)}{\varphi^2 - \omega^2}.$$

6. Determine the basic parameters of motion of a system, shown in Figure 6.3.1, moving on a horizontal frictionless surface while being subjected to a harmonic force. The mass of the system is connected to a non-movable support by a flexible link. The air resistance is negligible. The initial conditions of motion are as follows: for $t = 0$, $x = S$; $\frac{dx}{dt} = 0$.

Answer:

$$x = S \cos \omega t + \frac{a(\cos \omega t - \cos \varphi t)}{\varphi^2 - \omega^2};$$

$$\frac{dx}{dt} = -S\omega \sin \omega t + \frac{a(\varphi \sin \varphi t - \omega \sin \omega t)}{\varphi^2 - \omega^2};$$

$$\frac{d^2 x}{dt^2} = -S\omega^2 \cos \omega t + \frac{a(\varphi^2 \cos \varphi t - \omega^2 \cos \omega t)}{\varphi^2 - \omega^2}.$$

7. Determine the basic parameters of motion of a system, shown in Figure 6.3.1, moving on a horizontal frictionless surface while being subjected to a harmonic force. The mass of the system is connected to a non-movable support by a flexible link. The air resistance is negligible. The initial conditions of motion are as follows: for $t = 0$. $x = 0$; $\frac{dx}{dt} = V$.

Answer:

$$x = \frac{V}{\omega} \sin \omega t + \frac{a(\cos \omega t - \cos \varphi t)}{\varphi^2 - \omega^2};$$

$$\frac{dx}{dt} = V \cos \omega t + \frac{a(\varphi \sin \varphi t - \omega \sin \omega t)}{\varphi^2 - \omega^2};$$

$$\frac{d^2 x}{dt^2} = -V\omega \sin \omega t + \frac{a(\varphi^2 \cos \varphi t - \omega^2 \cos \omega t)}{\varphi^2 - \omega^2}.$$

8. Determine the basic parameters of motion of a system, shown in Figure 6.4.1, moving on a horizontal frictionless surface while being subjected to a constant active force and a harmonic force. The mass of the system is connected to a non-movable support by a flexible link. The air resistance is negligible.

The initial conditions of motion are as follows: for $t = 0$, $x = 0$; $\frac{dx}{dt} = 0$.

Answer:

$$x = \frac{p}{\omega^2}(1 - \cos \omega t) + \frac{a(\cos \omega t - \cos \varphi t)}{\varphi^2 - \omega^2};$$

$$\frac{dx}{dt} = \frac{p}{\omega}\sin \omega t + \frac{a(\varphi \sin \varphi t - \omega \sin \omega t)}{\varphi^2 - \omega^2};$$

$$\frac{d^2 x}{dt^2} = p \cos \omega t + \frac{a(\varphi^2 \cos \varphi t - \omega^2 \cos \omega t)}{\varphi^2 - \omega^2}.$$

9. Determine the basic parameters of motion of a system, shown in Figure 6.4.1, moving on a horizontal frictionless surface while being subjected to a constant active force and a harmonic force. The mass of the system is connected to a non-movable support by a flexible link. The air resistance is negligible. The initial conditions of motion are as follows: for $t = 0$, $x = S$; $\frac{dx}{dt} = 0$.

Answer:

$$x = S \cos \omega t + \frac{p}{\omega^2}(1 - \cos \omega t) + \frac{a(\cos \omega t - \cos \varphi t)}{\varphi^2 - \omega^2};$$

$$\frac{dx}{dt} = S\omega \sin \omega t + \frac{p}{\omega}\sin \omega t + \frac{a(\varphi \sin \varphi t - \omega \sin \omega t)}{\varphi^2 - \omega^2};$$

$$\frac{d^2 x}{dt^2} = S\omega^2 \cos \omega t + p \cos \omega t + \frac{a(\varphi^2 \cos \varphi t - \omega^2 \cos \omega t)}{\varphi^2 - \omega^2}.$$

10. Determine the basic parameters of motion of a system, shown in Figure 6.4.1, moving on a horizontal frictionless surface while being subjected to a constant active force and a harmonic force. The mass of the system is connected to a non-movable support by a flexible link. The air resistance is negligible. The initial conditions of motion are as follows: for $t = 0$, $x = 0$; $\frac{dx}{dt} = V$.

Answer:

$$x = \frac{V}{\omega}\sin\omega t + \frac{p}{\omega^2}(1 - \cos\omega t) + \frac{a(\cos\omega t - \cos\varphi t)}{\varphi^2 - \omega^2};$$

$$\frac{dx}{dt} = V\cos\omega t + \frac{p}{\omega}\sin\omega t + \frac{a(\varphi\sin\varphi t - \omega\sin\omega t)}{\varphi^2 - \omega^2};$$

$$\frac{d^2x}{dt^2} = V\sin\omega t + p\cos\omega t + \frac{a(\varphi^2\cos\varphi t - \omega^2\cos\omega t)}{\varphi^2 - \omega^2}.$$

CHAPTER 7

PROBLEMS

---

1. Determine the basic parameters of motion of a system, shown in
   Figure 7.1.1, moving on a horizontal frictionless surface due to
   the initial velocity of the system in the absence of active forces.
   The mass of the system is connected to a non-movable support
   by a flexible link and subjected to a constant resisting force.
   The air resistance is negligible. The initial conditions of motion
   are as follows: for $t = 0$, $x = 0$; $\frac{dx}{dt} = V$.

   Answer:

   $$x = \frac{V}{\omega}\sin\omega t - \frac{r}{\omega^2}(1 - \cos\omega t); \qquad \frac{dx}{dt} = V\cos\omega t - \frac{r}{\omega}\sin\omega t;$$

   $$\frac{d^2x}{dt^2} = -V\omega\sin\omega t - r\cos\omega t.$$

2. Determine the basic parameters of motion of a system, shown in
   Figure 7.2.1, moving on a horizontal frictionless surface while
   being subjected to a constant resisting force and a constant
   active force. The mass of the system is connected to a non-
   movable support by a flexible link. The air resistance is negligible.
   The initial conditions of motion are as follows: for $t = 0$, $x = 0$;
   $\frac{dx}{dt} = 0$.

Answer:

$$x = \frac{p-r}{\omega^2}(1 - \cos\omega t); \qquad \frac{dx}{dt} = \frac{p-r}{\omega}\sin\omega t;$$

$$\frac{d^2x}{dt^2} = (p - r)\cos\omega t.$$

3. Determine the basic parameters of motion of a system, shown in Figure 7.2.1, moving on a horizontal frictionless surface while being subjected to a constant resisting force and a harmonic force. The mass of the system is connected to a non-movable support by a flexible link. The air resistance is negligible. The initial conditions of motion are as follows: for $t = 0$, $x = S$; $\frac{dx}{dt} = 0$.

Answer:

$$x = S\cos\omega t - \frac{r}{\omega^2}(1 - \cos\omega t) + \frac{a(\cos\omega t - \cos\varphi t)}{\varphi^2 - \omega^2};$$

$$\frac{dx}{dt} = S\omega\sin\omega t - \frac{r}{\omega}\sin\omega t + \frac{a(\varphi\sin\varphi t - \omega\sin\omega t)}{\varphi^2 - \omega^2};$$

$$\frac{d^2x}{dt^2} = S\omega^2\cos\omega t - r\cos\omega t + \frac{a(\varphi^2\cos\varphi t - \omega^2\cos\omega t)}{\varphi^2 - \omega^2}.$$

4. Determine the basic parameters of motion of a system, shown in Figure 7.2.1, moving on a horizontal frictionless surface while being subjected to a constant resisting force, a constant active force, and a harmonic force. The mass of the system is connected to a non-movable support by a flexible link. The air resistance is negligible. The initial conditions of motion are as follows: for $t = 0$, $x = 0$; $\frac{dx}{dt} = V$.

Answer:

$$x = \frac{V}{\omega}\sin\omega t + \frac{p-r}{\omega^2}(1 - \cos\omega t) + \frac{a(\cos\omega t - \cos\varphi t)}{\varphi^2 - \omega^2};$$

$$\frac{dx}{dt} = V\cos\omega t + \frac{p-r}{\omega}\sin\omega t + \frac{a(\varphi\sin\varphi t - \omega\sin\omega t)}{\varphi^2 - \omega^2};$$

$$\frac{d^2x}{dt^2} = -V\omega\sin\omega t + (p - r)\cos\omega t + \frac{a(\varphi^2\cos\varphi t - \omega^2\cos\omega t)}{\varphi^2 - \omega^2}.$$

5. Determine the basic parameters of motion of a system, shown in Figure 7.3.1, moving on a horizontal frictionless surface while being subjected to a constant resisting force, a constant active, and a harmonic force. The mass of the system is connected to a non-movable support by a flexible link. The air resistance is negligible. The initial conditions of motion are as follows: for $t = 0$, $x = 0$; $\frac{dx}{dt} = 0$.

Answer:

$$x = \frac{p - r}{\omega^2}(1 - \cos\omega t) + \frac{a(\cos\omega t - \cos\varphi t)}{\varphi^2 - \omega^2};$$

$$\frac{dx}{dt} = \frac{p - r}{\omega}\sin\omega t + \frac{a(\varphi\sin\varphi t - \omega\sin\omega t)}{\varphi^2 - \omega^2};$$

$$\frac{d^2 x}{dt^2} = (p - r)\cos\omega t + \frac{a(\varphi^2\cos\varphi t - \omega^2\cos\omega t)}{\varphi^2 - \omega^2}.$$

6. Determine the basic parameters of motion of a system, shown in Figure 7.3.1, moving on a horizontal frictionless surface while being subjected to a constant resisting force, a constant active force, and a harmonic force. The mass of the system is connected to a non-movable support by a flexible link. The air resistance is negligible. The initial conditions of motion are as follows: for $t = 0$, $x = S$; $\frac{dx}{dt} = 0$.

Answer:

$$x = S\cos\omega t + \frac{p - r}{\omega^2}(1 - \cos\omega t) + \frac{a(\cos\omega t - \cos\varphi t)}{\varphi^2 - \omega^2};$$

$$\frac{dx}{dt} = -S\omega\sin\omega t + \frac{p - r}{\omega}\sin\omega t + \frac{a(\varphi\sin\varphi t - \omega\sin\omega t)}{\varphi^2 - \omega^2};$$

$$\frac{d^2 x}{dt^2} = -S\omega^2\cos\omega t + (p - r)\cos\omega t + \frac{a(\varphi^2\cos\varphi t - \omega^2\cos\omega t)}{\varphi^2 - \omega^2}.$$

7. Determine the basic parameters of motion of a system, shown in Figure 7.3.1, moving on a horizontal frictionless surface while being subjected to a constant resisting force, a constant active force, and a harmonic force. The mass of the system is connected

to a non-movable support by a flexible link. The air resistance is negligible. The initial conditions of motion are as follows: for $t = 0$, $x = 0$; $\frac{dx}{dt} = V$.

Answer:

$$x = \frac{V}{\omega}\sin \omega t + \frac{p-r}{\omega^2}(1 - \cos \omega t) + \frac{a(\cos \omega t - \cos \varphi t)}{\varphi^2 - \omega^2};$$

$$\frac{dx}{dt} = V \cos \omega t + \frac{p-r}{\omega}\sin \omega t + \frac{a(\varphi \sin \varphi t - \omega \sin \omega t)}{\varphi^2 - \omega^2};$$

$$\frac{d^2x}{dt^2} = -V\omega \sin \omega t + (p-r)\cos \omega t + \frac{a(\varphi^2 \cos \varphi t - \omega^2 \cos \omega t)}{\varphi^2 - \omega^2}.$$

8. Determine the basic parameters of motion of a system, shown in Figure 7.4.1, moving on a horizontal frictionless surface while being subjected to a constant active force and a harmonic force. The mass of the system is connected to a non-movable support by a flexible link. The air resistance is negligible. The initial conditions of motion are as follows: for $t = 0$, $x = 0$; $\frac{dx}{dt} = 0$.

   Answer:

$$x = \frac{p}{\omega^2}(1 - \cos \omega t) + \frac{a(\cos \omega t - \cos \varphi t)}{\varphi^2 - \omega^2};$$

$$\frac{dx}{dt} = \frac{p}{\omega}\sin \omega t + \frac{a(\varphi \sin \varphi t - \omega \sin \omega t)}{\varphi^2 - \omega^2};$$

$$\frac{d^2x}{dt^2} = p\cos \omega t + \frac{a(\varphi^2 \cos \varphi t - \omega^2 \cos \omega t)}{\varphi^2 - \omega^2}.$$

9. Determine the basic parameters of motion of a system, shown in Figure 7.4.1, moving on a horizontal frictional surface while being subjected to a constant active force and a harmonic force. The mass of the system is connected to a non-movable support by a flexible link. The air resistance is negligible. The initial conditions of motion are as follows: for $t = 0$, $x = S$; $\frac{dx}{dt} = 0$.

Answer:

$$x = S\cos\omega t + \frac{p}{\omega^2}(1 - \cos\omega t) + \frac{a(\cos\omega t - \cos\varphi t)}{\varphi^2 - \omega^2};$$

$$\frac{dx}{dt} = S\omega\sin\omega t + \frac{p}{\omega}\sin\omega t + \frac{a(\varphi\sin\varphi t - \omega\sin\omega t)}{\varphi^2 - \omega^2};$$

$$\frac{d^2x}{dt^2} = S\omega^2\cos\omega t + p\cos\omega t + \frac{a(\varphi^2\cos\varphi t - \omega^2\cos\omega t)}{\varphi^2 - \omega^2}.$$

10. Determine the basic parameters of motion of a system, shown in Figure 7.4.1, moving on a horizontal frictionless surface while being subjected to a constant active force and a harmonic force. The mass of the system is connected to a non-movable support by a flexible link. The air resistance is negligible. The initial conditions of motion are as follows: for $t = 0$, $x = 0$; $\frac{dx}{dt} = V$.

Answer:

$$x = \frac{V}{\omega}\sin\omega t + \frac{p}{\omega^2}(1 - \cos\omega t) + \frac{a(\cos\omega t - \cos\varphi t)}{\varphi^2 - \omega^2};$$

$$\frac{dx}{dt} = V\cos\omega t + \frac{p}{\omega}\sin\omega t + \frac{a(\varphi\sin\varphi t - \omega\sin\omega t)}{\varphi^2 - \omega^2};$$

$$\frac{d^2x}{dt^2} = V\sin\omega t + p\cos\omega t + \frac{a(\varphi^2\cos\varphi t - \omega^2\cos\omega t)}{\varphi^2 - \omega^2}.$$

CHAPTER 8

PROBLEMS

---

1. Determine the basic parameters of motion of a system, shown in Figure 8.1.1, moving on a horizontal surface due to the initial velocity of the system in the absence of active forces. The system is restricted by a flexible link and subjected to a dry friction force. The air resistance is negligible. The initial conditions of motion are as follows: for $t = 0$, $x = 0$; $\frac{dx}{dt} = V$.

   Answer:

$$x = \frac{V}{\omega} \sin \omega t - \frac{f}{\omega^2}(1 - \cos \omega t); \qquad \frac{dx}{dt} = V \cos \omega t - \frac{f}{\omega} \sin \omega t;$$

$$\frac{d^2 x}{dt^2} = -V\omega \sin \omega t - f \cos \omega t.$$

2. Determine the basic parameters of motion of a system, shown in Figure 8.2.1, moving on a horizontal surface and restricted by a flexible link, while being subjected to a dry friction force and a constant active force. The air resistance is negligible. The initial conditions of motion are as follows: for $t = 0$, $x = 0$; $\frac{dx}{dt} = 0$.

Answer:

$$x = \frac{p-f}{\omega^2}(1 - \cos\omega t); \qquad \frac{dx}{dt} = \frac{p-f}{\omega}\sin\omega t;$$

$$\frac{d^2x}{dt^2} = (p-f)\cos\omega t.$$

3. Determine the basic parameters of motion of a system, shown in Figure 8.2.1, moving on a horizontal surface and restricted by a flexible link while being subjected to a dry friction force and a constant active force. The air resistance is negligible. The initial conditions of motion are as follows: for $t = 0$, $x = S$; $\frac{dx}{dt} = 0$.

Answer:

$$x = S\cos\omega t + \frac{p-f}{\omega^2}(1 - \cos\omega t);$$

$$\frac{dx}{dt} = \frac{p-f}{\omega}\sin\omega t - S\omega\sin\omega t;$$

$$\frac{d^2x}{dt^2} = (p-f)\cos\omega t - S\omega^2\cos\omega t.$$

4. Determine the basic parameters of motion of a system, shown in Figure 8.2.1, moving on a horizontal surface while being restricted by a flexible link and subjected to a constant resisting force and a constant active force. The air resistance is negligible. The initial conditions of motion are as follows: for $t = 0$, $x = 0$; $\frac{dx}{dt} = V$.

Answer:

$$x = Vt - \frac{f}{\omega^2}(1 - \cos\omega t); \qquad \frac{dx}{dt} = V - f\sin\omega t;$$

$$\frac{d^2x}{dt^2} = -f\sin\omega t.$$

5. Determine the basic parameters of motion of a system, shown in Figure 8.3.1, moving on a horizontal surface while being restricted by a flexible link and subjected to a dry friction force and a harmonic force. The air resistance is negligible. The initial conditions of motion are as follows: for $t = 0$, $x = 0$; $\frac{dx}{dt} = 0$.

Answer:

$$x = \frac{a(\cos\omega t - \cos\varphi t)}{\varphi^2 - \omega^2} - \frac{f}{\omega^2}(1 - \cos\omega t);$$

$$\frac{dx}{dt} = \frac{a(\varphi\sin\varphi t - \omega\sin\omega t)}{\varphi^2 - \omega^2} - \frac{f}{\omega}\sin\omega t;$$

$$\frac{d^2 x}{dt^2} = \frac{a(\varphi^2\cos\varphi t - \omega^2\cos\omega t)}{\varphi^2 - \omega^2} - f\cos\omega t.$$

6. Determine the basic parameters of motion of a system, shown in Figure 8.3.1, moving on a horizontal surface while being restricted by a flexible link and subjected to a dry friction force and a harmonic force. The air resistance is negligible. The initial conditions of motion are as follows: for $t = 0$, $x = S$; $\frac{dx}{dt} = 0$.

Answer:

$$x = S\cos\omega t - \frac{f}{\omega^2}(1 - \cos\omega t) + \frac{a(\cos\omega t - \cos\varphi t)}{\varphi^2 - \omega^2};$$

$$\frac{dx}{dt} = \frac{a(\varphi\sin\varphi t - \omega\sin\omega t)}{\varphi^2 - \omega^2} - S\omega\sin\omega t - \frac{f}{\omega}\sin\omega t;$$

$$\frac{d^2 x}{dt^2} = \frac{a(\varphi^2\cos\varphi t - \omega^2\cos\omega t)}{\varphi^2 - \omega^2} - S\omega^2\cos\omega t - f\cos\omega t.$$

7. Determine the basic parameters of motion of a system, shown in Figure 8.3.1, moving on a horizontal surface while being restricted by a flexible link and subjected to a dry friction force and a harmonic force. The air resistance is negligible. The initial conditions of motion are as follows: for $t = 0$, $x = 0$; $\frac{dx}{dt} = V$.

Answer:

$$x = \frac{V}{\omega}\sin\omega t - \frac{f}{\omega^2}(1 - \cos\omega t) + \frac{a(\cos\omega t - \cos\varphi t)}{\varphi^2 - \omega^2};$$

$$\frac{dx}{dt} = V\cos\omega t - \frac{f}{\omega}\sin\omega t + \frac{a(\varphi\sin\varphi t - \omega\sin\omega t)}{\varphi^2 - \omega^2};$$

$$\frac{d^2 x}{dt^2} = -V\omega\sin\omega t - f\cos\omega t + \frac{a(\varphi^2\cos\varphi t - \omega^2\cos\omega t)}{\varphi^2 - \omega^2}.$$

8. Determine the basic parameters of motion of a system, shown in Figure 8.4.1, moving on a horizontal surface while being restricted by a flexible link and subjected to a dry friction force and a harmonic force. The air resistance is negligible. The initial conditions of motion are as follows: for $t = 0$, $x = 0$; $\frac{dx}{dt} = 0$.

Answer:

$$x = \frac{a(\cos \omega t - \cos \varphi t)}{\varphi^2 - \omega^2} - \frac{f}{\omega^2}(1 - \cos \omega t);$$

$$\frac{dx}{dt} = \frac{a(\varphi \sin \varphi t - \omega \sin \omega t)}{\varphi^2 - \omega^2} - \frac{f}{\omega}\sin \omega t;$$

$$\frac{d^2 x}{dt^2} = \frac{a(\varphi^2 \cos \varphi t - \omega^2 \cos \omega t)}{\varphi^2 - \omega^2} - f \cos \omega t.$$

9. Determine the basic parameters of motion of a system, shown in Figure 8.4.1, moving on a horizontal surface while being restricted by a flexible link and subjected to a dry friction force and a harmonic force. The air resistance is negligible. The initial conditions of motion are as follows: for $t = 0$, $x = S$; $\frac{dx}{dt} = 0$.

Answer:

$$x = S \cos \omega t - \frac{f}{\omega^2}(1 - \cos \omega t) + \frac{a(\cos \omega t - \cos \varphi t)}{\varphi^2 - \omega^2};$$

$$\frac{dx}{dt} = \frac{a(\varphi \sin \varphi t - \omega \sin \omega t)}{\varphi^2 - \omega^2} - S\omega \sin \omega t - \frac{r}{\omega}\sin \omega t;$$

$$\frac{d^2 x}{dt^2} = \frac{a(\varphi^2 \cos \varphi t - \omega^2 \cos \omega t)}{\varphi^2 - \omega^2} - S\omega^2 \cos \omega t - f \cos \omega t.$$

10. Determine the basic parameters of motion of a system, shown in Figure 8.4.1, moving on a horizontal surface while being restricted by a flexible link and subjected to a dry friction force and a harmonic force. The air resistance is negligible. The initial conditions of motion are as follows: for $t = 0$, $x = 0$; $\frac{dx}{dt} = V$.

Answer:

$$x = \frac{V}{\omega}\sin\omega t - \frac{f}{\omega^2}(1 - \cos\omega t) + \frac{a(\cos\omega t - \cos\varphi t)}{\varphi^2 - \omega^2};$$

$$\frac{dx}{dt} = V\cos\omega t - \frac{r}{\omega}\sin\omega t + \frac{a(\varphi\sin\varphi t - \omega\sin\omega t)}{\varphi^2 - \omega^2};$$

$$\frac{d^2x}{dt^2} = \frac{a(\varphi^2\cos\varphi t - \omega^2\cos\omega t)}{\varphi^2 - \omega^2} - V\omega\sin\omega t - f\cos\omega t.$$

CHAPTER 9

PROBLEMS

1. Determine the basic parameters of motion of the system shown in Figure 9.1.1. The system is restricted by a flexible link and moving on a horizontal surface due to its initial velocity in the absence of active forces. The system is subjected to a dry friction force and a constant resisting force. The air resistance is negligible. The initial conditions of motion are as follows: for $t = 0$, $x = 0$; $\frac{dx}{dt} = V$.

Answer:

$$x = \frac{V}{\omega} \sin \omega t - \frac{f + r}{\omega^2} (1 - \cos \omega t);$$

$$\frac{dx}{dt} = V \cos \omega t - \frac{f + r}{\omega} \sin \omega t;$$

$$\frac{d^2 x}{dt^2} = -V \omega \sin \omega t - (f + r) \cos \omega t.$$

2. Determine the basic parameters of motion of the system shown in Figure 9.2.1. The system is restricted by a flexible link and moving on a horizontal surface while being subjected to a dry friction force, a constant resisting force, and an active force. The air resistance is negligible. The initial conditions of motion are as follows: for $t = 0$, $x = 0$; $\frac{dx}{dt} = 0$.

Answer:

$$x = \frac{p - f - r}{\omega^2}(1 - \cos \omega t); \qquad \frac{dx}{dt} = \frac{p - f - r}{\omega}\sin \omega t;$$

$$\frac{d^2 x}{dt^2} = (p - f - r)\cos \omega t.$$

3. Determine the basic parameters of motion of a system, shown in Figure 9.2.1, moving on a horizontal surface and restricted by a flexible link while being subjected to a dry friction force, a constant resisting force, and a constant active force. The air resistance is negligible. The initial conditions of motion are as follows: for $t = 0$, $x = S$; $\frac{dx}{dt} = 0$.

Answer:

$$x = S \cos \omega t + \frac{p - f - r}{\omega^2}(1 - \cos \omega t);$$

$$\frac{dx}{dt} = \frac{p - f - r}{\omega}\sin \omega t - S\omega \sin \omega t;$$

$$\frac{d^2 x}{dt^2} = (p - f - r)\cos \omega t - S\omega^2 \cos \omega t.$$

4. Determine the basic parameters of motion of a system, shown in Figure 9.2.1, moving on a horizontal surface while being restricted by a flexible link and subjected to a dry friction force, a constant resisting force, and a constant active force. The air resistance is negligible. The initial conditions of motion are as follows: for $t = 0$, $x = 0$; $\frac{dx}{dt} = V$.

Answer:

$$x = \frac{p - f - r}{\omega^2}(1 - \cos \omega t) + \frac{V}{\omega}\sin \omega t;$$

$$\frac{dx}{dt} = \frac{p - f - r}{\omega}\sin \omega t + V \cos \omega t;$$

$$\frac{d^2 x}{dt^2} = (p - f - r)\cos \omega t - V\omega \sin \omega t.$$

5. Determine the basic parameters of motion of a system, shown in Figure 9.3.1, moving on a horizontal surface while being restricted by a flexible link and subjected to a dry friction force, a constant resisting force, and a harmonic force. The air resistance is negligible. The initial conditions of motion are as follows: for $t = 0$, $x = 0$; $\frac{dx}{dt} = 0$.

Answer:

$$x = \frac{a(\cos \omega t - \cos \varphi t)}{\varphi^2 - \omega^2} - \frac{f + r}{\omega^2}(1 - \cos \omega t);$$

$$\frac{dx}{dt} = \frac{a(\varphi \sin \varphi t - \omega \sin \omega t)}{\varphi^2 - \omega^2} - \frac{f + r}{\omega} \sin \omega t;$$

$$\frac{d^2 x}{dt^2} = \frac{a(\varphi^2 \cos \varphi t - \omega^2 \cos \omega t)}{\varphi^2 - \omega^2} - (f + r) \cos \omega t.$$

6. Determine the basic parameters of motion of a system, shown in Figure 9.3.1, moving on a horizontal surface while being restricted by a flexible link and subjected to a dry friction force, a constant resisting force, and a harmonic force. The air resistance is negligible. The initial conditions of motion are as follows: for $t = 0$, $x = S$; $\frac{dx}{dt} = 0$.

Answer:

$$x = \frac{a(\cos \omega t - \cos \varphi t)}{\varphi^2 - \omega^2} - \frac{f + r}{\omega^2}(1 - \cos \omega t) + S \cos \omega t;$$

$$\frac{dx}{dt} = \frac{a(\varphi \sin \varphi t - \omega \sin \omega t)}{\varphi^2 - \omega^2} - S\omega \sin \omega t - \frac{f + r}{\omega} \sin \omega t;$$

$$\frac{d^2 x}{dt^2} = \frac{a(\varphi^2 \cos \varphi t - \omega^2 \cos \omega t)}{\varphi^2 - \omega^2} - S\omega^2 \cos \omega t - (f + r) \cos \omega t.$$

7. Determine the basic parameters of motion of a system, shown in Figure 9.3.1, moving on a horizontal surface while being restricted by a flexible link and subjected to a dry friction force, a constant resisting force, and a harmonic force. The air resistance

is negligible. The initial conditions of motion are as follows: for $t = 0$, $x = 0$; $\frac{dx}{dt} = V$.

Answer:

$$x = \frac{V}{\omega}\sin\omega t - \frac{f+r}{\omega^2}(1 - \cos\omega t) + \frac{a(\cos\omega t - \cos\varphi t)}{\varphi^2 - \omega^2};$$

$$\frac{dx}{dt} = V\cos\omega t - \frac{f+r}{\omega}\sin\omega t + \frac{a(\varphi\sin\varphi t - \omega\sin\omega t)}{\varphi^2 - \omega^2};$$

$$\frac{d^2x}{dt^2} = -V\omega\sin\omega t - (f+r)\cos\omega t + \frac{a(\varphi^2\cos\varphi t - \omega^2\cos\omega t)}{\varphi^2 - \omega^2}.$$

8. Determine the basic parameters of motion of a system, shown in Figure 9.4.1, moving on a horizontal surface while being restricted by a flexible link and subjected to a dry friction force, a constant resisting force, a constant active force, and a harmonic force. The air resistance is negligible. The initial conditions of motion are as follows: for $t = 0$, $x = 0$; $\frac{dx}{dt} = 0$.

Answer:

$$x = \frac{a(\cos\omega t - \cos\varphi t)}{\varphi^2 - \omega^2} + \frac{p - f - r}{\omega^2}(1 - \cos\omega t);$$

$$\frac{dx}{dt} = \frac{a(\varphi\sin\varphi t - \omega\sin\omega t)}{\varphi^2 - \omega^2} + \frac{p - f - r}{\omega}\sin\omega t;$$

$$\frac{d^2x}{dt^2} = \frac{a(\varphi^2\cos\varphi t - \omega^2\cos\omega t)}{\varphi^2 - \omega^2} + (p - f - r)\cos\omega t.$$

9. Determine the basic parameters of motion of a system, shown in Figure 9.4.1, moving on a horizontal surface while being restricted by a flexible link and subjected to a dry friction force, a constant resisting force, a constant active force, and a harmonic force. The air resistance is negligible. The initial conditions of motion are as follows: for $t = 0$, $x = S$; $\frac{dx}{dt} = 0$.

Answer:

$$x = \frac{a(\cos\omega t - \cos\varphi t)}{\varphi^2 - \omega^2} + \frac{p - f - r}{\omega^2}(1 - \cos\omega t) + S\cos\omega t;$$

$$\frac{dx}{dt} = \frac{a(\varphi\sin\varphi t - \omega\sin\omega t)}{\varphi^2 - \omega^2} + \frac{(p - f - r)}{\omega}\sin\omega t - S\omega\sin\omega t;$$

$$\frac{d^2x}{dt^2} = \frac{a(\varphi^2\cos\varphi t - \omega^2\cos\omega t)}{\varphi^2 - \omega^2} + (p - f - r)\cos\omega t - S\omega^2\cos\omega t.$$

10. Determine the basic parameters of motion of a system, shown in Figure 9.4.1, moving on a horizontal surface while being restricted by a flexible link and subjected to a dry friction force, a constant resisting force, a constant active force, and a harmonic force. The air resistance is negligible. The initial conditions of motion are as follows: for $t = 0$, $x = 0$; $\frac{dx}{dt} = V$.

Answer:

$$x = \frac{V}{\omega}\sin\omega t + \frac{p - f - r}{\omega^2}(1 - \cos\omega t) + \frac{a(\cos\omega t - \cos\varphi t)}{\varphi^2 - \omega^2};$$

$$\frac{dx}{dt} = V\cos\omega t + \frac{(p - f - r)}{\omega}\sin\omega t + \frac{a(\varphi\sin\varphi t - \omega\sin\omega t)}{\varphi^2 - \omega^2};$$

$$\frac{d^2x}{dt^2} = \frac{a(\varphi^2\cos\varphi t - \omega^2\cos\omega t)}{\varphi^2 - \omega^2} - V\omega\sin\omega t + (p - f - r)\cos\omega t.$$

CHAPTER 10

PROBLEMS

---

1. Determine the basic parameters of motion of the system shown
   in Figure 10.1.1. The system is restricted by a fluid link and
   moving on a horizontal surface due to its initial velocity in
   the absence of active forces. The air resistance is negligible.
   The initial conditions of motion are as follows: for $t = 0$, $x = 0$;
   $\frac{dx}{dt} = V$.

   Answer:

   $$x = \frac{V}{2n}(1 - e^{-2nt}); \quad \frac{dx}{dt} = Ve^{-2nt}; \quad \frac{d^2x}{dt^2} = -2nVe^{-2nt}.$$

2. Determine the basic parameters of motion of the system shown in
   Figure 10.2.1. The system is restricted by a fluid link and moving
   on a horizontal surface while being subjected to a constant active
   force. The air resistance is negligible. The initial conditions of
   motion are as follows: for $t = 0$, $x = 0$; $\frac{dx}{dt} = 0$.

   Answer:

   $$x = \frac{p}{2n}\left[t + \frac{1}{2n}(e^{-2nt} - 1)\right]; \quad \frac{dx}{dt} = \frac{p}{2n}(1 - e^{-2nt});$$

   $$\frac{d^2x}{dt^2} = pe^{-2nt}.$$

3. Determine the basic parameters of motion of the system shown in Figure 10.2.1. The system is restricted by a fluid link and moving on a horizontal surface while being subjected to a constant active force. The air resistance is negligible. The initial conditions of motion are as follows: for $t = 0$, $x = S$; $\frac{dx}{dt} = 0$.

Answer:

$$x = S + \frac{p}{2n}\left[t + \frac{1}{2n}(e^{-2nt} - 1)\right];$$

$$\frac{dx}{dt} = \frac{p}{2n}(1 - e^{-2nt}); \quad \frac{d^2x}{dt^2} = pe^{-2nt}.$$

4. Determine the basic parameters of motion of a system, shown in Figure 10.2.1, moving on a horizontal surface. The system is restricted by a fluid link and moving on a horizontal surface while being subjected to a constant active force. The air resistance is negligible. The initial conditions of motion are as follows: for $t = 0$, $x = 0$; $\frac{dx}{dt} = V$.

Answer:

$$x = \frac{V}{2n}(1 - e^{-2nt}) + \frac{p}{2n}\left[t + \frac{1}{2n}(e^{-2nt} - 1)\right];$$

$$\frac{dx}{dt} = Ve^{-2nt} + \frac{p}{2n}(1 - e^{-2nt}); \quad \frac{d^2x}{dt^2} = e^{-2nt}(p - 2nV).$$

5. Determine the basic parameters of motion of a system, shown in Figure 10.3.1, moving on a horizontal surface. The system is restricted by a fluid link and moving on a horizontal surface while being subjected to a harmonic force. The air resistance is negligible. The initial conditions of motion are as follows: for $t = 0$, $x = 0$; $\frac{dx}{dt} = 0$.

Answer:

$$x = \frac{a}{\varphi^2 + 4n^2}\left(e^{-2nt} + \frac{2n}{\varphi}\sin\varphi t - \cos\varphi t\right);$$

$$\frac{dx}{dt} = \frac{a}{\varphi^2 + 4n^2}(2n\cos\varphi t - 2ne^{-2nt} + \varphi\sin\varphi t);$$

$$\frac{d^2x}{dt^2} = \frac{a}{\varphi^2 + 4n^2}(4n^2 e^{-2nt} - 2n\varphi\sin s\varphi t + \varphi^2\cos\varphi t).$$

6. Determine the basic parameters of motion of a system, shown in Figure 10.3.1, moving on a horizontal surface while being restricted by a fluid link and subjected to a harmonic force. The air resistance is negligible. The initial conditions of motion are as follows: for $t = 0$, $x = S$; $\frac{dx}{dt} = 0$.

Answer:

$$x = S + \frac{a}{\varphi^2 + 4n^2}\left(e^{-2nt} + \frac{2n}{\varphi}\sin\varphi t - \cos\varphi t\right);$$

$$\frac{dx}{dt} = \frac{a}{\varphi^2 + 4n^2}(2n\cos\varphi t - 2ne^{-2nt} + \varphi\sin\varphi t);$$

$$\frac{d^2x}{dt^2} = \frac{a}{\varphi^2 + 4n^2}(4n^2 e^{-2nt} - 2n\varphi\sin s\varphi t + \varphi^2\cos\varphi t).$$

7. Determine the basic parameters of motion of a system, shown in Figure 10.3.1, moving on a horizontal surface while being restricted by a fluid link and subjected to a harmonic force. The air resistance is negligible. The initial conditions of motion are as follows: for $t = 0$, $x = 0$; $\frac{dx}{dt} = V$.

Answer:

$$x = \frac{V}{2n}(1 - e^{-2nt}) + \frac{a}{\varphi^2 + 4n^2}\left(e^{-2nt} + \frac{2n}{\varphi}\sin\varphi t - \cos\varphi t\right);$$

$$\frac{dx}{dt} = Ve^{-2nt} + \frac{a}{\varphi^2 + 4n^2}(2n\cos\varphi t - 2ne^{-2nt} + \varphi\sin\varphi t);$$

$$\frac{d^2x}{dt^2} = -2nVe^{-2nt} + \frac{a}{\varphi^2 + 4n^2}$$
$$\times (4n^2 e^{-2nt} - 2n\varphi\sin s\varphi t + \varphi^2\cos\varphi t).$$

8. Determine the basic parameters of motion of a system, shown in Figure 10.4.1, moving on a horizontal surface while being restricted by a fluid link and subjected to a constant active force and a harmonic force. The air resistance is negligible. The initial conditions of motion are as follows: for $t = 0$, $x = 0$; $\frac{dx}{dt} = 0$.

Answer:

$$x = \frac{p}{2n}\left[t + \frac{1}{2n}(e^{-2nt} - 1)\right]$$

$$+ \frac{a}{\varphi^2 + 4n^2}\left(e^{-2nt} + \frac{2n}{\varphi}\sin\varphi t - \cos\varphi t\right);$$

$$\frac{dx}{dt} = \frac{p}{2n}(1 - e^{-2nt}) + \frac{a}{\varphi^2 + 4n^2}$$

$$\times (2n\cos\varphi t - 2ne^{-2nt} + \varphi\sin\varphi t);$$

$$\frac{d^2x}{dt^2} = pe^{-2nt} + \frac{a}{\varphi^2 + 4n^2}(4n^2 e^{-2nt} - 2n\varphi\sin s\varphi t + \varphi^2\cos\varphi t).$$

9. Determine the basic parameters of motion of a system, shown in Figure 10.4.1, moving on a horizontal surface while being restricted by a fluid link and subjected to a constant active force and a harmonic force. The air resistance is negligible. The initial conditions of motion are as follows: for $t = 0$, $x = S$; $\frac{dx}{dt} = 0$.

Answer:

$$x = S + \frac{p}{2n}\left[t + \frac{1}{2n}(e^{-2nt} - 1)\right]$$

$$+ \frac{a}{\varphi^2 + 4n^2}\left(e^{-2nt} + \frac{2n}{\varphi}\sin\varphi t - \cos\varphi t\right);$$

$$\frac{dx}{dt} = \frac{p}{2n}(1 - e^{-2nt}) + \frac{a}{\varphi^2 + 4n^2}$$

$$\times (2n\cos\varphi t - 2ne^{-2nt} + \varphi\sin\varphi t);$$

$$\frac{d^2x}{dt^2} = pe^{-2nt} + \frac{a}{\varphi^2 + 4n^2}(4n^2 e^{-2nt} - 2n\varphi\sin s\varphi t + \varphi^2\cos\varphi t).$$

10. Determine the basic parameters of motion of a system, shown in Figure 10.4.1, moving on a horizontal surface while being restricted by a fluid link and subjected to a constant active force and a harmonic force. The air resistance is negligible. The initial conditions of motion are as follows: for $t = 0$, $x = 0$; $\frac{dx}{dt} = V$.

Answer:

$$x = \frac{V}{2n}(1 - e^{-2nt}) + \frac{p}{2n}\left[t + \frac{1}{2n}(e^{-2nt} - 1)\right]$$

$$+ \frac{a}{\varphi^2 + 4n^2}\left(e^{-2nt} + \frac{2n}{\varphi}\sin\varphi t - \cos\varphi t\right);$$

$$\frac{dx}{dt} = Ve^{-2nt} + \frac{p}{2n}(1 - e^{-2nt})$$

$$+ \frac{a}{\varphi^2 + 4n^2}(2n\cos\varphi t - 2ne^{-2nt} + \varphi\sin\varphi t);$$

$$\frac{d^2x}{dt^2} = Ve^{-2nt} + pe^{-2nt} + \frac{a}{\varphi^2 + 4n^2}$$

$$\times (4n^2 e^{-2nt} - 2n\varphi\sin\varphi t + \varphi^2 \cos\varphi t).$$

CHAPTER 11

PROBLEMS

---

1. Determine the basic parameters of motion of the system shown in
   Figure 11.1.1. The system is restricted by a fluid link and moving
   on a horizontal surface due to its initial velocity in the absence
   of active forces. The system is subjected to a constant resisting
   force. The air resistance is negligible. The initial conditions of
   motion are as follows: for $t = 0$, $x = 0$; $\frac{dx}{dt} = V$.

   Answer:

   $$x = \frac{V}{2n}(1 - e^{-2nt}) - \frac{r}{2n}\left[t + \frac{1}{2n}(e^{-2nt} - 1)\right];$$

   $$\frac{dx}{dt} = Ve^{-2nt} - \frac{r}{2n}(1 - e^{-2nt}); \qquad \frac{d^2x}{dt^2} = -2nVe^{-2nt} - re^{-2nt}.$$

2. Determine the basic parameters of motion of the system shown in
   Figure 11.2.1. The system is restricted by a fluid link and moving
   on a horizontal surface while being subjected to a constant
   resisting force and a constant active force. The air resistance
   is negligible. The initial conditions of motion are as follows:
   for $t = 0$, $x = 0$; $\frac{dx}{dt} = 0$.

Answer:

$$x = \frac{p-r}{2n}\left[t + \frac{1}{2n}(e^{-2nt} - 1)\right]; \quad \frac{dx}{dt} = \frac{p-r}{2n}(1 - e^{-2nt});$$

$$\frac{d^2x}{dt^2} = (p-r)e^{-2nt}.$$

3. Determine the basic parameters of motion of the system shown in Figure 11.2.1. The system is restricted by a fluid link and moving on a horizontal surface while being subjected to a constant resisting force and a constant active force. The air resistance is negligible. The initial conditions of motion are as follows: for $t = 0$, $x = S$; $\frac{dx}{dt} = 0$.

Answer:

$$x = S + \frac{p-r}{2n}\left[t + \frac{1}{2n}(e^{-2nt} - 1)\right];$$

$$\frac{dx}{dt} = \frac{p-r}{2n}(1 - e^{-2nt}); \quad \frac{d^2x}{dt^2} = e^{-2nt}(p-r).$$

4. Determine the basic parameters of motion of a system, shown in Figure 11.2.1, moving on a horizontal surface. The system is restricted by a fluid link and moving on a horizontal surface while being subjected to a constant resisting force and a constant active force. The air resistance is negligible. The initial conditions of motion are as follows: for $t = 0$, $x = 0$; $\frac{dx}{dt} = V$.

Answer:

$$x = \frac{V}{2n}(1 - e^{-2nt}) + \frac{p-r}{2n}\left[t + \frac{1}{2n}(e^{-2nt} - 1)\right];$$

$$\frac{dx}{dt} = Ve^{-2nt} + \frac{p-r}{2n}(1 - e^{-2nt});$$

$$\frac{d^2x}{dt^2} = e^{-2nt}(p - r - 2nV).$$

5. Determine the basic parameters of motion of a system, shown in Figure 11.3.1, moving on a horizontal surface. The system is restricted by a fluid link and moving on a horizontal surface while being subjected to a constant resisting force and a harmonic force. The air resistance is negligible. The initial conditions of motion are as follows: for $t = 0$, $x = 0$; $\frac{dx}{dt} = 0$.

Answer:

$$x = \frac{a}{\varphi^2 + 4n^2}\left(e^{-2nt} + \frac{2n}{\varphi}\sin\varphi t - \cos\varphi t\right)$$
$$- \frac{r}{2n}\left[t + \frac{1}{2n}(e^{-2nt} - 1)\right];$$

$$\frac{dx}{dt} = \frac{a}{\varphi^2 + 4n^2}(\varphi\sin\varphi t + 2n\cos\varphi t - 2ne^{-2nt}) - \frac{r}{2n}(1 - e^{-2nt});$$

$$\frac{d^2x}{dt^2} = \frac{a}{\varphi^2 + 4n^2}(\varphi^2\cos\varphi t - 2n\varphi\sin s\varphi t + 4n^2 e^{-2nt}).$$

6. Determine the basic parameters of motion of a system, shown in Figure 11.3.1, moving on a horizontal surface while being restricted by a fluid link and subjected to a constant resisting force and a harmonic force. The air resistance is negligible. The initial conditions of motion are as follows: for $t = 0$, $x = S$; $\frac{dx}{dt} = 0$.

Answer:

$$x = S + \frac{a}{\varphi^2 + 4n^2}\left(e^{-2nt} + \frac{2n}{\varphi}\sin\varphi t - \cos\varphi t\right)$$
$$- \frac{r}{2n}\left[t + \frac{1}{2n}(e^{-2nt} - 1)\right];$$

$$\frac{dx}{dt} = \frac{a}{\varphi^2 + 4n^2}(\varphi\sin\varphi t + 2n\cos\varphi t - 2ne^{-2nt}) - \frac{r}{2n}(1 - e^{-2nt});$$

$$\frac{d^2x}{dt^2} = \frac{a}{\varphi^2 + 4n^2}(\varphi^2\cos\varphi t - 2n\varphi\sin\varphi t + 4n^2 e^{-2nt}) - re^{-2nt}.$$

7. Determine the basic parameters of motion of a system, shown in Figure 11.3.1, moving on a horizontal surface while being restricted by a fluid link and subjected to a constant resisting force and a harmonic force. The air resistance is negligible. The initial conditions of motion are as follows: for $t = 0$, $x = 0$; $\frac{dx}{dt} = V$.

Answer:

$$x = \frac{V}{2n}(1 - e^{-2nt}) - \frac{r}{2n}\left[t + \frac{1}{2n}(e^{-2nt} - 1)\right]$$

$$+ \frac{a}{\varphi^2 + 4n^2}\left(e^{-2nt} + \frac{2n}{\varphi}\sin\varphi t - \cos\varphi t\right);$$

$$\frac{dx}{dt} = Ve^{-2nt} + \frac{a}{\varphi^2 + 4n^2}(2n\cos\varphi t - 2ne^{-2nt} + \varphi\sin\varphi t)$$

$$- \frac{r}{2n}(1 - e^{-2nt});$$

$$\frac{d^2x}{dt^2} = -2nVe^{-2nt}$$

$$+ \frac{a}{\varphi^2 + 4n^2}(\varphi^2\cos\varphi t - 2n\varphi\sin\varphi t + 4n^2 e^{-2nt}) - re^{-2nt}.$$

8. Determine the basic parameters of motion of a system, shown in Figure 11.4.1, moving on a horizontal surface while being restricted by a fluid link and subjected to a constant resisting force, a constant active force, and a harmonic force. The air resistance is negligible. The initial conditions of motion are as follows: for $t = 0$, $x = 0$; $\frac{dx}{dt} = 0$.

Answer:

$$x = \frac{p - r}{2n}\left[t + \frac{1}{2n}(e^{-2nt} - 1)\right]$$

$$+ \frac{a}{\varphi^2 + 4n^2}\left(e^{-2nt} + \frac{2n}{\varphi}\sin\varphi t - \cos\varphi t\right);$$

$$\frac{dx}{dt} = \frac{p-r}{2n}(1 - e^{-2nt}) + \frac{a}{\varphi^2 + 4n^2}$$

$$\times (2n \cos \varphi t - 2ne^{-2nt} + \varphi \sin \varphi t);$$

$$\frac{d^2 x}{dt^2} = (p-r)e^{-2nt} + \frac{a}{\varphi^2 + 4n^2}$$

$$\times (4n^2 e^{-2nt} - 2n\varphi \sin s\varphi t + \varphi^2 \cos \varphi t).$$

9. Determine the basic parameters of motion of a system, shown in Figure 11.4.1, moving on a horizontal surface while being restricted by a fluid link and subjected to a constant resisting force, a constant active force, and a harmonic force. The air resistance is negligible. The initial conditions of motion are as follows: for $t = 0$, $x = S$; $\frac{dx}{dt} = 0$.

Answer:

$$x = S + \frac{p-r}{2n}\left[t + \frac{1}{2n}(e^{-2nt} - 1)\right]$$

$$+ \frac{a}{\varphi^2 + 4n^2}\left(e^{-2nt} + \frac{2n}{\varphi}\sin \varphi t - \cos \varphi t\right);$$

$$\frac{dx}{dt} = \frac{p-r}{2n}(1 - e^{-2nt}) + \frac{a}{\varphi^2 + 4n^2}$$

$$\times (2n \cos \varphi t - 2ne^{-2nt} + \varphi \sin \varphi t);$$

$$\frac{d^2 x}{dt^2} = (p-r)e^{-2nt} + \frac{a}{\varphi^2 + 4n^2}$$

$$\times (4n^2 e^{-2nt} - 2n\varphi \sin s\varphi t + \varphi^2 \cos \varphi t).$$

10. Determine the basic parameters of motion of a system, shown in Figure 11.4.1, moving on a horizontal surface while being restricted by a fluid link and subjected to a constant resisting force, a constant active force, and a harmonic force. The air

resistance is negligible. The initial conditions of motion are as follows: for $t = 0$, $x = 0$; $\frac{dx}{dt} = V$.

Answer:

$$x = \frac{V}{2n}(1 - e^{-2nt}) + \frac{p - r}{2n}\left[t + \frac{1}{2n}(e^{-2nt} - 1)\right]$$

$$+ \frac{a}{\varphi^2 + 4n^2}\left(e^{-2nt} + \frac{2n}{\varphi}\sin\varphi t - \cos\varphi t\right);$$

$$\frac{dx}{dt} = Ve^{-2nt} + \frac{p - r}{2n}(1 - e^{-2nt}) + \frac{a}{\varphi^2 + 4n^2}$$

$$\times (2n\cos\varphi t - 2ne^{-2nt} + \varphi\sin\varphi t);$$

$$\frac{d^2x}{dt^2} = -2nVe^{-2nt} + (p - r)e^{-2nt} + \frac{a}{\varphi^2 + 4n^2}$$

$$\times (4n^2 e^{-2nt} - 2n\varphi\sin s\varphi t + \varphi^2\cos\varphi t).$$

CHAPTER 12

PROBLEMS

---

1. Determine the basic parameters of motion of the system shown in
   Figure 12.1.1. The system is restricted by a fluid link and moving
   on a horizontal surface due to its initial velocity in the absence
   of active forces. The system is subjected to a dry friction force.
   The air resistance is negligible. The initial conditions of motion
   are as follows: for $t = 0$, $x = 0$; $\frac{dx}{dt} = V$.

   Answer:

   $$x = \frac{V}{2n}(1 - e^{-2nt}) - \frac{f}{2n}\left[t + \frac{1}{2n}(e^{-2nt} - 1)\right];$$

   $$\frac{dx}{dt} = Ve^{-2nt} - \frac{f}{2n}(1 - e^{-2nt}); \quad \frac{d^2x}{dt^2} = -2nVe^{-2nt} - fe^{-2nt}.$$

2. Determine the basic parameters of motion of the system shown in
   Figure 12.2.1. The system is restricted by a fluid link and moving
   on a horizontal surface while being subjected to a dry friction
   force and a constant active force. The air resistance is negligible.
   The initial conditions of motion are as follows: for $t = 0$, $x = 0$;
   $\frac{dx}{dt} = 0$.

Answer:

$$x = \frac{p-f}{2n}\left[t + \frac{1}{2n}(e^{-2nt} - 1)\right];$$

$$\frac{dx}{dt} = \frac{p-f}{2n}(1 - e^{-2nt}); \quad \frac{d^2x}{dt^2} = (p-f)e^{-2nt}.$$

3. Determine the basic parameters of motion of the system shown in Figure 12.2.1. The system is restricted by a fluid link and moving on a horizontal surface while being subjected to a dry friction force and a constant active force. The air resistance is negligible. The initial conditions of motion are as follows: for $t = 0$, $x = S$; $\frac{dx}{dt} = 0$.

Answer:

$$x = S + \frac{p-f}{2n}\left[t + \frac{1}{2n}(e^{-2nt} - 1)\right];$$

$$\frac{dx}{dt} = \frac{p-f}{2n}(1 - e^{-2nt}); \quad \frac{d^2x}{dt^2} = e^{-2nt}(p-f).$$

4. Determine the basic parameters of motion of a system, shown in Figure 12.2.1, moving on a horizontal surface. The system is restricted by a fluid link and moving on a horizontal surface while being subjected to a dry friction force and a constant active force. The air resistance is negligible. The initial conditions of motion are as follows: for $t = 0$, $x = 0$; $\frac{dx}{dt} = V$.

Answer:

$$x = \frac{V}{2n}(1 - e^{-2nt}) + \frac{p-f}{2n}\left[t + \frac{1}{2n}(e^{-2nt} - 1)\right];$$

$$\frac{dx}{dt} = Ve^{-2nt} + \frac{p-f}{2n}(1 - e^{-2nt}); \quad \frac{d^2x}{dt^2} = e^{-2nt}(p - f - 2nV).$$

5. Determine the basic parameters of motion of a system, shown in Figure 12.3.1, moving on a horizontal surface. The system

is restricted by a fluid link and moving on a horizontal surface while being subjected to a dry friction force and a harmonic force. The air resistance is negligible. The initial conditions of motion are as follows: for $t = 0$, $x = 0$; $\frac{dx}{dt} = 0$.

Answer:

$$x = \frac{a}{\varphi^2 + 4n^2}\left(e^{-2nt} + \frac{2n}{\varphi}\sin\varphi t - \cos\varphi t\right)$$

$$- f\left[t + \frac{1}{2n}(e^{-2nt} - 1)\right];$$

$$\frac{dx}{dt} = \frac{a}{\varphi^2 + 4n^2}(\varphi\sin\varphi t + 2n\cos\varphi t - 2ne^{-2nt}) - \frac{f}{2n}(1 - e^{-2nt});$$

$$\frac{d^2x}{dt^2} = \frac{a}{\varphi^2 + 4n^2}(\varphi^2\cos\varphi t - 2n\varphi\sin\varphi t + 4n^2 e^{-2nt}) - fe^{-2nt}.$$

6. Determine the basic parameters of motion of a system, shown in Figure 12.3.1, moving on a horizontal surface while being restricted by a fluid link and subjected to a dry friction force and a harmonic force. The air resistance is negligible. The initial conditions of motion are as follows: for $t = 0$, $x = S$; $\frac{dx}{dt} = 0$.

Answer:

$$x = S + \frac{a}{\varphi^2 + 4n^2}\left(e^{-2nt} + \frac{2n}{\varphi}\sin\varphi t - \cos\varphi t\right)$$

$$- \frac{f}{2n}\left[t + \frac{1}{2n}(e^{-2nt} - 1)\right];$$

$$\frac{dx}{dt} = \frac{a}{\varphi^2 + 4n^2}(\varphi\sin\varphi t + 2n\cos\varphi t - 2ne^{-2nt}) - \frac{f}{2n}(1 - e^{-2nt});$$

$$\frac{d^2x}{dt^2} = \frac{a}{\varphi^2 + 4n^2}(\varphi^2\cos\varphi t - 2n\varphi\sin\varphi t + 4n^2 e^{-2nt}) - fe^{-2nt}.$$

7. Determine the basic parameters of motion of a system, shown in Figure 12.3.1, moving on a horizontal surface while being restricted by a fluid link and subjected to a dry friction force and a harmonic force. The air resistance is negligible. The initial conditions of motion are as follows: for $t = 0$, $x = 0$; $\frac{dx}{dt} = V$.

Answer:

$$x = \frac{V}{2n}(1 - e^{-2nt}) + \frac{a}{\varphi^2 + 4n^2}\left(e^{-2nt} + \frac{2n}{\varphi}\sin\varphi t - \cos\varphi t\right)$$
$$- \frac{f}{2n}\left[t + \frac{1}{2n}(e^{-2nt} - 1)\right];$$

$$\frac{dx}{dt} = Ve^{-2nt} + \frac{a}{\varphi^2 + 4n^2}(2n\cos\varphi t - 2ne^{-2nt} + \varphi\sin\varphi t)$$
$$- \frac{f}{2n}(1 - e^{-2nt});$$

$$\frac{d^2x}{dt^2} = -2nVe^{-2nt} + \frac{a}{\varphi^2 + 4n^2}$$
$$\times (\varphi^2\cos\varphi t - 2n\varphi\sin\varphi t + 4n^2e^{-2nt}) - fe^{-2nt}.$$

8. Determine the basic parameters of motion of a system, shown in Figure 12.4.1, moving on a horizontal surface while being restricted by a fluid link and subjected to a dry friction force, a constant active force, and a harmonic force. The air resistance is negligible. The initial conditions of motion are as follows: for $t = 0$, $x = 0$; $\frac{dx}{dt} = 0$.

Answer:

$$x = \frac{p - f}{2n}\left[t + \frac{1}{2n}(e^{-2nt} - 1)\right] + \frac{a}{\varphi^2 + 4n^2}$$
$$\times \left(e^{-2nt} + \frac{2n}{\varphi}\sin\varphi t - \cos\varphi t\right);$$

$$\frac{dx}{dt} = \frac{p-f}{2n}(1 - e^{-2nt}) + \frac{a}{\varphi^2 + 4n^2}$$

$$\times\, (2n\cos\varphi t - 2ne^{-2nt} + \varphi\sin\varphi t);$$

$$\frac{d^2x}{dt^2} = (p-f)e^{-2nt} + \frac{a}{\varphi^2 + 4n^2}$$

$$\times\, (4n^2 e^{-2nt} - 2n\varphi\sin s\varphi t + \varphi^2\cos\varphi t).$$

9. Determine the basic parameters of motion of a system, shown in Figure 12.4.1, moving on a horizontal surface while being restricted by a fluid link and subjected to a dry friction force, a constant active force, and a harmonic force. The air resistance is negligible. The initial conditions of motion are as follows: for $t = 0$, $x = S$; $\frac{dx}{dt} = 0$.

Answer:

$$x = S + \frac{p-f}{2n}\left[t + \frac{1}{2n}(e^{-2nt} - 1)\right] + \frac{a}{\varphi^2 + 4n^2}$$

$$\times\, \left(e^{-2nt} + \frac{2n}{\varphi}\sin\varphi t - \cos\varphi t\right);$$

$$\frac{dx}{dt} = \frac{p-f}{2n}(1 - e^{-2nt}) + \frac{a}{\varphi^2 + 4n^2}$$

$$\times\, (2n\cos\varphi t - 2ne^{-2nt} + \varphi\sin\varphi t);$$

$$\frac{d^2x}{dt^2} = (p-f)e^{-2nt} + \frac{a}{\varphi^2 + 4n^2}$$

$$\times\, (4n^2 e^{-2nt} - 2n\varphi\sin s\varphi t + \varphi^2\cos\varphi t).$$

10. Determine the basic parameters of motion of a system shown in Figure 10.4.1, moving on a horizontal surface while being restricted by a fluid link and subjected to a dry friction force, a constant active force, and a harmonic force. The air resistance

is negligible. The initial conditions of motion are as follows: for $t = 0$, $x = 0$; $\frac{dx}{dt} = V$.

Answer:

$$x = \frac{V}{2n}(1 - e^{-2nt}) + \frac{p - f}{2n}\left[t + \frac{1}{2n}(e^{-2nt} - 1)\right]$$

$$+ \frac{a}{\varphi^2 + 4n^2}\left(e^{-2nt} + \frac{2n}{\varphi}\sin\varphi t - \cos\varphi t\right);$$

$$\frac{dx}{dt} = Ve^{-2nt} + \frac{p - f}{2n}(1 - e^{-2nt}) + \frac{a}{\varphi^2 + 4n^2}$$

$$\times (2n\cos\varphi t - 2ne^{-2nt} + \varphi\sin\varphi t);$$

$$\frac{d^2x}{dt^2} = -2nVe^{-2nt} + (p - f)e^{-2nt} + \frac{a}{\varphi^2 + 4n^2}$$

$$\times (4n^2 e^{-2nt} - 2n\varphi\sin s\varphi t + \varphi^2\cos\varphi t).$$

CHAPTER 13

PROBLEMS

---

1. Determine the basic parameters of motion of the system shown in Figure 13.1.1. The system is restricted by a fluid link and moving on a horizontal surface due to its initial velocity in the absence of active forces. The system is subjected to a dry friction force and a constant resisting force. The air resistance is negligible. The initial conditions of motion are as follows: for $t = 0$, $x = 0$; $\frac{dx}{dt} = V$.

   Answer:

   $$x = \frac{V}{2n}(1 - e^{-2nt}) - \frac{f + r}{2n}\left[t + \frac{1}{2n}(e^{-2nt} - 1)\right];$$

   $$\frac{dx}{dt} = Ve^{-2nt} - \frac{f + r}{2n}(1 - e^{-2nt});$$

   $$\frac{d^2 x}{dt^2} = -2nVe^{-2nt} - (f + r)e^{-2nt}.$$

2. Determine the basic parameters of motion of the system shown in Figure 13.2.1. The system is restricted by a fluid link and moving on a horizontal surface while being subjected to a dry friction force, constant resisting force, and a constant active force. The air resistance is negligible. The initial conditions of motion are as follows: for $t = 0$, $x = 0$; $\frac{dx}{dt} = 0$.

Answer:

$$x = \frac{p - f - r}{2n}\left[t + \frac{1}{2n}(e^{-2nt} - 1)\right];$$

$$\frac{dx}{dt} = \frac{p - f - r}{2n}(1 - e^{-2nt}); \quad \frac{d^2x}{dt^2} = (p - f - r)e^{-2nt}.$$

3. Determine the basic parameters of motion of the system shown in Figure 13.2.1. The system is restricted by a fluid link and moving on a horizontal surface while being subjected to a dry friction force, a constant resisting force, and a constant active force. The air resistance is negligible. The initial conditions of motion are as follows: for $t = 0$, $x = S$; $\frac{dx}{dt} = 0$.

Answer:

$$x = S + \frac{p - f - r}{2n}\left[t + \frac{1}{2n}(e^{-2nt} - 1)\right];$$

$$\frac{dx}{dt} = \frac{p - f - r}{2n}(1 - e^{-2nt}); \quad \frac{d^2x}{dt^2} = e^{-2nt}(p - f - r).$$

4. Determine the basic parameters of motion of a system, shown in Figure 13.2.1. The system is restricted by a fluid link and moving on a horizontal surface while being subjected to a dry friction force, a constant resisting force, and a constant active force. The air resistance is negligible. The initial conditions of motion are as follows: for $t = 0$, $x = 0$; $\frac{dx}{dt} = V$.

Answer:

$$x = \frac{V}{2n}(1 - e^{-2nt}) + \frac{p - f - r}{2n}\left[t + \frac{1}{2n}(e^{-2nt} - 1)\right];$$

$$\frac{dx}{dt} = Ve^{-2nt} + \frac{p - f - r}{2n}(1 - e^{-2nt});$$

$$\frac{d^2x}{dt^2} = e^{-2nt}(p - f - r - 2nV).$$

5. Determine the basic parameters of motion of a system, shown in Figure 13.3.1, moving on a horizontal surface. The system is restricted by a fluid link and moving on a horizontal surface while being subjected to a dry friction force, a constant resisting force, and a harmonic force. The air resistance is negligible. The initial conditions of motion are as follows: for $t = 0$, $x = 0$; $\frac{dx}{dt} = 0$.

Answer:

$$x = \frac{a}{\varphi^2 + 4n^2}\left(e^{-2nt} + \frac{2n}{\varphi}\sin\varphi t - \cos\varphi t\right)$$

$$- (f + r)\left[t + \frac{1}{2n}(e^{-2nt} - 1)\right];$$

$$\frac{dx}{dt} = \frac{a}{\varphi^2 + 4n^2}(\varphi\sin\varphi t + 2n\cos\varphi t - 2ne^{-2nt})$$

$$- \frac{f + r}{2n}(1 - e^{-2nt});$$

$$\frac{d^2x}{dt^2} = \frac{a}{\varphi^2 + 4n^2}(\varphi^2\cos\varphi t - 2n\varphi\sin\varphi t + 4n^2e^{-2nt})$$

$$- (f + r)e^{-2nt}.$$

6. Determine the basic parameters of motion of a system, shown in Figure 13.3.1, moving on a horizontal surface while being restricted by a fluid link and subjected to a dry friction force, a constant resisting force, and a harmonic force. The air resistance is negligible. The initial conditions of motion are as follows: for $t = 0$, $x = S$; $\frac{dx}{dt} = 0$.

Answer:

$$x = S + \frac{a}{\varphi^2 + 4n^2}\left(e^{-2nt} + \frac{2n}{\varphi}\sin\varphi t - \cos\varphi t\right) - \frac{f + r}{2n}$$

$$\times \left[t + \frac{1}{2n}(e^{-2nt} - 1)\right];$$

$$\frac{dx}{dt} = \frac{a}{\varphi^2 + 4n^2}(\varphi \sin \varphi t + 2n \cos \varphi t - 2ne^{-2nt})$$

$$- \frac{f+r}{2n}(1 - e^{-2nt});$$

$$\frac{d^2x}{dt^2} = \frac{a}{\varphi^2 + 4n^2}(\varphi^2 \cos \varphi t - 2n\varphi \sin \varphi t + 4n^2 e^{-2nt})$$

$$- (f+r)e^{-2nt}.$$

7. Determine the basic parameters of motion of a system, shown in Figure 13.3.1, moving on a horizontal surface while being restricted by a fluid link and subjected to a dry friction force, a constant resisting force, and a harmonic force. The air resistance is negligible. The initial conditions of motion are as follows: for $t = 0$, $x = 0$; $\frac{dx}{dt} = V$.

Answer:

$$x = \frac{V}{2n}(1 - e^{-2nt}) + \frac{a}{\varphi^2 + 4n^2}\left(e^{-2nt} + \frac{2n}{\varphi}\sin \varphi t - \cos \varphi t\right)$$

$$- \frac{f+r}{2n}\left[t + \frac{1}{2n}(e^{-2nt} - 1)\right];$$

$$\frac{dx}{dt} = Ve^{-2nt} + \frac{a}{\varphi^2 + 4n^2}(2n \cos \varphi t - 2ne^{-2nt} + \varphi \sin \varphi t)$$

$$- \frac{f+r}{2n}(1 - e^{-2nt});$$

$$\frac{d^2x}{dt^2} = -2nVe^{-2nt} + \frac{a}{\varphi^2 + 4n^2}$$

$$\times (\varphi^2 \cos \varphi t - 2n\varphi \sin \varphi t + 4n^2 e^{-2nt}) - (f+r)e^{-2nt}.$$

8. Determine the basic parameters of motion of a system, shown in Figure 13.4.1, moving on a horizontal surface while being restricted by a fluid link and subjected to a dry friction force, a constant resisting force, a constant active force, and a harmonic force. The air resistance is negligible. The initial conditions of motion are as follows: for $t = 0$, $x = 0$; $\frac{dx}{dt} = 0$.

Answer:

$$x = \frac{p-f-r}{2n}\left[t + \frac{1}{2n}(e^{-2nt} - 1)\right] + \frac{a}{\varphi^2 + 4n^2}$$

$$\times \left(e^{-2nt} + \frac{2n}{\varphi}\sin\varphi t - \cos\varphi t\right);$$

$$\frac{dx}{dt} = \frac{p-f-r}{2n}(1 - e^{-2nt}) + \frac{a}{\varphi^2 + 4n^2}$$

$$\times (2n\cos\varphi t - 2ne^{-2nt} + \varphi\sin\varphi t);$$

$$\frac{d^2x}{dt^2} = (p - f - r)e^{-2nt} + \frac{a}{\varphi^2 + 4n^2}$$

$$\times (4n^2 e^{-2nt} - 2n\varphi\sin s\varphi t + \varphi^2\cos\varphi t).$$

9. Determine the basic parameters of motion of a system, shown in Figure 13.4.1, moving on a horizontal surface while being restricted by a fluid link and subjected to a dry friction force, a constant active force, and a harmonic force. The air resistance is negligible. The initial conditions of motion are as follows: for $t = 0$, $x = S$; $\frac{dx}{dt} = 0$.

Answer:

$$x = S + \frac{p-f-r}{2n}\left[t + \frac{1}{2n}(e^{-2nt} - 1)\right] + \frac{a}{\varphi^2 + 4n^2}$$

$$\times \left(e^{-2nt} + \frac{2n}{\varphi}\sin\varphi t - \cos\varphi t\right);$$

$$\frac{dx}{dt} = \frac{p-f-r}{2n}(1 - e^{-2nt}) + \frac{a}{\varphi^2 + 4n^2}$$

$$\times (2n\cos\varphi t - 2ne^{-2nt} + \varphi\sin\varphi t);$$

$$\frac{d^2x}{dt^2} = (p - f - r)e^{-2nt} + \frac{a}{\varphi^2 + 4n^2}$$

$$\times (4n^2 e^{-2nt} - 2n\varphi\sin s\varphi t + \varphi^2\cos\varphi t).$$

10 Determine the basic parameters of motion of a system, shown in Figure 13.4.1, moving on a horizontal surface while being

restricted by a fluid link and subjected to a dry friction force, a constant active force, and a harmonic force. The air resistance is negligible. The initial conditions of motion are as follows: for $t = 0$, $x = 0$; $\frac{dx}{dt} = V$.

Answer:

$$x = \frac{V}{2n}(1 - e^{-2nt}) + \frac{p - f - r}{2n}\left[t + \frac{1}{2n}(e^{-2nt} - 1)\right]$$

$$+ \frac{a}{\varphi^2 + 4n^2}\left(e^{-2nt} + \frac{2n}{\varphi}\sin\varphi t - \cos\varphi t\right);$$

$$\frac{dx}{dt} = Ve^{-2nt} + \frac{p - f - r}{2n}(1 - e^{-2nt}) + \frac{a}{\varphi^2 + 4n^2}$$

$$\times (2n\cos\varphi t - 2ne^{-2nt} + \varphi\sin\varphi t);$$

$$\frac{d^2x}{dt^2} = -2nVe^{-2nt} + (p - f - r)e^{-2nt} + \frac{a}{\varphi^2 + 4n^2}$$

$$\times (4n^2 e^{-2nt} - 2n\varphi\sin s\varphi t + \varphi^2\cos\varphi t).$$

CHAPTER 14

PROBLEMS

---

1. Determine the basic parameters of motion of the system shown
   in Figure 14.1.1. The system is restricted by a pair of flexible
   and fluid links in parallel and is moving on a horizontal surface
   due to its initial velocity in the absence of active forces. The air
   resistance is negligible. The initial conditions of motion are as
   follows: for $t = 0$, $x = 0$; $\frac{dx}{dt} = V$.

   Answer:

   $$x = \frac{V}{\omega} e^{-nt} \sin \omega t; \qquad \frac{dx}{dt} = V e^{-nt} \left(\cos \omega t - \frac{n}{\omega} \sin \omega t \right);$$

   $$\frac{d^2 x}{dt^2} = \frac{V}{\omega} e^{-nt} [(n^2 - \omega^2) \sin \omega t - 2n\omega \cos \omega t].$$

2. Determine the basic parameters of motion of the system shown
   in Figure 14.2.1. The system is restricted by a pair of flexible
   and fluid links in parallel and is moving on a horizontal surface
   due to a constant active force. The air resistance is negligible.
   The initial conditions of motion are as follows: for $t = 0$, $x = 0$;
   $\frac{dx}{dt} = 0$.

Answer:

$$x = \frac{p}{\omega^2 + n^2}\left[1 - e^{-nt}(\cos\omega t + \frac{n}{\omega}\sin\omega t)\right];$$

$$\frac{dx}{dt} = \frac{p}{\omega}e^{-nt}\sin\omega t; \quad \frac{d^2x}{dt^2} = pe^{-nt}\left(\cos\omega t - \frac{n}{\omega}\sin\omega t\right).$$

3. Determine the basic parameters of motion of the system shown in Figure 14.2.1. The system is restricted by a pair of flexible and fluid links in parallel and is moving on a horizontal surface while being subjected to a constant active force. The air resistance is negligible. The initial conditions of motion are as follows: for $t = 0$, $x = S$; $\frac{dx}{dt} = 0$.

Answer:

$$x = Se^{-nt}\left(\frac{n}{\omega}\sin\omega t + \cos\omega t\right) + \frac{p}{\omega^2 + n^2}$$

$$\times \left[1 - e^{-nt}\left(\cos\omega t + \frac{n}{\omega}\sin\omega t\right)\right];$$

$$\frac{dx}{dt} = 2nSe^{-nt}\left(\cos\omega t - \frac{n}{\omega}\sin\omega t\right) + \frac{S}{\omega}e^{-nt}$$

$$\times \left[(n^2 - \omega^2)\sin\omega t - 2n\omega\cos\omega t\right] + \frac{p}{\omega}e^{-nt}\sin\omega t;$$

$$\frac{d^2x}{dt^2} = \frac{2nS}{\omega}e^{-nt}[(n^2 - \omega^2)\sin\omega t - 2n\omega\cos\omega t]$$

$$+ Se^{-nt}\left[\frac{n}{\omega}(3\omega^2 - n^2)\sin\omega t - (\omega^2 - 3n^2)\cos\omega t\right]$$

$$+ pe^{-nt}\left(\cos\omega t - \frac{n}{\omega}\sin\omega t\right).$$

4. Determine the basic parameters of motion of the system shown in Figure 14.2.1. The system is restricted by a pair of flexible and fluid links in parallel and is moving on a horizontal surface while being subjected to a constant active force. The air resistance is negligible. The initial conditions of motion are as follows: for $t = 0$, $x = 0$; $\frac{dx}{dt} = V$.

Answer:

$$x = \frac{V}{\omega} e^{-nt} \sin \omega t + \frac{p}{\omega^2 + n^2}\left[1 - e^{-nt}\left(\cos \omega t + \frac{n}{\omega} \sin \omega t\right)\right];$$

$$\frac{dx}{dt} = Ve^{-nt}\left(\cos \omega t - \frac{n}{\omega} \sin \omega t\right) + \frac{p}{\omega} e^{-nt} \sin \omega t;$$

$$\frac{d^2 x}{dt^2} = \frac{V}{\omega} e^{-nt}[(n^2 - \omega^2)\sin \omega t - 2n\omega \cos \omega t] + pe^{-nt}$$
$$\times \left(\cos \omega t - \frac{n}{\omega} \sin \omega t\right).$$

5. Determine the basic parameters of motion of the system shown in Figure 14.3.1. The system is restricted by a pair of flexible and fluid links in parallel and is moving on a horizontal surface while being subjected to a harmonic force. The air resistance is negligible. The initial conditions of motion are as follows: for $t = 0$, $x = 0$; $\frac{dx}{dt} = 0$.

Answer:

$$x = \frac{a}{4n^2\varphi^2 + (\omega^2 + n^2 - \varphi^2)^2}\left[(\omega^2 + n^2 - \varphi^2)\right.$$
$$\times (\cos \varphi t - e^{-nt} \cos \omega t) + 2n\varphi \sin \varphi t$$
$$\left. - \frac{n}{\omega}(\omega^2 + n^2 + \varphi^2)e^{-nt} \sin \omega t\right]$$

$$\frac{dx}{dt} = \frac{a}{4n^2\varphi^2 + (\omega^2 + n^2 - \varphi^2)^2}\left\{2n\varphi^2(\cos \varphi t - e^{-nt} \cos \omega t)\right.$$
$$- \varphi(\omega^2 + n^2 - \varphi^2)\sin \varphi t + \frac{1}{\omega}e^{-nt}[(\omega^2 + n^2)^2$$
$$\left. - \varphi^2(\omega^2 - n^2)]\sin \omega t\right\}$$

$$\frac{d^2 x}{dt^2} = \frac{a}{4n^2\varphi^2 + (\omega^2 + n^2 - \varphi^2)^2}\left\{[(n^2 + \omega^2)^2\right.$$
$$+ \varphi^2(3n^2 - \omega^2)]e^{-nt} \cos \omega t - \varphi^2(\omega^2 + n^2 - \varphi^2)\cos \varphi t$$
$$\left. - 2\varphi^3 n \sin \varphi t - \frac{n}{\omega}[(\omega^2 + n^2)^2 - \varphi^2(n^2 - 3\omega^2)]e^{-nt} \sin \omega t\right\}$$

6. Determine the basic parameters of motion of the system shown in Figure 14.3.1. The system is restricted by a pair of flexible and fluid links in parallel and is moving on a horizontal surface while being subjected to a harmonic force. The air resistance is negligible. The initial conditions of motion are as follows: for $t = 0$, $x = S$; $\frac{dx}{dt} = 0$.

Answer:

$$x = \frac{nS}{\omega}e^{-nt}\sin\omega t + Se^{-nt}\cos\omega t + \frac{a}{4n^2\varphi^2 + (\omega^2 + n^2 - \varphi^2)^2}$$

$$\times \left[(\omega^2 + n^2 - \varphi^2)(\cos\varphi t - e^{-nt}\cos\omega t) + 2n\varphi\sin\varphi t \right.$$

$$\left. - \frac{n}{\omega}(\omega^2 + n^2 + \varphi^2)e^{-nt}\sin\omega t \right]$$

$$\frac{dx}{dt} = 2nSe^{-nt}\left(\cos\omega t - \frac{n}{\omega}\sin\omega t\right) + \frac{S}{\omega}e^{-nt}[(n^2 - \omega^2)\sin\omega t$$

$$- 2n\omega\cos\omega t] + \frac{a}{4n^2\varphi^2 + (\omega^2 + n^2 - \varphi^2)^2}$$

$$\times \left\{ 2n\varphi^2(\cos\varphi t - e^{-nt}\cos\omega t) - \varphi(\omega^2 + n^2 - \varphi^2)\sin\varphi t \right.$$

$$\left. + \frac{1}{\omega}e^{-nt}[(\omega^2 + n^2)^2 - \varphi^2(\omega^2 - n^2)]\sin\omega t \right\}$$

$$\frac{d^2x}{dt^2} = \frac{2nS}{\omega}e^{-nt}[(n^2 - \omega^2)\sin\omega t - 2n\omega\cos\omega t]$$

$$+ Se^{-nt}\left[\frac{n}{\omega}(3\omega^2 - n^2)\sin\omega t - (\omega^2 - 3n^2)\cos\omega t \right]$$

$$+ \frac{a}{4n^2\varphi^2 + (\omega^2 + n^2 - \varphi^2)^2}\left\{ [(n^2 + \omega^2)^2 \right.$$

$$+ \varphi^2(3n^2 - \omega^2)]e^{-nt}\cos\omega t - \varphi^2(\omega^2 + n^2 - \varphi^2)\cos\varphi t$$

$$\left. - 2\varphi^3 n\sin\varphi t - \frac{n}{\omega}[(\omega^2 + n^2)^2 - \varphi^2(n^2 - 3\omega^2)]e^{-nt}\sin\omega t \right\}$$

7. Determine the basic parameters of motion of the system shown in Figure 14.3.1. The system is restricted by a pair of flexible

and fluid links in parallel and is moving on a horizontal surface while being subjected to a harmonic force. The air resistance is negligible. The initial conditions of motion are as follows: for $t = 0$, $x = 0$; $\frac{dx}{dt} = V$.

Answer:

$$x = \frac{V}{2n}(1 - e^{-2nt}) + \frac{a}{\varphi^2 + 4n^2}\left(e^{-2nt} + \frac{2n}{\varphi}\sin\varphi t - \cos\varphi t\right)$$

$$- \frac{f + r}{2n}\left[t + \frac{1}{2n}(e^{-2nt} - 1)\right];$$

$$\frac{dx}{dt} = Ve^{-2nt} + \frac{a}{\varphi^2 + 4n^2}(2n\cos\varphi t - 2ne^{-2nt} + \varphi\sin\varphi t)$$

$$- \frac{f + r}{2n}(1 - e^{-2nt});$$

$$\frac{d^2x}{dt^2} = -2nVe^{-2nt} + \frac{a}{\varphi^2 + 4n^2}$$

$$\times (\varphi^2\cos\varphi t - 2n\varphi\sin\varphi t + 4n^2 e^{-2nt}) - (f + r)e^{-2nt}.$$

8. Determine the basic parameters of motion of the system shown in Figure 14.4.1. The system is restricted by a pair of flexible and fluid links in parallel and is moving on a horizontal surface while being subjected to a constant active force and a harmonic force. The air resistance is negligible. The initial conditions of motion are as follows: for $t = 0$, $x = 0$; $\frac{dx}{dt} = 0$.

Answer:

$$x = \frac{p}{\omega^2 + n^2}\left[1 - e^{-nt}\left(\cos\omega t + \frac{n}{\omega}\sin\omega t\right)\right]$$

$$+ \frac{a}{4n^2\varphi^2 + (\omega^2 + n^2 - \varphi^2)^2}$$

$$\times \left[(\omega^2 + n^2 - \varphi^2)(\cos\varphi t - e^{-nt}\cos\omega t)\right.$$

$$\left. + 2n\varphi\sin\varphi t - \frac{n}{\omega}(\omega^2 + n^2 + \varphi^2)e^{-nt}\sin\omega t\right]$$

$$\frac{dx}{dt} = \frac{p}{\omega}e^{-nt}\sin\omega t + \frac{a}{4n^2\varphi^2 + (\omega^2 + n^2 - \varphi^2)^2}$$

$$\times \left\{ 2n\varphi^2(\cos\varphi t - e^{-nt}\cos\omega t) - \varphi(\omega^2 + n^2 - \varphi^2)\sin\varphi t \right.$$

$$\left. + \frac{1}{\omega}e^{-nt}[(\omega^2 + n^2)^2 - \varphi^2(\omega^2 - n^2)]\sin\omega t \right\}$$

$$\frac{d^2x}{dt^2} = pe^{-nt}(\cos\omega t - \frac{n}{\omega}\sin\omega t) + \frac{a}{4n^2\varphi^2 + (\omega^2 + n^2 - \varphi^2)^2}$$

$$\times \left\{ [(n^2 + \omega^2)^2 + \varphi^2(3n^2 - \omega^2)]e^{-nt}\cos\omega t \right.$$

$$-\varphi^2(\omega^2 + n^2 - \varphi^2)\cos\varphi t - 2\varphi^3 n\sin\varphi t$$

$$\left. - \frac{n}{\omega}[(\omega^2 + n^2)^2 - \varphi^2(n^2 - 3\omega^2)]e^{-nt}\sin\omega t \right\}.$$

9. Determine the basic parameters of motion of the system shown in Figure 14.4.1. The system is restricted by a pair of flexible and fluid links in parallel and is moving on a horizontal surface while being subjected to a constant active force and a harmonic force. The air resistance is negligible. The initial conditions of motion are as follows: for $t = 0$, $x = S$; $\frac{dx}{dt} = 0$.

Answer:

$$x = \frac{2nS}{\omega}e^{-nt}\sin\omega t + Se^{-nt}\left(\cos\omega t - \frac{n}{\omega}\sin\omega t\right)$$

$$+ \frac{p}{\omega^2 + n^2}\left[1 - e^{-nt}\left(\cos\omega t + \frac{n}{\omega}\sin\omega t\right)\right]$$

$$+ \frac{a}{4n^2\varphi^2 + (\omega^2 + n^2 - \varphi^2)^2}\left[(\omega^2 + n^2 - \varphi^2)\right.$$

$$\times (\cos\varphi t - e^{-nt}\cos\omega t) + 2n\varphi\sin\varphi t$$

$$\left. - \frac{n}{\omega}(\omega^2 + n^2 + \varphi^2)e^{-nt}\sin\omega t \right]$$

$$\frac{dx}{dt} = 2nSe^{-nt}\left(\cos \omega t - \frac{n}{\omega}\sin \omega t\right)$$

$$+ \frac{S}{\omega}e^{-nt}[(n^2 - \omega^2)\sin \omega t - 2n\omega \cos \omega t]$$

$$+ \frac{p}{\omega}e^{-nt}\sin \omega t + \frac{a}{4n^2\varphi^2 + (\omega^2 + n^2 - \varphi^2)^2}$$

$$\times \left\{ 2n\varphi^2(\cos \varphi t - e^{-nt}\cos \omega t) - \varphi(\omega^2 + n^2 - \varphi^2)\sin \varphi t \right.$$

$$\left. + \frac{1}{\omega}e^{-nt}[(\omega^2 + n^2)^2 - \varphi^2(\omega^2 - n^2)]\sin \omega t \right\}$$

$$\frac{d^2x}{dt^2} = \frac{2nS}{\omega}e^{-nt}[(n^2 - \omega^2)\sin \omega t - 2n\omega \cos \omega t]$$

$$+ Se^{-nt}\left[\frac{n}{\omega}(3\omega^2 - n^2)\sin \omega t - (\omega^2 - 3n^2)\cos \omega t\right]$$

$$+ pe^{-nt}(\cos \omega t - \frac{n}{\omega}\sin \omega t) + \frac{a}{4n^2\varphi^2 + (\omega^2 + n^2 - \varphi^2)^2}$$

$$\times \left\{ [(n^2 + \omega^2)^2 + \varphi^2(3n^2 - \omega^2)]e^{-nt}\cos \omega t \right.$$

$$- \varphi^2(\omega^2 + n^2 - \varphi^2)\cos \varphi t - 2\varphi^3 n \sin \varphi t$$

$$\left. - \frac{n}{\omega}[(\omega^2 + n^2)^2 - \varphi^2(n^2 - 3\omega^2)]e^{-nt}\sin \omega t \right\}.$$

10. Determine the basic parameters of motion of the system shown in Figure 14.4.1. The system is restricted by a pair of flexible and fluid links in parallel and is moving on a horizontal surface while being subjected to a constant active force and a harmonic force. The air resistance is negligible. The initial conditions of motion are as follows: for $t = 0$, $x = 0$; $\frac{dx}{dt} = V$.

Answer:

$$x = \frac{V}{\omega}e^{-nt}\sin \omega t + \frac{p}{\omega^2 + n^2}\left[1 - e^{-nt}\left(\cos \omega t + \frac{n}{\omega}\sin \omega t\right)\right]$$

$$+ \frac{a}{4n^2\varphi^2 + (\omega^2 + n^2 - \varphi^2)^2}\left[(\omega^2 + n^2 - \varphi^2)\right.$$

$$\times \left(\cos \varphi t - e^{-nt} \cos \omega t\right) + 2n\varphi \sin \varphi t$$

$$- \frac{n}{\omega}(\omega^2 + n^2 + \varphi^2)e^{-nt} \sin \omega t\Bigg]$$

$$\frac{dx}{dt} = Ve^{-nt}\left(\cos \omega t - \frac{n}{\omega}\sin \omega t\right) + \frac{p}{\omega}e^{-nt}\sin \omega t$$

$$+ \frac{a}{4n^2\varphi^2 + (\omega^2 + n^2 - \varphi^2)^2}\Big\{2n\varphi^2(\cos \varphi t - \dot{e}^{-nt}\cos \omega t)$$

$$- \varphi(\omega^2 + n^2 - \varphi^2)\sin \varphi t + \frac{1}{\omega}e^{-nt}[(\omega^2 + n^2)^2$$

$$- \varphi^2(\omega^2 - n^2)]\sin \omega t\Big\}$$

$$\frac{d^2x}{dt^2} = \frac{V}{\omega}e^{-nt}[(n^2 - \omega^2)\sin \omega t - 2n\omega \cos \omega t]$$

$$+ pe^{-nt}\left(\cos \omega t - \frac{n}{\omega}\sin \omega t\right) + \frac{a}{4n^2\varphi^2 + (\omega^2 + n^2 - \varphi^2)^2}$$

$$\times \Big\{[(n^2 + \omega^2)^2 + \varphi^2(3n^2 - \omega^2)]e^{-nt}\cos \omega t$$

$$- \varphi^2(\omega^2 + n^2 - \varphi^2)\cos \varphi t - 2\varphi^3 n \sin \varphi t$$

$$- \frac{n}{\omega}[(\omega^2 + n^2)^2 - \varphi^2(n^2 - 3\omega^2)]e^{-nt}\sin \omega t\Big\}.$$

CHAPTER 15

PROBLEMS

---

1. Determine the basic parameters of motion of the system shown
   in Figure 15.1.1. The system is restricted by a pair of flexible and
   fluid links in parallel, subjected to a constant resisting force, and
   moving on a horizontal surface due to the initial velocity of the
   mass. The air resistance is negligible. The initial conditions of
   motion are as follows: for $t = 0$, $x = 0$; $\frac{dx}{dt} = V$.

   Answer:

   $$x = \frac{V}{\omega} e^{-nt} \sin \omega t - \frac{r}{\omega^2 + n^2}\left[1 - e^{-nt}\left(\cos \omega t + \frac{n}{\omega}\sin \omega t\right)\right];$$

   $$\frac{dx}{dt} = V e^{-nt}\left(\cos \omega t - \frac{n}{\omega} e^{-nt}\sin \omega t\right) - \frac{r}{\omega} e^{-nt}\sin \omega t;$$

   $$\frac{d^2 x}{dt^2} = \frac{V}{\omega} e^{-nt}[(n^2 - \omega^2)\sin \omega t - 2n\omega \cos \omega t]$$

   $$- r e^{-nt}\left(\cos \omega t - \frac{n}{\omega}\sin \omega t\right).$$

2. Determine the basic parameters of motion of the system shown
   in Figure 15.2.1. The system is restricted by a pair of flexible
   and fluid links in parallel and is moving on a horizontal surface
   while being subjected to a constant resisting force and a constant
   active force. The air resistance is negligible. The initial conditions
   of motion are as follows: for $t = 0$, $x = 0$; $\frac{dx}{dt} = 0$.

Answer:

$$x = \frac{p - r}{\omega^2 + n^2}\left[1 - e^{-nt}\left(\cos \omega t + \frac{n}{\omega}\sin \omega t\right)\right];$$

$$\frac{dx}{dt} = \frac{p - r}{\omega}e^{-nt}\sin \omega t; \quad \frac{d^2x}{dt^2} = (p - r)e^{-nt}\left(\cos \omega t - \frac{n}{\omega}\sin \omega t\right).$$

3. Determine the basic parameters of motion of the system shown in Figure 15.2.1. The system is restricted by a pair of flexible and fluid links in parallel and is moving on a horizontal surface while being subjected to a constant resisting and constant active force. The air resistance is negligible. The initial conditions of motion are as follows: for $t = 0$, $x = S$; $\frac{dx}{dt} = 0$.

Answer:

$$x = \frac{p - r}{\omega^2 + n^2}\left[1 - e^{-nt}\left(\cos \omega t + \frac{n}{\omega}\sin \omega t\right)\right];$$

$$\frac{dx}{dt} = \frac{p - r}{\omega}e^{-nt}\sin \omega t; \quad \frac{d^2x}{dt^2} = (p - r)e^{-nt}\left(\cos \omega t - \frac{n}{\omega}\sin \omega t\right).$$

4. Determine the basic parameters of motion of the system shown in Figure 15.2.1. The system is restricted by a pair of flexible and fluid links in parallel and is moving on a horizontal surface while being subjected to a constant resisting force and a constant active force. The air resistance is negligible. The initial conditions of motion are as follows: for $t = 0$, $x = 0$; $\frac{dx}{dt} = V$.

Answer:

$$x = \frac{V}{\omega}e^{-nt}\sin \omega t + \frac{p - r}{\omega^2 + n^2}\left[1 - e^{-nt}\left(\cos \omega t + \frac{n}{\omega}\sin \omega t\right)\right];$$

$$\frac{dx}{dt} = Ve^{-nt}\left(\cos \omega t - \frac{n}{\omega}\sin \omega t\right) + \frac{p - r}{\omega}e^{-nt}\sin \omega t;$$

$$\frac{d^2x}{dt^2} = \frac{V}{\omega}e^{-nt}[(n^2 - \omega^2)\sin \omega t - 2n\omega \cos \omega t] + (p - r)e^{-nt}$$

$$\times \left(\cos \omega t - \frac{n}{\omega}\sin \omega t\right).$$

5. Determine the basic parameters of motion of the system shown in Figure 15.3.1. The system is restricted by a pair of flexible and fluid links in parallel and is moving on a horizontal surface while being subjected to a constant resisting force and a harmonic force. The air resistance is negligible. The initial conditions of motion are as follows: for $t = 0$, $x = 0$; $\frac{dx}{dt} = 0$.

Answer:

$$
x = \frac{a}{4n^2\varphi^2 + (\omega^2 + n^2 - \varphi^2)^2} \left[(\omega^2 + n^2 - \varphi^2) \right.
$$

$$
\times (\cos\varphi t - e^{-nt}\cos\omega t) + 2n\varphi\sin\varphi t
$$

$$
\left. - \frac{n}{\omega}(\omega^2 + n^2 + \varphi^2)e^{-nt}\sin\omega t \right]
$$

$$
- \frac{r}{\omega^2 + n^2}[1 - e^{-nt}\left(\cos\omega t + \frac{n}{\omega}\sin\omega t\right)
$$

$$
\frac{dx}{dt} = \frac{a}{4n^2\varphi^2 + (\omega^2 + n^2 - \varphi^2)^2} \left\{ 2n\varphi^2(\cos\varphi t - e^{-nt}\cos\omega t) \right.
$$

$$
- \varphi(\omega^2 + n^2 - \varphi^2)\sin\varphi t + \frac{1}{\omega}e^{-nt}[(\omega^2 + n^2)^2
$$

$$
\left. - \varphi^2(\omega^2 - n^2)]\sin\omega t \right\} - \frac{r}{\omega}e^{-nt}\sin\omega t
$$

$$
\frac{d^2x}{dt^2} = \frac{a}{4n^2\varphi^2 + (\omega^2 + n^2 - \varphi^2)^2} \left\{ [(n^2 + \omega^2)^2 \right.
$$

$$
+ \varphi^2(3n^2 - \omega^2)]e^{-nt}\cos\omega t - \varphi^2(\omega^2 + n^2 - \varphi^2)\cos\varphi t
$$

$$
\left. - 2\varphi^3 n\sin\varphi t - \frac{n}{\omega}[(\omega^2 + n^2)^2 - \varphi^2(n^2 - 3\omega^2)]e^{-nt}\sin\omega t \right\}
$$

$$
- re^{-nt}\left(\cos\varphi t - \frac{n}{\varphi}\sin\varphi t\right).
$$

6. Determine the basic parameters of motion of the system shown in Figure 15.3.1. The system is restricted by a pair of flexible and fluid links in parallel and is moving on a horizontal surface while being subjected to a constant resisting force and a harmonic force. The air resistance is negligible. The initial conditions of motion are as follows: for $t = 0$, $x = S$; $\frac{dx}{dt} = 0$.

Answer:

$$x = \frac{2nS}{\omega} e^{-nt} \sin \omega t + S e^{-nt} \left(\cos \omega t - \frac{n}{\omega} \sin \omega t \right)$$

$$+ \frac{a}{4n^2 \varphi^2 + (\omega^2 + n^2 - \varphi^2)^2}$$

$$\times \left[(\omega^2 + n^2 - \varphi^2)(\cos \varphi t - e^{-nt} \cos \omega t) \right.$$

$$\left. + 2n\varphi \sin \varphi t - \frac{n}{\omega}(\omega^2 + n^2 + \varphi^2) e^{-nt} \sin \omega t \right]$$

$$- \frac{r}{\omega^2 + n^2}[1 - e^{-nt} \left(\cos \omega t + \frac{n}{\omega} \sin \omega t \right)$$

$$\frac{dx}{dt} = 2nS e^{-nt} \left(\cos \omega t - \frac{n}{\omega} \sin \omega t \right)$$

$$+ \frac{S}{\omega} e^{-nt}[(n^2 - \omega^2) \sin \omega t - 2n\omega \cos \omega t]$$

$$+ \frac{a}{4n^2 \varphi^2 + (\omega^2 + n^2 - \varphi^2)^2} \left\{ 2n\varphi^2 (\cos \varphi t - e^{-nt} \cos \omega t) \right.$$

$$- \varphi(\omega^2 + n^2 - \varphi^2) \sin \varphi t + \frac{1}{\omega} e^{-nt}[(\omega^2 + n^2)^2$$

$$\left. - \varphi^2 (\omega^2 - n^2)] \sin \omega t \right\} - \frac{r}{\omega} e^{-nt} \sin \omega t$$

$$\frac{d^2 x}{dt^2} = \frac{2nS}{\omega} e^{-nt}[(n^2 - \omega^2) \sin \omega t - 2n\omega \cos \omega t]$$

$$+ S e^{-nt} \left[\frac{n}{\omega}(3\omega^2 - n^2) \sin \omega t - (\omega^2 - 3n^2) \cos \omega t \right]$$

$$+ \frac{a}{4n^2 \varphi^2 + (\omega^2 + n^2 - \varphi^2)^2} \left\{ [(n^2 + \omega^2)^2 \right.$$

$$+ \varphi^2 (3n^2 - \omega^2)] e^{-nt} \cos \omega t - \varphi^2 (\omega^2 + n^2 - \varphi^2) \cos \varphi t$$

$$\left. - 2\varphi^3 n \sin \varphi t - \frac{n}{\omega}[(\omega^2 + n^2)^2 - \varphi^2 (n^2 - 3\omega^2)] e^{-nt} \sin \omega t \right\}$$

$$- r e^{-nt} \left(\cos \varphi t - \frac{n}{\varphi} \sin \varphi t \right)$$

7. Determine the basic parameters of motion of the system shown in Figure 15.3.1. The system is restricted by a pair of flexible and fluid links in parallel and is moving on a horizontal surface while being subjected to a constant resisting force and a harmonic force. The air resistance is negligible. The initial conditions of motion are as follows: for $t = 0$, $x = 0$; $\frac{dx}{dt} = V$.

Answer:

$$x = \frac{V}{\omega} e^{-nt} \sin \omega t + \frac{a}{4n^2\varphi^2 + (\omega^2 + n^2 - \varphi^2)^2}$$

$$\times \left[(\omega^2 + n^2 - \varphi^2)(\cos \varphi t - e^{-nt} \cos \omega t) + 2n\varphi \sin \varphi t \right.$$

$$\left. - \frac{n}{\omega}(\omega^2 + n^2 + \varphi^2)e^{-nt} \sin \omega t \right] - \frac{r}{\omega^2 + n^2}[1 - e^{-nt}$$

$$\times \left(\cos \omega t + \frac{n}{\omega} \sin \omega t \right)$$

$$\frac{dx}{dt} = V e^{-nt} \left(\cos \omega t - \frac{n}{\omega} \sin \omega t \right) + \frac{a}{4n^2\varphi^2 + (\omega^2 + n^2 - \varphi^2)^2}$$

$$\times \left\{ 2n\varphi^2 (\cos \varphi t - e^{-nt} \cos \omega t) - \varphi(\omega^2 + n^2 - \varphi^2) \sin \varphi t \right.$$

$$\left. + \frac{1}{\omega} e^{-nt}[(\omega^2 + n^2)^2 - \varphi^2(\omega^2 - n^2)] \sin \omega t \right\} - \frac{r}{\omega} e^{-nt} \sin \omega t$$

$$\frac{d^2x}{dt^2} = \frac{V}{\omega} e^{-nt}[(n^2 - \omega^2) \sin \omega t - 2n\omega \cos \omega t]$$

$$+ \frac{a}{4n^2\varphi^2 + (\omega^2 + n^2 - \varphi^2)^2} \left\{ [(n^2 + \omega^2)^2 \right.$$

$$+ \varphi^2(3n^2 - \omega^2)]e^{-nt} \cos \omega t - \varphi^2(\omega^2 + n^2 - \varphi^2) \cos \varphi t$$

$$\left. - 2\varphi^3 n \sin \varphi t - \frac{n}{\omega}[(\omega^2 + n^2)^2 - \varphi^2(n^2 - 3\omega^2)]e^{-nt} \sin \omega t \right\}$$

$$- re^{-nt} \left(\cos \varphi t - \frac{n}{\varphi} \sin \varphi t \right)$$

8. Determine the basic parameters of motion of the system shown in Figure 15.4.1. The system is restricted by a pair of flexible and fluid links in parallel and is moving on a horizontal surface

while being subjected to a constant resisting force, a constant active force, and a harmonic force. The air resistance is negligible. The initial conditions of motion are as follows: for $t = 0$, $x = 0$; $\frac{dx}{dt} = 0$.

Answer:

$$x = \frac{p-r}{\omega^2 + n^2}\left[1 - e^{-nt}\left(\cos\omega t + \frac{n}{\omega}\sin\omega t\right)\right]$$

$$+ \frac{a}{4n^2\varphi^2 + (\omega^2 + n^2 - \varphi^2)^2}$$

$$\times \left[(\omega^2 + n^2 - \varphi^2)(\cos\varphi t - e^{-nt}\cos\omega t)\right.$$

$$\left. + 2n\varphi\sin\varphi t - \frac{n}{\omega}(\omega^2 + n^2 + \varphi^2)e^{-nt}\sin\omega t\right]$$

$$\frac{dx}{dt} = \frac{p-r}{\omega}e^{-nt}\sin\omega t + \frac{a}{4n^2\varphi^2 + (\omega^2 + n^2 - \varphi^2)^2}$$

$$\times \left\{2n\varphi^2(\cos\varphi t - e^{-nt}\cos\omega t) - \varphi(\omega^2 + n^2 - \varphi^2)\sin\varphi t\right.$$

$$\left. + \frac{1}{\omega}e^{-nt}[(\omega^2 + n^2)^2 - \varphi^2(\omega^2 - n^2)]\sin\omega t\right\}$$

$$\frac{d^2x}{dt^2} = (p-r)e^{-nt}\left(\cos\omega t - \frac{n}{\omega}\sin\omega t\right) + \frac{a}{4n^2\varphi^2 + (\omega^2 + n^2 - \varphi^2)^2}$$

$$\times \left\{[(n^2 + \omega^2)^2 + \varphi^2(3n^2 - \omega^2)]e^{-nt}\cos\omega t\right.$$

$$- \varphi^2(\omega^2 + n^2 - \varphi^2)\cos\varphi t - 2\varphi^3 n\sin\varphi t$$

$$\left. - \frac{n}{\omega}[(\omega^2 + n^2)^2 - \varphi^2(n^2 - 3\omega^2)]e^{-nt}\sin\omega t\right\}$$

9. Determine the basic parameters of motion of the system shown in Figure 15.4.1. The system is restricted by a pair of flexible and fluid links in parallel and is moving on a horizontal surface while being subjected to a constant resisting force, a constant active force, and a harmonic force. The air resistance is negligible. The initial conditions of motion are as follows: for $t = 0$, $x = S$; $\frac{dx}{dt} = 0$.

Answer:

$$x = \frac{2nS}{\omega}e^{-nt}\sin\omega t + Se^{-nt}\left(\cos\omega t - \frac{n}{\omega}\sin\omega t\right)$$

$$+ \frac{p-r}{\omega^2 + n^2}\left[1 - e^{-nt}\left(\cos\omega t + \frac{n}{\omega}\sin\omega t\right)\right]$$

$$+ \frac{a}{4n^2\varphi^2 + (\omega^2 + n^2 - \varphi^2)^2}$$

$$\times \left[(\omega^2 + n^2 - \varphi^2)(\cos\varphi t - e^{-nt}\cos\omega t)\right.$$

$$\left. + 2n\varphi\sin\varphi t - \frac{n}{\omega}(\omega^2 + n^2 + \varphi^2)e^{-nt}\sin\omega t\right]$$

$$\frac{dx}{dt} = 2nSe^{-nt}\left(\cos\omega t - \frac{n}{\omega}\sin\omega t\right)$$

$$+ \frac{S}{\omega}e^{-nt}[(n^2 - \omega^2)\sin\omega t - 2n\omega\cos\omega t] + \frac{p-r}{\omega}e^{-nt}\sin\omega t$$

$$+ \frac{a}{4n^2\varphi^2 + (\omega^2 + n^2 - \varphi^2)^2}\left\{2n\varphi^2(\cos\varphi t - e^{-nt}\cos\omega t)\right.$$

$$- \varphi(\omega^2 + n^2 - \varphi^2)\sin\varphi t + \frac{1}{\omega}e^{-nt}[(\omega^2 + n^2)^2$$

$$\left. - \varphi^2(\omega^2 - n^2)]\sin\omega t\right\}$$

$$\frac{d^2x}{dt^2} = \frac{2nS}{\omega}e^{-nt}[(n^2 - \omega^2)\sin\omega t - 2n\omega\cos\omega t]$$

$$+ Se^{-nt}\left[\frac{n}{\omega}(3\omega^2 - n^2)\sin\omega t - (\omega^2 - 3n^2)\cos\omega t\right]$$

$$+ (p-r)e^{-nt}\left(\cos\omega t - \frac{n}{\omega}\sin\omega t\right)$$

$$+ \frac{a}{4n^2\varphi^2 + (\omega^2 + n^2 - \varphi^2)^2}$$

$$\times \left\{[(n^2 + \omega^2)^2 + \varphi^2(3n^2 - \omega^2)]e^{-nt}\cos\omega t\right.$$

$$- \varphi^2(\omega^2 + n^2 - \varphi^2)\cos\varphi t - 2\varphi^3 n\sin\varphi t$$

$$\left. - \frac{n}{\omega}[(\omega^2 + n^2)^2 - \varphi^2(n^2 - 3\omega^2)]e^{-nt}\sin\omega t\right\}$$

10. Determine the basic parameters of motion of the system shown in Figure 15.4.1. The system is restricted by a pair of flexible and fluid links in parallel and is moving on a horizontal surface while being subjected to a constant resisting force, a constant active force, and a harmonic force. The air resistance is negligible. The initial conditions of motion are as follows: for $t = 0$, $x = 0$; $\frac{dx}{dt} = V$.

Answer:

$$x = \frac{V}{\omega} e^{-nt} \sin \omega t + \frac{p - r}{\omega^2 + n^2} \left[1 - e^{-nt} \left(\cos \omega t + \frac{n}{\omega} \sin \omega t \right) \right]$$

$$+ \frac{a}{4n^2\varphi^2 + (\omega^2 + n^2 - \varphi^2)^2}$$

$$\times \left[(\omega^2 + n^2 - \varphi^2)(\cos \varphi t - e^{-nt} \cos \omega t) \right.$$

$$\left. + 2n\varphi \sin \varphi t - \frac{n}{\omega}(\omega^2 + n^2 + \varphi^2) e^{-nt} \sin \omega t \right]$$

$$\frac{dx}{dt} = V e^{-nt} \left(\cos \omega t - \frac{n}{\omega} \sin \omega t \right) + \frac{p - r}{\omega} e^{-nt} \sin \omega t$$

$$+ \frac{a}{4n^2\varphi^2 + (\omega^2 + n^2 - \varphi^2)^2} \left\{ 2n\varphi^2 (\cos \varphi t - e^{-nt} \cos \omega t) \right.$$

$$- \varphi(\omega^2 + n^2 - \varphi^2) \sin \varphi t + \frac{1}{\omega} e^{-nt}[(\omega^2 + n^2)^2$$

$$\left. - \varphi^2(\omega^2 - n^2)] \sin \omega t \right\}$$

$$\frac{d^2x}{dt^2} = \frac{V}{\omega} e^{-nt}[(n^2 - \omega^2) \sin \omega t - 2n\omega \cos \omega t]$$

$$+ (p - r)e^{-nt} \left(\cos \omega t - \frac{n}{\omega} \sin \omega t \right)$$

$$+ \frac{a}{4n^2\varphi^2 + (\omega^2 + n^2 - \varphi^2)^2}$$

$$\times \left\{ [(n^2 + \omega^2)^2 + \varphi^2(3n^2 - \omega^2)]e^{-nt} \cos \omega t \right.$$

$$- \varphi^2(\omega^2 + n^2 - \varphi^2) \cos \varphi t - 2\varphi^3 n \sin \varphi t$$

$$\left. - \frac{n}{\omega}[(\omega^2 + n^2)^2 - \varphi^2(n^2 - 3\omega^2)]e^{-nt} \sin \omega t \right\}$$

CHAPTER 16

PROBLEMS

---

1. Determine the basic parameters of motion of the system shown in Figure 16.1.1. The system is restricted by a pair of flexible and fluid links in parallel and is subjected to a dry friction force while moving on a horizontal surface due to the initial velocity of the mass. The motion of the system represents critical damping. The air resistance is negligible. The initial conditions of motion are as follows: for $t = 0$, $x = 0$; $\frac{dx}{dt} = V$.

   Answer:

$$x = Vte^{-nt} - \frac{f}{n^2}[1 - (1 + nt)e^{-nt}];$$

$$\frac{dx}{dt} = Ve^{-nt}(1 - nt) - fte^{-nt};$$

$$\frac{d^2x}{dt^2} = Vne^{-nt}(nt - 2) - fe^{-nt}(1 - nt).$$

2. Determine the basic parameters of motion of the system shown in Figure 16.2.1. The system is restricted by a pair of flexible and fluid links in parallel and is moving on a horizontal surface while being subjected to a dry friction force and to a constant active force. The motion of the system represents critical damping. The air resistance is negligible. The initial conditions of motion are as follows: for $t = 0$, $x = 0$; $\frac{dx}{dt} = 0$.

Answer:

$$x = \frac{p-f}{n^2}[1 - (1+nt)e^{-nt}]; \quad \frac{dx}{dt} = (p-f)te^{-nt};$$

$$\frac{d^2x}{dt^2} = (p-f)e^{-nt}(1-nt).$$

3. Determine the basic parameters of motion of the system shown in Figure 16.2.1. The system is restricted by a pair of flexible and fluid links in parallel and is moving on a horizontal surface while being subjected to a dry friction force and a constant active force. The motion of the system represents critical damping. The air resistance is negligible. The initial conditions of motion are as follows: for $t = 0$, $x = S$; $\frac{dx}{dt} = 0$.

Answer:

$$x = S(1+nt)e^{-nt} + \frac{p-f}{n^2}[1 - e^{-nt}(1+nt)];$$

$$\frac{dx}{dt} = te^{-nt}(p - f - Sn^2);$$

$$\frac{d^2x}{dt^2} = e^{-nt}(1-nt)(p - f - n^2 S).$$

4. Determine the basic parameters of motion of the system shown in Figure 16.2.1. The system is restricted by a pair of flexible and fluid links in parallel and is moving on a horizontal surface while being subjected to a constant resisting force and a constant active force. The motion of the system represents critical damping. The air resistance is negligible. The initial conditions of motion are as follows: for $t = 0$, $x = 0$; $\frac{dx}{dt} = V$.

5. Determine the basic parameters of motion of the system shown in Figure 16.3.1. The system is restricted by a pair of flexible and fluid links in parallel and is moving on a horizontal surface while being subjected to a dry friction force and a harmonic force. The motion of the system represents critical damping. The air resistance is negligible. The initial conditions of motion are as follows: for $t = 0$, $x = 0$; $\frac{dx}{dt} = 0$.

Answer:

$$x = \frac{a}{(\varphi^2 + n^2)^2}\{2n\varphi \sin \varphi t + e^{-nt}[\varphi^2 - n^2 - nt(\varphi^2 + n^2)]$$

$$- (\varphi^2 - n^2)\cos \varphi t\} - \frac{f}{n^2}[1 - e^{-nt}(1 + nt)];$$

$$\frac{dx}{dt} = \frac{a}{(\varphi^2 + n^2)^2}[(\varphi^2 - n^2)(\varphi \sin \varphi t + n^2 t e^{-nt})$$

$$- 2\varphi^2 n(\cos \varphi t - e^{-nt})] - f t e^{-nt};$$

$$\frac{d^2x}{dt^2} = \frac{1}{\varphi^2 + n^2}[\varphi^2(\varphi^2 - n^2)\cos \varphi t - 2n\varphi^3 \sin \varphi t$$

$$- \varphi^2 n^2 e^{-nt}(nt - 3) - n^4 e^{-nt}(nt - 1) - f e^{-nt}(1 - nt).$$

6. Determine the basic parameters of motion of the system shown in Figure 16.3.1. The system is restricted by a pair of flexible and fluid links in parallel and is moving on a horizontal surface while being subjected to a dry friction force and a harmonic force. The motion of the system represents critical damping. The air resistance is negligible. The initial conditions of motion are as follows: for $t = 0$, $x = S$; $\frac{dx}{dt} = 0$.

7. Determine the basic parameters of motion of the system shown in Figure 16.3.1. The system is restricted by a pair of flexible and fluid links in parallel and is moving on a horizontal surface while being subjected to a constant resisting force and a harmonic force. The motion of the system represents critical damping. The air resistance is negligible. The initial conditions of motion are as follows: for $t = 0$, $x = 0$; $\frac{dx}{dt} = V$.

Answer:

$$x = \frac{V}{\omega}e^{-nt}\sin \omega t + \frac{a}{4n^2\varphi^2 + (\omega^2 + n^2 - \varphi^2)^2}$$

$$\times \left[(\omega^2 + n^2 - \varphi^2)(\cos \varphi t - e^{-nt}\cos \omega t)\right.$$

$$\left. + 2n\varphi \sin \varphi t - \frac{n}{\omega}(\omega^2 + n^2 + \varphi^2)\, e^{-nt}\sin \omega t\right]$$

$$- \frac{r}{\omega^2 + n^2} \left[1 - e^{-nt} \left(\cos \omega t + \frac{n}{\omega} \sin \omega t \right) \right.$$

$$\frac{dx}{dt} = V e^{-nt} \left(\cos \omega t - \frac{n}{\omega} \sin \omega t \right) + \frac{a}{4n^2 \varphi^2 + (\omega^2 + n^2 - \varphi^2)^2}$$

$$\times \left\{ 2n\varphi^2 (\cos \varphi t - e^{-nt} \cos \omega t) - \varphi(\omega^2 + n^2 - \varphi^2) \sin \varphi t \right.$$

$$\left. + \frac{1}{\omega} e^{-nt} [(\omega^2 + n^2)^2 - \varphi^2(\omega^2 - n^2)] \sin \omega t \right\} - \frac{r}{\omega} e^{-nt} \sin \omega t$$

$$\frac{d^2 x}{dt^2} = \frac{V}{\omega} e^{-nt} [(n^2 - \omega^2) \sin \omega t - 2n\omega \cos \omega t]$$

$$+ \frac{a}{4n^2 \varphi^2 + (\omega^2 + n^2 - \varphi^2)^2}$$

$$\times \left\{ [(n^2 + \omega^2)^2 + \varphi^2(3n^2 - \omega^2)] e^{-nt} \cos \omega t \right.$$

$$- \varphi^2(\omega^2 + n^2 - \varphi^2) \cos \varphi t - 2\varphi^3 n \sin \varphi t$$

$$\left. - \frac{n}{\omega} [(\omega^2 + n^2)^2 - \varphi^2(n^2 - 3\omega^2)] e^{-nt} \sin \omega t \right\}$$

$$- r e^{-nt} \left(\cos \varphi t - \frac{n}{\varphi} \sin \varphi t \right)$$

8. Determine the basic parameters of motion of the system shown in Figure 16.4.1. The system is restricted by a pair of flexible and fluid links in parallel and is moving on a horizontal surface while being subjected to a constant resisting force, a constant active force, and a harmonic force. The air resistance is negligible. The initial conditions of motion are as follows: for $t = 0$, $x = 0$; $\frac{dx}{dt} = 0$.

Answer:

$$x = \frac{p - r}{\omega^2 + n^2} \left[1 - e^{-nt} \left(\cos \omega t + \frac{n}{\omega} \sin \omega t \right) \right]$$

$$+ \frac{a}{4n^2 \varphi^2 + (\omega^2 + n^2 - \varphi^2)^2} \left[(\omega^2 + n^2 - \varphi^2)(\cos \varphi t \right.$$

$$\left. - e^{-nt} \cos \omega t) + 2n\varphi \sin \varphi t - \frac{n}{\omega}(\omega^2 + n^2 + \varphi^2) e^{-nt} \sin \omega t \right]$$

$$\frac{dx}{dt} = \frac{p-r}{\omega} e^{-nt} \sin \omega t + \frac{a}{4n^2 \varphi^2 + (\omega^2 + n^2 - \varphi^2)^2}$$

$$\times \left\{ 2n\varphi^2 (\cos \varphi t - e^{-nt} \cos \omega t) - \varphi(\omega^2 + n^2 - \varphi^2) \sin \varphi t \right.$$

$$\left. + \frac{1}{\omega} e^{-nt} [(\omega^2 + n^2)^2 - \varphi^2(\omega^2 - n^2)] \sin \omega t \right\}$$

$$\frac{d^2 x}{dt^2} = (p-r) e^{-nt} \left(\cos \omega t - \frac{n}{\omega} \sin \omega t \right)$$

$$+ \frac{a}{4n^2 \varphi^2 + (\omega^2 + n^2 - \varphi^2)^2}$$

$$\times \left\{ [(n^2 + \omega^2)^2 + \varphi^2 (3n^2 - \omega^2)] e^{-nt} \cos \omega t \right.$$

$$- \varphi^2 (\omega^2 + n^2 - \varphi^2) \cos \varphi t - 2\varphi^3 n \sin \varphi t$$

$$\left. - \frac{n}{\omega} [(\omega^2 + n^2)^2 - \varphi^2 (n^2 - 3\omega^2)] e^{-nt} \sin \omega t \right\}$$

9. Determine the basic parameters of motion of the system shown in Figure 16.4.1. The system is restricted by a pair of flexible and fluid links in parallel and is moving on a horizontal surface while being subjected to a constant resisting force, a constant active force, and a harmonic force. The air resistance is negligible. The initial conditions of motion are as follows: for $t = 0$, $x = S$; $\frac{dx}{dt} = 0$.

10. Determine the basic parameters of motion of the system shown in Figure 16.4.1. The system is restricted by a pair of flexible and fluid links in parallel and is moving on a horizontal surface while being subjected to a constant resisting force, a constant active force, and a harmonic force. The air resistance is negligible. The initial conditions of motion are as follows: for $t = 0$, $x = 0$; $\frac{dx}{dt} = V$.

Answer:

$$x = \frac{V}{\omega} e^{-nt} \sin \omega t + \frac{p-r}{\omega^2 + n^2} \left[1 - e^{-nt} \left(\cos \omega t + \frac{n}{\omega} \sin \omega t \right) \right]$$

$$+ \frac{a}{4n^2\varphi^2 + (\omega^2 + n^2 - \varphi^2)^2}$$

$$\times \left[(\omega^2 + n^2 - \varphi^2)(\cos\varphi t - e^{-nt}\cos\omega t) + 2n\varphi\sin\varphi t \right.$$

$$\left. - \frac{n}{\omega}(\omega^2 + n^2 + \varphi^2)e^{-nt}\sin\omega t \right]$$

$$\frac{dx}{dt} = V e^{-nt}\left(\cos\omega t - \frac{n}{\omega}\sin\omega t\right) + \frac{p - r}{\omega}e^{-nt}\sin\omega t$$

$$+ \frac{a}{4n^2\varphi^2 + (\omega^2 + n^2 - \varphi^2)^2}\left\{ 2n\varphi^2(\cos\varphi t - e^{-nt}\cos\omega t) \right.$$

$$- \varphi(\omega^2 + n^2 - \varphi^2)\sin\varphi t + \frac{1}{\omega}e^{-nt}[(\omega^2 + n^2)^2$$

$$\left. - \varphi^2(\omega^2 - n^2)]\sin\omega t \right\}$$

$$\frac{d^2x}{dt^2} = \frac{V}{\omega}e^{-nt}[(n^2 - \omega^2)\sin\omega t - 2n\omega\cos\omega t] + (p - r)e^{-nt}$$

$$\times \left(\cos\omega t - \frac{n}{\omega}\sin\omega t\right) + \frac{a}{4n^2\varphi^2 + (\omega^2 + n^2 - \varphi^2)^2}$$

$$\times \left\{ [(n^2 + \omega^2)^2 + \varphi^2(3n^2 - \omega^2)]e^{-nt}\cos\omega t \right.$$

$$- \varphi^2(\omega^2 + n^2 - \varphi^2)\cos\varphi t - 2\varphi^3 n\sin\varphi t$$

$$\left. - \frac{n}{\omega}[(\omega^2 + n^2)^2 - \varphi^2(n^2 - 3\omega^2)]e^{-nt}\sin\omega t \right\}$$

CHAPTER 17

PROBLEMS

---

1. Determine the basic parameters of motion of the system shown
   in Figure 17.1.1. The system is restricted by a pair of flexible and
   fluid links in parallel and is subjected to a dry friction force and
   a constant resisting force while moving on a horizontal surface
   due to the initial velocity of the mass. The system performs
   overdamped motion. The air resistance is negligible. The initial
   conditions of motion are as follows: for $t = 0$, $x = 0$; $\frac{dx}{dt} = V$.

   Answer:

$$
x = \frac{V}{\omega} e^{-nt} \sinh \omega t - \frac{f + r}{n^2 - \omega^2}
$$

$$
\times \left[1 - e^{-nt} \left(\cosh \omega t + \frac{n}{\omega} \sinh \omega t \right) \right] ;
$$

$$
\frac{d^2 x}{dt^2} = \frac{V}{\omega} e^{-nt} [(n^2 - \omega^2) \sinh \omega t - 2n\omega \cosh \omega t]
$$

$$
- (f + r) e^{-nt} \left(\cosh \omega t - \frac{n}{\omega} \sinh \omega t \right) .
$$

2. Determine the basic parameters of motion of the system shown
   in Figure 17.2.1. The system is restricted by a pair of flexible and
   fluid links in parallel and is moving on a horizontal surface while
   being subjected to a dry friction force, a constant resisting force,
   and a constant active force. The system performs overdamped

motion. The air resistance is negligible. The initial conditions of motion are as follows: for $t = 0$, $x = 0$; $\frac{dx}{dt} = 0$.

Answer:

$$x = \frac{p - f - r}{n^2 - \omega^2} \left[1 - e^{-nt} \left(\cosh \omega t - \frac{n}{\omega} \sinh \omega t \right) \right] ;$$

$$\frac{dx}{dt} = \frac{p - f - r}{\omega} e^{-nt} \sinh \omega t;$$

$$\frac{d^2 x}{dt^2} = (p - f - r)e^{-nt} \left(\cosh \omega t - \frac{n}{\omega} \sinh \omega t \right) .$$

3. Determine the basic parameters of motion of the system shown in Figure 17.2.1. The system is restricted by a pair of flexible and fluid links in parallel and is moving on a horizontal surface while being subjected to a dry friction force, a constant resisting force, and a constant active force. The system performs overdamped motion. The air resistance is negligible. The initial conditions of motion are as follows: for $t = 0$, $x = S$; $\frac{dx}{dt} = 0$.

   Answer:

$$x = \frac{nS}{\omega} e^{-nt} \sinh \omega t + S e^{-nt} \cosh \omega t$$

$$+ \frac{p - f - r}{n^2 - \omega^2} \left[1 - e^{-nt} \left(\cosh \omega t - \frac{n}{\omega} \sinh \omega t \right) \right] ;$$

$$\frac{dx}{dt} = \frac{1}{\omega} S e^{-nt} (\omega^2 - n^2) \sinh \omega t + \frac{p - f - r}{\omega} e^{-nt} \sinh \omega t;$$

$$\frac{d^2 x}{dt^2} = \frac{2nS}{\omega} e^{-nt} [(n^2 - \omega^2) \sin \omega t - 2n\omega \cos \omega t]$$

$$+ S e^{-nt} \left[\frac{n}{\omega} (3\omega^2 - n^2) \sin \omega t - (\omega^2 - 3n^2) \cos \omega t \right]$$

$$+ (p - f - r)e^{-nt} \left(\cos \omega t - \frac{n}{\omega} \sin \omega t \right) .$$

4. Determine the basic parameters of motion of the system shown in Figure 17.2.1. The system is restricted by a pair of flexible and fluid links in parallel and is moving on a horizontal surface while

being subjected to a dry friction force, a constant resisting force, and a constant active force. The system performs overdamped motion. The air resistance is negligible. The initial conditions of motion are as follows: for $t = 0$, $x = 0$; $\frac{dx}{dt} = V$.

Answer:

$$x = \frac{V}{\omega}e^{-nt}\sin\omega t + \frac{p-f-r}{\omega^2+n^2}\left[1 - e^{-nt}\left(\cos\omega t + \frac{n}{\omega}\sin\omega t\right)\right];$$

$$\frac{dx}{dt} = Ve^{-nt}\left(\cos\omega t - \frac{n}{\omega}\sin\omega t\right) + \frac{p-f-r}{\omega}e^{-nt}\sin\omega t;$$

$$\frac{d^2x}{dt^2} = \frac{V}{\omega}e^{-nt}[(n^2-\omega^2)\sin\omega t - 2n\omega\cos\omega t]$$

$$+ (p-f-r)e^{-nt}\left(\cos\omega t - \frac{n}{\omega}\sin\omega t\right).$$

5. Determine the basic parameters of motion of the system shown in Figure 17.3.1. The system is restricted by a pair of flexible and fluid links in parallel and is moving on a horizontal surface while being subjected to a dry friction force, a constant resisting force, and a harmonic force. The system performs overdamped motion. The air resistance is negligible. The initial conditions of motion are as follows: for $t = 0$, $x = 0$; $\frac{dx}{dt} = 0$.

Answer:

$$x = \frac{V}{\omega}e^{-nt}\sinh\omega t + \frac{a}{4n^2\varphi^2 + (n^2-\omega^2-\varphi^2)^2}$$

$$\times [(n^2-\omega^2-\varphi^2)\left(\cos\varphi t - e^{-nt}\cosh\omega t\right)$$

$$+ 2n\varphi\sin\varphi t - \frac{n}{\omega}(n^2-\omega^2+\varphi^2)e^{-nt}\sinh\omega t]$$

$$- \frac{f+r}{n^2-\omega^2}\left[1 - e^{-nt}\left(\cosh\omega t - \frac{n}{\omega}\sinh\omega t\right)\right];$$

$$\frac{dx}{dt} = Ve^{-nt}\left(\cos\omega t - \frac{n}{\omega}\sinh\omega t\right) + \frac{a}{4n^2\varphi^2 + (n^2-\omega^2-\varphi^2)^2}$$

$$\times \left\{2n\varphi^2(\cos\varphi t - e^{-nt}\cosh\omega t) - \varphi(n^2-\omega^2-\varphi^2)\sin\varphi t\right.$$

$$+\frac{1}{\omega}e^{-nt}[(n^2-\omega^2)^2+\varphi^2(\omega^2+n^2)]\sinh\omega t\Big\}$$

$$-\frac{f+r}{\omega}e^{-nt}\sinh\omega t;$$

$$\frac{d^2x}{dt^2}=\frac{V}{\omega}e^{-nt}[(n^2+\omega^2)\sinh\omega t-2n\omega\cosh\omega t]$$

$$+\frac{a}{4n^2\varphi^2+(n^2-\omega^2-\varphi^2)^2}\{[(n^2-\omega^2)^2$$

$$+\varphi^2(3n^2+\omega^2)]e^{-nt}\cosh\omega t-\varphi^2(n^2-\omega^2-\varphi^2)\cos\varphi t$$

$$-2\varphi^3 n\sin\varphi t-\frac{n}{\omega}[(n^2-\omega^2)^2-\varphi^2(n^2+3\omega^2)]e^{-nt}\sinh\omega t\}$$

$$-(f+r)\left(\cosh\omega t-\frac{n}{\omega}\sinh\omega t\right).$$

6. Determine the basic parameters of motion of the system shown in Figure 17.3.1. The system is restricted by a pair of flexible and fluid links in parallel and is moving on a horizontal surface while being subjected to a dry friction force and a harmonic force. The system performs overdamped motion. The air resistance is negligible. The initial conditions of motion are as follows: for $t=0$, $x=S$; $\frac{dx}{dt}=0$.

7. Determine the basic parameters of motion of the system shown in Figure 17.3.1. The system is restricted by a pair of flexible and fluid links in parallel and is moving on a horizontal surface while being subjected to a dry friction force, constant resisting force, and a harmonic force. The system performs overdamped motion. The air resistance is negligible. The initial conditions of motion are as follows: for $t=0$, $x=0$; $\frac{dx}{dt}=V$.

8. Determine the basic parameters of motion of the system shown in Figure 17.4.1. The system is restricted by a pair of flexible and fluid links in parallel and is moving on a horizontal surface while being subjected to a dry friction force, a constant resisting

force, a constant active force, and a harmonic force. The system performs overdamped motion. The air resistance is negligible. The initial conditions of motion are as follows: for $t = 0$, $x = 0$; $\frac{dx}{dt} = 0$.

Answer:

$$x = \frac{a}{4n^2\varphi^2 + (n^2 - \omega^2 - \varphi^2)^2}\left[(n^2 - \omega^2 - \varphi^2)\right.$$

$$\times (\cos\varphi t - e^{-nt}\cosh\omega t) + 2n\varphi\sin\varphi t - \frac{n}{\omega}(n^2 - \omega^2 + \varphi^2)$$

$$\left. \times\ e^{-nt}\sinh\omega t\right] + \frac{p - f - r}{n^2 - \omega^2}$$

$$\times \left[1 - e^{-nt}\left(\cosh\omega t - \frac{n}{\omega}\sinh\omega t\right)\right]$$

$$\frac{dx}{dt} = \frac{a}{4n^2\varphi^2 + (n^2 - \omega^2 - \varphi^2)^2}\left\{2n\varphi^2(\cos\varphi t - e^{-nt}\cosh\omega t)\right.$$

$$-\varphi(n^2 - \omega^2 - \varphi^2)\sin\varphi t + \frac{1}{\omega}e^{-nt}[(n^2 - \omega^2)^2$$

$$\left. +\varphi^2(\omega^2 + n^2)]\sinh\omega t\right\} + \frac{p - f - r}{\omega}e^{-nt}\sinh\omega t$$

$$\frac{d^2x}{dt^2} = \frac{a}{4n^2\varphi^2 + (n^2 - \omega^2 - \varphi^2)^2}\left\{[(n^2 - \omega^2)^2 + \varphi^2(3n^2 + \omega^2)]\right.$$

$$\times e^{-nt}\cosh\omega t - \varphi^2(n^2 - \omega^2 + -\varphi^2)\cos\varphi t - 2\varphi^3 n\sin\varphi t$$

$$\left. -\frac{n}{\omega}[(n^2 - \omega^2)^2 - \varphi^2(n^2 + 3\omega^2)]e^{-nt}\sinh\omega t\right\}$$

$$+(p - f - r)e^{-nt}\left(\cosh\omega t - \frac{n}{\omega}\sinh\omega t\right).$$

9. Determine the basic parameters of motion of the system shown in Figure 17.4.1. The system is restricted by a pair of flexible and fluid links in parallel and is moving on a horizontal surface while being subjected to a dry friction force, a constant resisting force, a constant active force, and a harmonic force. The system performs overdamped motion. The air resistance is negligible.

The initial conditions of motion are as follows: for $t = 0$, $x = S$; $\frac{dx}{dt} = 0$.

10. Determine the basic parameters of motion of the system shown in Figure 17.4.1. The system is restricted by a pair of flexible and fluid links in parallel and is moving on a horizontal surface while being subjected to a dry friction force, constant resisting force, constant active force, and a harmonic force. The system performs overdamped motion. The air resistance is negligible. The initial conditions of motion are as follows: for $t = 0$, $x = 0$; $\frac{dx}{dt} = V$.

Answer:

$$x = \frac{V}{\omega} e^{-nt} \sinh \omega t + \frac{p - f - r}{\omega^2 + n^2}$$

$$\times \left[1 - e^{-nt} \left(\cosh \omega t + \frac{n}{\omega} \sinh \omega t \right) \right]$$

$$+ \frac{a}{4n^2 \varphi^2 + (n^2 - \omega^2 - \varphi^2)^2}$$

$$\times \left[(n^2 - \omega^2 - \varphi^2)(\cos \varphi t - e^{-nt} \cosh \omega t) + 2n\varphi \sin \varphi t \right.$$

$$\left. - \frac{n}{\omega}(n^2 - \omega^2 + \varphi^2) e^{-nt} \sinh \omega t \right]$$

$$\frac{dx}{dt} = V e^{-nt} \left(\cosh \omega t - \frac{n}{\omega} \sinh \omega t \right) + \frac{p - f - r}{\omega} e^{-nt} \sinh \omega t$$

$$+ \frac{a}{4n^2 \varphi^2 + (n^2 - \omega^2 - \varphi^2)^2} \left\{ 2n\varphi^2 (\cos \varphi t - e^{-nt} \cosh \omega t) \right.$$

$$- \varphi(n^2 - \omega^2 - \varphi^2) \sin \varphi t + \frac{1}{\omega} e^{-nt} [(n^2 - \omega^2)^2$$

$$\left. + \varphi^2(\omega^2 + n^2)] \sinh \omega t \right\}$$

$$\frac{d^2x}{dt^2} = \frac{V}{\omega} e^{-nt} [(n^2 + \omega^2) \sinh \omega t - 2n\omega \cosh \omega t]$$

$$+ (p - f - r) e^{-nt} \left(\cosh \omega t - \frac{n}{\omega} \sinh \omega t \right)$$

$$+ \frac{a}{4n^2\varphi^2 + (\omega^2 + n^2 - \varphi^2)^2}$$

$$\times \left\{ [(n^2 - \omega^2)^2 + \varphi^2(3n^2 + \omega^2)] e^{-nt} \cosh \omega t \right.$$

$$- \varphi^2(n^2 - \omega^2 - \varphi^2) \cos \varphi t - 2\varphi^3 n \sin \varphi t$$

$$\left. - \frac{n}{\omega} [(n^2 - \omega^2)^2 - \varphi^2(n^2 + 3\omega^2)] e^{-nt} \sinh \omega t \right\}$$

CHAPTER 19

PROBLEMS

---

1. Determine the basic parameters of motion of the masses
   of the unrestricted two-degree-of-freedom system shown in
   Figure 19.1.1. The masses are connected to each other by a
   flexible link and are moving on a horizontal surface due to the
   initial conditions of motion of the masses. The air resistance is
   negligible. The initial conditions of motion are as follows:

$$x_1 = S_1; \quad \frac{dx_1}{dt} = 0;$$

for $t = 0$

$$x_2 = S_2; \quad \frac{dx_2}{dt} = 0$$

Answer:

$$x_1 = S_1 \cos \omega t + \frac{S_1 \omega_2^2 + S_2 \omega_1^2}{\omega^2}(1 - \cos \omega t)$$

$$x_2 = S_2 \cos \omega t + \frac{S_2 \omega_1^2 + S_1 \omega_2^2}{\omega^2}(1 - \cos \omega t)$$

2. Determine the basic parameters of motion of the masses
   of the unrestricted two-degree-of-freedom system shown in
   Figure 19.1.1. The masses are connected to each other by a
   flexible link and are moving on a horizontal surface due to the

initial conditions of motion the masses. The air resistance is negligible. The initial conditions of motion are as follows:

$$x_1 = 0; \quad \frac{dx_1}{dt} = V_1;$$

for $t = 0$

$$x_2 = 0; \quad \frac{dx_2}{dt} = V_2$$

Answer:

$$x_1 = \frac{V_1}{\omega} \sin \omega t + \frac{V_1 \omega_2^2 + V_2 \omega_1^2}{\omega^2} \left(t - \frac{1}{\omega} \sin \omega t \right)$$

$$x_2 = \frac{V_2}{\omega} \sin \omega t + \frac{V_2 \omega_1^2 + V_1 \omega_2^2}{\omega^2} \left(t - \frac{1}{\omega} \sin \omega t \right)$$

3. Determine the basic parameters of motion of the masses of the unrestricted two-degree-of-freedom system shown in Figure 19.1.1. The masses are connected to each other by a flexible link and are moving on a horizontal surface due to the initial conditions of motion of the masses. The air resistance is negligible. The initial conditions of motion are as follows:

$$x_1 = S_1; \quad \frac{dx_1}{dt} = 0;$$

for $t = 0$

$$x_2 = 0; \quad \frac{dx_2}{dt} = V_2$$

Answer:

$$x_1 = \frac{V_2 \omega_1^2}{\omega^2} \left(t - \frac{1}{\omega} \sin \omega t \right) + S_1 \cos \omega t + \frac{S_1 \omega_2^2}{\omega^2} (1 - \cos \omega t)$$

$$x_2 = \frac{V_2}{\omega} \sin \omega t + \frac{V_2 \omega_1^2}{\omega^2} \left(t - \frac{1}{\omega} \sin \omega t \right) + \frac{S_1 \omega_2^2}{\omega^2} (1 - \cos \omega t)$$

4. Determine the basic parameters of motion of the masses of the unrestricted two-degree-of-freedom system shown in Figure 19.1.1. The masses are connected to each other by a flexible link and are moving on a horizontal surface due to the initial conditions of motion of the masses. The air resistance is negligible. The initial conditions of motion are as follows:

$$x_1 = S_1; \quad \frac{dx_1}{dt} = V_1;$$

for $t = 0$

$$x_2 = 0; \quad \frac{dx_2}{dt} = 0$$

Answer:

$$x_1 = \frac{V_1}{\omega}\sin\omega t + \frac{V_1\omega_2^2}{\omega^2}\left(t - \frac{1}{\omega}\sin\omega t\right) + S_1\cos\omega t$$

$$+ \frac{S_1\omega_2^2}{\omega^2}(1 - \cos\omega t)$$

$$x_2 = \left(t - \frac{1}{\omega}\sin\omega t\right) + \frac{S_1\omega_2^2}{\omega^2}(1 - \cos\omega t)$$

5. Determine the basic parameters of motion of the masses of the unrestricted two-degree-of-freedom system shown in Figure 19.2.1. The masses are connected to each other by a flexible link and are moving on a horizontal surface while being subjected to dry friction, constant resisting, and constant active force. The air resistance is negligible. The initial conditions of motion are as follows:

$$x_1 = S_1; \quad \frac{dx_1}{dt} = 0;$$

for $t = 0$

$$x_2 = S_2; \quad \frac{dx_2}{dt} = 0$$

Answer:

$$x_1 = S_1 \cos \omega t + \frac{S_1 \omega_2^2 + S_2 \omega_1^2}{\omega^2}(1 - \cos \omega t) + \frac{p_1 - f_1 - r_1}{\omega^2}$$

$$\times (1 - \cos \omega t) + [\omega_2^2(p_1 - f_1 - r_1) + \omega_1^2(p_2 - f_2 - r_2)]$$

$$\times \left[\frac{1}{\omega^4}(\cos \omega t - 1) + \frac{1}{2\omega^2}t^2\right]$$

$$x_2 = S_2 \cos \omega t + \frac{S_2 \omega_1^2 + S_1 \omega_2^2}{\omega^2}(1 - \cos \omega t) + \frac{p_2 - f_2 - r_2}{\omega^2}$$

$$\times (1 - \cos \omega t) + [\omega_1^2(p_2 - f_2 - r_2) + \omega_2^2(p_1 - f_1 - r_1)]$$

$$\times \left[\frac{1}{\omega^4}(\cos \omega t - 1) + \frac{1}{2\omega^2}t^2\right]$$

6. Determine the basic parameters of motion of the masses of the unrestricted two-degree-of-freedom system shown in Figure 19.2.1. The masses are connected to each other by a flexible link and are moving on a horizontal surface while being subjected to dry friction, constant resisting, and constant active forces. The air resistance is negligible. The initial conditions of motion are as follows:

$$x_1 = 0; \quad \frac{dx_1}{dt} = V_1;$$

for $t = 0$

$$x_2 = 0; \quad \frac{dx_2}{dt} = V_2$$

Answer:

$$x_1 = \frac{V_1}{\omega}\sin \omega t + \frac{V_1\omega_2^2 + V_2\omega_1^2}{\omega^2}\left(t - \frac{1}{\omega}\sin \omega t\right) + \frac{p_1 - f_1 - r_1}{\omega^2}$$

$$\times (1 - \cos \omega t) + [\omega_2^2(p_1 - f_1 - r_1) + \omega_1^2(p_2 - f_2 - r_2)]$$

$$\times \left[\frac{1}{\omega^4}(\cos \omega t - 1) + \frac{1}{2\omega^2}t^2\right]$$

$$x_2 = \frac{V_2}{\omega}\sin \omega t + \frac{V_2\omega_1^2 + V_1\omega_2^2}{\omega^2}\left(t - \frac{1}{\omega}\sin \omega t\right) + \frac{p_2 - f_2 - r_2}{\omega^2}$$

$$\times (1 - \cos \omega t) + [\omega_1^2(p_2 - f_2 - r_2) + \omega_2^2(p_1 - f_1 - r_1)]$$

$$\times \left[\frac{1}{\omega^4}(\cos \omega t - 1) + \frac{1}{2\omega^2}t^2 \right]$$

7. Determine the basic parameters of motion of the masses of the unrestricted two-degree-of-freedom system shown in Figure 19.2.1. The masses are connected to each other by a flexible link and are moving on a horizontal surface while being subjected to dry friction, constant resisting, and constant active forces. The air resistance is negligible. The initial conditions of motion are as follows:

$$x_1 = S_1; \quad \frac{dx_1}{dt} = 0;$$

for $t = 0$

$$x_2 = C; \quad \frac{dx_2}{dt} = V_2$$

Answer:

$$x_1 = \frac{V_2\omega_1^2}{\omega^2}\left(t - \frac{1}{\omega}\sin \omega t\right) + S_1 \cos \omega t + \frac{S_1\omega_2^2 + S_2\omega_1^2}{\omega^2}$$

$$\times (1 - \cos \omega t) + \frac{p_1 - f_1 - r_1}{\omega^2}(1 - \cos \omega t)$$

$$+ [\omega_2^2(p_1 - f_1 - r_1) + \omega_1^2(p_2 - f_2 - r_2)]$$

$$\times \left[\frac{1}{\omega^4}(\cos \omega t - 1) + \frac{1}{2\omega^2}t^2 \right]$$

$$x_2 = \frac{V_2}{\omega}\sin \omega t + \frac{V_2\omega_1^2}{\omega^2}\left(t - \frac{1}{\omega}\sin \omega t\right) + \frac{S_1\omega_2^2}{\omega^2}(1 - \cos \omega t)$$

$$+ \frac{p_2 - f_2 - r_2}{\omega^2}(1 - \cos \omega t) + [\omega_1^2(p_2 - f_2 - r_2)$$

$$+ \omega_2^2(p_1 - f_1 - r_1)] \left[\frac{1}{\omega^4}(\cos \omega t - 1) + \frac{1}{2\omega^2}t^2 \right]$$

8. Determine the basic parameters of motion of the masses of the unrestricted two-degree-of-freedom system shown in Figure 19.3.1. The masses are connected to each other by a flexible link and are moving on a horizontal surface while being subjected to harmonic forces. The air resistance is negligible. The initial conditions of motion are as follows:

$$x_1 = S_1; \quad \frac{dx_1}{dt} = V_1;$$

for $t = 0$

$$x_2 = 0; \quad \frac{dx_2}{dt} = 0$$

Answer:

$$x_1 = \frac{V_1}{\omega}\sin\omega t + \frac{V_1\omega_2^2 + V_2\omega_1^2}{\omega^2}\left(t - \frac{1}{\omega}\sin\omega t\right) + S_1\cos\omega t$$

$$+ \frac{S_1\omega_2^2 + S_2\omega_1^2}{\omega^2}(1 - \cos\omega t) + a_1\frac{\cos\omega t - \cos\varphi_1 t}{\varphi_1^2 - \omega^2}$$

$$+ \frac{a_1\omega_2^2}{\varphi_1^2 - \omega^2}\left[\frac{1}{\omega^2}(1 - \cos\omega t) - \frac{1}{\varphi_1^2}(1 - \cos\varphi_1 t)\right] + \frac{a_2\omega_1^2}{\varphi_2^2 - \omega^2}$$

$$\times \left[\frac{1}{\omega^2}(1 - \cos\omega t) - \frac{1}{\varphi_2^2}(1 - \cos\varphi_2 t)\right]$$

$$x_2 = \frac{V_1\omega_2^2}{\omega^2}\left(t - \frac{1}{\omega}\sin\omega t\right) + \frac{S_1\omega_2^2}{\omega^2}(1 - \cos\omega t)$$

$$+ a_2\frac{\cos\omega t - \cos\varphi_2 t}{\varphi_2^2 - \omega^2} + \frac{a_2\omega_1^2}{\varphi_2^2 - \omega^2}$$

$$\times \left[\frac{1}{\omega^2}(1 - \cos\omega t) - \frac{1}{\varphi_2^2}(1 - \cos\varphi_2 t)\right]$$

$$+ \frac{a_1\omega_2^2}{\varphi_1^2 - \omega^2}\left[\frac{1}{\omega^2}(1 - \cos\omega t) - \frac{1}{\varphi_1^2}(1 - \cos\varphi_1 t)\right]$$

9. Determine the basic parameters of motion of the masses of the unrestricted two-degree-of-freedom system shown in Figure 19.3.1. The masses are connected to each other by a flexible link and are moving on a horizontal surface while being

subjected to harmonic forces. The air resistance is negligible. The initial conditions of motion are as follows:

$$x_1 = S_1; \quad \frac{dx_1}{dt} = 0;$$

for $t = 0$

$$x_2 = S_2; \quad \frac{dx_2}{dt} = 0$$

Answer:

$$x_1 = S_1 \cos \omega t + \frac{S_1 \omega_2^2 + S_2 \omega_1^2}{\omega^2}(1 - \cos \omega t) + a_1 \frac{\cos \omega t - \cos \varphi_1 t}{\varphi_1^2 - \omega^2}$$

$$+ \frac{a_1 \omega_2^2}{\varphi_1^2 - \omega^2}\left[\frac{1}{\omega^2}(1 - \cos \omega t) - \frac{1}{\varphi_1^2}(1 - \cos \varphi_1 t)\right]$$

$$+ \frac{a_2 \omega_1^2}{\varphi_2^2 - \omega^2}\left[\frac{1}{\omega^2}(1 - \cos \omega t) - \frac{1}{\varphi_2^2}(1 - \cos \varphi_2 t)\right]$$

$$x_2 = S_2 \cos \omega t + \frac{S_2 \omega_1^2 + S_1 \omega_2^2}{\omega^2}(1 - \cos \omega t) + a_2 \frac{\cos \omega t - \cos \varphi_2 t}{\varphi_2^2 - \omega^2}$$

$$+ \frac{a_2 \omega_1^2}{\varphi_2^2 - \omega^2}\left[\frac{1}{\omega^2}(1 - \cos \omega t) - \frac{1}{\varphi_2^2}(1 - \cos \varphi_2 t)\right]$$

$$+ \frac{a_1 \omega_2^2}{\varphi_1^2 - \omega^2}\left[\frac{1}{\omega^2}(1 - \cos \omega t) - \frac{1}{\varphi_1^2}(1 - \cos \varphi_1 t)\right]$$

10. Determine the basic parameters of motion of the masses of the unrestricted two-degree-of-freedom system shown in Figure 19.3.1. The masses are connected to each other by a flexible link and are moving on a horizontal surface while being subjected to harmonic forces. The air resistance is negligible. The initial conditions of motion are as follows:

$$x_1 = 0; \quad \frac{dx_1}{dt} = V_1;$$

for $t = 0$

$$x_2 = 0; \quad \frac{dx_2}{dt} = V_2$$

Answer:

$$x_1 = \frac{V_1}{\omega}\sin\omega t + \frac{V_1\omega_2^2 + V_2\omega_1^2}{\omega^2}\left(t - \frac{1}{\omega}\sin\omega t\right)$$

$$+ a_1\frac{\cos\omega t - \cos\varphi_1 t}{\varphi_1^2 - \omega^2} + \frac{a_1\omega_2^2}{\varphi_1^2 - \omega^2}$$

$$\times\left[\frac{1}{\omega^2}(1 - \cos\omega t) - \frac{1}{\varphi_1^2}(1 - \cos\varphi_1 t)\right]$$

$$+ \frac{a_2\omega_1^2}{\varphi_2^2 - \omega^2}\left[\frac{1}{\omega^2}(1 - \cos\omega t) - \frac{1}{\varphi_2^2}(1 - \cos\varphi_2 t)\right]$$

$$x_2 = \frac{V_2}{\omega}\sin\omega t + \frac{V_2\omega_1^2 + V_1\omega_2^2}{\omega^2}\left(t - \frac{1}{\omega}\sin\omega t\right)$$

$$+ a_2\frac{\cos\omega t - \cos\varphi_2 t}{\varphi_2^2 - \omega^2} + \frac{a_2\omega_1^2}{\varphi_2^2 - \omega^2}$$

$$\times\left[\frac{1}{\omega^2}(1 - \cos\omega t) - \frac{1}{\varphi_2^2}(1 - \cos\varphi_2 t)\right]$$

$$+ \frac{a_1\omega_2^2}{\varphi_1^2 - \omega^2}\left[\frac{1}{\omega^2}(1 - \cos\omega t) - \frac{1}{\varphi_1^2}(1 - \cos\varphi_1 t)\right]$$

CHAPTER 20

PROBLEMS

---

1. Determine the basic parameters of motion of the masses
   of the unrestricted two-degree-of-freedom system shown in
   Figure 20.1.1. The masses are connected to each other by a fluid
   link and are moving on a horizontal surface due to the initial
   velocity of the masses. The air resistance is negligible. The initial
   conditions of motion are as follows:

   $$x_1 = 0; \quad \frac{dx_1}{dt} = V_1;$$

   for $t = 0$

   $$x_2 = 0; \quad \frac{dx_2}{dt} = V_2$$

   Answer:

   $$x_1 = \frac{V_1}{2n}(1 - e^{-2nt}) + \frac{V_1 n_2 + V_2 n_1}{n}\left[t + \frac{1}{2n}(e^{-2nt} - 1)\right]$$

   $$x_2 = \frac{V_2}{2n}(1 - e^{-2nt}) + \frac{V_2 n_1 + V_1 n_2}{n}\left[t + \frac{1}{2n}(e^{-2nt} - 1)\right]$$

2. Determine the basic parameters of motion of the masses
   of the unrestricted two-degree-of-freedom system shown in
   Figure 20.1.1. The masses are connected to each other by a fluid
   link and are moving on a horizontal surface while being subjected

to constant resisting, dry friction, and constant active forces. The air resistance is negligible. The initial conditions of motion are as follows:

$$x_1 = 0; \quad \frac{dx_1}{dt} = V_1;$$

for $t = 0$

$$x_2 = 0; \quad \frac{dx_2}{dt} = 0$$

3. Determine the basic parameters of motion of the masses of the unrestricted two-degree-of-freedom system shown in Figure 20.2.1. The masses are connected to each other by a fluid link and are moving on a horizontal surface while being subjected to constant resisting, dry friction, and constant active forces. The air resistance is negligible. The initial conditions of motion are as follows:

$$x_1 = S_1; \quad \frac{dx_1}{dt} = 0;$$

for $t = 0$

$$x_2 = 0; \quad \frac{dx_2}{dt} = V_2$$

Answer:

$$x_1 = \frac{2nS_1}{2n}(1 - e^{-2nt}) + \frac{2n_1V_2 + p_1 - r_1 - f_1}{2n}$$

$$\times \left[t + \frac{1}{2n}(e^{-2nt} - 1)\right] + S_1 e^{-2nt} + \frac{1}{4n}2[n_2(p_1 - r_1 - f_1)$$

$$+ n_1(p_2 - r_2 - f_2)]\left[t^2 - \frac{1}{n}t - \frac{1}{2n^2}(e^{-2nt} - 1)\right]$$

$$x_2 = \frac{V_2}{2n}(1 - e^{-2nt}) + \frac{2n_1V_2 + p_2 - r_2 - f_2}{2n}\left[t + \frac{1}{2n}(e^{-2nt} - 1)\right]$$

$$+ \frac{1}{4n}2[n_1(p_2 - r_2 - f_2) + n_2(p_1 - r_1 - f_1)]$$

$$\times \left[t^2 - \frac{1}{n}t - \frac{1}{2n^2}(e^{-2nt} - 1)\right]$$

4. Determine the basic parameters of motion of the masses of the unrestricted two-degree-of-freedom system shown in Figure 20.2.1. The masses are connected to each other by a fluid link and are moving on a horizontal surface while being subjected to constant resisting, dry friction, and constant active forces. The air resistance is negligible. The initial conditions of motion are as follows:

$$x_1 = S_1; \quad \frac{dx_1}{dt} = V_1;$$

for $t = 0$

$$x_2 = 0; \quad \frac{dx_2}{dt} = 0$$

Answer:

$$x_1 = \frac{V_1 + 2nS_1}{2n}(1 - e^{-2nt}) + \frac{2n_2V_1 + p_1 - r_1 - f_1}{2n}$$

$$\times \left[t + \frac{1}{2n}(e^{-2nt} - 1) \right] + S_1 e^{-2nt} + \frac{1}{4n}2[n_2(p_1 - r_1 - f_1)$$

$$+ n_1(p_2 - r_2 - f_2)]\left[t^2 - \frac{1}{n}t - \frac{1}{2n^2}(e^{-2nt} - 1) \right]$$

$$x_2 = \frac{2n_2V_1 + p_2 - r_2 - f_2}{2n}\left[t + \frac{1}{2n}(e^{-2nt} - 1) \right]$$

$$+ \frac{1}{4n}2[n_1(p_2 - r_2 - f_2) + n_2(p_1 - r_1 - f_1)]$$

$$\times \left[t^2 - \frac{1}{n}t - \frac{1}{2n^2}(e^{-2nt} - 1) \right]$$

5. Determine the basic parameters of motion of the masses of the unrestricted two-degree-of-freedom system shown in Figure 20.2.1. The masses are connected to each other by a fluid link and are moving on a horizontal surface while being subjected to constant resisting, dry friction, and active forces. The air resistance is negligible. The initial conditions of motion are as

follows:

$$x_1 = S_1; \quad \frac{dx_1}{dt} = 0;$$

for $t = 0$

$$x_2 = 0; \quad \frac{dx_2}{dt} = 0$$

Answer:

$$x_1 = S_1(1 - e^{-2nt}) + \frac{p_1 - r_1 - f_1}{2n}\left[t + \frac{1}{2n}(e^{-2nt} - 1)\right]$$

$$+ S_1 e^{-2nt} + \frac{1}{4n}2[n_2(p_1 - r_1 - f_1) + n_1(p_2 - r_2 - f_2)]$$

$$\times \left[t^2 - \frac{1}{n}t - \frac{1}{2n^2}(e^{-2nt} - 1)\right]$$

$$x_2 = \frac{p_2 - r_2 - f_2}{2n}\left[t + \frac{1}{2n}(e^{-2nt} - 1)\right] + \frac{1}{4n}2[n_1(p_2 - r_2 - f_2)$$

$$+ n_2(p_1 - r_1 - f_1)]\left[t^2 - \frac{1}{n}t - \frac{1}{2n^2}(e^{-2nt} - 1)\right]$$

6. Determine the basic parameters of motion of the masses of the unrestricted two-degree-of-freedom system shown in Figure 20.2.1. The masses are connected to each other by a fluid link and are moving on a horizontal surface while being subjected to constant resisting, dry friction, and active forces. The air resistance is negligible. The initial conditions of motion are as follows:

$$x_1 = 0; \quad \frac{dx_1}{dt} = V_1;$$

for $t = 0$

$$x_2 = 0; \quad \frac{dx_2}{dt} = 0$$

Answer:

$$x_1 = \frac{V_1}{2n}(1 - e^{-2nt}) + \frac{2n_2 V_1 + p_1 - r_1 - f_1}{2n}\left[t + \frac{1}{2n}(e^{-2nt} - 1)\right]$$

$$+ \frac{1}{4n}2[n_2(p_1 - r_1 - f_1) + n_1(p_2 - r_2 - f_2)]$$

$$\times \left[t^2 - \frac{1}{n}t - \frac{1}{2n^2}(e^{-2nt} - 1)\right]$$

$$x_2 = \frac{2n_2 V_1 + p_2 - r_2 - f_2}{2n}\left[t + \frac{1}{2n}(e^{-2nt} - 1)\right]$$

$$+ \frac{1}{4n}2[n_1(p_2 - r_2 - f_2) + n_2(p_1 - r_1 - f_1)]$$

$$\times \left[t^2 - \frac{1}{n}t - \frac{1}{2n^2}(e^{-2nt} - 1)\right]$$

7. Determine the basic parameters of motion of the masses of the unrestricted two-degree-of-freedom system shown in Figure 20.2.1. The masses are connected to each other by a fluid link and are moving on a horizontal surface while being subjected to constant resisting, dry friction, and constant active forces. The air resistance is negligible. The initial conditions of motion are as follows:

$$x_1 = 0; \quad \frac{dx_1}{dt} = V_1;$$

for $t = 0$

$$x_2 = 0; \quad \frac{dx_2}{dt} = V_2$$

Answer:

$$x_1 = \frac{V_1}{2n}(1 - e^{-2nt}) + \frac{2(n_2 V_1 + n_1 V_2) + p_1 - r_1 - f_1}{2n}$$

$$\times \left[t + \frac{1}{2n}(e^{-2nt} - 1)\right] + \frac{1}{4n}2[n_2(p_1 - r_1 - f_1)$$

$$+ n_1(p_2 - r_2 - f_2)]\left[t^2 - \frac{1}{n}t - \frac{1}{2n^2}(e^{-2nt} - 1)\right]$$

$$x_2 = \frac{V_2}{2n}(1 - e^{-2nt}) + \frac{2(n_1 V_2 + n_2 V_1) + p_2 - r_2 - f_2}{2n}$$

$$\times \left[t + \frac{1}{2n}(e^{-2nt} - 1)\right] + \frac{1}{4n}2[n_1(p_2 - r_2 - f_2)$$

$$+ n_2(p_1 - r_1 - f_1)]\left[t^2 - \frac{1}{n}t - \frac{1}{2n^2}(e^{-2nt} - 1)\right]$$

8. Determine the basic parameters of motion of the masses of the unrestricted two-degree-of-freedom system shown in Figure 20.3.1. The masses are connected to each other by a fluid link and are moving on a horizontal surface while being subjected to harmonic forces. The air resistance is negligible. The initial conditions of motion are as follows:

$$x_1 = 0; \quad \frac{dx_1}{dt} = V_1;$$

for $t = 0$

$$x_2 = 0; \quad \frac{dx_2}{dt} = V_2$$

Answer:

$$x_1 = \frac{V_1}{2n}(1 - e^{-2nt}) + \frac{V_1 n_2 + V_2 n_1}{n}\left[t + \frac{1}{2n}(e^{-2nt} - 1)\right]$$

$$+ \frac{a_1}{\varphi_1^2 + 4n^2}\left(e^{-2nt} + \frac{2n}{\varphi_1}\sin\varphi_1 t - \cos\varphi_1 t\right) + \frac{2n_2 a_1}{\varphi_1^2 + 4n^2}$$

$$\times \left[\frac{2n}{\varphi_1^2}(1 - \cos\varphi_1 t) - \frac{1}{\varphi_1}\sin\varphi_1 t + \frac{1}{2n}(1 - e^{-2nt})\right]$$

$$+ \frac{2n_1 a_2}{\varphi_2^2 + 4n^2}\left[\frac{2n}{\varphi_2^2}(1 - \cos\varphi_2 t) - \frac{1}{\varphi_2}\sin\varphi_2 t + \frac{1}{2n}(1 - e^{-2nt})\right]$$

$$x_2 = \frac{V_2}{2n}(1 - e^{-2nt}) + \frac{V_2 n_1 + V_1 n_2}{n}\left[t + \frac{1}{2n}(e^{-2nt} - 1)\right]$$

$$+ \frac{a_2}{\varphi_2^2 + 4n^2}\left(e^{-2nt} + \frac{2n}{\varphi_2}\sin\varphi_2 t - \cos\varphi_2 t\right) + \frac{2n_1 a_2}{\varphi_2^2 + 4n^2}$$

$$\times \left[\frac{2n}{\varphi_2^2}(1 - \cos \varphi_2 t) - \frac{1}{\varphi_2}\sin \varphi_2 t + \frac{1}{2n}(1 - e^{-2nt}) \right]$$

$$+ \frac{2n_2 a_1}{\varphi_1^2 + 4n^2}\left[\frac{2n}{\varphi_1^2}(1 - \cos \varphi_1 t) - \frac{1}{\varphi_1}\sin \varphi_1 t + \frac{1}{2n}(1 - e^{-2nt}) \right]$$

9. Determine the basic parameters of motion of the masses of the unrestricted two-degree-of-freedom system shown in Figure 20.3.1. The masses are connected to each other by a fluid link and are moving on a horizontal surface while being subjected to harmonic forces. The air resistance is negligible. The initial conditions of motion are as follows:

$$x_1 = S_1; \quad \frac{dx_1}{dt} = 0;$$

for $t = 0$

$$x_2 = S_2; \quad \frac{dx_2}{dt} = 0$$

Answer:

$$x_1 = S_1 e^{-2nt} + S_1(1 - e^{-2nt}) + \frac{a_1}{\varphi_1^2 + 4n^2}$$

$$\times \left(e^{-2nt} + \frac{2n}{\varphi_1}\sin \varphi_1 t - \cos \varphi_1 t \right) + \frac{2n_2 a_1}{\varphi_1^2 + 4n^2}$$

$$\times \left[\frac{2n}{\varphi_1^2}(1 - \cos \varphi_1 t) - \frac{1}{\varphi_1}\sin \varphi_1 t + \frac{1}{2n}(1 - e^{-2nt}) \right]$$

$$+ \frac{2n_1 a_2}{\varphi_2^2 + 4n^2}\left[\frac{2n}{\varphi_2^2}(1 - \cos \varphi_2 t) - \frac{1}{\varphi_2}\sin \varphi_2 t + \frac{1}{2n}(1 - e^{-2nt}) \right]$$

$$x_2 = S_2 e^{-2nt} + S_2(1 - e^{-2nt}) + \frac{a_2}{\varphi_2^2 + 4n^2}$$

$$\times \left(e^{-2nt} + \frac{2n}{\varphi_2}\sin \varphi_2 t - \cos \varphi_2 t \right) + \frac{2n_1 a_2}{\varphi_2^2 + 4n^2}$$

$$\times \left[\frac{2n}{\varphi_2^2}(1 - \cos \varphi_2 t) - \frac{1}{\varphi_2} \sin \varphi_2 t + \frac{1}{2n}(1 - e^{-2nt}) \right]$$

$$+ \frac{2n_2 a_1}{\varphi_1^2 + 4n^2} \left[\frac{2n}{\varphi_1^2}(1 - \cos \varphi_1 t) - \frac{1}{\varphi_1} \sin \varphi_1 t + \frac{1}{2n}(1 - e^{-2nt}) \right]$$

10. Determine the basic parameters of motion of the masses of the unrestricted two-degree-of-freedom system shown in Figure 20.3.1. The masses are connected to each other by a fluid link and are moving on a horizontal surface while being subjected to harmonic forces. The air resistance is negligible. The initial conditions of motion are as follows:

$$x_1 = 0; \quad \frac{dx_1}{dt} = V_1;$$

for $t = 0$

$$x_2 = S_2; \quad \frac{dx_2}{dt} = 0$$

Answer:

$$x_1 = \frac{V_1}{2n}(1 - e^{-2nt}) + \frac{V_1 n_2}{n}\left[t + \frac{1}{2n}(e^{-2nt} - 1) \right] + \frac{a_1}{\varphi_1^2 + 4n^2}$$

$$\times \left(e^{-2nt} + \frac{2n}{\varphi_1} \sin \varphi_1 t - \cos \varphi_1 t \right) + \frac{2n_2 a_1}{\varphi_1^2 + 4n^2}$$

$$\times \left[\frac{2n}{\varphi_1^2}(1 - \cos \varphi_1 t) - \frac{1}{\varphi_1} \sin \varphi_1 t + \frac{1}{2n}(1 - e^{-2nt}) \right]$$

$$+ \frac{2n_1 a_2}{\varphi_2^2 + 4n^2} \left[\frac{2n}{\varphi_2^2}(1 - \cos \varphi_2 t) - \frac{1}{\varphi_2} \sin \varphi_2 t + \frac{1}{2n}(1 - e^{-2nt}) \right]$$

$$x_2 = S_2 e^{-2nt} + S_2(1 - e^{-2nt}) + \frac{V_1 n_2}{n}\left[t + \frac{1}{2n}(e^{-2nt} - 1) \right]$$

$$+ \frac{a_2}{\varphi_2^2 + 4n^2} \left(e^{-2nt} + \frac{2n}{\varphi_2} \sin \varphi_2 t - \cos \varphi_2 t \right) + \frac{2n_1 a_2}{\varphi_2^2 + 4n^2}$$

$$\times \left[\frac{2n}{\varphi_2^2}(1 - \cos \varphi_2 t) - \frac{1}{\varphi_2} \sin \varphi_2 t + \frac{1}{2n}(1 - e^{-2nt}) \right]$$

$$+ \frac{2n_2 a_1}{\varphi_1^2 + 4n^2} \left[\frac{2n}{\varphi_1^2}(1 - \cos \varphi_1 t) - \frac{1}{\varphi_1} \sin \varphi_1 t + \frac{1}{2n}(1 - e^{-2nt}) \right]$$

CHAPTER 21

PROBLEMS

---

1. Determine the basic parameters of motion of the masses
   of the unrestricted two-degree-of-freedom system shown in
   Figure 21.1.1. The masses are connected to each other by
   a flexible link and a fluid link in parallel and are moving
   on a horizontal surface, performing underdamped vibrations.
   The initial conditions of motion are as follows:

$$x_1 = 0; \quad \frac{dx_1}{dt} = V_1;$$

for $t = 0$

$$x_2 = 0; \quad \frac{dx_2}{dt} = V_2$$

Answer:

$$x_1 = \frac{V_1}{\omega} e^{-nt} \sin \omega t + \frac{2(V_1 n_2 + V_2 n_1)}{\omega^2 + n^2}$$

$$\times \left[1 - e^{-nt} \left(\cos \omega t + \frac{n}{\omega} \sin \omega t \right) \right] + \frac{V_1 \omega_2^2 + V_2 \omega_1^2}{\omega(\omega^2 + n^2)^2}$$

$$\times \{ (\omega + n^2)\omega t - 2n\omega - e^{-nt}[(\omega^2 - n^2) \sin \omega t - 2n\omega \cos \omega t] \}$$

$$x_2 = \frac{V_2}{\omega}e^{-nt}\sin\omega t + \frac{2(V_2 n_1 + V_1 n_2)}{\omega^2 + n^2}$$

$$\times \left[1 - e^{-nt}\left(\cos\omega t + \frac{n}{\omega}\sin\omega t\right)\right] + \frac{V_2\omega_1^2 + V_1\omega_2^2}{\omega(\omega^2 + n^2)^2}$$

$$\times \{(\omega^2 + n^2)\omega t - 2n\omega - e^{-nt}[(\omega^2 - n^2)\sin\omega t - 2n\omega\cos\omega t]\}$$

2. Determine the basic parameters of motion of the masses of the unrestricted two-degree-of-freedom system shown in Figure 21.1.1. The masses are connected to each other by a flexible link and a fluid link in parallel and are moving on a horizontal surface, performing critically damped motion. The initial conditions of motion are as follows:

$$x_1 = S_1; \quad \frac{dx_1}{dt} = V_1;$$

for $t = 0$

$$x_2 = 0; \quad \frac{dx_2}{dt} = 0$$

Answer:

$$x_1 = S_1(1 - nt)e^{-nt} + 2nS_1 te^{-nt} + \frac{S_1\omega_2^2}{n^2}[1 - e^{-nt}(1 + nt)]$$

$$+ \frac{2}{n^2}S_1(n_1\omega_2^2 - n_2\omega_1^2)\left[t - \frac{2}{n} + e^{-nt}\left(\frac{2}{n} + t\right)\right] + V_1 te^{-nt}$$

$$+ \frac{2V_1 n_2}{n^2}[1 - e^{-nt}(1 + nt)] + \frac{V_1\omega_2^2}{n^2}\left[t - \frac{2}{n} + e^{-nt}\left(\frac{2}{n} + t\right)\right]$$

$$x_2 = \frac{S_1\omega_2^2}{n^2}[1 - e^{-nt}(1 + nt)]$$

$$- \frac{2}{n^2}S_1(n_2\omega_1^2 - n_1\omega_2^2)\left[t - \frac{2}{n} + e^{-nt}\left(\frac{2}{n} + t\right)\right]$$

$$+ \frac{2V_1 n_2}{n^2}[1 - e^{-nt}(1 + nt)] + \frac{V_1\omega_2^2}{n^2}\left[t - \frac{2}{n} + e^{-nt}\left(\frac{2}{n} + t\right)\right]$$

3. Determine the basic parameters of motion of the masses of the unrestricted two-degree-of-freedom system shown in Figure 21.1.1. The masses are connected to each other by a flexible link and a fluid link in parallel and are moving on a horizontal surface, performing overdamped motion. The initial conditions of motion are as follows:

$$x_1 = S_1; \quad \frac{dx_1}{dt} = 0;$$

for $t = 0$

$$x_2 = 0; \quad \frac{dx_2}{dt} = V_2$$

Answer:

$$x_1 = S_1 e^{-nt}\left(\cosh \omega t - \frac{n}{\omega}\sinh \omega t\right) + 2nS_1 e^{-nt}\sinh \omega t$$

$$+ \frac{S_1\omega_2^2}{n^2 - \omega^2}\left[1 - e^{-nt}\left(\cosh \omega t + \frac{n}{\omega}\sinh \omega t\right)\right]$$

$$+ \frac{2S_1(n_1\omega_2^2 - n_2\omega_1^2)}{\omega(n^2 - \omega^2)^2}\{(n^2 - \omega^2)\omega t - 2n\omega$$

$$+ e^{-nt}[(\omega^2 + n^2)\sinh \omega t + 2n\omega \cosh \omega t]\}$$

$$+ \frac{2V_2 n_1}{n^2 - \omega^2}\left[1 - e^{-nt}\left(\cosh \omega t + \frac{n}{\omega}\sinh \omega t\right)\right]$$

$$+ \frac{V_2\omega_1^2}{\omega(n^2 - \omega^2)^2}\{(n^2 - \omega^2)\omega t - 2n\omega$$

$$+ e^{-nt}[(\omega^2 + n^2)\sinh \omega t + 2n\omega \cosh \omega t]\}$$

$$x_2 = \frac{S_1\omega_2^2}{n^2 - \omega^2}\left[1 - e^{-nt}\left(\cosh \omega t + \frac{n}{\omega}\sinh \omega t\right)\right]$$

$$- \frac{2S_1(n_2\omega_1^2 - n_1\omega_2^2)}{\omega(n^2 - \omega^2)^2}\{(n^2 - \omega^2)\omega t - 2n\omega$$

$$+ e^{-nt}[(\omega^2 + n^2)\sinh \omega t + 2n\omega \cosh \omega t]\}$$

$$+ \frac{V_2}{\omega}e^{-nt}\sinh \omega t + \frac{2V_2 n_1}{n^2 - \omega^2}$$

$$\times \left[1 - e^{-nt}\left(\cosh \omega t + \frac{n}{\omega}\sinh \omega t\right)\right]$$

$$+ \frac{V_2\omega_1^2}{\omega(n^2 - \omega^2)^2}\{(n^2 - \omega^2)\omega t - 2n\omega$$

$$+ e^{-nt}[(\omega^2 + n^2)\sinh \omega t + 2n\omega \cosh \omega t]\}$$

4. Determine the basic parameters of motion of the masses of the unrestricted two-degree-of-freedom system shown in Figure 21.2.1. The masses are connected to each other by a flexible link and a fluid link in parallel and are moving on a horizontal surface, performing underdamped motion. The masses are subjected to constant resisting, dry friction, and constant active forces. The initial conditions of motion are as follows:

$$x_1 = S_1; \quad \frac{dx_1}{dt} = 0;$$

for $t = 0$

$$x_2 = S_2; \quad \frac{dx_2}{dt} = 0$$

Answer:

$$x_1 = S_1 e^{-nt}\left(\cos \omega t - \frac{n}{\omega}\sin \omega t\right) + \frac{p_1 - r_1 - f_1}{\omega^2 + n^2}$$

$$\times \left[1 - e^{-nt}\left(\cos \omega t + \frac{n}{\omega}\sin \omega t\right)\right]$$

$$+ \frac{2[n_2(p_1 - r_1 - f_1) + n_1(p_2 - r_2 - f_2)}{\omega(\omega^2 + n^2)^2}$$

$$\times \{(\omega^2 + n^2)\omega t - 2n\omega - e^{-nt}[(\omega^2 - n^2)\sin \omega t - 2n\omega \cos \omega t]\}$$

$$+ \frac{\omega_2^2(p_1 - r_1 - f_1) + \omega_1^2(p_2 - r_2 - f_2)}{\omega(\omega^2 + n^2)^2}\left\{\frac{\omega t^2}{2}(\omega^2 + n^2)\right.$$

$$- 2n\omega t - \frac{\omega(\omega^2 - n^2)}{\omega^2 + n^2}\left[1 - e^{-nt}\left(\cos \omega t + \frac{n}{\omega}\sin \omega t\right)\right]$$

$$\left. - \frac{2n^2\omega}{\omega^2 + n^2}\left[1 + e^{-nt}\left(\frac{\omega}{n}\sin \omega t - \cos \omega t\right)\right]\right\}$$

$$x_2 = S_2 e^{-nt}\left(\cos\omega t - \frac{n}{\omega}\sin\omega t\right) + \frac{p_2 - r_2 - f_2}{\omega^2 + n^2}$$

$$\times \left[1 - e^{-nt}\left(\cos\omega t + \frac{n}{\omega}\sin\omega t\right)\right]$$

$$+ \frac{2[n_1(p_2 - r_2 - f_2) + n_2(p_1 - r_1 - f_1)]}{\omega(\omega^2 + n^2)^2}\{(\omega^2 + n^2)\omega t$$

$$- 2n\omega - e^{-nt}[(\omega^2 - n^2)\sin\omega t - 2n\omega\cos\omega t]\}$$

$$+ \frac{\omega_1^2(p_2 - r_2 - f_2) + \omega_2^2(p_1 - r_1 - f_1)}{\omega(\omega^2 + n^2)^2}\left\{\frac{\omega t^2}{2}(\omega^2 + n^2)\right.$$

$$- 2n\omega t - \frac{\omega(\omega^2 - n^2)}{\omega^2 + n^2}\left[1 - e^{-nt}\left(\cos\omega t + \frac{n}{\omega}\sin\omega t\right)\right]$$

$$\left. - \frac{2n^2\omega}{\omega^2 + n^2}\left[1 + e^{-nt}\left(\frac{\omega}{n}\sin\omega t - \cos\omega t\right)\right]\right\}$$

5. Determine the basic parameters of motion of the masses of the unrestricted two-degree-of-freedom system shown in Figure 21.2.1. The masses are connected to each other by a flexible link and a fluid link in parallel and are moving on a horizontal surface, performing underdamped motion while being subjected to constant resisting, dry friction, and constant active forces. The initial conditions of motion are as follows:

$$x_1 = S_1; \quad \frac{dx_1}{dt} = 0;$$

for $t = 0$

$$x_2 = S_2; \quad \frac{dx_2}{dt} = 0$$

6. Determine the basic parameters of motion of the masses of the unrestricted two-degree-of-freedom system shown in Figure 21.2.1. The masses are connected to each other by a flexible link and a fluid link and are moving on a horizontal surface while being subjected to constant resisting, dry friction, and active forces and performing critically damped motion. The initial

conditions of motion are as follows:

$$x_1 = S_1; \quad \frac{dx_1}{dt} = 0;$$

for $t = 0$

$$x_2 = S_2; \quad \frac{dx_2}{dt} = 0$$

7. Determine the basic parameters of motion of the masses of the unrestricted two-degree-of-freedom system shown in Figure 21.2.1. The masses are connected to each other by a flexible link and a fluid link and are moving on a horizontal surface while being subjected to constant resisting, dry friction, and constant active forces and performing critically damped motion. The initial conditions of motion are as follows:

$$x_1 = 0; \quad \frac{dx_1}{dt} = V_1;$$

for $t = 0$

$$x_2 = 0; \quad \frac{dx_2}{dt} = V_2$$

Answer:

$$x_1 = \frac{V_1}{2n}(1 - e^{-2nt}) + \frac{2(n_2 V_1 + n_1 V_2) + p_1 - r_1 - f_1}{2n}$$
$$\times \left[t + \frac{1}{2n}(e^{-2nt} - 1) \right] + \frac{1}{4n} 2[n_2(p_1 - r_1 - f_1)$$
$$+ n_1(p_2 - r_2 - f_2)] \left[t^2 - \frac{1}{n}t - \frac{1}{2n^2}(e^{-2nt} - 1) \right]$$

$$x_2 = \frac{V_2}{2n}(1 - e^{-2nt}) + \frac{2(n_1 V_2 + n_2 V_1) + p_2 - r_2 - f_2}{2n}$$
$$\times \left[t + \frac{1}{2n}(e^{-2nt} - 1) \right] + \frac{1}{4n} 2[n_1(p_2 - r_2 - f_2)$$
$$+ n_2(p_1 - r_1 - f_1)] \left[t^2 - \frac{1}{n}t - \frac{1}{2n^2}(e^{-2nt} - 1) \right]$$

8. Determine the basic parameters of motion of the masses of the unrestricted two-degree-of-freedom system shown in Figure 21.3.1. The masses are connected to each other by a flexible link and a fluid link in parallel and are moving on a horizontal surface while being subjected to harmonic forces. The initial conditions of motion are as follows:

$$x_1 = 0; \quad \frac{dx_1}{dt} = V_1;$$

for $t = 0$

$$x_2 = 0; \quad \frac{dx_2}{dt} = V_2$$

Answer:

$$x_1 = \frac{V_1}{2n}(1 - e^{-2nt}) + \frac{V_1 n_2 + V_2 n_1}{n}\left[t + \frac{1}{2n}(e^{-2nt} - 1)\right]$$

$$+ \frac{a_1}{\varphi_1^2 + 4n^2}\left(e^{-2nt} + \frac{2n}{\varphi_1}\sin\varphi_1 t - \cos\varphi_1 t\right)$$

$$+ \frac{2n_2 a_1}{\varphi_1^2 + 4n^2}\left[\frac{2n}{\varphi_1^2}(1 - \cos\varphi_1 t) - \frac{1}{\varphi_1}\sin\varphi_1 t + \frac{1}{2n}(1 - e^{-2nt})\right]$$

$$+ \frac{2n_1 a_2}{\varphi_2^2 + 4n^2}\left[\frac{2n}{\varphi_2^2}(1 - \cos\varphi_2 t) - \frac{1}{\varphi_2}\sin\varphi_2 t + \frac{1}{2n}(1 - e^{-2nt})\right]$$

$$x_2 = \frac{V_2}{2n}(1 - e^{-2nt}) + \frac{V_2 n_1 + V_1 n_2}{n}\left[t + \frac{1}{2n}(e^{-2nt} - 1)\right]$$

$$+ \frac{a_2}{\varphi_2^2 + 4n^2}\left(e^{-2nt} + \frac{2n}{\varphi_2}\sin\varphi_2 t - \cos\varphi_2 t\right)$$

$$+ \frac{2n_1 a_2}{\varphi_2^2 + 4n^2}\left[\frac{2n}{\varphi_2^2}(1 - \cos\varphi_2 t) - \frac{1}{\varphi_2}\sin\varphi_2 t + \frac{1}{2n}(1 - e^{-2nt})\right]$$

$$+ \frac{2n_2 a_1}{\varphi_1^2 + 4n^2}\left[\frac{2n}{\varphi_1^2}(1 - \cos\varphi_1 t) - \frac{1}{\varphi_1}\sin\varphi_1 t + \frac{1}{2n}(1 - e^{-2nt})\right]$$

9. Determine the basic parameters of motion of the masses of the unrestricted two-degree-of-freedom system shown in

Figure 21.3.1. The masses are connected to each other by a flexible link and a fluid link and are moving on a horizontal surface while being subjected to harmonic forces. The initial conditions of motion are as follows:

$$x_1 = S_1; \quad \frac{dx_1}{dt} = 0;$$

for $t = 0$

$$x_2 = S_2; \quad \frac{dx_2}{dt} = 0$$

Answer:

$$x_1 = S_1 e^{-2nt} + S_1(1 - e^{-2nt})$$

$$+ \frac{a_1}{\varphi_1^2 + 4n^2} \left(e^{-2nt} + \frac{2n}{\varphi_1} \sin \varphi_1 t - \cos \varphi_1 t \right)$$

$$+ \frac{2n_2 a_1}{\varphi_1^2 + 4n^2} \left[\frac{2n}{\varphi_1^2} (1 - \cos \varphi_1 t) - \frac{1}{\varphi_1} \sin \varphi_1 t + \frac{1}{2n} (1 - e^{-2nt}) \right]$$

$$+ \frac{2n_1 a_2}{\varphi_2^2 + 4n^2} \left[\frac{2n}{\varphi_2^2} (1 - \cos \varphi_2 t) - \frac{1}{\varphi_2} \sin \varphi_2 t + \frac{1}{2n} (1 - e^{-2nt}) \right]$$

$$x_2 = S_2 e^{-2nt} + S_2(1 - e^{-2nt})$$

$$+ \frac{a_2}{\varphi_2^2 + 4n^2} \left(e^{-2nt} + \frac{2n}{\varphi_2} \sin \varphi_2 t - \cos \varphi_2 t \right)$$

$$+ \frac{2n_1 a_2}{\varphi_2^2 + 4n^2} \left[\frac{2n}{\varphi_2^2} (1 - \cos \varphi_2 t) - \frac{1}{\varphi_2} \sin \varphi_2 t + \frac{1}{2n} (1 - e^{-2nt}) \right]$$

$$+ \frac{2n_2 a_1}{\varphi_1^2 + 4n^2} \left[\frac{2n}{\varphi_1^2} (1 - \cos \varphi_1 t) - \frac{1}{\varphi_1} \sin \varphi_1 t + \frac{1}{2n} (1 - e^{-2nt}) \right]$$

10. Determine the basic parameters of motion of the masses of the unrestricted two-degree-of-freedom system shown in Figure 21.3.1. The masses are connected to each other by a flexible link and a fluid link and are moving on a horizontal surface while being subjected to harmonic forces. The initial

conditions of motion are as follows:

$$x_1 = 0; \quad \frac{dx_1}{dt} = V_1;$$

for $t = 0$

$$x_2 = S_2; \quad \frac{dx_2}{dt} = 0$$

Answer:

$$x_1 = \frac{V_1}{2n}(1 - e^{-2nt}) + \frac{V_1 n_2}{n}\left[t + \frac{1}{2n}(e^{-2nt} - 1)\right]$$

$$+ \frac{a_1}{\varphi_1^2 + 4n^2}\left(e^{-2nt} + \frac{2n}{\varphi_1}\sin\varphi_1 t - \cos\varphi_1 t\right)$$

$$+ \frac{2n_2 a_1}{\varphi_1^2 + 4n^2}\left[\frac{2n}{\varphi_1^2}(1 - \cos\varphi_1 t) - \frac{1}{\varphi_1}\sin\varphi_1 t + \frac{1}{2n}(1 - e^{-2nt})\right]$$

$$+ \frac{2n_1 a_2}{\varphi_2^2 + 4n^2}\left[\frac{2n}{\varphi_2^2}(1 - \cos\varphi_2 t) - \frac{1}{\varphi_2}\sin\varphi_2 t + \frac{1}{2n}(1 - e^{-2nt})\right]$$

$$x_2 = S_2 e^{-2nt} + S_2(1 - e^{-2nt}) + \frac{V_1 n_2}{n}\left[t + \frac{1}{2n}(e^{-2nt} - 1)\right]$$

$$+ \frac{a_2}{\varphi_2^2 + 4n^2}\left(e^{-2nt} + \frac{2n}{\varphi_2}\sin\varphi_2 t - \cos\varphi_2 t\right)$$

$$+ \frac{2n_1 a_2}{\varphi_2^2 + 4n^2}\left[\frac{2n}{\varphi_2^2}(1 - \cos\varphi_2 t) - \frac{1}{\varphi_2}\sin\varphi_2 t + \frac{1}{2n}(1 - e^{-2nt})\right]$$

$$+ \frac{2n_2 a_1}{\varphi_1^2 + 4n^2}\left[\frac{2n}{\varphi_1^2}(1 - \cos\varphi_1 t) - \frac{1}{\varphi_1}\sin\varphi_1 t + \frac{1}{2n}(1 - e^{-2nt})\right]$$

PROBLEMS

---

1. Determine the basic parameters of motion of the masses of the restricted two-degree-of-freedom system shown in Figure 22.1.1. The masses are connected to each other by a flexible link and each mass is attached to a non-movable support by an identical link. The masses are moving on a horizontal surface. The initial conditions of motion are as follows:

$$x_1 = 0; \quad \frac{dx_1}{dt} = V_1;$$

for $t = 0$

$$x_2 = 0; \quad \frac{dx_2}{dt} = V_2$$

Answer:

$$x_1 = \frac{V_1}{\omega} \sin \omega t + \frac{1}{\omega^2}[V_1(\omega_{12}^2 + \omega_{22}^2 + \omega_{32}^2)$$

$$+ V_2(\omega_{11}^2 + \omega_{21}^2 + \omega_{31}^2)]\left(t - \frac{1}{\omega}\sin \omega t\right)$$

$$x_2 = \frac{V_2}{\omega} \sin \omega t + \frac{1}{\omega^2}[V_2(\omega_{11}^2 + \omega_{21}^2 + \omega_{31}^2)$$

$$+ V_1(\omega_{12}^2 + \omega_{22}^2 + \omega_{32}^2)]\left(t - \frac{1}{\omega}\sin \omega t\right)$$

2. Determine the basic parameters of motion of the masses of the restricted two-degree-of-freedom system shown in Figure 22.1.1. The masses are connected to each other by a flexible link and each mass is attached to a non-movable support by an identical link. The masses are moving on a horizontal surface. The initial conditions of motion are as follows:

$$x_1 = S_1; \quad \frac{dx_1}{dt} = 0;$$

for $t = 0$

$$x_2 = S_2; \quad \frac{dx_2}{dt} = 0$$

Answer:

$$x_1 = S_1 \cos \omega t + \frac{1}{\omega^2}[S_1(\omega_{12}^2 + \omega_{22}^2 + \omega_{32}^2)$$

$$+ S_2(\omega_{11}^2 + \omega_{21}^2 + \omega_{31}^2)](1 - \cos \omega t)$$

$$x_2 = S_2 \cos \omega t + \frac{1}{\omega^2}[S_2(\omega_{11}^2 + \omega_{21}^2 + \omega_{31}^2)$$

$$+ S_1(\omega_{12}^2 + \omega_{22}^2 + \omega_{32}^2)](1 - \cos \omega t)$$

3. Determine the basic parameters of motion of the masses of the restricted two-degree-of-freedom system shown in Figure 22.1.1. The masses are connected to each other by a flexible link and each mass is attached to a non-movable support by an identical link. The masses are moving on a horizontal surface. The initial conditions of motion are as follows:

$$x_1 = S_1; \quad \frac{dx_1}{dt} = 0;$$

for $t = 0$

$$x_2 = 0; \quad \frac{dx_2}{dt} = V_2$$

Answer:

$$x_1 = \frac{1}{\omega^2} V_2(\omega_{11}^2 + \omega_{21}^2 + \omega_{31}^2)\left(t - \frac{1}{\omega}\sin\omega t\right)$$

$$+ S_1 \cos\omega t + \frac{1}{\omega^2} S_1(\omega_{12}^2 + \omega_{22}^2 + \omega_{32}^2)(1 - \cos\omega t)$$

$$x_2 = \frac{V_2}{\omega}\sin\omega t + \frac{1}{\omega^2} V_2(\omega_{11}^2 + \omega_{21}^2 + \omega_{31}^2)\left(t - \frac{1}{\omega}\sin\omega t\right)$$

$$+ \frac{1}{\omega^2} S_1(\omega_{12}^2 + \omega_{22}^2 + \omega_{32}^2)(1 - \cos\omega t)$$

4. Determine the basic parameters of motion of the masses of the restricted two-degree-of-freedom system shown in Figure 22.1.1. The masses are connected to each other by a flexible link and each mass is attached to a non-movable support by an identical link. The masses are moving on a horizontal surface. The initial conditions of motion are as follows:

$$x_1 = 0; \quad \frac{dx_1}{dt} = V_1;$$

for $t = 0$

$$x_2 = S_2; \quad \frac{dx_2}{dt} = 0$$

Answer:

$$x_1 = \frac{V_1}{\omega}\sin\omega t + \frac{1}{\omega^2} V_1(\omega_{12}^2 + \omega_{22}^2 + \omega_{32}^2)]\left(t - \frac{1}{\omega}\sin\omega t\right)$$

$$+ \frac{1}{\omega^2} S_2(\omega_{11}^2 + \omega_{21}^2 + \omega_{31}^2)(1 - \cos\omega t)$$

$$x_2 = V_1(\omega_{12}^2 + \omega_{22}^2 + \omega_{32}^2)\left(t - \frac{1}{\omega}\sin\omega t\right)$$

$$+ S_2 \cos\omega t + \frac{1}{\omega^2} S_2(\omega_{11}^2 + \omega_{21}^2 + \omega_{31}^2)(1 - \cos\omega t)$$

5. Determine the basic parameters of motion of the masses of the restricted two-degree-of-freedom system shown in Figure 22.1.1. The masses are connected to each other by a flexible link and each mass is attached to a non-movable support by an identical link. The masses are moving on a horizontal surface. The initial conditions of motion are as follows:

$$x_1 = S_1; \quad \frac{dx_1}{dt} = 0;$$

for $t = 0$

$$x_2 = S_2; \quad \frac{dx_2}{dt} = 0$$

Answer:

$$x_1 = S_1 \cos \omega t + \frac{1}{\omega^2}[S_1(\omega_{12}^2 + \omega_{22}^2 + \omega_{32}^2)$$
$$+ S_2(\omega_{11}^2 + \omega_{21}^2 + \omega_{31}^2)](1 - \cos \omega t)$$
$$x_2 = S_2 \cos \omega t + \frac{1}{\omega^2}[S_2(\omega_{11}^2 + \omega_{21}^2 + \omega_{31}^2)$$
$$+ S_1(\omega_{12}^2 + \omega_{22}^2 + \omega_{32}^2)](1 - \cos \omega t)$$

6. Determine the basic parameters of motion of the masses of the restricted two-degree-of-freedom system shown in Figure 22.1.1. The masses are connected to each other by a flexible link and each mass is attached to a non-movable support by an identical link. The masses are moving on a horizontal surface. The initial conditions of motion are as follows

$$x_1 = S_1; \quad \frac{dx_1}{dt} = 0;$$

for $t = 0$

$$x_2 = S_2; \quad \frac{dx_2}{dt} = 0$$

Answer:

$$x_1 = S_1 \cos \omega t + \frac{1}{\omega^2}[S_1(\omega_{12}^2 + \omega_{22}^2 + \omega_{32}^2)$$

$$+ S_2(\omega_{11}^2 + \omega_{21}^2 + \omega_{31}^2)](1 - \cos \omega t)$$

$$x_2 = S_2 \cos \omega t + \frac{1}{\omega^2}[S_2(\omega_{11}^2 + \omega_{21}^2 + \omega_{31}^2)$$

$$+ S_1(\omega_{12}^2 + \omega_{22}^2 + \omega_{32}^2)](1 - \cos \omega t)$$

7. Determine the basic parameters of motion of the masses of the restricted two-degree-of-freedom system shown in Figure 22.2.1. The masses are connected to each other by a flexible link and each mass is attached to a non-movable support by an identical link. The masses are moving on a horizontal surface, while only the first mass subjected to a constant resisting and a constant active force. The initial conditions of motion are as follows

$$x_1 = 0; \quad \frac{dx_1}{dt} = 0;$$

for $t = 0$

$$x_2 = 0; \quad \frac{dx_2}{dt} = 0$$

Answer:

$$x_1 = \frac{p_1 - r_1}{\omega^2}(1 - \cos \omega t) + [(\omega_{12}^2 + \omega_{22}^2 + \omega_{32}^2)(p_1 - r_1)$$

$$\times \left[\frac{1}{\omega^4}(\cos \omega t - 1) + \frac{1}{2\omega^2}t^2\right]$$

$$x_2 = (\omega_{12}^2 + \omega_{22}^2 + \omega_{32}^2)(p_1 - r_1)\left[\frac{1}{\omega^4}(\cos \omega t - 1) + \frac{1}{2\omega^2}t^2\right]$$

8. Determine the basic parameters of motion of the masses of the restricted two-degree-of-freedom system shown in Figure 22.2.1. The masses are connected to each other by a flexible link and each

mass is attached to a non-movable support by an identical link. The masses are moving on a horizontal surface, while only the second mass is subjected to a constant resisting and a constant active force. The initial conditions of motion are as follows

$$x_1 = 0; \quad \frac{dx_1}{dt} = 0;$$

for $t = 0$

$$x_2 = 0; \quad \frac{dx_2}{dt} = 0$$

Answer:

$$x_1 = \frac{p_1 - r_1}{\omega^2}(1 - \cos \omega t) + (\omega_{12}^2 + \omega_{22}^2 + \omega_{32}^2)(p_1 - r_1)$$

$$\times \left[\frac{1}{\omega^4}(\cos \omega t - 1) + \frac{1}{2\omega^2}t^2 \right]$$

$$x_2 = (\omega_{12}^2 + \omega_{22}^2 + \omega_{32}^2)(p_1 - r_1)] \left[\frac{1}{\omega^4}(\cos \omega t - 1) + \frac{1}{2\omega^2}t^2 \right]$$

9. Determine the basic parameters of motion of the masses of the restricted two-degree-of-freedom system shown in Figure 22.3.1. The masses are connected to each other by a flexible link and each mass is attached to a non-movable support by an identical link. The masses are moving on a horizontal surface, while only the first mass is subjected to a harmonic force. The initial conditions of motion are as follows

$$x_1 = 0; \quad \frac{dx_1}{dt} = 0;$$

for $t = 0$

$$x_2 = 0; \quad \frac{dx_2}{dt} = 0$$

Answer:

$$x_1 = a_1 \frac{\cos \omega t - \cos \varphi_1 t}{\varphi_1^2 - \omega^2} + \frac{a_1 (\omega_{12}^2 + \omega_{22}^2 + \omega_{32}^2)}{\varphi_1^2 - \omega^2}$$

$$\times \left[\frac{1}{\omega^2} (1 - \cos \omega t) - \frac{1}{\varphi_1^2} (1 - \cos \varphi_1 t) \right]$$

$$x_2 = \frac{a_1 (\omega_{12}^2 + \omega_{22}^2 + \omega_{32}^2)}{\varphi_1^2 - \omega^2} \left[\frac{1}{\omega^2} (1 - \cos \omega t) - \frac{1}{\varphi_1^2} (1 - \cos \varphi_1 t) \right]$$

10. Determine the basic parameters of motion of the masses of the restricted two-degree-of-freedom system shown in Figure 22.3.1. The masses are connected to each other by a flexible link and each mass is attached to a non-movable support by an identical link. The masses are moving on a horizontal surface, while only the second mass is subjected to a harmonic force. The initial conditions of motion are as follows:

$$x_1 = 0; \quad \frac{dx_1}{dt} = 0;$$

for $t = 0$

$$x_2 = 0; \quad \frac{dx_2}{dt} = 0$$

Answer:

$$x_1 = \frac{a_2 (\omega_{11}^2 + \omega_{21}^2 + \omega_{31}^2)}{\varphi_2^2 - \omega^2} \left[\frac{1}{\omega^2} (1 - \cos \omega t) - \frac{1}{\varphi_2^2} (1 - \cos \varphi_2 t) \right]$$

$$x_2 = a_2 \frac{\cos \omega t - \cos \varphi_2 t}{\varphi_2^2 - \omega^2} + \frac{a_2 (\omega_{11}^2 + \omega_{21}^2 + \omega_{31}^2)}{\varphi_2^2 - \omega^2}$$

$$\times \left[\frac{1}{\omega^2} (1 - \cos \omega t) - \frac{1}{\varphi_2^2} (1 - \cos \varphi_2 t) \right]$$

CHAPTER 23

PROBLEMS

---

1. Determine the basic parameters of motion of the masses of the restricted two-degree-of-freedom system shown in Figure 23.1.1. The masses are connected to each other by a fluid link and are moving on a horizontal surface. The initial conditions of motion are as follows:

$$x_1 = 0; \quad \frac{dx_1}{dt} = V_1;$$

for $t = 0$

$$x_2 = 0; \quad \frac{dx_2}{dt} = V_2$$

Answer:

$$x_1 = \frac{V_1}{2n}(1 - e^{-2nt}) + \frac{V_1(n_{12} + n_{22}) + V_2(n_{11} + n_{21})}{n}$$

$$\times \left[t + \frac{1}{2n}(e^{-2nt} - 1) \right]$$

$$x_2 = \frac{V_2}{2n}(1 - e^{-2nt}) + \frac{V_2(n_{11} + n_{21}) + V_1(n_{12} + n_{22})}{n}$$

$$\times \left[t + \frac{1}{2n}(e^{-2nt} - 1) \right]$$

2. Determine the basic parameters of motion of the masses of the restricted two-degree-of-freedom system shown in Figure 23.1.1. The masses are connected to each other by a fluid link and are moving on a horizontal surface. The initial conditions of motion are as follows:

$$x_1 = S_1; \quad \frac{dx_1}{dt} = V_1;$$

for $t = 0$

$$x_2 = 0; \quad \frac{dx_2}{dt} = 0$$

Answer:

$$x_1 = S_1 + \frac{V_1}{2n}(1 - e^{-2nt}) + \frac{V_1(n_{12} + n_{22})}{n}\left[t + \frac{1}{2n}(e^{-2nt} - 1)\right]$$

$$x_2 = \frac{V_1(n_{12} + n_{22})}{n}\left[t + \frac{1}{2n}(e^{-2nt} - 1)\right]$$

3. Determine the basic parameters of motion of the masses of the restricted two-degree-of-freedom system shown in Figure 23.1.1. The masses are connected to each other by a fluid link and are moving on a horizontal surface. The initial conditions of motion are as follows:

$$x_1 = S_1; \quad \frac{dx_1}{dt} = 0;$$

for $t = 0$

$$x_2 = 0; \quad \frac{dx_2}{dt} = V_2$$

Answer:

$$x_1 = S_1 + \frac{V_2(n_{11} + n_{21})}{n}\left[t + \frac{1}{2n}(e^{-2nt} - 1)\right]$$

$$x_2 = \frac{V_2}{2n}(1 - e^{-2nt}) + \frac{V_2(n_{11} + n_{21})}{n}\left[t + \frac{1}{2n}(e^{-2nt} - 1)\right]$$

4. Determine the basic parameters of motion of the masses of the restricted two-degree-of-freedom system shown in Figure 23.1.1. The masses are connected to each other by a fluid link and are moving on a horizontal surface. The initial conditions of motion are as follows:

$$x_1 = 0; \quad \frac{dx_1}{dt} = V_1;$$

for $t = 0$

$$x_2 = S_2; \quad \frac{dx_2}{dt} = 0$$

Answer:

$$x_1 = \frac{V_1}{2n}(1 - e^{-2nt}) + \frac{V_1(n_{12} + n_{22})}{n}\left[t + \frac{1}{2n}(e^{-2nt} - 1)\right]$$

$$x_2 = S_2 + \frac{V_1(n_{12} + n_{22})}{n}\left[t + \frac{1}{2n}(e^{-2nt} - 1)\right]$$

5. Determine the basic parameters of motion of the masses of the restricted two-degree-of-freedom system shown in Figure 23.2.1. The masses are connected to each other by a fluid link and are moving on a horizontal surface while being subjected to constant resisting, dry friction, and constant active forces. The initial conditions of motion are as follows:

$$x_1 = S_1; \quad \frac{dx_1}{dt} = 0;$$

for $t = 0$

$$x_2 = S_2; \quad \frac{dx_2}{dt} = 0$$

Answer:

$$x_1 = \frac{2nS_1}{2n}(1 - e^{-2nt}) + S_1 e^{-2nt}$$

$$+ \frac{(n_{12} + n_{22})(p_1 - r_1 - f_1) + (n_{11} + n_{21})(p_2 - r_2 - f_2)}{2n}$$

$$\times \left[t^2 - \frac{t}{n} - \frac{1}{2n^2}(e^{-2nt} - 1) \right]$$

$$x_2 = \frac{2nS_2}{2n}(1 - e^{-2nt}) + S_2 e^{-2nt}$$

$$+ \frac{(n_{11} + n_{21})(p_2 - r_2 - f_2) + (n_{12} + n_{22})(p_1 - r_1 - f_1)}{2n}$$

$$\times \left[t^2 - \frac{t}{n} - \frac{1}{2n^2}(e^{-2nt} - 1) \right]$$

6. Determine the basic parameters of motion of the masses of the restricted two-degree-of-freedom system shown in Figure 23.2.1. The masses are connected to each other by a fluid link and are moving on a horizontal surface while being subjected to constant resisting, dry friction, and constant active forces. The initial conditions of motion are as follows:

$$x_1 = S_1; \quad \frac{dx_1}{dt} = 0;$$

for $t = 0$

$$x_2 = S_2; \quad \frac{dx_2}{dt} = 0$$

Answer:

$$x_1 = S_1 + \frac{p_1 - r_1 - f_1}{2n}\left[t + \frac{1}{2n}(e^{-2nt} - 1) \right]$$

$$+ \frac{(n_{12} + n_{22})(p_1 - r_1 - f_1) + (n_{11} + n_{21})(p_2 - r_2 - f_2)}{2n}$$

$$\times \left[t^2 - \frac{t}{n} - \frac{1}{2n^2}(e^{-2nt} - 1) \right]$$

$$x_2 = S_2 + \frac{p_2 - r_2 - f_2}{2n}\left[t + \frac{1}{2n}(e^{-2nt} - 1)\right]$$

$$+ \frac{(n_{11} + n_{21})(p_2 - r_2 - f_2) + (n_{12} + n_{22})(p_1 - r_1 - f_1)}{2n}$$

$$\times \left[t^2 - \frac{t}{n} - \frac{1}{2n^2}(e^{-2nt} - 1)\right]$$

7. Determine the basic parameters of motion of the masses of the restricted two-degree-of-freedom system shown in Figure 23.2.1. The masses are connected to each other by a flexible link and are moving on a horizontal surface while being subjected to constant resisting, dry friction, and constant active forces. The initial conditions of motion are as follows:

$$x_1 = 0; \quad \frac{dx_1}{dt} = V_1;$$

for $t = 0$

$$x_2 = 0; \quad \frac{dx_2}{dt} = V_2$$

8. Determine the basic parameters of motion of the masses of the restricted two-degree-of-freedom system shown in Figure 23.2.1. The masses are connected to each other by a fluid link and are moving on a horizontal surface while being subjected to constant resisting, dry friction, and constant active forces. The initial conditions of motion are as follows:

$$x_1 = S_1; \quad \frac{dx_1}{dt} = V_1;$$

for $t = 0$

$$x_2 = 0; \quad \frac{dx_2}{dt} = 0$$

Answer:

$$x_1 = S_1 + \frac{2[V_1(n_{12} + n_{22})] + p_1 - r_1 - f_1}{2n}$$

$$\times \left[t + \frac{1}{2n}(e^{-2nt} - 1)\right]$$

$$+ \frac{(n_{12} + n_{22})(p_1 - r_1 - f_1) + (n_{11} + n_{21})(p_2 - r_2 - f_2)}{2n}$$

$$\times \left[t^2 - \frac{t}{n} - \frac{1}{2n^2}(e^{-2nt} - 1)\right]$$

$$x_2 = \frac{2V_1(n_{12} + n_{22}) + p_2 - r_2 - f_2}{2n}\left[t + \frac{1}{2n}(e^{-2nt} - 1)\right]$$

$$+ \frac{(n_{11} + n_{21})(p_2 - r_2 - f_2) + (n_{12} + n_{22})(p_1 - r_1 - f_1)}{2n}$$

$$\times \left[t^2 - \frac{t}{n} - \frac{1}{2n^2}(e^{-2nt} - 1)\right]$$

9. Determine the basic parameters of motion of the masses of the restricted two-degree-of-freedom system shown in Figure 23.3.1. The masses are connected to each other by a fluid link and are moving on a horizontal surface while being subjected to harmonic forces. The initial conditions of motion are as follows:

$$x_1 = 0; \quad \frac{dx_1}{dt} = V_1;$$

for $t = 0$

$$x_2 = S_2; \quad \frac{dx_2}{dt} = 0$$

Answer:

$$x_1 = \frac{V_1}{2n}(1 - e^{-2nt}) + \frac{V_1(n_{12} + n_{22})}{n}\left[t + \frac{1}{2n}(1 - e^{-2nt})\right]$$

$$+ \frac{a_1}{\varphi_1^2 + 4n^2}\left(e^{-2nt} + \frac{2n}{\varphi_1}\sin \varphi_1 t - \cos \varphi_1 t\right)$$

$$+ \frac{2a_1(n_{12}+n_{22})}{\varphi_1^2+4n^2}\left[\frac{2n}{\varphi_1^2}(1-\cos\varphi_1 t)-\frac{1}{\varphi_1}\sin\varphi_1 t\right.$$

$$\left.+\frac{1}{2n}(1-e^{-2nt})\right]+\frac{2a_2(n_{11}+n_{21})}{\varphi_2^2+4n^2}\left[\frac{2n}{\varphi_2^2}(1-\cos\varphi_2 t)\right.$$

$$\left.-\frac{1}{\varphi_2}\sin\varphi_2 t+\frac{1}{2n}(1-e^{-2nt})\right]$$

$$x_2 = S_2 + \frac{V_1}{2n}(1-e^{-2nt})+\frac{V_1(n_{11}+n_{21})}{n}\left[t+\frac{1}{2n}(1-e^{-2nt})\right]$$

$$+\frac{a_2}{\varphi_2^2+4n^2}\left(e^{-2nt}+\frac{2n}{\varphi_2}\sin\varphi_2 t-\cos\varphi_2 t\right)$$

$$+\frac{2a_2(n_{11}+n_{21})}{\varphi_2^2+4n^2}\left[\frac{2n}{\varphi_2^2}(1-\cos\varphi_2 t)-\frac{1}{\varphi_2}\sin\varphi_2 t\right.$$

$$\left.+\frac{1}{2n}(1-e^{-2nt})\right]+\frac{2a_1(n_{12}+n_{22})}{\varphi_1^2+4n^2}\left[\frac{2n}{\varphi_1^2}(1-\cos\varphi_1 t)\right.$$

$$\left.-\frac{1}{\varphi_1}\sin\varphi_1 t+\frac{1}{2n}(1-e^{-2nt})\right]$$

10. Determine the basic parameters of motion of the masses of the restricted two-degree-of-freedom system shown in Figure 23.3.1. The masses are connected to each other by a fluid link and are moving on a horizontal surface while being respectively subjected to harmonic forces. The initial conditions of motion are as follows:

$$x_1 = S_1; \quad \frac{dx_1}{dt}=0;$$

for $t=0$

$$x_2 = 0; \quad \frac{dx_2}{dt}=V_2$$

Answer:

$$x_1 = S_1 + \frac{V_2(n_{11} + n_{21})}{n}\left[t + \frac{1}{2n}(1 - e^{-2nt})\right] + \frac{a_1}{\varphi_1^2 + 4n^2}$$

$$\times \left(e^{-2nt} + \frac{2n}{\varphi_1}\sin\varphi_1 t - \cos\varphi_1 t\right) + \frac{2a_1(n_{12} + n_{22})}{\varphi_1^2 + 4n^2}$$

$$\times \left[\frac{2n}{\varphi_1^2}(1 - \cos\varphi_1 t) - \frac{1}{\varphi_1}\sin\varphi_1 t + \frac{1}{2n}(1 - e^{-2nt})\right]$$

$$+ \frac{2a_2(n_{11} + n_{21})}{\varphi_2^2 + 4n^2}\left[\frac{2n}{\varphi_2^2}(1 - \cos\varphi_2 t) - \frac{1}{\varphi_2}\sin\varphi_2 t\right.$$

$$\left. + \frac{1}{2n}(1 - e^{-2nt})\right]$$

$$x_2 = \frac{V_2(n_{12} + n_{22})}{n}\left[t + \frac{1}{2n}(1 - e^{-2nt})\right] + \frac{a_2}{\varphi_2^2 + 4n^2}$$

$$\times \left(e^{-2nt} + \frac{2n}{\varphi_2}\sin\varphi_2 t - \cos\varphi_2 t\right) + \frac{2a_2(n_{11} + n_{21})}{\varphi_2^2 + 4n^2}$$

$$\times \left[\frac{2n}{\varphi_2^2}(1 - \cos\varphi_2 t) - \frac{1}{\varphi_2}\sin\varphi_2 t + \frac{1}{2n}(1 - e^{-2nt})\right]$$

$$+ \frac{2a_1(n_{12} + n_{22})}{\varphi_1^2 + 4n^2}\left[\frac{2n}{\varphi_1^2}(1 - \cos\varphi_1 t) - \frac{1}{\varphi_1}\sin\varphi_1 t\right.$$

$$\left. + \frac{1}{2n}(1 - e^{-2nt})\right]$$

PROBLEMS

---

1. Determine the basic parameters of motion of the masses of the
restricted two-degree-of-freedom system shown in Figure 24.1.1.
The masses are connected to each other by a fluid link and a
flexible link in parallel, while one of the masses is attached to
a non-movable support by a similar pair of links. The masses
are moving on a horizontal surface, performing underdamped
vibratory motion. The initial conditions of motion are as
follows:

$$x_1 = 0; \quad \frac{dx_1}{dt} = V_1;$$

for $t = 0$

$$x_2 = 0; \quad \frac{dx_2}{dt} = V_2$$

Answer:

$$x_1 = \frac{V_1}{\omega} e^{-nt} \sin \omega t + \frac{2V_1(n_{12} + n_{22})}{\omega^2 + n^2} \left[1 - e^{-nt} \left(\cos \omega t + \frac{n}{\omega} \sin \omega t \right) \right.$$

$$+ \frac{V_1(\omega_{12}^2 + \omega_{22}^2)}{\omega(\omega^2 + n^2)^2} \{ (\omega^2 + n^2)\omega t - 2n\omega - e^{-nt}$$

$$\times [(\omega^2 - n^2) \sin \omega t - 2n\omega \cos \omega t] \}$$

$$x_2 = \frac{2V_1(n_{12} + n_{22})}{\omega^2 + n^2}\left[1 - e^{-nt}\left(\cos\omega t + \frac{n}{\omega}\sin\omega t\right)\right]$$

$$+ \frac{V_1(\omega_{12}^2 + \omega_{22}^2)}{\omega(\omega^2 + n^2)^2}\{(\omega^2 + n^2)\omega t - 2n\omega - e^{-nt}$$

$$\times [(\omega^2 - n^2)\sin\omega t - 2n\omega\cos\omega t]\}$$

2. Determine the basic parameters of motion of the masses of the restricted two-degree-of-freedom system shown in Figure 24.1.1. The masses are connected to each other by a fluid link and a flexible link in parallel, while one of the masses is attached to a non-movable support by a similar pair of links. The masses are moving on a horizontal surface, performing underdamped vibratory motion. The initial conditions of motion are as follows:

$$x_1 = 0; \quad \frac{dx_1}{dt} = 0;$$

for $t = 0$

$$x_2 = 0; \quad \frac{dx_2}{dt} = V_2$$

Answer:

$$x_1 = \frac{V_2(n_{11} + n_{21})}{\omega^2 + n^2}\left[1 - e^{-nt}\left(\cos\omega t + \frac{n}{\omega}\sin\omega t\right)\right]$$

$$+ \frac{V_2(\omega_{11}^2 + \omega_{21}^2)}{\omega(\omega^2 + n^2)^2}\{(\omega^2 + n^2)\omega t - 2n\omega - e^{-nt}$$

$$\times [(\omega^2 - n^2)\sin\omega t - 2n\omega\cos\omega t]\}$$

$$x_2 = \frac{V_2}{\omega}e^{-nt}\sin\omega t + \frac{2V_2(n_{11} + n_{21})}{\omega^2 + n^2}$$

$$\times \left[1 - e^{-nt}\left(\cos\omega t + \frac{n}{\omega}\sin\omega t\right)\right] + \frac{V_2(\omega_{11}^2 + \omega_{21}^2)}{\omega(\omega^2 + n^2)^2}$$

$$\times \{(\omega^2 + n^2)\omega t - 2n\omega - e^{-nt}$$

$$\times [(\omega^2 - n^2)\sin\omega t - 2n\omega\cos\omega t]\}$$

3. Determine the basic parameters of motion of the masses of the restricted two-degree-of-freedom system shown in Figure 24.1.1. The masses are connected to each other by a fluid link and a flexible link in parallel, while one of the masses is attached to a non-movable support by a similar pair of links. The masses are moving on a horizontal surface, performing critically damped motion. The initial conditions of motion are as follows:

$$x_1 = 0; \quad \frac{dx_1}{dt} = V_1;$$

for $t = 0$

$$x_2 = 0; \quad \frac{dx_2}{dt} = 0$$

Answer:

$$x_1 = V_1 t e^{-nt} + \frac{2V_1(n_{12} + n_{22})}{n^2}[1 - e^{-nt}(1 + nt)]$$
$$+ \frac{V_1(\omega_{12}^2 + \omega_{22}^2)}{n^2}\left[t - \frac{2}{n} + e^{-nt}\left(\frac{2}{n} + t\right)\right]$$
$$x_2 = \frac{2V_1(n_{12} + n_{22})}{n^2}[1 - e^{-nt}(1 + nt)]$$
$$+ \frac{V_1(\omega_{12}^2 + \omega_{22}^2)}{n^2}\left[t - \frac{2}{n} + e^{-nt}\left(\frac{2}{n} + t\right)\right]$$

4. Determine the basic parameters of motion of the masses of the restricted two-degree-of-freedom system shown in Figure 24.1.1. The masses are connected to each other by a fluid link and a flexible link in parallel, while one of the masses is attached to a non-movable support by a similar pair of links. The masses are moving on a horizontal surface, performing critically

damped motion. The initial conditions of motion are as follows:

$$x_1 = 0; \quad \frac{dx_1}{dt} = 0;$$

for $t = 0$

$$x_2 = 0; \quad \frac{dx_2}{dt} = V_2$$

Answer:

$$x_1 = \frac{2V_2(n_{11} + n_{21})}{n^2}[1 - e^{-nt}(1 + nt)]$$
$$+ \frac{V_2(\omega_{11}^2 + \omega_{21}^2)}{n^2}\left[t - \frac{2}{n} + e^{-nt}\left(\frac{2}{n} + t\right)\right]$$

$$x_2 = V_2 t e^{-nt} + \frac{2V_2(n_{11} + n_{21})}{n^2}[1 - e^{-nt}(1 + nt)]$$
$$+ \frac{V_2(\omega_{11}^2 + \omega_{21}^2)}{n^2}\left[t - \frac{2}{n} + e^{-nt}\left(\frac{2}{n} + t\right) = \right]$$

5. Determine the basic parameters of motion of the masses of the restricted two-degree-of-freedom system shown in Figure 24.1.1. The masses are connected to each other by a fluid link and a flexible link in parallel, while one of the masses is attached to a non-movable support by a similar pair of links. The masses are moving on a horizontal surface, performing overdamped motion. The initial conditions of motion are as follows:

$$x_1 = 0; \quad \frac{dx_1}{dt} = V_1;$$

for $t = 0$

$$x_2 = 0; \quad \frac{dx_2}{dt} = V_2$$

Answer:

$$x_1 = \frac{V_1}{\omega} e^{-nt} \sinh \omega t + \frac{2V_1(n_{12} + n_{22})}{n^2 - \omega^2}$$
$$\times \left[1 - e^{-nt}\left(\cosh \omega t + \frac{n}{\omega} \sinh \omega t\right)\right]$$

$$+ \frac{V_1(\omega_{12}^2 + \omega_{22}^2)}{\omega(n^2 - \omega^2)^2}\{(n^2 - \omega^2)\omega t - 2n\omega$$

$$+ e^{-nt}[(\omega^2 + n^2)\sinh \omega t + 2n\omega \cosh \omega t]\}$$

$$x_2 = \frac{2V_1(n_{12} + n_{22})}{n^2 - \omega^2}\left[1 - e^{-nt}\left(\cosh \omega t + \frac{n}{\omega}\sinh \omega t\right)\right]$$

$$+ \frac{V_1(\omega_{12}^2 + \omega_{22}^2)}{\omega(n^2 - \omega^2)^2}\{(n^2 - \omega^2)\omega t - 2n\omega$$

$$+ e^{-nt}[(\omega^2 + n^2)\sinh \omega t + 2n\omega \cosh \omega t]\}$$

6. Determine the basic parameters of motion of the masses of the restricted two-degree-of-freedom system shown in Figure 24.1.1. The masses are connected to each other by a fluid link and a flexible link in parallel, while one of the masses is attached to a non-movable support by an identical pair of links. The masses are moving on a horizontal surface, performing overdamped motion. The initial conditions of motion are as follows:

$$x_1 = 0; \quad \frac{dx_1}{dt} = 0;$$

for $t = 0$

$$x_2 = 0; \quad \frac{dx_2}{dt} = 0$$

Answer:

$$x_1 = \frac{2V_2(n_{11} + n_{21})]}{n^2 - \omega^2}\left[1 - e^{-nt}\left(\cosh \omega t + \frac{n}{\omega}\sinh \omega t\right)\right]$$

$$+ \frac{V_2(\omega_{11}^2 + \omega_{21}^2)}{\omega(n^2 - \omega^2)^2}\{(n^2 - \omega^2)\omega t - 2n\omega$$

$$+ e^{-nt}[(\omega^2 + n^2)\sinh \omega t + 2n\omega \cosh \omega t]\}$$

$$x_2 = \frac{V_2}{\omega}e^{-nt}\sinh \omega t + \frac{2V_2(n_{11} + n_{21})}{n^2 - \omega^2}$$

$$\times \left[1 - e^{-nt}\left(\cosh \omega t + \frac{n}{\omega}\sinh \omega t\right)\right] + \frac{V_2(\omega_{11}^2 + \omega_{21}^2)}{\omega(n^2 - \omega^2)^2}$$

$$\times \{(n^2 - \omega^2)\omega t - 2n\omega + e^{-nt}$$

$$\times [(\omega^2 + n^2)\sin h\omega t + 2n\omega \cosh \omega t]\}$$

7. Determine the basic parameters of motion of the masses of the restricted two-degree-of-freedom system shown in Figure 24.2.1. The masses are connected to each other by a fluid link and a flexible link, while one of the masses is attached to a non-movable support by an identical pair of links. The masses are moving on a horizontal surface, while only the first mass is subjected to a constant active force. The masses are performing underdamped vibrations. The initial conditions of motion are as follows:

$$x_1 = 0; \quad \frac{dx_1}{dt} = 0;$$

for $t = 0$

$$x_2 = 0; \quad \frac{dx_2}{dt} = 0$$

Answer:

$$x_1 = \frac{p_1}{\omega^2 + n^2}\left[1 - e^{-nt}\left(\cos\omega t + \frac{n}{\omega}\sin\omega t\right)\right] + \frac{2p_1(n_{12} + n_{22})}{\omega(\omega^2 + n^2)^2}$$

$$\times \{(\omega^2 + n^2)\omega t - 2n\omega - e^{-nt}$$

$$\times [(\omega^2 - n^2)\sin\omega t - 2n\omega\cos\omega t]\}$$

$$+ \frac{p_1(\omega_{12}^2 + \omega_{22}^2)}{\omega(\omega^2 + n^2)^2}\left\{\frac{\omega t^2}{2}(\omega^2 + n^2) - 2n\omega t - \frac{\omega(\omega^2 - n^2)}{\omega^2 + n^2}\right.$$

$$\times \left[1 - e^{-nt}\left(\cos\omega t + \frac{n}{\omega}\sin\omega t\right)\right] - \frac{2n^2\omega}{\omega^2 + n^2}$$

$$\times \left.\left[1 + e^{-nt}\left(\frac{\omega}{n}\sin\omega t - \cos\omega t\right)\right]\right\}$$

$$x_2 = \frac{p_1(n_{12} + n_{22})}{\omega(\omega^2 + n^2)^2}\{(\omega^2 + n^2)\omega t - 2n\omega - e^{-nt}[(\omega^2 - n^2)\sin\omega t$$

$$- 2n\omega\cos\omega t]\} + \frac{p_1(\omega_{12}^2 + \omega_{22}^2)}{\omega(\omega^2 + n^2)^2}\left\{\frac{\omega t^2}{2}(\omega^2 + n^2) - 2n\omega t\right.$$

$$-\frac{\omega(\omega^2 - n^2)}{\omega^2 + n^2}\left[1 - e^{-nt}\left(\cos\omega t + \frac{n}{\omega}\sin\omega t\right)\right] - \frac{2n^2\omega}{\omega^2 + n^2}$$

$$\times\left[1 + e^{-nt}\left(\frac{\omega}{n}\sin\omega t - \cos\omega t\right)\right]\bigg\}$$

8. Determine the basic parameters of motion of the masses of the restricted two-degree-of-freedom system shown in Figure 24.2.1. The masses are connected to each other by a fluid link and a flexible link, while one of the masses is attached to a non-movable support by an identical pair of links. The masses are moving on a horizontal surface, while only the second mass is subjected to a constant active force. The masses are performing underdamped vibrations. The initial conditions of motion are as follows:

$$x_1 = S_1; \quad \frac{dx_1}{dt} = 0;$$

for $t = 0$

$$x_2 = 0; \quad \frac{dx_2}{dt} = 0$$

9. Determine the basic parameters of motion of the masses of the restricted two-degree-of-freedom system shown in Figure 24.2.1. The masses are connected to each other by a fluid link and a flexible link, while one of the masses is attached to a non-movable support by an identical pair of links. The masses are moving on a horizontal surface, while only the first mass is subjected to a constant resisting force and a constant active force. The masses are performing critical damping. The initial conditions of motion are as follows:

$$x_1 = 0; \quad \frac{dx_1}{dt} = 0;$$

for $t = 0$

$$x_2 = S_2; \quad \frac{dx_2}{dt} = 0$$

10. Determine the basic parameters of motion of the masses of the restricted two-degree-of-freedom system shown in Figure 24.2.1. The masses are connected to each other by a fluid link and a flexible link, while one of the masses is attached to a non-movable support by an identical pair of links. The masses are moving on a horizontal surface, while only the first mass is subjected to a constant resisting force and a constant active force. The masses are performing overdamped motion. The initial conditions of motion are as follows:

$$x_1 = 0; \quad \frac{dx_1}{dt} = 0;$$

for $t = 0$

$$x_2 = 0; \quad \frac{dx_2}{dt} = 0$$

PROBLEMS

---

1. Determine the basic parameters of motion of the masses of the restricted two-degree-of-freedom system shown in Figure 25.1.1. The masses are connected to each other by a flexible link, and each mass is attached to a non-movable support by an identical link. The masses are moving on a horizontal surface. The initial conditions of motion are as follows:

$$x_1 = 0; \quad \frac{dx_1}{dt} = V_1;$$

for $t = 0$

$$x_2 = 0; \quad \frac{dx_2}{dt} = V_2$$

Answer:

$$x_1 = \frac{V_1}{\omega} \sin \omega t + \frac{1}{\omega^2} [V_1(\omega_{12}^2 + \omega_{22}^2 + \omega_{32}^2)$$

$$+ V_2(\omega_{11}^2 + \omega_{21}^2 + \omega_{31}^2)] \left(t - \frac{1}{\omega} \sin \omega t \right)$$

$$x_2 = \frac{V_2}{\omega} \sin \omega t + \frac{1}{\omega^2} [V_2(\omega_{11}^2 + \omega_{21}^2 + \omega_{31}^2)$$

$$+ V_1(\omega_{12}^2 + \omega_{22}^2 + \omega_{32}^2)] \left(t - \frac{1}{\omega} \sin \omega t \right)$$

2. Determine the basic parameters of motion of the masses of the restricted two-degree-of-freedom system shown in Figure 25.1.1. The masses are connected to each other by a flexible link, and each mass is attached to a non-movable support by an identical link. The masses are moving on a horizontal surface. The initial conditions of motion are as follows:

$$x_1 = S_1; \quad \frac{dx_1}{dt} = 0;$$

for $t = 0$

$$x_2 = S_2; \quad \frac{dx_2}{dt} = 0$$

Answer:

$$x_1 = S_1 \cos \omega t + \frac{1}{\omega^2}[S_1(\omega_{12}^2 + \omega_{22}^2 + \omega_{32}^2)$$

$$+ S_2(\omega_{11}^2 + \omega_{21}^2 + \omega_{31}^2)](1 - \cos \omega t)$$

$$x_2 = S_2 \cos \omega t + \frac{1}{\omega^2}[S_2(\omega_{11}^2 + \omega_{21}^2 + \omega_{31}^2)$$

$$+ S_1(\omega_{12}^2 + \omega_{22}^2 + \omega_{32}^2)](1 - \cos \omega t)$$

3. Determine the basic parameters of motion of the masses of the restricted two-degree-of-freedom system shown in Figure 25.1.1. The masses are connected to each other by a flexible link, and each mass is attached to a non-movable support by an identical link. The masses are moving on a horizontal surface. The initial conditions of motion are as follows:

$$x_1 = S_1; \quad \frac{dx_1}{dt} = 0;$$

for $t = 0$

$$x_2 = 0; \quad \frac{dx_2}{dt} = V_2$$

Answer:

$$x_1 = \frac{1}{\omega^2} V_2(\omega_{11}^2 + \omega_{21}^2 + \omega_{31}^2)\left(t - \frac{1}{\omega}\sin\omega t\right) + S_1 \cos\omega t$$

$$+ \frac{1}{\omega^2} S_1(\omega_{12}^2 + \omega_{22}^2 + \omega_{32}^2)(1 - \cos\omega t)$$

$$x_2 = \frac{V_2}{\omega}\sin\omega t + \frac{1}{\omega^2} V_2(\omega_{11}^2 + \omega_{21}^2 + \omega_{31}^2)\left(t - \frac{1}{\omega}\sin\omega t\right)$$

$$+ \frac{1}{\omega^2} S_1(\omega_{12}^2 + \omega_{22}^2 + \omega_{32}^2)(1 - \cos\omega t)$$

4. Determine the basic parameters of motion of the masses of the restricted two-degree-of-freedom system shown in Figure 25.1.1. The masses are connected to each other by a flexible link, and each mass is attached to a non-movable support by an identical link. The masses are moving on a horizontal surface. The initial conditions of motion are as follows:

$$x_1 = 0; \quad \frac{dx_1}{dt} = V_1;$$

for $t = 0$

$$x_2 = S_2; \quad \frac{dx_2}{dt} = 0$$

Answer:

$$x_1 = \frac{V_1}{\omega}\sin\omega t + \frac{1}{\omega^2} V_1(\omega_{12}^2 + \omega_{22}^2 + \omega_{32}^2)]\left(t - \frac{1}{\omega}\sin\omega t\right)$$

$$+ \frac{1}{\omega^2} S_2(\omega_{11}^2 + \omega_{21}^2 + \omega_{31}^2)(1 - \cos\omega t)$$

$$x_2 = V_1(\omega_{12}^2 + \omega_{22}^2 + \omega_{32}^2)\left(t - \frac{1}{\omega}\sin\omega t\right) + S_2 \cos\omega t$$

$$+ \frac{1}{\omega^2} S_2(\omega_{11}^2 + \omega_{21}^2 + \omega_{31}^2)(1 - \cos\omega t)$$

5. Determine the basic parameters of motion of the masses of the restricted two-degree-of-freedom system shown in Figure 25.1.1. The masses are connected to each other by a flexible link, and each mass is attached to a non-movable support by an identical link. The masses are moving on a horizontal surface. The initial conditions of motion are as follows:

$$x_1 = S_1; \quad \frac{dx_1}{dt} = 0;$$

for $t = 0$

$$x_2 = S_2; \quad \frac{dx_2}{dt} = 0$$

Answer:

$$x_1 = S_1 \cos \omega t + \frac{1}{\omega^2}[S_1(\omega_{12}^2 + \omega_{22}^2 + \omega_{32}^2) + S_2(\omega_{11}^2 + \omega_{21}^2 + \omega_{31}^2)]$$
$$\times (1 - \cos \omega t)$$

$$x_2 = S_2 \cos \omega t + \frac{1}{\omega^2}[S_2(\omega_{11}^2 + \omega_{21}^2 + \omega_{31}^2) + S_1(\omega_{12}^2 + \omega_{22}^2 + \omega_{32}^2)]$$
$$\times (1 - \cos \omega t)$$

6. Determine the basic parameters of motion of the masses of the restricted two-degree-of-freedom system shown in Figure 25.1.1. The masses are connected to each other by a flexible link, and each mass is attached to a non-movable support by an identical link. The masses are moving on a horizontal surface. The initial conditions of motion are as follows:

$$x_1 = S_1; \quad \frac{dx_1}{dt} = 0;$$

for $t = 0$

$$x_2 = S_2; \quad \frac{dx_2}{dt} = 0$$

Answer:

$$x_1 = S_1 \cos \omega t + \frac{1}{\omega^2}[S_1(\omega_{12}^2 + \omega_{22}^2 + \omega_{32}^2)$$
$$+ S_2(\omega_{11}^2 + \omega_{21}^2 + \omega_{31}^2)](1 - \cos \omega t)$$

$$x_2 = S_2 \cos \omega t + \frac{1}{\omega^2}[S_2(\omega_{11}^2 + \omega_{21}^2 + \omega_{31}^2)$$
$$+ S_1(\omega_{12}^2 + \omega_{22}^2 + \omega_{32}^2)](1 - \cos \omega t)$$

7. Determine the basic parameters of motion of the masses of the restricted two-degree-of-freedom system shown in Figure 25.2.1. The masses are connected to each other by a flexible link, and each mass is attached to a non-movable support by an identical link. The masses are moving on a horizontal surface, while only the first mass is subjected to a constant resisting force and a constant active force. The initial conditions of motion are as follows:

$$x_1 = 0; \quad \frac{dx_1}{dt} = 0;$$

for $t = 0$

$$x_2 = 0; \quad \frac{dx_2}{dt} = 0$$

Answer:

$$x_1 = \frac{p_1 - r_1}{\omega^2}(1 - \cos \omega t) + [(\omega_{12}^2 + \omega_{22}^2 + \omega_{32}^2)(p_1 - r_1)$$
$$\times \left[\frac{1}{\omega^4}(\cos \omega t - 1) + \frac{1}{2\omega^2}t^2\right]$$

$$x_2 = (\omega_{12}^2 + \omega_{22}^2 + \omega_{32}^2)(p_1 - r_1)\left[\frac{1}{\omega^4}(\cos \omega t - 1) + \frac{1}{2\omega^2}t^2\right]$$

8. Determine the basic parameters of motion of the masses of the restricted two-degree-of-freedom system shown in Figure 25.2.1. The masses are connected to each other by a flexible link, and each mass is attached to a non-movable support by an identical link. The masses are moving on a horizontal surface, while only the second mass is subjected to a constant resisting force and a constant active force. The initial conditions of motion are as follows:

$$x_1 = 0; \quad \frac{dx_1}{dt} = 0;$$

for $t = 0$

$$x_2 = 0; \quad \frac{dx_2}{dt} = 0$$

Answer:

$$x_1 = \frac{p_1 - r_1}{\omega^2}(1 - \cos\omega t) + (\omega_{12}^2 + \omega_{22}^2 + \omega_{32}^2)(p_1 - r_1)$$

$$\times \left[\frac{1}{\omega^4}(\cos\omega t - 1) + \frac{1}{2\omega^2}t^2\right]$$

$$x_2 = (\omega_{12}^2 + \omega_{22}^2 + \omega_{32}^2)(p_1 - r_1)]\left[\frac{1}{\omega^4}(\cos\omega t - 1) + \frac{1}{2\omega^2}t^2\right]$$

9. Determine the basic parameters of motion of the masses of the restricted two-degree-of-freedom system shown in Figure 25.3.1. The masses are connected to each other by a flexible link, and each mass is attached to a non-movable support by an identical link. The masses are moving on a horizontal surface, while only the first mass is subjected to a harmonic force. The initial conditions of motion are as follows:

$$x_1 = 0; \quad \frac{dx_1}{dt} = 0;$$

for $t = 0$

$$x_2 = 0; \quad \frac{dx_2}{dt} = 0$$

Answer:

$$x_1 = a_1\frac{\cos\omega t - \cos\varphi_1 t}{\varphi_1^2 - \omega^2} + \frac{a_1(\omega_{12}^2 + \omega_{22}^2 + \omega_{32}^2)}{\varphi_1^2 - \omega^2}$$

$$\times \left[\frac{1}{\omega^2}(1 - \cos\omega t) - \frac{1}{\varphi_1^2}(1 - \cos\varphi_1 t)\right]$$

$$x_2 = \frac{a_1(\omega_{12}^2 + \omega_{22}^2 + \omega_{32}^2)}{\varphi_1^2 - \omega^2}\left[\frac{1}{\omega^2}(1 - \cos\omega t) - \frac{1}{\varphi_1^2}(1 - \cos\varphi_1 t)\right]$$

10. Determine the basic parameters of motion of the masses of the restricted two-degree-of-freedom system shown in Figure 25.3.1. The masses are connected to each other by a flexible link, and each mass is attached to a non-movable support by an identical link. The masses are moving on a horizontal surface, while only the second mass is subjected to a harmonic force. The initial conditions of motion are as follows:

$$x_1 = 0; \quad \frac{dx_1}{dt} = 0;$$

for $t = 0$

$$x_2 = 0; \quad \frac{dx_2}{dt} = 0$$

Answer:

$$x_1 = \frac{a_2(\omega_{11}^2 + \omega_{21}^2 + \omega_{31}^2)}{\varphi_2^2 - \omega^2} \left[\frac{1}{\omega^2}(1 - \cos \omega t) - \frac{1}{\varphi_2^2}(1 - \cos \varphi_2 t) \right]$$

$$x_2 = a_2 \frac{\cos \omega t - \cos \varphi_2 t}{\varphi_2^2 - \omega^2} + \frac{a_2(\omega_{11}^2 + \omega_{21}^2 + \omega_{31}^2)}{\varphi_2^2 - \omega^2}$$

$$\times \left[\frac{1}{\omega^2}(1 - \cos \omega t) - \frac{1}{\varphi_2^2}(1 - \cos \varphi_2 t) \right]$$

CHAPTER 26

PROBLEMS

---

1. Determine the basic parameters of motion of the masses of the
   restricted two-degree-of-freedom system shown in Figure 26.1.1.
   The masses are connected to each other by a fluid link, and
   each mass is attached to a non-movable support by an identical
   link. The masses are moving on a horizontal surface. The initial
   conditions of motion are as follows:

$$x_1 = 0; \quad \frac{dx_1}{dt} = V_1;$$

for $t = 0$

$$x_2 = 0; \quad \frac{dx_2}{dt} = 0$$

Answer:

$$x_1 = \frac{V_1}{2n}(1 - e^{-2nt}) + \frac{V_1(n_{12} + n_{22} + n_{32})}{n}\left[t + \frac{1}{2n}(e^{-2nt} - 1)\right]$$

$$x_2 = \frac{V_1(n_{12} + n_{22} + n_{32})}{n}\left[t + \frac{1}{2n}(e^{-2nt} - 1)\right]$$

2. Determine the basic parameters of motion of the masses of the
   restricted two-degree-of-freedom system shown in Figure 26.1.1.
   The masses are connected to each other by a fluid link, and

each mass is attached to a non-movable support by an identical link. The masses are moving on a horizontal surface. The initial conditions of motion are as follows:

$$x_1 = 0; \quad \frac{dx_1}{dt} = 0;$$

for $t = 0$

$$x_2 = 0; \quad \frac{dx_2}{dt} = V_2$$

Answer:

$$x_1 = \frac{V_2(n_{11} + n_{21} + n_{31})}{n}\left[t + \frac{1}{2n}(e^{-2nt} - 1)\right]$$

$$x_2 = \frac{V_2}{2n}(1 - e^{-2nt}) + \frac{V_2(n_{11} + n_{21} + n_{31})}{n}\left[t + \frac{1}{2n}(e^{-2nt} - 1)\right]$$

3. Determine the basic parameters of motion of the masses of the restricted two-degree-of-freedom system shown in Figure 26.1.1. The masses are connected to each other by a fluid link, and each mass is attached to a non-movable support by an identical link. The masses are moving on a horizontal surface. The initial conditions of motion are as follows:

$$x_1 = S_1; \quad \frac{dx_1}{dt} = 0;$$

for $t = 0$

$$x_2 = 0; \quad \frac{dx_2}{dt} = V_2$$

Answer:

$$x_1 = S_1 + \frac{V_2(n_{11} + n_{21} + n_{31})}{n}\left[t + \frac{1}{2n}(e^{-2nt} - 1)\right]$$

$$x_2 = \frac{V_2(n_{11} + n_{21} + n_{31})}{n}\left[t + \frac{1}{2n}(e^{-2nt} - 1)\right]$$

4. Determine the basic parameters of motion of the masses of the restricted two-degree-of-freedom system shown in Figure 26.1.1. The masses are connected to each other by a fluid link, and each mass is attached to a non-movable support by an identical link. The masses are moving on a horizontal surface. The initial conditions of motion are as follows:

$$x_1 = 0; \quad \frac{dx_1}{dt} = V_1;$$

for $t = 0$

$$x_2 = S_2; \quad \frac{dx_2}{dt} = 0$$

Answer:

$$x_1 = \frac{V_1}{2n}(1 - e^{-2nt}) + \frac{V_1(n_{12} + n_{22} + n_{32})}{n}\left[t + \frac{1}{2n}(e^{-2nt} - 1)\right]$$

$$x_2 = S_2 + \frac{V_1(n_{12} + n_{22} + n_{32})}{n}\left[t + \frac{1}{2n}(e^{-2nt} - 1)\right]$$

5. Determine the basic parameters of motion of the masses of the restricted two-degree-of-freedom system shown in Figure 26.2.1. The masses are connected to each other by a fluid link, and each mass is attached to a non-movable support by an identical link. The masses are moving on a horizontal surface while being subjected to constant resisting, dry friction, and constant active forces. The initial conditions of motion are as follows:

$$x_1 = 0; \quad \frac{dx_1}{dt} = V_1;$$

for $t = 0$

$$x_2 = 0; \quad \frac{dx_2}{dt} = 0$$

6. Determine the basic parameters of motion of the masses of the restricted two-degree-of-freedom system shown in Figure 26.2.1. The masses are connected to each other by a fluid link, and each mass is attached to a non-movable support by an identical link. The masses are moving on a horizontal surface while being subjected to constant resisting, dry friction, and constant active forces. The initial conditions of motion are as follows:

$$x_1 = 0; \quad \frac{dx_1}{dt} = 0;$$

for $t = 0$

$$x_2 = 0; \quad \frac{dx_2}{dt} = V_2$$

7. Determine the basic parameters of motion of the masses of the restricted two-degree-of-freedom system shown in Figure 26.2.1. The masses are connected to each other by a fluid link, and each mass is attached to a non-movable support by an identical link. The masses are moving on a horizontal surface, while only the first mass is subjected to a constant resisting force and a constant active force. The initial conditions of motion are as follows:

$$x_1 = 0; \quad \frac{dx_1}{dt} = 0;$$

for $t = 0$

$$x_2 = 0; \quad \frac{dx_2}{dt} = 0$$

Answer:

$$x_1 = \frac{p_1 - r_1}{\omega^2}(1 - \cos \omega t) + [(\omega_{12}^2 + \omega_{22}^2 + \omega_{32}^2)(p_1 - r_1)$$

$$\times \left[\frac{1}{\omega^4}(\cos \omega t - 1) + \frac{1}{2\omega^2}t^2 \right]$$

$$x_2 = (\omega_{12}^2 + \omega_{22}^2 + \omega_{32}^2)(p_1 - r_1) \left[\frac{1}{\omega^4}(\cos \omega t - 1) + \frac{1}{2\omega^2}t^2 \right]$$

8. Determine the basic parameters of motion of the masses of the restricted two-degree-of-freedom system shown in Figure 26.2.1. The masses are connected to each other by a fluid link, and each mass is attached to a non-movable support by an identical link. The masses are moving on a horizontal surface, while only the second mass is subjected to a constant resisting force and a constant active force. The initial conditions of motion are as follows:

$$x_1 = 0; \quad \frac{dx_1}{dt} = 0;$$

for $t = 0$

$$x_2 = 0; \quad \frac{dx_2}{dt} = 0$$

Answer:

$$x_1 = \frac{(n_{11} + n_{21} + n_{31})(p_2 - r_2 - f_2)}{2n} \left[t^2 - \frac{t}{n} - \frac{1}{2n^2}(e^{-2nt} - 1) \right]$$

$$x_2 = \frac{p_2 - r_2 - f_2}{2n} \left[t + \frac{1}{2n}(e^{-2nt} - 1) \right]$$

$$+ \frac{(n_{11} + n_{21} + n_{31})(p_2 - r_2 - f_2)}{2n}$$

$$\times \left[t^2 - \frac{t}{n} - \frac{1}{2n^2}(e^{-2nt} - 1) \right]$$

9. Determine the basic parameters of motion of the masses of the restricted two-degree-of-freedom system shown in Figure 26.3.1. The masses are connected to each other by a fluid link, and each mass is attached to a non-movable support by an identical link. The masses are moving on a horizontal surface, while only the first mass is subjected to a harmonic force. The initial conditions of motion are as follows:

$$x_1 = 0; \quad \frac{dx_1}{dt} = 0;$$

for $t = 0$

$$x_2 = 0; \quad \frac{dx_2}{dt} = 0$$

Answer:

$$x_1 = +\frac{a_1}{\varphi_1^2 + 4n^2}\left(e^{-2nt} + \frac{2n}{\varphi_1}\sin\varphi_1 t - \cos\varphi_1 t\right)$$

$$+ \frac{2a_1(n_{12} + n_{22} + n_{32})}{\varphi_1^2 + 4n^2}$$

$$\times \left[\frac{2n}{\varphi_1^2}(1 - \cos\varphi_1 t) - \frac{1}{\varphi_1}\sin\varphi_1 t + \frac{1}{2n}(1 - e^{-2nt})\right]$$

$$x_2 = \frac{2a_1(n_{12} + n_{22} + n_{32})}{\varphi_1^2 + 4n^2}$$

$$\times \left[\frac{2n}{\varphi_1^2}(1 - \cos\varphi_1 t) - \frac{1}{\varphi_1}\sin\varphi_1 t + \frac{1}{2n}(1 - e^{-2nt})\right]$$

10. Determine the basic parameters of motion of the masses of the restricted two-degree-of-freedom system shown in Figure 26.3.1. The masses are connected to each other by a fluid link, and each mass is attached to a non-movable support by an identical link. The masses are moving on a horizontal surface, while only the second mass is subjected to a harmonic force. The initial conditions of motion are as follows:

$$x_1 = 0; \quad \frac{dx_1}{dt} = 0;$$

for $t = 0$

$$x_2 = 0; \quad \frac{dx_2}{dt} = 0$$

Answer:

$$x_1 = \frac{2a_2(n_{11} + n_{21} + n_{31})}{\varphi_2^2 + 4n^2}$$

$$\times \left[\frac{2n}{\varphi_2^2}(1 - \cos\varphi_2 t) - \frac{1}{\varphi_2}\sin\varphi_2 t + \frac{1}{2n}(1 - e^{-2nt})\right]$$

$$x_2 = \frac{a_2}{\varphi_2^2 + 4n^2}\left(e^{-2nt} + \frac{2n}{\varphi_2}\sin\varphi_2 t - \cos\varphi_2 t\right)$$

$$+ \frac{2a_2(n_{11} + n_{21} + n_{31})}{\varphi_2^2 + 4n^2}$$

$$\times \left[\frac{2n}{\varphi_2^2}(1 - \cos\varphi_2 t) - \frac{1}{\varphi_2}\sin\varphi_2 t + \frac{1}{2n}(1 - e^{-2nt})\right]$$

PROBLEMS

---

1. Determine the basic parameters of motion of the masses of the restricted two-degree-of-freedom system shown in Figure 27.1.1. The masses are connected to each other by a fluid link and a flexible link in parallel, while both masses are attached to non-movable supports by identical pairs of links. The masses are moving on a horizontal surface, performing underdamped vibratory motion. The initial conditions of motion are as follows:

$$x_1 = 0; \quad \frac{dx_1}{dt} = V_1;$$

for $t = 0$

$$x_2 = 0; \quad \frac{dx_2}{dt} = 0$$

Answer:

$$x_1 = \frac{V_1}{\omega} e^{-nt} \sin \omega t + \frac{2V_1(n_{12} + n_{22} + n_{32})}{\omega^2 + n^2}$$

$$\times \left[1 - e^{-nt} \left(\cos \omega t + \frac{n}{\omega} \sin \omega t \right) \right]$$

$$+ \frac{V_1(\omega_{12}^2 + \omega_{22}^2 + \omega_{32}^2)}{\omega(\omega^2 + n^2)^2} \{ (\omega^2 + n^2)\omega t - 2n\omega - e^{-nt}$$

$$\times [(\omega^2 - n^2) \sin \omega t - 2n\omega \cos \omega t] \}$$

$$x_2 = \frac{2V_1(n_{12} + n_{22} + n_{32})}{\omega^2 + n^2}\left[1 - e^{-nt}\left(\cos\omega t + \frac{n}{\omega}\sin\omega t\right)\right]$$

$$+ \frac{V_1(\omega_{12}^2 + \omega_{22}^2 + \omega_{32}^2)}{\omega(\omega^2 + n^2)^2}\{(\omega^2 + n^2)\omega t - 2n\omega$$

$$- e^{-nt}[(\omega^2 - n^2)\sin\omega t - 2n\omega\cos\omega t]\}$$

2. Determine the basic parameters of motion of the masses of the restricted two-degree-of-freedom system shown in Figure 27.1.1. The masses are connected to each other by a fluid link and a flexible link in parallel, while both masses are attached to non-movable supports by identical pairs of links. The masses are moving on a horizontal surface, performing underdamped vibratory motion. The initial conditions of motion are as follows:

$$x_1 = 0; \quad \frac{dx_1}{dt} = 0;$$

for $t = 0$

$$x_2 = 0; \quad \frac{dx_2}{dt} = V_2$$

Answer:

$$x_1 = \frac{2V_2(n_{11} + n_{21} + n_{31})}{\omega^2 + n^2}\left[1 - e^{-nt}\left(\cos\omega t + \frac{n}{\omega}\sin\omega t\right)\right.$$

$$+ \frac{V_2(\omega_{11}^2 + \omega_{21}^2 + \omega_{31}^2)}{\omega(\omega^2 + n^2)^2}\{(\omega^2 + n^2)\omega t - 2n\omega$$

$$- e^{-nt}[(\omega^2 - n^2)\sin\omega t - 2n\omega\cos\omega t]\}$$

$$x_2 = \frac{V_2}{\omega}e^{-nt}\sin\omega t + \frac{2[V_2(n_{11} + n_{21} + n_{31})]}{\omega^2 + n^2}$$

$$\times \left[1 - e^{-nt}\left(\cos\omega t + \frac{n}{\omega}\sin\omega t\right)\right]$$

$$+ \frac{V_2(\omega_{11}^2 + \omega_{21}^2 + \omega_{31}^2)}{\omega(\omega^2 + n^2)^2}\{(\omega^2 + n^2)\omega t - 2n\omega$$

$$- e^{-nt}[(\omega^2 - n^2)\sin\omega t - 2n\omega\cos\omega t]\}$$

3. Determine the basic parameters of motion of the masses of the restricted two-degree-of-freedom system shown in Figure 27.1.1. The masses are connected to each other by a fluid link and a flexible link in parallel, while both masses are attached to non-movable supports by identical pairs of links. The masses are moving on a horizontal surface, performing critically damped motion. The initial conditions of motion are as follows:

$$x_1 = 0; \quad \frac{dx_1}{dt} = V_1;$$

for $t = 0$

$$x_2 = 0; \quad \frac{dx_2}{dt} = 0$$

Answer:

$$x_1 = V_1 t e^{-nt} + \frac{2V_1(n_{12} + n_{22} + n_{32})}{n^2}\left[1 - e^{-nt}(1 + nt)\right]$$

$$+ \frac{V_1(\omega_{12}^2 + \omega_{22}^2 + \omega_{32}^2)}{n^2}\left[t - \frac{2}{n} + e^{-nt}\left(\frac{2}{n} + t\right)\right]$$

$$x_2 = \frac{2V_1(n_{12} + n_{22} + n_{32})}{n^2}\left[1 - e^{-nt}(1 + nt)\right]$$

$$+ \frac{V_1(\omega_{12}^2 + \omega_{22}^2 + \omega_{32}^2)}{n^2}\left[t - \frac{2}{n} + e^{-nt}\left(\frac{2}{n} + t\right)\right]$$

4. Determine the basic parameters of motion of the masses of the restricted two-degree-of-freedom system shown in Figure 27.1.1. The masses are connected to each other by a fluid link and a flexible link in parallel, while both masses are attached to non-movable supports by identical pairs of links. The masses are moving on a horizontal surface, performing critically damped motion. The initial conditions of motion are as follows:

$$x_1 = 0; \quad \frac{dx_1}{dt} = 0;$$

for $t = 0$

$$x_2 = 0; \quad \frac{dx_2}{dt} = V_2$$

Answer:

$$x_1 = \frac{2V_2(n_{11} + n_{21} + n_{31})}{n^2}[1 - e^{-nt}(1 + nt)]$$

$$+ \frac{V_2(\omega_{11}^2 + \omega_{21}^2 + \omega_{31}^2)}{n^2}\left[t - \frac{2}{n} + e^{-nt}\left(\frac{2}{n} + t\right)\right]$$

$$x_2 = V_2 t e^{-nt} + \frac{2V_2(n_{11} + n_{21} + n_{31})}{n^2}[1 - e^{-nt}(1 + nt)]$$

$$+ \frac{V_2(\omega_{11}^2 + \omega_{21}^2 + \omega_{31}^2)}{n^2}\left[t - \frac{2}{n} + e^{-nt}\left(\frac{2}{n} + t\right)\right]$$

5. Determine the basic parameters of motion of the masses of the restricted two-degree-of-freedom system shown in Figure 27.1.1. The masses are connected to each other by a fluid link and a flexible link in parallel, while both masses are attached to non-movable supports by identical pairs of links. The masses are moving on a horizontal surface, performing overdamped motion. The initial conditions of motion are as follows:

$$x_1 = 0; \quad \frac{dx_1}{dt} = V_1;$$

for $t = 0$

$$x_2 = S_2; \quad \frac{dx_2}{dt} = 0$$

Answer:

$$x_1 = \frac{V_1}{\omega}e^{-nt}\sinh\omega t + \frac{2[V_1(n_{12} + n_{22} + n_{32})]}{n^2 - \omega^2}$$

$$\times \left[1 - e^{-nt}\left(\cosh\omega t + \frac{n}{\omega}\sinh\omega t\right)\right]$$

$$+ \frac{V_1(\omega_{12}^2 + \omega_{22}^2 + \omega_{32}^2)}{\omega(n^2 - \omega^2)^2}\{(n^2 - \omega^2)\omega t - 2n\omega$$

$$+ e^{-nt}[(\omega^2 + n^2)\sinh\omega t + 2n\omega\cosh\omega t]\}$$

$$x_2 = \frac{2V_1(n_{12} + n_{22} + n_{32})}{n^2 - \omega^2}\left[1 - e^{-nt}\left(\cosh\omega t + \frac{n}{\omega}\sinh\omega t\right)\right]$$

$$+ \frac{V_1(\omega_{12}^2 + \omega_{22}^2 + \omega_{32}^2)}{\omega(n^2 - \omega^2)^2}\{(n^2 - \omega^2)\omega t - 2n\omega$$

$$+ e^{-nt}[(\omega^2 + n^2)\sinh\omega t + 2n\omega\cosh\omega t]\}$$

6. Determine the basic parameters of motion of the masses of the restricted two-degree-of-freedom system shown in Figure 27.1.1. The masses are connected to each other by a fluid link and a flexible link in parallel, while both masses are attached to non-movable supports by an identical pair of links. The masses are moving on a horizontal surface, performing overdamped motion. The initial conditions of motion are as follows:

$$x_1 = 0; \quad \frac{dx_1}{dt} = 0;$$

for $t = 0$

$$x_2 = 0; \quad \frac{dx_2}{dt} = V_2$$

Answer:

$$x_1 = \frac{2V_2(n_{11} + n_{21} + n_{31})}{n^2 - \omega^2}\left[1 - e^{-nt}\left(\cosh\omega t + \frac{n}{\omega}\sinh\omega t\right)\right]$$

$$+ \frac{V_2(\omega_{11}^2 + \omega_{21}^2 + \omega_{31}^2)}{\omega(n^2 - \omega^2)^2}\{(n^2 - \omega^2)\omega t - 2n\omega$$

$$+ e^{-nt}[(\omega^2 + n^2)\sinh\omega t + 2n\omega\cosh\omega t]\}$$

$$x_2 = \frac{V_2}{\omega}e^{-nt}\sinh\omega t + \frac{2V_2(n_{11} + n_{21} + n_{31})}{n^2 - \omega^2}$$

$$\times \left[1 - e^{-nt}\left(\cosh\omega t + \frac{n}{\omega}\sinh\omega t\right)\right]$$

$$+ \frac{V_2(\omega_{11}^2 + \omega_{21}^2 + \omega_{31}^2)}{\omega(n^2 - \omega^2)^2}\{(n^2 - \omega^2)\omega t - 2n\omega$$

$$+ e^{-nt}[(\omega^2 + n^2)\sinh\omega t + 2n\omega\cosh\omega t]\}$$

7. Determine the basic parameters of motion of the masses of the restricted two-degree-of-freedom system shown in Figure 27.2.1. The masses are connected to each other by a fluid link and a flexible link, while both masses are attached to non-movable supports by an identical pair of links. The masses are moving on a horizontal surface, while only the first mass is subjected to a constant active force. The masses are performing under-damped vibrations. The initial conditions of motion are as follows:

$$x_1 = 0; \quad \frac{dx_1}{dt} = 0;$$

for $t = 0$

$$x_2 = 0; \quad \frac{dx_2}{dt} = 0$$

Answer:

$$x_1 = \frac{p_1}{\omega^2 + n^2}\left[1 - e^{-nt}\left(\cos\omega t + \frac{n}{\omega}\sin\omega t\right)\right]$$

$$+ \frac{2p_1(n_{12} + n_{22} + n_{32})}{\omega(\omega^2 + n^2)^2}\{(\omega^2 + n^2)\omega t - 2n\omega$$

$$- e^{-nt}[(\omega^2 - n^2)\sin\omega t - 2n\omega\cos\omega t]\}$$

$$+ \frac{p_1(\omega_{12}^2 + \omega_{22}^2 + \omega_{32}^2)}{\omega(\omega^2 + n^2)^2}\left\{\frac{\omega t^2}{2}(\omega^2 + n^2) - 2n\omega t\right.$$

$$- \frac{\omega(\omega^2 - n^2)}{\omega^2 + n^2}\left[1 - e^{-nt}\left(\cos\omega t + \frac{n}{\omega}\sin\omega t\right)\right]$$

$$\left. - \frac{2n^2\omega}{\omega^2 + n^2}\left[1 + e^{-nt}\left(\frac{\omega}{n}\sin\omega t - \cos\omega t\right)\right]\right\}$$

$$x_2 = \frac{p_1(n_{12} + n_{22} + n_{32})}{\omega(\omega^2 + n^2)^2}\{(\omega^2 + n^2)\omega t - 2n\omega$$

$$- e^{-nt}[(\omega^2 - n^2)\sin\omega t - 2n\omega\cos\omega t]\}$$

$$+ \frac{p_1(\omega_{12}^2 + \omega_{22}^2 + \omega_{32}^2)}{\omega(\omega^2 + n^2)^2}\left\{\frac{\omega t^2}{2}(\omega^2 + n^2) - 2n\omega t\right.$$

$$-\frac{\omega(\omega^2 - n^2)}{\omega^2 + n^2}\left[1 - e^{-nt}\left(\cos\omega t + \frac{n}{\omega}\sin\omega t\right)\right]$$

$$\left. -\frac{2n^2\omega}{\omega^2 + n^2}\left[1 + e^{-nt}\left(\frac{\omega}{n}\sin\omega t - \cos\omega t\right)\right]\right\}$$

8. Determine the basic parameters of motion of the masses of the restricted two-degree-of-freedom system shown in Figure 24.2.1. The masses are connected to each other by a fluid link and a flexible link, while both are attached to non-movable supports by an identical pair of links. The masses are moving on a horizontal surface, while only the second mass is subjected to a constant active force. The masses are performing under-damped vibrations. The initial conditions of motion are as follows:

$$x_1 = S_1; \quad \frac{dx_1}{dt} = 0;$$

for $t = 0$

$$x_2 = 0; \quad \frac{dx_2}{dt} = 0$$

9. Determine the basic parameters of motion of the masses of the restricted two-degree-of-freedom system shown in Figure 24.2.1. The masses are connected to each other by a fluid link and a flexible link, while both masses are attached to non-movable supports by an identical pair of links. The masses are moving on a horizontal surface, while only the first mass is subjected to a constant resisting force and a constant active force. The masses are performing critical damping. The initial conditions of motion are as follows:

$$x_1 = 0; \quad \frac{dx_1}{dt} = 0;$$

for $t = 0$

$$x_2 = S_2; \quad \frac{dx_2}{dt} = 0$$

10. Determine the basic parameters of motion of the masses of the restricted two-degree-of-freedom system shown in Figure 27.2.1. The masses are connected to each other by a fluid link and a flexible link, while both masses are attached to non-movable supports by an identical pair of links. The masses are moving on a horizontal surface, while only the first mass is subjected to a constant resisting force and a constant active force. The masses are performing overdamped motion. The initial conditions of motion are as follows:

$$x_1 = 0; \quad \frac{dx_1}{dt} = 0;$$

for $t = 0$

$$x_2 = 0; \quad \frac{dx_2}{dt} = 0$$

Answer:

$$x_1 = \frac{p_1 - r_1}{n^2 - \omega^2} \left[1 - e^{-nt} \left(\cosh \omega t + \frac{n}{\omega} \sinh \omega t \right) \right]$$

$$+ \frac{2(p_1 - r_1)(n_{12} + n_{22} + n_{32})}{\omega(n^2 - \omega^2)^2} \{ \omega t(n^2 - \omega^2) - 2n\omega$$

$$+ e^{-nt}[\omega t(n^2 + \omega^2) \sin \omega t + 2n\omega \cosh \omega t] \}$$

$$+ \frac{(p_1 - r_1)(\omega_{12}^2 + \omega_{22}^2 + \omega_{32}^2)}{\omega(n^2 - \omega^2)^2} \left\{ \frac{\omega t^2}{2}(n^2 - \omega^2) - 2n\omega t \right.$$

$$+ \frac{\omega(n^2 + \omega^2)}{n^2 - \omega^2} \left[1 - e^{-nt} \left(\cosh \omega t + \frac{n}{\omega} \sinh \omega t \right) \right]$$

$$\left. - \frac{2n^2\omega}{n^2 - \omega^2} \left[1 + e^{-nt} \left(\frac{n}{\omega} \sinh \omega t - \cosh \omega t \right) \right] \right\}$$

$$x_2 = \frac{2(p_1 - r_1)(n_{12} + n_{22} + n_{32})}{\omega(n^2 - \omega^2)^2}\{\omega t(n^2 - \omega^2) - 2n\omega$$

$$+ e^{-nt}[\omega t(n^2 + \omega^2)\sin\omega t + 2n\omega\cosh\omega t]\}$$

$$+ \frac{(p_1 - r_1)(\omega_{12}^2 + \omega_{22}^2 + \omega_{32}^2)}{\omega(n^2 - \omega^2)^2}\left\{\frac{\omega t^2}{2}(n^2 - \omega^2) - 2n\omega t\right.$$

$$+ \frac{\omega(n^2 + \omega^2)}{n^2 - \omega^2}\left[1 - e^{-nt}\left(\cosh\omega t + \frac{n}{\omega}\sinh\omega t\right)\right]$$

$$\left. - \frac{2n^2\omega}{n^2 - \omega^2}\left[1 + e^{-nt}\left(\frac{n}{\omega}\sinh\omega t - \cosh\omega t\right)\right]\right\}$$

APPENDIX A-1

Laplace Transform Pairs

$$f(t) \qquad\qquad F(s) = s \int_0^\infty e^{-st} f(t)\,dt$$

1. $u(t)$ 1. $u(s)$

2. $Constant$ 2. $Constant$

3. t 3. $\dfrac{1}{s}$

4. $\dfrac{du}{dt}$ 4. $su(s) - su(0)$

5. $\dfrac{d^2u}{dt^2}$ 5. $s^2 u(s) - s\dfrac{du(0)}{dt} - s^2 u(0)$

6. $\dfrac{t^i}{i!}$ 6. $\dfrac{1}{s^i}$ (i is a positive integer)

7. $\dfrac{1}{n}(1 - e^{-nt})$ 7. $\dfrac{1}{s+n}$ (n is a positive number)

8. $\dfrac{1}{2n}(1 - e^{-2nt})$ 8. $\dfrac{1}{s+2n}$

9. $\dfrac{1}{n}(e^{-nt} - 1)$ 9. $\dfrac{1}{s-n}$

10. $\dfrac{1}{2n}(e^{-2nt} - 1)$ 10. $\dfrac{s}{s-2n}$

11. e^{nt} 11. $\dfrac{s}{s-n}$

12. e^{2nt} 12. $\dfrac{s}{s-2n}$

13. e^{-nt} 13. $\dfrac{s}{s+n}$

14	e^{-2nt}	14.	$\dfrac{s}{s+2n}$
15.	$\dfrac{1}{n}\left[t+\dfrac{1}{n}(e^{-nt}-1)\right]$	15.	$\dfrac{1}{s(s+n)}$
16.	$\dfrac{1}{2n}\left[t+\dfrac{1}{2n}(e^{-2nt}-1)\right]$	16.	$\dfrac{1}{s(s+2n)}$
17.	$\dfrac{1}{\omega^2}(1-\cos\omega t)$	17.	$\dfrac{1}{s^2+\omega^2}$
18.	$\dfrac{1}{\omega^2}(\cosh\omega t-1)$	18.	$\dfrac{1}{s^2-\omega^2}$
19.	$\dfrac{1}{n^2}[1-e^{-nt}(1+nt)]$	19.	$\dfrac{1}{(s+n)^2}$
20.	$\dfrac{1}{4n^2}[1-e^{-2nt}(1+2nt)]$	20.	$\dfrac{1}{(s+2n)^2}$
21.	$\dfrac{1}{\omega^2+n^2}\left[1-e^{-nt}\left(\cos\omega t+\dfrac{n}{\omega}\sin\omega t\right)\right]$	21.	$\dfrac{1}{(s+n)^2+\omega^2}$
22.	$\dfrac{1}{n^2-\omega^2}\left[1-e^{-nt}\left(\cosh\omega t+\dfrac{n}{\omega}\sinh\omega t\right)\right]$	22.	$\dfrac{1}{(s+n)^2-\omega^2}$
23.	$\sin\omega t$	23.	$\dfrac{\omega s}{s^2+\omega^2}$
24.	$\dfrac{1}{\omega}\sinh\omega t$	24.	$\dfrac{s}{s^2-\omega^2}$
25.	te^{-nt}	25.	$\dfrac{s}{(s+n)^2}$
26.	te^{-2nt}	26.	$\dfrac{s}{(s+2n)^2}$
27.	$\dfrac{1}{\omega}e^{-nt}\sin\omega t$	27.	$\dfrac{s}{(s+n)^2+\omega^2}$
28.	$\dfrac{1}{\omega}e^{-nt}\sinh\omega t$	28.	$\dfrac{s}{(s+n)^2-\omega^2}$
29.	$\cos\omega t$	29.	$\dfrac{s^2}{s^2+\omega^2}$
30.	$\cosh\omega t$	30.	$\dfrac{s^2}{s^2-\omega^2}$
31.	$(1-nt)e^{-nt}$	31.	$\dfrac{s^2}{(s+n)^2}$
32.	$(1-2nt)e^{-2nt}$	32.	$\dfrac{s^2}{(s+2n)^2}$

33. $e^{-nt}\cos\omega t$

33. $\dfrac{s(s+n)}{(s+n)^2+\omega^2}$

34. $e^{-nt}\cosh\omega t$

34. $\dfrac{s(s+n)}{(s+n)^2-\omega^2}$

35. $e^{-nt}\left(\cos\omega t-\dfrac{n}{\omega}\sin\omega t\right)$

35. $\dfrac{s^2}{(s+n)^2+\omega^2}$

36. $e^{-nt}\left(\cosh\omega t-\dfrac{n}{\omega}\sinh\omega t\right)$

36. $\dfrac{s^2}{(s+n)^2-\omega^2}$

37. $\dfrac{1}{n^2}\left[t-\dfrac{2}{n}+e^{-nt}\left(\dfrac{2}{n}+t\right)\right]$

37. $\dfrac{1}{s(s+n)^2}$

38. $\dfrac{1}{4n^2}\left[t-\dfrac{1}{n}+e^{-2nt}\left(\dfrac{1}{n}+t\right)\right]$

38. $\dfrac{1}{s(s+2n)^2}$

39. $\dfrac{1}{\omega^2}\left(t-\dfrac{1}{\omega}\sin\omega t\right)$

39. $\dfrac{1}{s(s^2+\omega^2)}$

40. $\dfrac{1}{\omega^2}\left(\dfrac{1}{\omega}\sinh\omega t-t\right)$

40. $\dfrac{1}{s(s^2-\omega^2)}$

41. $\dfrac{1}{\omega^2+n^2}\left[\dfrac{n}{\omega^2}(1-\cos\omega t)\right.$
$\left.-\dfrac{1}{\omega}\sin\omega t+\dfrac{1}{n}(1-e^{-nt})\right]$

41. $\dfrac{1}{(s+n)(s^2+\omega^2)}$

42. $\dfrac{1}{\omega^2+4n^2}\left[\dfrac{2n}{\omega^2}(1-\cos\omega t)\right.$
$\left.-\dfrac{1}{\omega}\sin\omega t+\dfrac{1}{2n}(1-e^{-2nt})\right]$

42. $\dfrac{1}{(s+2n)(s^2+\omega^2)}$

43. $\dfrac{1}{\omega^2+n^2}\left(1-\cos\omega t+\dfrac{n}{\omega}\sin\omega t\right.$
$\left.-1+e^{-nt}\right)$

43. $\dfrac{s}{(s+n)(s^2+\omega^2)}$

44. $\dfrac{1}{\omega^2+4n^2}\left(e^{-2nt}+\dfrac{2n}{\omega}\sin\omega t-\cos\omega t\right)$

44. $\dfrac{s}{(s+2n)(s^2+\omega^2)}$

45. $\dfrac{1}{\omega(\omega^2+n^2)^2}\{(\omega^2+n^2)\omega t-2n\omega$
$\quad-e^{-nt}[(\omega^2-n^2)\sin\omega t-2n\omega\cos\omega t]\}$

45. $\dfrac{1}{s[(s+n)^2+\omega^2]}$

46. $\dfrac{1}{\omega(n^2-\omega^2)^2}\{\omega t(n^2-\omega^2)-2n\omega$
$\quad+e^{-nt}[(n^2+\omega^2)\sinh\omega t+2n\omega\cosh\omega t]\}$

46. $\dfrac{1}{s[(s+n)^2-\omega^2]}$

47. $\dfrac{1}{2n}\left[t^2 - \dfrac{2}{n}t - \dfrac{2}{n^2}(e^{-nt}-1)\right]$ 47. $\dfrac{1}{s^2(s+n)}$

48. $\dfrac{1}{4n}\left[t^2 - \dfrac{1}{n}t - \dfrac{1}{2n^2}(e^{-2nt}-1)\right]$ 48. $\dfrac{1}{s^2(s+2n)}$

49. $\dfrac{t}{n\omega^2} - \dfrac{1}{\omega^2+n^2}\left(\dfrac{1-e^{-nt}}{n^2}\right.$ 49. $\dfrac{1}{s(s+n)(s^2+\omega^2)}$

$\left.+\dfrac{1-\cos\omega t}{\omega^2}+\dfrac{n\sin\omega t}{\omega^3}\right)$

50. $\dfrac{t}{2n\omega^2} - \dfrac{1}{\omega^2+4n^2}\left(\dfrac{1-e^{-2nt}}{4n^2}\right.$ 50. $\dfrac{1}{s(s+2n)(s^2+\omega^2)}$

$\left.+\dfrac{1-\cos\omega t}{\omega^2}+\dfrac{2n\sin\omega t}{\omega^3}\right)$

51. $\dfrac{1}{\omega^4}(\cos\omega t - 1) + \dfrac{1}{2\omega^2}t^2$ 51. $\dfrac{1}{s^2(s^2+\omega^2)}$

52. $\dfrac{1}{n}\left[\dfrac{t^3}{12} - \dfrac{t^2}{2n} - \dfrac{t}{n^2} - \dfrac{1}{n^3}(1-e^{-nt})\right]$ 52. $\dfrac{1}{s^3(s+n)}$

53. $\dfrac{1}{8n}\left[\dfrac{t^3}{3} - \dfrac{t^2}{n} + \dfrac{t}{n^2} - \dfrac{1}{2n^3}(1-e^{-2nt})\right]$ 53. $\dfrac{1}{s^3(s+2n)}$

54. $\dfrac{1}{n^2}\left[\dfrac{t^2}{2} - \dfrac{2t}{n} - \dfrac{3}{n^2} + \dfrac{e^{-nt}}{n^2}(3+nt)\right]$ 54. $\dfrac{1}{s^2(s+n)^2}$

55. $\dfrac{1}{\omega^2-\varphi^2}\left[\dfrac{1}{\varphi^2}(1-\cos\varphi t) - \dfrac{1}{\omega^2}(1-\cos\omega t)\right]$ 55. $\dfrac{1}{(s^2+\omega^2)(s^2+\varphi^2)}$

56. $\dfrac{1}{\omega(\omega^2+n^2)^2}\left\{\dfrac{\omega t^2}{2}(\omega^2+n^2) - 2n\omega t\right.$ 56. $\dfrac{1}{s^2[(s+n)^2+\omega^2]}$

$-\dfrac{\omega(\omega^2-n^2)}{\omega^2+n^2}\left[1 - e^{-nt}\left(\cos\omega t + \dfrac{n}{\omega}\sin\omega t\right)\right]$

$\left.-\dfrac{2n^2\omega}{\omega^2+n^2}\left[1 + e^{-nt}\left(\dfrac{\omega}{n}\sin\omega t - \cos\omega t\right)\right]\right\}$

57. $\dfrac{1}{\omega(n^2-\omega^2)^2}\left\{\dfrac{\omega t^2}{2}(n^2-\omega^2) - 2n\omega t\right.$ 57. $\dfrac{1}{s^2[(s+n)^2-\omega^2]}$

$+\dfrac{\omega(\omega^2+n^2)}{n^2-\omega^2}\left[1 - e^{-nt}\left(\cosh\omega t + \dfrac{n}{\omega}\sinh\omega t\right)\right]$

$\left.-\dfrac{2n^2\omega}{n^2-\omega^2}\left[1 + e^{-nt}\left(\dfrac{\omega}{n}\sinh\omega t - \cosh\omega t\right)\right]\right\}$

58.
$$\frac{1}{4n^2\varepsilon^2 + (\omega^2 + n^2 - \varepsilon^2)^2}$$
$$\times \left\{ \frac{1}{\varepsilon^2}(\omega^2 + n^2 - \varepsilon^2)(1 - \cos\varepsilon t) \right.$$
$$-2n\left(\frac{1}{\varepsilon}\sin\varepsilon t - \frac{1}{\omega}e^{-nt}\sin\omega t\right)$$
$$+\frac{(3n^2 + \varepsilon^2 - \omega^2)}{\omega^2 + n^2}\left[1 - e^{-nt}\left(\cos\omega t\right.\right.$$
$$\left.\left.\left. +\frac{n}{\omega}\sin\omega t\right)\right]\right\}$$

58.
$$\frac{1}{(s^2 + \varepsilon^2)[(s + n)^2 + \omega^2]}$$

59.
$$\frac{1}{4n^2\varepsilon^2 + (n^2 - \omega^2 - \varepsilon^2)^2}$$
$$\times \left\{ \frac{(n^2 - \omega^2 - \varepsilon^2)}{\varepsilon^2}(1 - \cos\varepsilon t) \right.$$
$$-2n\left(\frac{1}{\varepsilon}\sin\varepsilon t - \frac{1}{\omega}e^{-nt}\sinh\omega t\right)$$
$$+\frac{(3n^2 + \varepsilon^2 + \omega^2)}{n^2 - \omega^2}\left[1 - e^{-nt}\left(\cosh\omega t\right.\right.$$
$$\left.\left.\left. +\frac{n}{\omega}\sinh\omega t\right)\right]\right\}$$

59.
$$\frac{1}{(s^2 + \varepsilon^2)[(s + n)^2 - \omega^2]}$$

60.
$$\frac{1}{(n^2 + \varepsilon^2)^2}\left\{ \frac{n^2 - \varepsilon^2}{\varepsilon^2}(1 - \cos\varepsilon t) \right.$$
$$-\frac{2n}{\varepsilon}\sin\varepsilon t + \frac{3n^2 + \varepsilon^2}{\varepsilon^2}[1 - e^{-nt}(1 + nt)]$$
$$\left. +2nte^{-nt}\right\}$$

60.
$$\frac{1}{(s^2 + \varepsilon^2)(s + n)^2}$$

61.
$$\frac{1}{\omega^4}(1 - \cos\omega t) - \frac{1}{2\omega^3}t\sin\omega t$$

61.
$$\frac{1}{(s^2 + \omega^2)^2}$$

62.
$$\frac{\sin\omega t}{2\omega^3} - \frac{\omega t\cos\omega t}{2\omega^3}$$

62.
$$\frac{s}{(s^2 + \omega^2)^2}$$

63.
$$\frac{1}{(\varepsilon^2 + n^2)^2}\left\{ 2n\varepsilon[e^{-nt}(1 + nt) - \cos\varepsilon t] \right.$$
$$\left. -(\varepsilon^2 - n^2)(\sin\varepsilon t - \varepsilon te^{-nt})\right\}$$

63.
$$\frac{s}{(s^2 + \varepsilon^2)(s + n)^2}$$

64.
$$\frac{\omega\sin\varphi t - \varphi\sin\omega t}{\omega\varphi(\omega^2 - \varphi^2)}$$

64.
$$\frac{s}{(s^2 + \omega^2)(s^2 + \varphi^2)}$$

65. $\dfrac{1}{4n^2\varepsilon^2 + (\omega^2 + n^2 - \varepsilon^2)^2}\Big[2n(e^{-nt}\cos\omega t$

$\qquad - \cos\varepsilon t) + \dfrac{1}{\varepsilon}(\omega^2 + n^2 - \varepsilon^2)\sin\varepsilon t$

$\qquad - \dfrac{1}{\omega}(\omega^2 - n^2 - \varepsilon^2)e^{-nt}\sin\omega t\Big]$

65. $\dfrac{s}{(s^2 + \varepsilon^2)[(s + n)^2 + \omega^2]}$

66. $\dfrac{1}{4n^2\varepsilon^2 + (n^2 - \omega^2 - \varepsilon^2)^2}\Big[2n(e^{-nt}\cosh\omega t$

$\qquad - \cos\varepsilon t) + \dfrac{1}{\varepsilon}(n^2 - \omega^2 - \varepsilon^2)\sin\varepsilon t$

$\qquad + \dfrac{1}{\omega}(\omega^2 + n^2 + \varepsilon^2)e^{-nt}\sinh\omega t\Big]$

66. $\dfrac{s}{(s^2 + \varepsilon^2)[(s + n)^2 - \omega^2]}$

67. $\dfrac{\omega t\sin\omega t}{2\omega^2}$

67. $\dfrac{s^2}{(s^2 + \omega^2)^2}$

68. $\dfrac{1}{4n^2\varepsilon^2 + (\omega^2 + n^2 - \varepsilon^2)^2}\Big[(\omega^2 + n^2 - \varepsilon^2)$

$\qquad \times(\cos\varepsilon t - e^{-nt}\cos\omega t) + 2n\varepsilon\sin\varepsilon t$

$\qquad - \dfrac{n}{\omega}(\omega^2 + n^2 + \varepsilon^2)e^{-nt}\sin\omega t\Big]$

68. $\dfrac{s^2}{(s^2 + \varepsilon^2)[(s + n)^2 + \omega^2]}$

69. $\dfrac{1}{4n^2\varepsilon^2 + (n^2 - \omega^2 - \varepsilon^2)^2}\Big[(n^2 - \omega^2 - \varepsilon^2)$

$\qquad \times (\cos\varepsilon t - e^{-nt}\cosh\omega t) + 2n\varepsilon\sin\varepsilon t$

$\qquad - \dfrac{n}{\omega}(n^2 - \omega^2 + \varepsilon^2)e^{-nt}\sinh\omega t\Big]$

69. $\dfrac{s^2}{(s^2 + \varepsilon^2)[(s + n)^2 - \omega^2]}$

70. $\dfrac{\cos\varphi t - \cos\omega t}{\omega^2 - \varphi^2}$

70. $\dfrac{s^2}{(s^2 + \omega^2)(s^2 + \varphi^2)}$

71. $\dfrac{1}{(\varepsilon^2 + n^2)^2}\{2n\varepsilon\sin\varepsilon t + e^{-nt}[\varepsilon^2 - n^2$

$\qquad - nt(\varepsilon^2 + n^2)] - (\varepsilon^2 - n^2)\cos\varepsilon t\}$

71. $\dfrac{s^2}{(s^2 + \varepsilon^2)(s + n)^2}$

72. $\dfrac{\sin\omega t + \omega t\cos\omega t}{2\omega}$

72. $\dfrac{s^3}{(s^2 + \omega^2)^2}$

73. $\dfrac{1}{(\varepsilon^2 + n^2)^2}[nt(3\varepsilon^2 - n^2)e^{-nt}$

$\qquad + 2n\varepsilon^2(\cos\varepsilon t - e^{-nt}) - \varepsilon(\varepsilon^2 - n^2)\sin\varepsilon t]$

73. $\dfrac{s^3}{(s^2 + \varepsilon^2)(s + n)^2}$

74.
$$\frac{\omega \sin \omega t - \varphi \sin \varphi t}{\omega^2 - \varphi^2}$$

74.
$$\frac{s^3}{(s^2 + \omega^2)(s^2 + \varphi^2)}$$

75.
$$\frac{1}{4n^2\varepsilon^2 + (\omega^2 + n^2 - \varepsilon^2)^2}\left\{2n\varepsilon^2(\cos \varepsilon t \right.$$
$$- e^{-nt}\cos \omega t) - \varepsilon(\omega^2 + n^2 - \varepsilon^2)\sin \varepsilon t$$
$$\left. + \frac{1}{\omega}[(n^2 + \omega^2)^2 + \varepsilon^2(n^2 - \omega^2)]e^{-nt}\sin \omega t\right\}$$

75.
$$\frac{s^3}{(s^2 + \varepsilon^2)[(s + n)^2 + \omega^2]}$$

76.
$$\frac{1}{4n^2\varepsilon^2 + (n^2 - \omega^2 - \varepsilon^2)^2}\left\{2n\varepsilon^2(\cos \varepsilon t \right.$$
$$- e^{-nt}\cosh \omega t) - \varepsilon(n^2 - \omega^2 - \varepsilon^2)\sin \varepsilon t$$
$$\left. + \frac{1}{\omega}[(n^2 + \omega^2) + \varepsilon^2(n^2 - \omega^2)]e^{-nt}\sinh \omega t\right\}$$

76.
$$\frac{s^3}{(s^2 + \varepsilon^2)[(s + n)^2 - \omega^2]}$$

77.
$$\frac{1}{4n^2\varepsilon^2 + (\varepsilon^2 - n^2)^2}\{n^2 e^{-nt}[nt(n^2 - \varepsilon^2)$$
$$- 5\varepsilon^2] - \varepsilon^3(\varepsilon \cos \varepsilon t + 2n \sin \varepsilon t)\}$$

77.
$$\frac{s^4}{(s^2 + \varepsilon^2)(s + n)^2}$$

78.
$$\frac{1}{4n^2\varepsilon^2 + (\omega^2 + n^2 - \varepsilon^2)^2}\left\{[(n^2 + \omega^2)^2 \right.$$
$$+ \varepsilon^2(3n^2 - \omega^2)]e^{-nt}\cos \omega t$$
$$- \varepsilon^2(\omega^2 + n^2 - \varepsilon^2)\cos \varepsilon t$$
$$- \frac{n}{\omega}[(n^2 + \omega^2)^2 + \varepsilon^2(n^2 - 3\omega^2)]e^{-nt}$$
$$\left. \times \sin \omega t - 2\varepsilon^3 n \sin \varepsilon t\right\}$$

78.
$$\frac{s^4}{(s^2 + \varepsilon^2)[(s + n)^2 + \omega^2]}$$

79.
$$\frac{1}{4n^2\varepsilon^2 + (n^2 - \omega^2 - \varepsilon^2)^2}\left\{[(n^2 - \omega^2)^2 \right.$$
$$+ \varepsilon^2(3n^2 + \omega^2)]e^{-nt}\cosh \omega t$$
$$- \varepsilon^2(n^2 - \omega^2 - \varepsilon^2)\cos \varepsilon t$$
$$- \frac{n}{\omega}[(n^2 - \omega^2)^2 + \varepsilon^2(n^2 + 3\omega^2)]e^{-nt}\sinh \omega t$$
$$\left. - 2\varepsilon^3 n \sin \varepsilon t\right\}$$

79.
$$\frac{s^4}{(s^2 + \varepsilon^2)[(s + n)^2 - \omega^2]}$$

80. $\dfrac{\omega^2 \cos \omega t - \varphi^2 \cos \varphi t}{\omega^2 - \varphi^2}$

80. $\dfrac{s^4}{(s^2 + \omega^2)(s^2 + \varphi^2)}$

81. $\dfrac{1}{\omega^2}\left(\dfrac{t^3}{6} - \dfrac{t}{\omega^2} + \dfrac{1}{\omega^3}\sin \omega t\right)$

81. $\dfrac{1}{s^3(s^2 + \omega^2)}$

82. $\dfrac{1}{\varphi^2(\omega^2 + n^2)[4n^2\varphi^2 + (\varphi^2 - \omega^2 - n^2)^2]}$

$$\times \left\{ t[4n^2\varphi^2 + (\varphi^2 - \omega^2 - n^2)^2] \right.$$

$$- 2n(\omega^2 + n^2)(1 - \cos \varphi t)$$

$$+ \frac{1}{\varphi}(\omega^2 + n^2)(\varphi^2 - \omega^2 - n^2)$$

$$\times \sin \varphi t - \frac{2n\varphi^2(2n^2 - 2\omega^2 + \varphi^2)}{\omega^2 + n^2}$$

$$\times \left[1 - e^{-nt}\left(\cos \omega t + \frac{n}{\omega}\sin \omega t\right)\right]$$

$$\left. - \frac{\varphi^2}{\omega}(3n^2 - \omega^2 + \varphi^2)e^{-nt}\sin \omega t \right\}$$

82. $\dfrac{1}{s(s^2 + \varphi^2)[(s + n)^2 + \omega^2]}$

83. $\dfrac{1}{\varphi^2(n^2 - \omega^2)[4n^2\varphi^2 + (\varphi^2 + \omega^2 - n^2)^2]}$

$$\times \left\{ t[4n^2\varphi^2 + (\varphi^2 + \omega^2 - n^2)^2] \right.$$

$$- 2n(n^2 - \omega^2)(1 - \cos \varphi t) + \frac{1}{\varphi}$$

$$\times (n^2 - \omega^2)(\varphi^2 + \omega^2 - n^2)\sin \varphi t$$

$$- \frac{2n\varphi^2(2n^2 + 2\omega^2 + \varphi^2)}{n^2 - \omega^2}$$

$$\times \left[1 - e^{-nt}\left(\cosh \omega t + \frac{n}{\omega}\sinh \omega t\right)\right]$$

$$\left. - \frac{\varphi^2}{\omega}(3n^2 + \omega^2 + \varphi^2)e^{-nt}\sinh \omega t \right.$$

83. $\dfrac{1}{s(s^2 + \varphi^2)[(s + n)^2 - \omega^2]}$

84. $\dfrac{t}{\omega^2\varphi^2} - \dfrac{1}{\varphi^2 - \omega^2}\left(\dfrac{1}{\omega^3}\sin \omega t - \dfrac{1}{\varphi^3}\sin \varphi t\right)$

84. $\dfrac{1}{s(s^2 + \omega^2)(s^2 + \varphi^2)}$

85. $\dfrac{1}{(\varepsilon^2 + n^2)^2}\left\{\dfrac{t(\varepsilon^2 + n^2)^2}{\varepsilon^2 n^2} - \dfrac{2n}{\varepsilon^2}(1 - \cos \varepsilon t)\right.$

$$- \frac{\varepsilon^2 - n^2}{\varepsilon^3}\sin \varepsilon t + \frac{2(\varepsilon^2 + 2n^2)}{n^3}$$

$$\left. \times [1 - e^{-nt}(1 + nt)] + \frac{\varepsilon^2 + 3n^2}{n^2}te^{-nt} \right\}$$

85. $\dfrac{1}{s(s^2 + \varepsilon^2)(s + n)^2}$

86. $\dfrac{t}{2n^2\omega^2}(nt-2) + \dfrac{1}{\omega^2+n^2}\left[\dfrac{1-e^{-nt}}{n^3}\right.$

$\left. - \dfrac{n(1-\cos\omega t)}{\omega^4} + \dfrac{\sin\omega t}{\omega^3}\right]$

86. $\dfrac{1}{s^2(s+n)(s^2+\omega^2)}$

87. $\dfrac{t}{4n\omega^2}\left(t-\dfrac{1}{n}\right) + \dfrac{1}{\omega^2+4n^2}\left[\dfrac{1-e^{-2nt}}{8n^3}\right.$

$\left. - \dfrac{2n}{\omega^4}(1-\cos\omega t) + \dfrac{1}{\omega^3}\sin\omega t\right]$

87. $\dfrac{1}{s^2(s+2n)(s^2+\omega^2)}$

88. $\dfrac{1}{\omega^2+n^2}\left\{\dfrac{t^3}{6} - \dfrac{nt^2}{\omega^2+n^2} - \dfrac{t(\omega^2-3n^2)}{(\omega^2+n^2)^2}\right.$

$+ \dfrac{4n(\omega^2-n^2)}{(\omega^2+n^2)^3}\left[1-e^{-nt}(\cos\omega t\right.$

$\left.+ \dfrac{n}{\omega}\sin\omega t)\right] + \dfrac{\omega^2-3n^2}{\omega(\omega^2+n^2)^2}e^{-nt}\sin\omega t\Big\}$

88. $\dfrac{1}{s^3[(s+n)^2+\omega^2]}$

89. $\dfrac{1}{n^2-\omega^2}\left\{\dfrac{t^3}{6} - \dfrac{nt^2}{n^2-\omega^2} + \dfrac{t(\omega^2+3n^2)}{(n^2-\omega^2)^2}\right.$

$- \dfrac{4n(\omega^2+n^2)}{(n^2-\omega^2)^3}\left[1-e^{-nt}(\cosh\omega t\right.$

$\left.+ \dfrac{n}{\omega}\sin\omega t)\right] - \dfrac{\omega^2+3n^2}{\omega(n^2-\omega^2)^2}e^{-nt}\sinh\omega t\Big\}$

89. $\dfrac{1}{s^3[(s+n)^2-\omega^2]}$

90. $\dfrac{t^3}{6n^2} - \dfrac{t^2}{n^3} + \dfrac{t}{n^4}(3+e^{-nt}) - \dfrac{4}{n^5}(1-e^{-nt})$

90. $\dfrac{1}{s^3(s+n)^2}$

91. $\dfrac{t^3}{24n^2} - \dfrac{t^2}{8n^3} + \dfrac{t}{16n^4}(3+e^{-2nt})$

$- \dfrac{1}{8n^5}(1-e^{-2nt})$

91. $\dfrac{1}{s^3(s+2n)^2}$

92. $\dfrac{1}{n}\left[\dfrac{t^4}{24} - \dfrac{t^3}{6n} + \dfrac{t^2}{2n^2} + \dfrac{1}{n^4}(1-e^{-nt})\right]$

92. $\dfrac{1}{s^4(s+n)}$

93. $\dfrac{1}{8n}\left[\dfrac{t^4}{6} - \dfrac{t^3}{3n} + \dfrac{t^2}{2n^2} - \dfrac{t}{2n^3}\right.$

$\left. - \dfrac{1}{4n^4}(e^{-2nt}-1)\right]$

93. $\dfrac{1}{s^4(s+2n)}$

94. $\dfrac{t^4}{24n^2} - \dfrac{t^3}{3n^2} + \dfrac{3t^2}{2n^4} - \dfrac{t}{n^5}(4+e^{-nt})$

$+ \dfrac{5}{n^6}(1-e^{-nt})$

94. $\dfrac{1}{s^4(s+n)^2}$

95.
$$\frac{t^4}{96n^2} - \frac{t^3}{24n^2} + \frac{3t^2}{32n^4} - \frac{t}{32n^5}(4 + e^{-nt})$$
$$+ \frac{5}{64n^6}(1 - e^{-nt})$$

95.
$$\frac{1}{s^4(s + 2n)^2}$$

96.
$$\frac{t^2}{2\varphi^2\omega^2} - \frac{1}{\omega^2 - \varphi^2}\left[\frac{1}{\varphi^4}(1 - \cos\varphi t)\right.$$
$$\left. + \frac{1}{\omega^4}(1 - \cos\omega t)\right]$$

96.
$$\frac{1}{s^2(s^2 + \varphi^2)(s^2 + \omega^2)}$$

97.
$$\frac{1}{(\varepsilon^2 - \omega^2)(\varphi^2 - \varepsilon^2)}\left\{\frac{\varphi^2 + \omega^2 - \varepsilon^2}{\omega^2 - \varphi^2}\right.$$
$$\times \left[\frac{1}{\varphi^2}(1 - \cos\varphi t) - \frac{1}{\omega^2}(1 - \cos\omega t)\right]$$
$$\left. + \frac{\cos\varphi t - \cos\omega t}{\omega^2 - \varphi^2} - \frac{1}{\varepsilon^2}(1 - \cos\varepsilon t)\right\}$$

97.
$$\frac{1}{(s^2 + \varepsilon^2)(s^2 + \omega^2)(s^2 + \varphi^2)}$$

98.
$$\frac{1}{(\varepsilon^2 - \omega^2)(\varphi^2 - \varepsilon^2)}\left[\frac{\varphi^2 + \omega^2 - \varepsilon^2}{\omega^2 - \varphi^2}\right.$$
$$\times \left(\frac{1}{\varphi}\sin\varphi t - \frac{1}{\omega}\sin\omega t\right)$$
$$\left. + \frac{\omega\sin\omega t - \varphi\sin\varphi t}{\omega^2 - \varphi^2} - \frac{1}{\varepsilon}\sin\varepsilon t\right]$$

98.
$$\frac{s}{(s^2 + \varepsilon^2)(s^2 + \omega^2)(s^2 + \varphi^2)}$$

99.
$$\frac{1}{(\varepsilon^2 - \omega^2)(\varphi^2 - \varepsilon^2)}\left[\frac{\varphi^2 + \omega^2 - \varepsilon^2}{\omega^2 - \varphi^2}\right.$$
$$\times (\cos\varphi t - \cos\omega t)$$
$$\left. + \frac{\omega^2\cos\omega t - \varphi^2\cos\varphi t}{\omega^2 - \varphi^2} - \cos\varepsilon t\right]$$

99.
$$\frac{s^2}{(s^2 + \varepsilon^2)(s^2 + \omega^2)(s^2 + \varphi^2)}$$

100.
$$\frac{1}{(\varepsilon^2 - \omega^2)(\varphi^2 - \varepsilon^2)}\left[\frac{\varphi^2 + \omega^2 - \varepsilon^2}{\omega^2 - \varphi^2}\right.$$
$$\times (\omega\sin\omega t - \varphi\sin\varphi t)$$
$$\left. + \frac{\varphi^3\sin\varphi t - \omega^3\sin\omega t}{\omega^2 - \varphi^2} + \varepsilon\sin\varepsilon t\right]$$

100.
$$\frac{s^3}{(s^2 + \varepsilon^2)(s^2 + \omega^2)(s^2 + \varphi^2)}$$

101.
$$\frac{1}{(\varepsilon^2 - \omega^2)(\varphi^2 - \varepsilon^2)}\left[\frac{\varphi^2 + \omega^2 - \varepsilon^2}{\omega^2 - \varphi^2}\right.$$
$$\times (\omega^2\cos\omega t - \varphi^2\cos\varphi t)$$
$$\left. + \frac{\varphi^4\cos\varphi t - \omega^4\cos\omega t}{\omega^2 - \varphi^2} + \varepsilon^2\cos\varepsilon t\right]$$

101.
$$\frac{s^4}{(s^2 + \varepsilon^2)(s^2 + \omega^2)(s^2 + \varphi^2)}$$

INDEX

Printed in the USA
CPSIA information can be obtained
at www.ICGtesting.com
LVHW011734091223
764860LV00002BA/3